INTERNATIONAL UNION OF THEORETICAL
AND APPLIED MECHANICS

RHEOLOGY AND SOIL MECHANICS

SYMPOSIUM GRENOBLE, APRIL 1–8, 1964

RHÉOLOGIE ET MÉCANIQUE
DES SOLS

SYMPOSIUM GRENOBLE, 1ER–8 AVRIL 1964

EDITORS

J. KRAVTCHENKO
UNIVERSITÉ DE GRENOBLE

P. M. SIRIEYS
UNIVERSITÉ DE GRENOBLE

WITH 325 FIGURES

SPRINGER-VERLAG BERLIN HEIDELBERG GMBH
1966

© Springer-Verlag Berlin Heidelberg 1966
Ursprünglich erschienen bei Springer-Verlag Berlin Heidelberg New York 1966
Softcover reprint of the hardcover 1st edition 1966

ISBN 978-3-662-38599-9   ISBN 978-3-662-39449-6 (eBook)
DOI 10.1007/978-3-662-39449-6

Titel Nr. 1300

Preface

For a long period Soil Mechanics has remained at the semi-empirica
stage, and only a few decades ago it has shown a tendency to become a
fundamental science. However, this evolution is taking place slowly;
in spite of the efforts of numerous research scientists, the very complex
rheological laws of soils are still not well known. Even if these laws
were elucidated, it would take a long time still to deduce simple rules
from them for reliable and convenient use in current practical engineering.

In the pursuit of these distant aims — and of others more immediate — fundamental research and applied research are very active,
both in Rheology and Soil Mechanics. The complexity of the problems
to be solved should incite the laboratory researchers and the engineers to
a continuous collaboration. Everyone acknowledges the advantage of
these connections although aware of the difficulty of realizing this
wish. However, contacts are being made little by little between the representatives of the different branches of Rheology and Soil Mechanics,
to the great benefit of science.

The bureau of the International Union of Theoretical and Applied
Mechanics (IUTAM), aware of the importance of these two associat-
ed fields of mechanics, considered it possible to accelerate the natural
and necessary processus of their interpenetration by organizing in
Grenoble, from 1st to 8th April 1964 an International Symposium on
Rheology and Soil Mechanics.

Therefore the specialists of these two sciences were seen together —
apparently for the first time — endeavouring to attain the following
aims:

To increase the connections between the engineers and the laboratory researchers;

To study the practical range of theoretical methods of calculus of
works in soils and of the recent measurement methods (in the laboratory and in situ) of the rheological constants of soils;

To bring out the interest, from the point of view of fundamental
research, of the observations and measures accumulated on the building sites;

To promote fundamental research in some fields of practical interest.

On account of the aim pursued, the organizers attached a major importance to discussions and seminars which were to serve as a starting point to the dialogue between practicians, experimental and theoretical. It was to be feared that in this respect IUTAM's appeal would not be heard. However, thanks to the good will of all and the procedure of those responsible for the debates, these apprehensions proved groundless. The exchanges views were lively from the first session; they were prolonged beyond the duration of the session, and increased in the course of the symposium. The editors have endeavoured to recreate in this volume the atmosphere of the debates. They are aware that this has only been done imperfectly, in spite of the eager cooperation of the participants.

It is not for us to draw up a scientific balance sheet of the symposium, but we may say that the members of the congress separated with the wish to meet again and take up anew the thread of their discussions. That is all the organizers could expect within reason; and it is their reward.

The Bureau of the IUTAM has entrusted the responsibility of the organization of the symposium to three committees, the constitution of which is to be found on pp. X and XI. The International Committee looked after the scientific organization. The Patronal Committee helped to obtain the indispensable financial aid and contributed to the organization of the receptions given in honour of the participants and their wives. Monsieur le Préfet de l'Isère wished particularly to welcome them in the reception rooms of the Préfecture de Grenoble. Monsieur le Recteur TRÉHIN welcomed them in the new buildings of the Université de Grenoble. Mr. NEEL put several rooms of the Institut Polytechnique at the disposal of the participants. Mr. BARBIER, assessor to the dean, welcomed the participants in the Faculty of Science in his own name and in the name of Dean WEIL who was absent at the time on a mission abroad.

The preparation of the symposium was entrusted to the Service de Mécanique de l'Université de Grenoble. A National Committee was formed to carry out this duty. As well as the scientific personalities, the Committee associated Mr. FRAPPAT, General Secretary of "l'Association des Amis de l'Université de Grenoble", and his assistant Mr. ESPIC, who together with Messrs. SANTON and BIAREZ, assumed a large part of the responsibilities of the organization.

Several public and private bodies have granted subventions for the organization charges of the symposium:

La Direction de la Coopération avec la Communauté et l'Étranger du Ministère de l'Éducation Nationale.

La Délégation Générale à la Recherche Scientifique et Technique (DGRST).

Le Bureau de l'IUTAM.

Le Comité Français de Mécanique des Sols et Fondations.

Les Fédérations Nationales de Bâtiment et des Travaux Publics et leurs organismes techniques: CEBTP, SOCOTEC, UTI.

Le Bureau Veritas.

The Comité Francais de Mécanique des Sols patronized the symposium. Its General Secretary Mr. BUISSON, acted as an intermediary to the representatives of professional bodies. The President Mr. CAQUOT was unable to come to Grenoble due to mourning. The editors wish to mention here that messages of condolence and deferential sympathy were sent to him by all the participants.

The Bureau de l'IUTAM was represented by its president, Mr. TEMPLE, its general secretary, Mr. ROY, who took up the duty of acting presidents of the symposium, and its treasurer, Mr. KOITER. We must recall here the part taken by Mr. ROY, in the scientific and material organization of the meeting; Mr. KOITER also contributed to this double duty.

The presidents of the Sections and Sub-Sections and the general commentators were responsible for the working sessions, a list of which is to be found on pp. IX and X. Mr. SOKOLOVSKI, commentator of the second Sub-Section was unable to come to Grenoble for reasons of health; Mr. RADENKOVIC agreed to take his place. Finally Messrs. BIAREZ and ROSCOE looked after the organization of the two seminars; the works are reproduced in the "Discussions".

Grenoble, April 1966

J. Kravtchenko and **P. M. Sirieys**

Avant-Propos

Pendant une longue période, la Mécanique des Sols est restée au stade du semi-empirisme. Depuis quelques dizaines d'années seulement, elle a tendance à devenir une discipline fondamentale. Mais cette évolution s'accomplit avec lenteur; les lois rhéologiques des sols, très complexes, sont encore mal connues, en dépit des efforts de nombreux chercheurs. Et ces lois seraient-elles élucidées, qu'il faudrait encore de longs délais pour en tirer des règles simples, d'un emploi sûr et commode dans la pratique courante du Génie Civil.

Poursuivant ces objectifs lointains — et d'autres plus immédiats — la recherche fondamentale et la recherche appliquée sont fort actives, tant en Rhéologie qu'en Mécanique des Sols. La complexité des problèmes à résoudre devrait inciter les chercheurs de Laboratoire et les ingénieurs à pratiquer une collaboration continue. Tout le monde d'ailleurs reconnaît l'utilité de ces liaisons, tout en sachant qu'il est difficile de faire de ce vœu une réalité. Cependant, des contacts s'établissent peu à peu entre les représentants des différentes branches de la Rhéologie et de la Mécanique des Sols, pour le plus grand profit de la science.

Le bureau de l'Union Internationale de Mécanique Théorique et Appliquée (IUTAM), conscient de l'importance de ces deux domaines associés de la Mécanique, a jugé possible d'accélérer le processus naturel et nécessaire de leur interpénétration, en organisant à Grenoble, du 1er au 8 Avril 1964, un Symposium International de Rhéologie et de Mécanique des Sols.

Ainsi a-t-on vu réunis — pour la première fois, semble-t-il — les spécialistes de ces deux disciplines, pour essayer d'atteindre les objectifs suivants:

multiplier les liaisons entre les ingénieurs-constructeurs et les chercheurs de Laboratoire;

étudier la portée pratique des méthodes théoriques de calculs d'ouvrages en terre et des méthodes récentes de mesure (en Laboratoire et in situ) des constantes rhéologiques des sols;

dégager l'intérêt, au point de vue de la recherche fondamentale, des observations et des mesures accumulées sur les chantiers;

promouvoir la recherche fondamentale dans quelques domaines présentant un intérêt pratique.

Eu égard au but poursuivi, les organisateurs ont attaché une importance majeure aux discussions et séminaires qui devaient servir de point de départ au dialogue entre praticiens, expérimentateurs et théoriciens. On pouvait craindre qu'à cet égard l'appel de l'IUTAM ne fût pas entendu. Grâce à la bonne volonté de tous et à l'action des responsables des débats, ces appréhensions se sont révélées vaines. Les échanges de vue ont été animés dès la première séance; ils se sont prolongés hors séance et n'ont fait que s'amplifier au cours du Symposium. Les Éditeurs ont fait un effort pour restituer dans ce volume l'atmosphère des débats. Ils ont conscience de n'y avoir que très imparfaitement réussi, en dépit du concours empressé des participants.

Il ne nous appartient pas de dresser le bilan scientifique du Symposium. Mais nous croyons pouvoir dire que les congressistes se sont séparés avec le désir de se retrouver pour reprendre le dialogue interrompu. C'est tout ce que les organisateurs pouvaient raisonnablement espérer; et c'est là leur récompense.

Le Bureau de l'IUTAM a confié la responsabilité de l'organisation du Symposium à trois Comités dont on trouvera la composition aux pp. X et XI. Le Comité International avait la charge de l'organisation scientifique. Le Comité de Patronage a apporté son aide pour obtenir les concours financiers indispensables et a contribué à l'organisation des réceptions en l'honneur des participants et de leurs épouses, que M. le Préfet de l'Isère a tenu, en particulier, à accueillir dans les salons de la Préfecture de Grenoble. M. le Recteur TRÉHIN les a accueillis dans les nouveaux locaux de l'Université de Grenoble. M. NEEL a mis à la disposition des Congressistes plusieurs salles de l'Institut Polytechnique. M. BARBIER, assesseur du Doyen, a accueilli les participants à la Faculté des Sciences en son nom personnel et au nom de M. le Doyen WEIL, à l'époque en mission à l'étranger.

La préparation du Symposium a été confiée au Service de Mécanique de l'Université de Grenoble. Un Comité National a été constitué pour assumer en fait cette charge. En plus des personnalités scientifiques, le Comité s'est adjoint MM. FRAPPAT, Secrétaire Général de l'Association des Amis de l'Université de Grenoble, et son Adjoint, M. ESPIC, qui, avec MM. SANTON et BIAREZ, ont assumé une grande part de responsabilités de l'organisation.

Plusieurs organismes publics et privés ont accordé des subventions pour faire face aux frais d'organisation du Symposium:

La Direction de la Coopération avec la Communauté et l'Étranger du Ministère de l'Éducation Nationale.

La Délégation Générale à la Recherche Scientifique et Technique (DGRST).

Le Bureau de l'IUTAM.

Le Comité Français de Mécanique des Sols et des Fondations.

Les Fédérations Nationales du Bâtiment et des Travaux Publics et leurs organismes techniques: CEBTP, SOCOTEC, UTI.

Le Bureau Veritas.

Le Comité Français de Mécanique des Sols a accordé au Symposium son patronage. Son Secrétaire Général, M. BUISSON, a servi d'intermédiaire auprès des représentants d'organismes professionnels. Son Président, M. CAQUOT, n'a pu se rendre à Grenoble en raison d'un deuil. Les Éditeurs se doivent de rappeler ici les messages de condoléances et de déférente sympathie qui lui ont été adressés par l'ensemble des participants.

Le Bureau de l'IUTAM était représenté par son Président, M. TEMPLE, son Secrétaire Général, M. ROY, qui ont assumé la charge de Présidents effectifs du Symposium, et son Trésorier, M. KOITER. Nous devons rappeler ici la part prise par M. ROY, à l'organisation scientifique et matérielle du Colloque; M. KOITER a également apporté sa contribution à cette double tâche.

La responsabilité des séances de travail était confiée aux Présidents des Sections et des Sous-Sections et aux Rapporteurs Généraux dont on trouvera la liste aux pp. IX et X. M. SOKOLOVSKI, Rapporteur de la deuxième Sous-Section n'a pas pu, pour des raisons de santé, se rendre à Grenoble. M. RADENKOVIC a accepté de le remplacer dans cette tâche. Enfin MM. BIAREZ et ROSCOE se sont chargés de l'organisation de deux séminaires, dont les travaux ont été reproduits aux ,,Discussions".

Grenoble, Avril 1966

J. Kravtchenko et **P. M. Sirieys**

SYMPOSIUM INTERNATIONAL
DE L'UNION INTERNATIONALE DE MÉCANIQUE THÉORIQUE ET APPLIQUÉE

Avec le Concours du Comité Français de Mécanique des Sols

et des Travaux de Fondations

RHÉOLOGIE ET MÉCANIQUE DES SOLS

Grenoble, du 1er au 8 avril 1964

Président d'honneur: M. ALBERT CAQUOT
Membre de l'Institut de France

Présidents: M. MAURICE ROY
Membre de l'Institut de France

M. GEORGE TEMPLE
Sedleian Professor of Natural
Philosophy University of Oxford

PREMIÈRE SECTION: SECTION THÉORIQUE

President: M. WACLAW OLSZAK
Membre de l'Académie Polonaise des Sciences

Première Sous-Section: Rhéologie Théorique

Président: M. JEAN MANDEL
Professeur à l'Ecole Polytechnique

Rapporteur: M. DRAGOS RADENKOVIC
Professeur à l'Université de Belgrade

Deuxieme Sous-Section:
Études des Équations de la Mécanique des Sols

Président: M. WILLIAM PRAGER
Professeur à Brown University

Rapporteur: M. VADIM SOKOLOVSKY
Membre Correspondant de l'Académie des Sciences de l'U.R.S.S.

DEUXIÈME SECTION: SECTION EXPÉRIMENTALE

Président: M. MAURICE ROY
             Membre de l'Institut

Troisième Sous-Section:
Études Expérimentales des Propriétés Mécaniques des Sols

Président: M. JEAN KÉRISEL
             Professeur à l'École Nationale des Ponts et Chaussées

Rapporteur: M. JEAN BIAREZ
             Maître de Conférences à la Faculté des Sciences de Grenoble

Quatrième Sous-Section:
Modèles Réduits et Mesures "in Situ"

Président: M. RAYMOND PELTIER
             Directeur du Laboratoire Central des Ponts et Chaussées

Rapporteur: M. PIERRE-MARCEL SIRIEYS
             Chargé de Cours à la Faculté des Sciences de Grenoble

COMITÉ SCIENTIFIQUE

MM. KRAVTCHENKO   (France) Président
    BISHOP   (Grande-Bretagne)
    FERRANDON   (France)
    MANDEL   (France)
    OLSZAK   (Pologne)
    PRAGER   (États-Unis)
    RIVLIN   (États-Unis)
    ROY   (France)
    SKEMPTON   (Grande-Bretagne)
    SOKOLOVSKY   (U.R.S.S.)
    TSYTOVITCH   (U.R.S.S.)

COMITÉ NATIONAL D'ORGANISATION

MM. CAQUOT   (Paris) Président d'honneur
    ROY   (Paris) Président
    ANGLES D'AURIAC   (Grenoble)
    BIAREZ   (Grenoble)
    ESPIC   (Genoble) Secrétaire administratif
    FRAPPAT   (Grenoble) Commissaire général
    GERMAIN   (Paris)
    KERISEL   (Paris)
    KRAVTCHENKO   (Grenoble)
    MANDEL   (Paris)
    PELTIER   (Paris)
    RADENKOVIC   (Paris)
    SANTON   (Grenoble)
    SIRIEYS   (Grenoble) Secrétaire général

COMITÉ DE PATRONAGE

MM. Tréhin	Recteur de l'Université de Grenoble, Président
Néel	Membre de l'Institut, Directeur de l'Institut Polytechnique de Grenoble
Weil	Doyen de la Faculté des Sciences de Grenoble
Merlin	Président de l'Association des Amis de l'Université de Grenoble

Authors of the Articles Contained in the Book and of the Contributions to the Discussions

Auteurs des Articles Contenus dans le Livre et des Contributions aux Discussions

Anglès d'Auriac, Paul, Professeur, Université de Grenoble, Laboratoires de Mécanique des Fluides, 44—46, Avenue Félix-Viallet, Grenoble, France.

de Beer, Edward, Professeur, Rijksinstituut voor Grondmechanica, Laboratorium, St.-Pietersnieuwstraat, 41, Gent, Belgium.

Biarez, Jean, Professeur, Université de Grenoble, Laboratoires de Mécanique des Fluides, 44—46, Avenue Félix-Viallet, Grenoble, France.

Belot, A., Université de Grenoble, Laboratoires de Mécanique des Fluides, 44—46, Avenue Félix-Viallet, Grenoble, France.

Bjerrum, Laurits, Dr. techn., Director Norwegian Geotechnical Institute, Forskningsveien 1, Oslo 3, Norway.

Boucherie, M., Université de Grenoble, Laboratoires de Mécanique des Fluides. 44—46, Avenue Félix-Viallet, Grenoble, France.

Boucraut, L. M., Université de Grenoble, Laboratoires de Mécanique des Fluides, 44—46, Avenue Félix-Viallet, Grenoble, France.

Brinch Hansen, Jørgen, Professor Dr. techn., Dr. Ir. h. c., Geoteknisk Institut, 10, Østervoldgade, København K, Denmark.

Buisson, Maurice, Secrétaire Général du Comité Français de Mécanique des Sols, 31, rue H. Rochefort, Paris 17e, France.

Dayre, Michel, Université de Grenoble, Laboratoires de Mécanique des Fluides, 44—46, Avenue Félix-Viallet, Grenoble, France.

Drucker, Daniel, L. Herbert Ballou University Professor, Brown University, Division of Engineering, Barus-Holley Building, Providence, R. I. 02912, U. S. A.

Freudenthal, A. M., Professor, Columbia University in the City of New York, New York 27, N. Y., U. S. A.

FUJIMOTO, HIROSHI, Soil Mechanics Laboratory, Faculty of Engineering, Miyazaki University, 118, Nishy-Maruyama-Cho, Miyazaki-City, Japan.

GEUZE, E. C. W. A., Professor of Soil Mechanics and Foundation Engineering, Rensselaer Polytechnic Institute, Troy, N. Y., U. S. A.

GREEN, A. E., Professor of Applied Mathematics, The University of Newcastle-upon-Tyne, Newcastle-upon-Tyne 1, England

HABIB, PIERRE, Laboratoire de Mécanique des Solides, École Polytechnique, 17, rue Descartes, Paris 5e, France.

HAERINGER, J., Université de Grenoble, Laboratoires de Mécanique des Fluides, 44—46, Avenue Félix-Viallet, Grenoble, France

HANSEN, BENT, Civil Engineer, Geoteknisk Institut, 10, Østervoldgade, København K, Denmark.

HASEGAWA, HIROSHI, Professor of Civil Engineering, National Gunma Technical College, 580, Toriba-Cho, Maebashi-Shi, Gunma-Ken, Japan

HAYTHORNTHWAITE, ROBERT M., Professor of Engineering Science, The University of Michigan, Ann Arbor, Mich., U. S. A.

IRMAY, SHRAGGA, Professor, Hydraulic Laboratory, Technion-Israel, Institute of Technology, Haifa, Israel.

ITO, K., Staff of Building Research Institute, Ministry of Construction, 4, Chome Hyakunin-cho, Shinjuku-ku, Tokyo, Japan.

DE JOSSELIN DE JONG, G., Professor of Soil Mechanics, Technological University, Oostplantsoen 25, Delft, Netherlands.

KAWAKAMI, F., Dr.-Ing., Professor of Civil Engineering, Faculty of Engineering, Tohoku University, Sendai, Japan

KÉRISEL, JEAN, Professeur, SIMECSOL, 115, rue Saint-Dominique, Paris 7e, France.

KÉZDI, ÁRPÁD, Professor Dr. techn., Müegyetem rakpart 3, Budapest 11, Hungary.

KOIZUMI, YASUNORI, Dr. of Eng., Staff of Building Research Institute, Ministry of Construction, 4, Chome Hyakunin-cho, Shinjuku-ku, Tokyo, Japan.

KONDNER, ROBERT L., Professor of Engineering Physics, Loyola College, Baltimore, Md. 21210 U. S. A.

KRAUSE, J., Technische Hochschule Aachen, 51 Aachen, Mies-van-der-Rohe-Straße, Germany.

KRAVTCHENKO, JULIEN, Professeur, Université de Grenoble, Laboratoires de Mécanique des Fluides, 44—46, Avenue Félix-Viallet, Grenoble, France.

LITWINISZYN, JERZY, Professor, Polska Akademia Nauk, Zakład Mechaniki Górotworu, Mickiewicza 30, Kraków, Poland.

LONDE, PIERRE, Directeur, Coyne & Bellier, 19, rue Alphonse de Neuville, Paris 17e, France.

MANDEL, JEAN, Professeur, École Polytechnique, 17, rue Descartes, Paris 5e, France.

MARTIN, D., Université de Grenoble, Laboratoires de Mécanique des Fluides, 44—46, Avenue Félix-Viallet, Grenoble, France.

MONTEL, B., Université de Grenoble, Laboratoires de Mécanique des Fluides, 44—46, Avenue Félix-Viallet, Grenoble, France.

MURAYAMA, SAKURO, Professor, Dr. Engg., Kyoto University, Kita-Ku, Shichiku Kamiu- menoki-cho 26, Kyoto, Japan.

NÈGRE, R., Université de Grenoble, Laboratoires de Mécanique des Fluides, 44—46, Avenue Félix-Viallet, Grenoble, France.

OLSZAK, WACŁAW, Professor, Polish Academy of Sciences, ul. Swietokrzyska 21, Warszawa, Poland.

PELTIER, RAYMOND, Directeur du Laboratoire Central des Ponts et Chaussées, 58, Bd. Lefebvre, Paris 15e, France.

PERZYNA, P., Polish Academy of Sciences, ul. Swietokrzyska 21, Warszawa, Poland

PIERRARD, JEAN-MARIE, Université de Grenoble, Laboratoires de Mécanique des Fluides, 44—46, Avenue Félix-Viallet, Grenoble, France.

POTTIER, JEAN, Institut Français du Pétrole, 1 & 4, Avenue de Bois-Préau, Rueil-Malmaison, S.-&-O., France.

RADENKOVIC, DRAGOS, Directeur de Recherches à l'École Polytechnique, Laboratoire de Mécanique des Solides, 17, rue Descartes, Paris 5e, France

RIVLIN, RONALD S. L. Herbert Ballou University, Professor, Brown University Providence, R. I. 02912 U. S. A.

ROSCOE, KENNETH H., University Engineering Department, Trumpington Street, Cambridge, England.

SCHIFFMAN, ROBERT L., Professor, Rensselaer Polytechnic Institute, Troy, N. Y., and Lecturer, Massachusetts Institute of Technology, Cambridge, Mass., U. S. A.

SCHULTZE, EDGAR, Professor Dr.-Ing., Technische Hochschule Aachen, 51 Aachen, Mies-van-der-Rohe-Straße, Germany.

SIBILLE, ROBERT, Université de Grenoble, Laboratoires de Mécanique des Fluides, 44—46, Avenue Félix-Viallet, Grenoble, France.

SIRIEYS, PIERRE MARCEL, Professeur, Université de Grenoble, Laboratoires de Mécanique des Fluides, 44—46, Avenue Félix-Viallet, Grenoble, France.

SKEMPTON, ALEC, Professor of Civil Engineering, Imperial College, South Kensington, London S. W. 7, England.

SOBOTKA, Z., Doc. Dr.-Ing., Švédská 10, Praha XVI-Smíchov, ČSSR.

STROGANOV, A. S., Institut des Fondations et des Constructions Souterraines, 2 Institutskaja 6, Moscow Ж-389, USSR.

STUTZ, P., Université de Grenoble, Laboratoires de Mécanique des Fluides, 44—46, Avenue Félix-Viallet, Grenoble, France.

ŠUKLJE, LUJO, Professeur, Université de Ljubljana, Lepi pot 11, Ljubljana, Jugoslavia.

TAN TJONG-KIE, Professor Dr.-Ir., Institute for Rock and Soil Mechanics, Academia Sinica, Wuchang, Wuhan, China.

TSYTOVITCH, N. A., Professor Dr., Gosstroi, Prosp. Marxa 4, Moscow K—25, USSR.

VERDEYEN, JACQUES, Professeur ordinaire, Université Libre de Bruxelles, École Polytechnique, Faculté des Sciences Appliquées, Laboratoire de Mécanique des Sols, 87, Avenue Adolphe Buyl, Bruxelles 5, Belgium.

VERIGIN, N. N., Scientific Research Institute of Foundations and Underground Structures, Gosstroi, Prosp. Marxa 4, Moscow K—25, USSR.

VYALOV, S. S., c/o Professor Dr. N. A. TSYTOVITCH, Gosstroi, Prosp. Marxa 4, Moscow K—25. USSR,

WACK, BERNARD, Université de Grenoble, Laboratoires de Mécanique des Fluides, 44—46, Avenue Félix-Viallet, Grenoble, France.

WHITMAN, ROBERT, Professor of Civil Engineering, Massachusetts Institute of Technology, Cambridge, Mass., U. S. A.

WIENDIECK, K., Université de Grenoble, Laboratoires de Mécanique des Fluides, 44—46, Avenue Félix-Viallet, Grenoble, France.

ZARETSKY, J. K., c/o Professor Dr. N. A. TSYTOVITCH, Gosstroi, Prosp. Marxa 4, Moscow K—25, USSR.

Contents - Sommaire

Section I (Theoretical Section)

Sub-Section 1

Contents - Sommaire

Sub-Section 2

Section II (Experimental Section)
Sub-Section 3

Sub-Section 4

1.1 The Mechanics of Materials with Structure

By

A. E. Green and R. S. Rivlin

1. Introduction

In classical continuum mechanics, we consistently ignore any structure which the material may possess. The predictions of the theory, whether they take the form of calculated forces, stresses, displacements, or velocities, are then valid only as average values over regions the dimensions of which are large on the scale of the structure. There are many circumstances in which such average values do not represent adequately, or even meaningfully, the response of the body. We may cite some examples.

Consider a crystal of sodium chloride, which consists of positive sodium ions and negative chlorine ions arranged alternately in a cubic lattice. If a uniform electric field is applied, the forces on the sodium ions and those on the chlorine ions will be of equal magnitudes but of opposite directions and the displacements of the sodium and chlorine ions will also be oppositely directed. It is plainly not possible to represent the force system acting, or the resulting displacements, in a meaningful manner by average values taken over volumes large compared with the crystalline cell.

Again, if we consider a polycrystalline material, or a material consisting of particles embedded in a matrix with different mechanical properties, we cannot, by the usual procedures of classical continuum mechanics, obtain meaningful information regarding the effect of force systems which vary significantly in distances comparable with the lengths characterizing the structure, except by invoking a detailed physical model, of the system considered.

It is the object of this paper to draw attention to the manner in which the methods of classical continuum mechanics may be extended to render them applicable to systems and situations which heretofore were outside the range of continuum mechanics.

The bulk of the material presented in this paper is discussed more fully and in greater generality in papers pending publication elsewhere [1], [2], [3]. The idea of describing a deformation by tensor fields of various orders and of defining the associated forces as the coefficients of these tensors in a work formula was apparently first suggested by Truesdell and Toupin [4]. Particular cases of the theory presented here have been developed by Grioli [5], by Toupin [6], by Mindlin and Tiersten [7], and by Mindlin [8].

2. The Physical Background

In order to be explicit, we shall present the underlying physical notions of the present theory in terms of a specific physical situation — that of an elastic matrix in which elastic particles with different mechanical properties are distributed in a random manner. We consider that a force system is applied to the body and that a vector displacement field results. We divide the body into regions each of which contains many particles and has volume V in the undeformed state. The vector displacement field in each of these regions will, in general, vary throughout the region in a complicated manner. We may, however, describe it also by a vector and by tensors of various orders bound at the center of mass of the region. The bound vector will be the average of the vector displacement field in the region and the bound tensors may be moments of various types of the displacement field, or spatial derivatives of it of various orders. Together they provide a more detailed description of the displacement field in the region than would be provided by the bound vector alone. (In classical continuum mechanics, we would take the bound vector as describing the displacement field in the region.) It may, in certain circumstances, be desirable to describe the displacement field in the region by more than one vector bound at the center of mass and by more than one bound tensor of the same order. Indeed, many formally different, but physically equivalent, descriptions of the same displacement field are possible. In the interests of explicitness we shall limit our discussion, in the present paper, to the case of a single bound vector displacement and a single bound tensor of each order $2, \ldots, \mu$.

We now replace the vectors and tensors bound at the centers of mass of the various regions into which the body was divided and, assuming that the applied force system is such that they vary smoothly, we may replace them by continuous vector and tensor fields possessing as many spatial derivatives as may be required in the analysis. These are then taken as a description of the deformation of the body. The tensor field of order $\alpha + 1$ is called a 2^{α}-pole displacement field. Generically they are called multipolar displacement fields.

3. Fundamental Concepts

In accordance with the ideas expressed in the previous section, we describe the deformation in a body in a rectangular Cartesian coordinate system x. We assume that a generic point of the body initially at X_A, at some reference time, moves to $x_i(\tau)$ at time τ. In this way we define a vector displacement field $u_i(\tau)$ by

$$u_i(\tau) = x_i(\tau) - X_i. \qquad (3.1)$$

We assume that to complete the description of the deformation we require also tensor fields $x_{iA_1...A_\alpha}(\tau)$, $(\alpha = 1, ..., \mu)$. As a particular case it has been considered [1] that the tensor fields $x_{iA_1...A_\alpha}(\tau)$ are given by the spatial derivatives of $u_i(\tau)$ with respect to the coordinates X_A, so that

$$x_{iA_1...A_\alpha}(\tau) = x_{i,A_1...A_\alpha}(\tau), \quad (\alpha \geq 1). \qquad (3.2)$$

In this case it is clear that if we superpose on the deformation of the body a rigid time-dependent rotation, which takes the point $x_i(\tau)$ to $\bar{x}_i(\tau)$ so that

$$\bar{x}_i(\tau) = Q_{ij}(\tau) x_j(\tau), \qquad (3.3)$$

where $Q_{ij}(\tau)$ is a proper orthogonal tensor, then the tensor field $x_{i,A_1...A_\alpha}(\tau)$ is changed to $\bar{x}_{i,A_1...A_\alpha}(\tau)$, where

$$\bar{x}_{i,A_1...A_\alpha}(\tau) = Q_{ij}(\tau) x_{j,A_1...A_\alpha}(\tau). \qquad (3.4)$$

Motivated by this consideration, we *assume* that a superposed rigid rotation $Q_{ij}(\tau)$ changes the tensor field $x_{iA_1...A_\alpha}(\tau)$ to $\bar{x}_{iA_1...A_\alpha}(\tau)$, where

$$\bar{x}_{iA_1...A_\alpha}(\tau) = Q_{ij}(\tau) x_{jA_1...A_\alpha}(\tau). \qquad (3.5)$$

Other laws expressing the effect of a superposed rigid rotation on the multipolar displacement fields may be used and have been discussed elsewhere, but in the present paper we shall confine our attention to the law expressed by (3.5). The multipolar displacement fields are assumed to be unaltered by a rigid translation in accord with the fact that they are considered to express the detailed local deformation in a region of the body relative to the center of mass of that region.

We now have to define the forces associated with multipolar displacements. This is done with the aid of the time derivatives of these displacements. We therefore define multipolar velocity fields $v_{iA_1...A_\alpha}(\tau)$ by

$$v_{iA_1...A_\alpha}(\tau) = \dot{x}_{iA_1...A_\alpha}(\tau) \qquad (3.6)$$

and the ordinary (monopolar) velocity field by

$$v_i(\tau) = \dot{x}_i(\tau). \qquad (3.7)$$

We assume that the rate at which work is done by the system of body and surface forces acting on the body is given by

$$\int_V \varrho F_i v_i \, dV + \int_S p_i v_i \, dS$$
$$+ \sum_{\alpha=1}^{\mu} \left[\int_V \varrho F_{iA_1 \ldots A_\alpha} v_{iA_1 \ldots A_\alpha} \, dV + \int_S p_{iA_1 \ldots A_\alpha} v_{iA_1 \ldots A_\alpha} \, dS \right], \tag{3.8}$$

where V is the volume and S is the surface of the body in its unde-formed state and ϱ is the density of the material in this state.

Then, $F_{iA_1 \ldots A_\alpha}$ is called a *body force per unit mass of the* $(\alpha + 1)^{\text{th}}$ *kind* and $p_{iA_1 \ldots A_\alpha}$ is called a *surface traction of the* $(\alpha + 1)^{\text{th}}$ *kind per unit surface area*. Since the work is a scalar it follows that these forces are tensors.

We shall denote by $\Pi_{KiA_1 \ldots A_\alpha}$ a surface force of the $(\alpha + 1)^{\text{th}}$ kind acting on a plane perpendicular to the x_K-axis, per unit area measured in the undeformed state. When $\alpha = 0$, this becomes the KIRCHOFF stress tensor Π_{Ki}. For $\alpha > 0$, $\Pi_{KiA_1 \ldots A_\alpha}$ is, from its definition, a tensor with regard to the indices $i A_1 \ldots A_\alpha$, but it is not, in general, a tensor with regard to the index K.

4. The Energy Balance Equation and its Consequences

The expression (3.8) for the rate at which work is done by the system of forces acting on the body may be used to construct an energy balance equation. Here we shall do this only in the case when the body is elastic and is deformed isothermally. The more general case is discussed in [2]. For isothermal elastic deformations, the rate at which work is done by the forces acting (including inertial forces) is equal to the rate of increase in the HELMHOLTZ free-energy of the body. Then, if $\mathcal{A}$ is the HELMHOLTZ free-energy per unit mass and the forces F_i and $F_{iA_1 \ldots A_\alpha}$ in (3.8) are assumed to include the inertial forces, we have

$$\int_V \varrho F_i v_i \, dV + \int_S p_i v_i \, dS + \sum_{\alpha=1}^{\mu} \left[\int_V \varrho F_{iA_1 \ldots A_\alpha} v_{iA_1 \ldots A_\alpha} \, dV \right.$$
$$\left. + \int_S p_{iA_1 \ldots A_\alpha} v_{iA_1 \ldots A_\alpha} \, dS \right] = \int_V \varrho \dot{\mathcal{A}} \, dV . \tag{4.1}$$

We have already assumed that a rigid translation of the body leaves the multipolar displacement fields unaltered. It follows that a rigid translational velocity (i.e. a change of v_i to $v_i + a_i$, where a_i is constant) leaves the multipolar velocity fields unaltered. This change plainly leaves the HELMHOLTZ free-energy $\mathcal{A}$ and the forces p_i and F_i unaltered. We shall also assume that it leaves the multipolar forces $F_{iA_1 \ldots A_\alpha}$ and $p_{iA_1 \ldots A_\alpha}$ unaltered. Then, it follows from (4.1) that

$$\int_V \varrho F_i \, dV + \int_S p_i \, dS = 0 . \tag{4.2}$$

Bearing in mind that F_i includes the inertial forces, this is essentially the usual expression of NEWTON's second law in integral form. By considering an elementary tetrahedron in the deformed body, which has edges parallel to the coordinate axes, we obtain the classical KIRCHOFF equations

$$\Pi_{Ki,K} + \varrho F_i = 0$$

and

$$p_i = \Pi_{Ki} N_K,$$

(4.3)

where N_K is the unit vector normal to the surface considered in the undeformed state.

Introducing the relations (4.3) into (4.1), we obtain, using the Divergence Theorem,

$$\int_V \Pi_{Ai} v_{i,A}\, dV + \sum_{\alpha=1}^{\mu} \left[\int_V \varrho F_{iA_1\ldots A_\alpha} v_{iA_1\ldots A_\alpha}\, dV \right.$$
$$\left. + \int_S p_{iA_1\ldots A_\alpha} v_{iA_1\ldots A_\alpha}\, dS \right] = \int_V \varrho \dot{\mathcal{A}}\, dV.$$

(4.4)

Applying this to an elementary tetrahedron, we obtain

$$\sum_{\alpha=1}^{\mu} \left(p_{iA_1\ldots A_\alpha} - N_K \Pi_{KiA_1\ldots A_\alpha} \right) v_{iA_1\ldots A_\alpha} = 0.$$

(4.5)

Introducing (4.5) into (4.4) and using the Divergence Theorem, we obtain

$$\Pi_{Ai} v_{i,A} + \sum_{\alpha=1}^{\mu} \left(\varrho F_{iA_1\ldots A_\alpha} + \Pi_{KiA_1\ldots A_\alpha, K} \right) v_{iA_1\ldots A_\alpha}$$
$$+ \sum_{\alpha=1}^{\mu} \Pi_{KiA_1\ldots A_\alpha} v_{iA_1\ldots A_\alpha, K} = \varrho \dot{\mathcal{A}}.$$

(4.6)

It is evident that before we can proceed further, we must make some assumption as to the manner in which $\mathcal{A}$ depends on the monopolar and multipolar deformation fields and the manner in which $\Pi_{KiA_1\ldots A_\alpha}$ depends on the monopolar and multipolar deformation fields. In other words we require constitutive equations for these quantities. We have already remarked that $\Pi_{KiA_1\ldots A_\alpha}$ is not necessarily a tensor with respect to the index K (except for $\alpha = 0$). It will be, however, if we assume that $p_{iA_1\ldots A_\alpha} - N_K \Pi_{KiA_1\ldots A_\alpha}$ $(\alpha \geq 1)$ is independent of $v_{iA_1\ldots A_\beta}$ $(\beta = 1, 2, \ldots, \mu)$, although it may depend on the spatial derivatives of these quantities. In this case, it follows from (4.5) that

$$p_{iA_1\ldots A_\alpha} - N_K \Pi_{KiA_1\ldots A_\alpha} = 0,$$

(4.7)

from which the tensor character of $\Pi_{KiA_1\ldots A_\alpha}$ with respect to the index K follows immediately. If this is not the case, then we require a constitutive equation for the quantity $p_{iA_1\ldots A_\alpha} - N_K \Pi_{KiA_1\ldots A_\alpha}$ $(\alpha \geq 1)$, but the material cannot be elastic.

5. Constitutive Equations for Elastic Materials

We shall now assume that the Helmholtz free-energy $\mathcal{A}$ depends only on the deformation gradients $x_{i,A}$ and on the multipolar displacement fields and their first derivatives, i.e., on $x_{iA_1...A_\alpha}$ and $x_{iA_1...A_\alpha,K}$ ($\alpha = 1, ..., \mu$). Thus,

$$\mathcal{A} = \mathcal{A}(x_{i,A}, x_{iA_1...A_\alpha}, x_{iA_1...A_\alpha,K}). \tag{5.1}$$

Introducing (5.1) into (4.5), we have

$$\Pi_{Ai}v_{i,A} + \sum_{\alpha=1}^{\mu} (\varrho F_{iA_1...A_\alpha} + \Pi_{KiA_1...A_\alpha,K}) v_{iA_1...A_\alpha}$$

$$+ \sum_{\alpha=1}^{\mu} \Pi_{KiA_1...A_\alpha} v_{iA_1...A_\alpha,K} = \varrho \frac{\partial \mathcal{A}}{\partial x_{i,A}} v_{i,A} \tag{5.2}$$

$$+ \varrho \sum_{\alpha=1}^{\mu} \left[\frac{\partial \mathcal{A}}{\partial x_{iA_1...A_\alpha}} v_{iA_1...A_\alpha} + \frac{\partial \mathcal{A}}{\partial x_{iA_1...A_\alpha,K}} v_{iA_1...A_\alpha,K} \right].$$

Since the coefficients of $v_{i,A}$, $v_{iA_1...A_\alpha}$ and $v_{iA_1...A_\alpha,K}$ in Eq. (5.2) are independent of these quantities, we may equate them to zero.

We then obtain

$$\Pi_{Ai} = \varrho \frac{\partial \mathcal{A}}{\partial x_{i,A}},$$

$$\Pi_{KiA_1...A_\alpha,K} + \varrho F_{iA_1...A_\alpha} = \varrho \frac{\partial \mathcal{A}}{\partial x_{iA_1...A_\alpha}}, \tag{5.3}$$

$$\Pi_{KiA_1...A_\alpha} = \varrho \frac{\partial \mathcal{A}}{\partial x_{iA_1...A_\alpha,K}}.$$

In this case, we note that the conditions for (4.7) to be valid are satisfied, so that no constitutive equation is required for the quantity on the left of (4.7) and $\Pi_{KiA_1...A_\alpha}$ is a tensor with respect to all indices.

Acknowledgement

The work described here was carried out under Office of Naval Research Contract Nonr 562(10) with Brown University.

References

[1] Green, A. E., and R. S. Rivlin: Arch. Rational Mech. Anal. 16, 325 (1964).
[2] Green, A. E., and R. S. Rivlin: Arch. Rational Mech. Anal. 17, 113 (1964).
[3] Green, A. E., P. Naghdi and R. S. Rivlin: Int. J. Engineering Science 2, 611 (1965).
[4] Truesdell, C., and R. A. Toupin: The Classical Field Theories. Handbuch der Physik, Vol. III/1, Berlin/Göttingen/Heidelberg: Springer 1960.
[5] Grioli, G.: Ann. di Mat. Pura ed App. 50, 389 (1960).
[6] Toupin, R. A.: Arch. Rational Mech. Anal. 11, 385 (1962).
[7] Mindlin, R. D., and H. F. Tiersten: Arch. Rational Mech. Anal. 11, 415 (1962).
[8] Mindlin, R. D.: Arch. Rational Mech. Anal. 16, 51 (1964).

Discussion

Question posée par W. OLSZAK: The author showed a very interesting way of formulating the fundamental relations for media "with structure". An important role is played by the scalar function representing the work. The general theory holds for any kind of material, independently of its physical (mechanical) properties. As an example elastic bodies were considered.

I would like to ask how we should proceed to introduce into the theory different other properties of materials as, for instance, anisotropy (of any kind), time dependence, etc.

Réponse de R. S. RIVLIN: We may introduce other properties of the materials, such as time-dependence, through constitutive equations in which the multipolar stresses depend not only on the displacement gradients, the multipolar displacements and their spatial derivatives, but also on the time derivatives of these. Material symmetry of the various types can be introduced into these relations in a manner similar to that employed when multipolar displacement and stresses are not present, except insofar as the stresses need not be tensors in certain indices.

1.2 Définitions et Principes en Rhéologie Tensorielle

Par

Paul Anglès d'Auriac

Cette étude ne concerne que les corps assimilables à un milieu continu, unique et homogène, au moins localement.

Deux sortes de lois régissent ces corps:

Des lois *générales* (cinématique, statique et dynamique) communes à tous les milieux, et tirées de la Mécanique Rationnelle.

Des lois *spécifiques* propres à tel ou tel corps, et qu'on appelle lois de comportement.

Bien entendu, il existe des formules mixtes combinant des lois des deux types. Par exemple, les équations de l'Elasticité de LAMÉ résultent de la loi générale « équations de l'équilibre » combinée avec la loi de comportement « solide élastique isotrope de HOOKE ».

Nous ne nous intéressons pas, ici, à ces lois mixtes, mais uniquement aux lois de comportement *à l'etat pur*.

Nous cherchons une classification des comportements, et, par suite, des corps. A cette fin, nous *proposons* certains principes. Ces principes, bien évidemment, ne sont pas universels comme ceux de la Mécanique Rationnelle. Les cas où ils s'appliquent sont plus ou moins généraux.

Par ailleurs, le vocabulaire courant de la Rhéologie est actuellement imprécis. Par exemple, la viscosité newtonnienne est parfaitement définie, mais le mot viscosité tout court ne l'est pas.

En présence d'un mauvais vocabulaire, on peut décider de l'abandonner complètement et d'en créer un nouveau. Nous avons essayé une autre solution: à savoir de conserver les vieux mots en les justifiant par des définitions rigoureuses.

Notations (en axes Cartésiens)

Soit

$$u^j(x^i, t),$$

u étant la vitesse du point matériel qui se trouve à l'instant t au point x de l'espace.

Soit

$$\mathscr{T}^{ij} = \frac{\partial u^j}{\partial x^i},$$

appelé vitesse de transformation.

Pendant le temps dt, le point x va en

$$\xi^i = x^i + u^i dt.$$

La transformation de x en ξ est dite *transformation infinitésimale* et s'écrit :

$$\frac{\partial \xi^j}{\partial x^i} = \delta^{ij} + \mathscr{T}^{ij} dt = \delta^{ij} + dT^{ij}.$$

$\mathscr{T}^{ij}$ est la somme de la *vitesse de déformation* et de la *vitesse de rotation* respectivement définies par

$$\mathscr{D}^{ij} = \frac{1}{2} \left(\frac{\partial u^j}{\partial x^i} + \frac{\partial u^i}{\partial x^j} \right),$$

$$\mathscr{R}^{ij} = \frac{1}{2} \left(\frac{\partial u^j}{\partial x^i} - \frac{\partial u^i}{\partial x^j} \right).$$

On pose :

$$\mathscr{D}^{ij} dt = dD^{ij},$$

$$\mathscr{R}^{ij} dt = dR^{ij}.$$

La transformation infinitésimale est le produit (commutatif) de la *déformation infinitésimale*

$$\delta^{ij} + dD^{ij}$$

par la *rotation infinitésimale*

$$\delta^{ij} + dR^{ij}.$$

Pour des transformations finies, $\frac{\partial \xi^j}{\partial x^i}$ peut toujours se mettre, et d'une seule façon, sous la forme

$$T^{ij} = \frac{\partial \xi^j}{\partial x^i} = D^{ik} R^{kj},$$

$D^{ik} = D^{ki}$ est la déformation.

R^{kj} est la rotation des axes principeux.

Dans le cas particulier où $\frac{\partial \xi^j}{\partial x^i} = \frac{\partial \xi^i}{\partial x^j}$, on a simplement

$$D^{ij} = \frac{\partial \xi^j}{\partial x^i}.$$

Dans le cas général, on a toujours le carré de D par

$$\frac{\partial \xi^k}{\partial x^i} \frac{\partial \xi^k}{\partial x^j} = D^{ik} D^{kj}.$$

Principe I (d'homogénéité)

Il affirme que l'on peut exprimer les lois de comportement en supposant constants dans l'espace (mais non dans le temps) les différents tenseurs intervenant $(D, \mathcal{D}, \mathcal{J}, \sigma, \ldots)$.

En d'autres termes, le principe exclut que la loi de comportement fasse intervenir des tenseurs à 3 indices représentant la dérivée par rapport à l'espace de $D, \mathcal{D}, \mathcal{J}, \sigma$ etc. ...

Repère Rhéologique

La fonction (I) étant définie, $\mathcal{R}^{ij}$ est défini à tout instant. Soit un repère $\bar{x}$ animé à tout instant de la vitesse de rotation $\mathcal{R}$ par rapport au repère x. Le repère $\bar{x}$ est dit lié au corps ou repère rhéologique. Il est caractérisé par

$$\mathcal{J} = \mathcal{D}, \qquad \mathcal{R} = 0.$$

Toute transformation infinitésimale est une déformation pure. Bien entendu, sur un intervalle de temps fini Δt, le produit de ces déformations infinitésimales n'est pas une déformation pure.

S'il y a transformation finie instantanée, on la suppose déformation pure.

Tout repère fixe par rapport à un repère rhéologique est lui-même rhéologique.

Principe II (isotropie de l'espace)

Une loi de comportement exprimée en repère rhéologique ne fait jamais intervenir la vitesse de rotation de ce repère (par rapport à quoi que ce soit).

Il s'en suit que, dans un repère non rhéologique, interviendra la vitesse de rotation de ce repère par rapport à un repère rhéologique.

C'est donc dans les repères rhéologiques que les lois s'expriment le plus simplement. Ce sont ceux utilisés ci-dessous.

Sollicitation et Réponse

Les lois de comportement posent en général des relations entre contraintes et déformations telles que, si l'on se donne les unes, les autres sont déterminées. On peut indifféremment considérer les unes ou les autres comme cause ou comme effet.

Par pure convention, nous appelons les contraintes « sollicitation », et nous appelons les déformations « réponse ».

La sollicitation est donc une fonction $\sigma^{ij}(t)$, c'est-à-dire six fonctions scalaires. Nous en représentons une fig. 1. σ peut faire des sauts instantanés à certains instants.

La réponse comporte d'une part les six fonctions scalaires $\mathcal{D}^{ij}(t)$ et éventuellement les déformations D^{ij} correspondant aux sauts de contraintes.

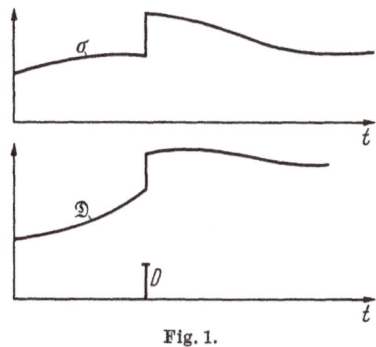

Fig. 1.

Principe III

Pour un corps donné, si l'on se donne la vitesse de déformation à l'instant 0 $\mathcal{D}^{ij}(0)$ et la sollicitation $\sigma^{ij}(t)$, la réponse est déterminée (ce serait évidemment faux en axes non rhéologiques).

Deux cas sont à distinguer : la réponse ne dépend que de la sollicitation elle-même, c'est-à-dire de facteurs mécaniques.

La réponse dépend, en outre, de facteurs non mécaniques (chimiques ou autres) pouvant varier dans le temps.

Nous appelons lois strictement mécaniques celles dans lesquelles tous les paramètres non mécaniques sont supposés constants dans le temps. C'est ce que nous supposons en général dans ce qui suit.

Propriétés d'un Corps

On appelle propriétés d'un corps l'ensemble des réponses correspondant aux différentes sollicitations.

Si l'on considère deux sollicitations pour un même corps, il est bien entendu qu'il ne faut pas les faire l'une après l'autre sur la même éprouvette, car la première sollicitation peut changer les propriétés du corps.

Il faut faire les deux sollicitations simultanément sur deux échantillons identiques par définition.

On dit donc que deux corps ont mêmes propriétés si toute sollicitation s'accompagne de réponses identiques sur l'un et l'autre.

Lois Complète ou Incomplètes

Il est bien évident qu'une sollicitation quelconque n'est pas forcément possible. Il faut donc considérer le domaine des sollicitations possibles.

Nous pouvons représenter une contrainte par un point dans un espace à six dimensions E_6. Nous supposons qu'il existe en général dans E_6 un certain volume à six dimensions où les sollicitations sont possibles.

Dans ce cas, nous admettons que, pour une sollicitation donnée, la réponse est en général déterminée. C'est ce que nous appelons une loi complète.

Toutefois, nous envisagerons plus loin le cas où, pour une certaine valeur de la contrainte, la réponse n'est pas entièrement déterminée mais contient un paramètre arbitraire. Nous disons, dans ce cas, que la loi est incomplète. Ce n'est qu'un cas limite (plasticité parfaite par exemple).

Il y a enfin le cas où le domaine des sollicitations possibles n'est pas à six dimensions. Dans ce cas, la loi ne peut être qu'incomplète. (Fluide parfait, par exemple.)

Principe IV

Si, pendant un intervalle de temps fini, on a, à la fois, $\mathcal{D} = 0$ et $\sigma = cte$ (notamment $\sigma = 0$), les propriétés du corps ne changent pas.

Comportement Solide

Si, sous contrainte indéfiniment constante dans le temps, un corps tend vers une forme limite (atteinte ou non), on dit que le corps a un comportement solide pour cette sollicitation. Ceci comporte que $\mathcal{D}$ tende vers zéro pour $t \, \infty$.

Cas particulier de comportement solide: sous contrainte constante, le corps ne se déforme pas.

Comportement Fluide

Si, sous contrainte indéfiniment constante dans le temps, $\mathcal{D}$ tend vers une limite non nulle (atteinte ou non), on dit que le corps a un comportement fluide pour cette contrainte.

Cas particulier: sous contrainte constante, $\mathcal{D} = cte$.

Principe V

Un corps étant pris dans certaines conditions initiales de contrainte et de vitesse de déformation, si on lui applique instantanément une contrainte maintenue indéfiniment constante, il n'a que trois comportements possibles:

le comportement solide;

le comportement fluide;

la rupture, c'est-à-dire la destruction nécessaire de l'homogénéité.

Ce que ce principe exclut, c'est par exemple que, sous contrainte constante, $\mathcal{D}$ oscille perpétuellement entre deux valeurs.

Les Trois Domaines

Un corps étant pris dans certaines conditions initiales, on peut considérer l'ensemble des sollicitations possibles $\sigma = cte$. Chacune d'elles sera représentée par un point dans E_6.

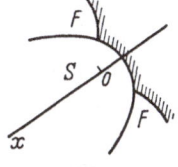

Ces points remplissent un domaine à 6 dimensions dans E_6. On suppose que ces 3 sortes de points peuvent se grouper en 3 domaines (fig. 2).

Le domaine de rupture R est la limite de l'ensemble formé par les domaines de comportement solide S et fluide F.

Fig. 2.

R est donc une surface (hypersurface à 5 dimensions au maximum) de l'autre coté de laquelle sont les sollicitations impossibles.

Solide et Fluide — Principe VI

Tous les corps, même ceux vulgairement appelés liquides, ont un comportement solide sur la demi-droite représentant les contraintes isotropes négatives (compression) (Ox, fig. 2). Ceci est un principe.

Le domaine de comportement solide comporte donc toujours au moins cette demi-droite. Compte tenu de ce fait, on appellera solide un corps dont le domaine de comportement solide est à 6 dimensions — et fluide un corps dont le domaine de comportement solide n'est qu'à une dimension.

Ceci semble exclure les cas intermédiaires où le domaine de comportement solide aurait un nombre de dimensions compris entre 1 et 6.

Or, ce cas intermédiaire peut se produire, mais n'est point étudié ici.

Viscosité

Soit un corps dont l'état de contrainte est représenté fig. 3 par le vecteur OA dans l'espace E_6 des contraintes.

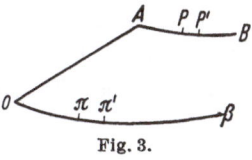

Considérons une sollicitation représentée par le trajet AB parcouru avec une certaine loi des temps satisfaisant à la condition suivante : le point contrainte peut se deplacer de A vers B (ou s'arrêter) mais pas de B vers A.

Fig. 3.

Si nous prenons comme forme de référence la forme initiale correspondant au point A, l'évolution de la forme dans le temps définit une fonction $T(t)$ qui est une courbe dans un espace à 9 dimensions E_9.

Aux points P et P' dans E_6 correspondent π et π' dans E_9. La transformation finie $O\pi$ n'est pas forcément une déformation, tandis que toute transformation infinitésimale telle que $\pi\pi'$ est une déformation (puisque nous sommes en axes rhéologiques).

Autrement dit, quand on se trouve dans l'espace E_9 au point π, les directions possibles pour $\pi\pi'$ ne dépendent que de 6 paramètres.

Un trajet quelconque $O\beta$ dans E_9 n'est pas forcément possible.

Malgré cela, on peut toujours aller de O en un point quelconque β par un trajet possible et même par une infinité.

Pour représenter l'évolution de la forme comparée à la forme initiale de référence, nous avons effectivement besoin d'un espace à 9 dimensions. Mais, à tout point de cet espace E_9 correspond localement un sous-espace E_6 représentant l'ensemble des directions de départ possibles.

En général, le trajet-transformation $O\beta$ dépend non seulement du trajet-contrainte AB, mais encore de la loi des temps sur AB.

Si la loi des temps n'intervient pas, on dit que la sollicitation est non visqueuse (trajet non visqueux).

Si, autour du point A, existe un certain domaine dans lequel tous les trajets sont non visqueux, on dira que le corps est non visqueux (à partir du point A) dans ce domaine.

Il est évident qu'un fluide ne peut pas être non visqueux. En effet, si l'on s'arrête au point P, la vitesse de déformation $\mathcal{D}$ tend vers une limite finie; donc, T ne peut rester constant.

Bien entendu, ce raisonnement ne vaudrait pas si le trajet AB était sur la droite des contraintes isotropes. Mais, on sait précisément que sur cette droite, fluide et solide ont même comportement.

Nous n'avons donc que trois cas à étudier: solide non visqueux, solide visqueux, fluide visqueux.

Solide non Visqueux

Vitesse de Déformation. Puisqu il y a correspondance (pour un trajet donné) entre le point contrainte et le point transformation, la vitesse de déformation $\mathcal{D}$ est déterminée quand on se donne la loi des temps. Il est donc inutile d'en parler.

Dans le cas particulier où la contrainte reste constante, $\mathcal{D}$ s'annule et *les propriétés du corps ne changent pas dans le temps* ainsi qu'il résulte de la définition même du trajet non visqueux.

Trajet Aller-Retour

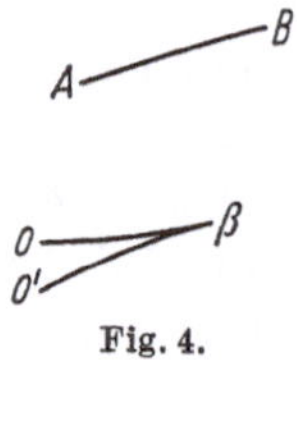

Supposons que le point contrainte trace le trajet aller-retour $AB\,BA$ (fig. 4).

Chacun de ces trajets satisfaisant aux conventions ci-dessus.

Fig. 4.

Il en résulte, en transformation, le trajet $O\beta\beta O'$.

1. Si le trajet retour $\beta O'$ est confondu avec le trajet aller O *avec la même correspondance contrainte — transformation*, on dit le trajet réversible.

2. Dans le cas contraire, le trajet est irréversible.

Transformation Élastique — Domaine Élastique

Si un trajet tel que AB est réversible, la transformation y correspondant est dite élastique.

S'il existe autour du point A une région telle que tout trajet partant de A et restant à l'intérieur soit réversible, cette région est dite domaine élastique.

Il est facile de voir que tout trajet partant d'un autre point du domaine jouit de la même propriété. Si bien que le point A ne joue pas de rôle privilégié.

Si l'origine O fait partie du domaine élastique, il est naturel ne la prendre pour définir la forme de référence. On appelle celle-ci *forme neutre*.

On peut démontrer qu'à l'intérieur du domaine élastique existe une fonction scalaire W (du point contrainte) appelée potentiel élastique et jouissant de la propriété suivante: l'énergie à fournir au contour pour aller du point contrainte A au point B est $W_B - W_A$.

Si, partant du point O, on trace un trajet contrainte fermé $OBCO$ restant à l'intérieur du domaine élastique, le trajet T correspondant n'est pas forcément fermé et peut être $O\beta\gamma O'$, avec OO' représentant une rotation.

Cette complication peut paraître fâcheuse. Heureusement, on peut l'éliminer par la convention suivante, qui, bien entendu, n'est intéressante qu'à l'intérieur du domaine élastique.

Convention Spéciale

Au lieu de considérer la transformation comme résultant de la contrainte, nous allons considérer la contrainte comme résultant de la transformation.

Par ailleurs, en fait de trajets transformations, nous ne considérons que des trajets rectilignes passant par l'origine, tels que $O\alpha$ et $O\beta$ (fig. 5).

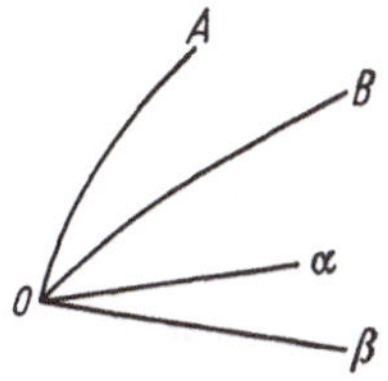

Tous ces trajets se faisant à directions invariantes constantes, le produit de déformations infinitésimales donnera une déformation finie. L'ensemble des points tels que α, β constitue donc un espace à 6 dimensions.

Fig. 5.

La correspondance entre les points-déformations α, β et les points-contraintes A, B suffit à définir la loi élastique. A un trajet rectiligne $O\alpha$ correspond un trajet contrainte OA non rectiligne, mais bien déterminé.

Si donc, nous voulons aller de A en B, nous ne le ferons que par le trajet privilégié $AO\ OB$. Pour tout autre trajet AB, le point figuratif de la transformation pourrait sortir de l'espace à 6 dimensions.

Par ailleurs, nous n'avons pas le droit de faire des trajets $\alpha\beta$ quelconques dans E_6, car nous ne serions plus en axes rhéologiques.

A l'intérieur du domaine élastique, la relation déformation contrainte est biunivoque et doit satisfaire à une relation d'intégrabilité, exprimant l'existence du scalaire W.

Tout trajet rectiligne aller-retour $O\alpha O$ ne change évidemment pas les propriétés du corps pour tout trajet futur à l'intérieur du domaine élastique. Et l'on pose en principe (VII) que cela ne change pas les propriétés du corps, dans toute l'acception du terme.

La limite du domaine élastique est représentée par une hypersurface à 5 dimensions dans E_6, lequel est un sous-espace de E_9.

Transformation Élasto-Plastique

Si, au trajet contrainte $ABBA$, correspond le trajet transformation $O\beta\beta O'$, on dit que le trajet AB est irréversible et que la transformation $O\beta$ est élasto-plastique.

Le trajet retour BA peut lui-même être réversible ou non. Pour le savoir il faut le faire suivre d'un nouveau trajet AB. Ainsi, dans le trajet transformation $O\beta\beta O'$, $O\beta$ est élasto-plastique et $\beta O'$ est soit élastique, soit élasto-plastique.

Si $\beta O'$ est élastique, on conviendra de dire que la transformation élasto-plastique $O\beta$ est la somme de la transformation plastique OO' et de la transformation élastique $O'\beta$. Mais, si $\beta O'$ est lui-même élasto-plastique, on ne peut pas, dans la transformation $O\beta$, chiffrer la part d'élasticité et de plasticité.

Le mot élasto-plastique doit s'entendre alors de façon purement qualitative.

Il y a toutefois deux cas où l'on peut, en quelque sorte, « chiffrer » la plasticité; ce sont les 2 cas extrêmes de la plasticité infiniment petite et infiniment grande.

Plasticité Infiniment Petite

Si le trajet en contrainte AB est infinitésimal, les trajets en transformation $O\beta$ et $\beta O'$ le seront aussi. Si OO' est du second ordre par rapport à $O\beta$ et $O'\beta$, on dit que la plasticité est infinitésimale ou encore que la loi élasto-plastique est tangente à une loi élastique.

La limite élastique que nous avons définie plus haut peut encore s'appeler seuil de plasticité. Elle portera l'un ou l'autre nom, selon que l'on regarde d'un coté ou de l'autre de la surface.

Et quand on sort du domaine élastique en franchissant le seuil de plasticité, on peut soit rencontrer brusquement une plasticité finie, soit rencontrer une plasticité croissant progressivement à partir de zéro.

Dans ce dernier cas, si l'on se place sur la limite élastique, on a d'un côté de la surface une loi rigoureusement élastique et, de l'autre côté, une loi tangente à une loi élastique.

Plasticité Infinie

Supposons qu'au trajet fini en contrainte AB corresponde le trajet infini en transformation $O\beta$ (fig. 6).

Traçons sur $O\beta$ des intervalles de longueurs égales $\beta_1\beta_2\beta_3\ldots\beta_n\ldots$

Il y correspond sur AB des intervalles $B_1B_2B_3\ldots B_n\ldots$ tendant vers zéro.

On dit que la plasticité augmente de plus en plus et tend vers l'∞ en B.

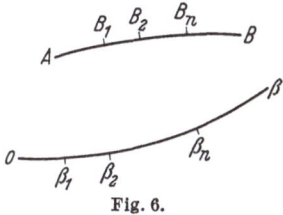

Fig. 6.

Le point B ne peut être dépassé si on l'atteint par le trajet AB. Mais rien ne prouve que par un autre chemin on n'aurait pas pu atteindre un point dans le prolongement de la courbe AB.

Quand on est suffisamment près du point B, on peut considérer la déformation comme arbitraire, à une constante multiplicative près. Plus exactement, on se donne une certaine vitesse de déformation $\mathcal{D}$ constante et l'on peut prendre la transformation T résultant de l'application de $\mathcal{D}$ pendant un temps quelconque.

Un point de plasticité infinie suppose une déformation à volume constant car il est physiquement absurde d'envisager une variation de volume infinie soit en augmentation, soit en diminution.

On posera, (ou on ne posera pas), le principe VIII suivant : Pendant la transformation (arbitraire à un paramètre près) correspondant à un point de plasticité infinie, les propriétés du corps ne changent pas.

Cas Schématiques

En principe, on considère une plasticité variant progressivement, c'est-à-dire tangente à l'élasticité au voisinage du seuil et tendant vers l'∞ au voisinage d'un point de plasticité infinie, c'est-à-dire que ce point est une limite inaccessible.

Mais ce cas général peut être simplifié de deux façons: on peut supposer que, au franchissement du seuil de plasticité, il y a saut brusque faisant passer la plasticité de O à une valeur finie.

De même, au point de plasticité infinie, on peut supposer un saut brusque faisant passer d'une valeur finie à l'infini. Dans ce cas, ce point peut être effectivement atteint.

La plus grande simplification consiste à faire sauter la plasticité de O à l'∞. Dans ce cas, d'un côté de la limite élastique, on a l'élasticité et, de l'autre côté, plasticité infinie. La limite élastique est alors infranchissable (dans l'espace des contraintes). Ce cas est appelé plasticité parfaite ou idéale.

Ecrouissage

D'une façon générale, tout trajet irréversible peut changer les propriétés du corps et notamment sa limite élastique. Si, partant d'un point A sur la limite élastique, on effectue le trajet AB dans le domaine élasto-plastique, il peut se faire que la nouvelle limite élastique passe par le point B. Quand ce phénomène se produit, on l'appelle écrouissage.

Mais il peut aussi se faire qu'en parcourant le trajet AB, on ait aboli le domaine élastique, c'est-à-dire que tous les trajets partant de B sont élasto-plastiques.

Solide Visqueux

Post-Effet. Partant du point-contrainte A (fig. 7), (auquel nous sommes restés suffisamment longtemps pour que la forme finale d'équilibre soit atteinte), nous effectuons le trajet AB à vitesse uniforme dans le temps Δt.

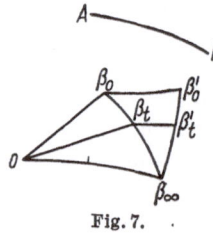

Fig. 7.

Après quoi, nous restons indéfiniment au point B jusqu'à $t = \infty$.

Au trajet AB correspond le trajet $O\beta_t$. Ensuite, la contrainte restant constante, le corps, puisque solide, doit tendre vers une forme limite, d'où résulte le trajet $\beta_t \beta_t'$ appelé post-effet.

Le trajet $O\beta_t \beta_t'$ dépend du paramètre Δt. Nous admettons que, quand t tend vers O, il existe le trajet limite $O\beta_0 \beta_0'$.

$O\beta_0$ est la déformation instantanée pour une variation de contrainte instantanée, $\beta_0 \beta_0'$ est le post-effet, toujours pour une variation de contrainte instantanée.

Quand le paramètre Δt tend vers l'∞, on admet que le post-effet tend vers O. Les points β_t et β_t' tendent vers la même limite β_∞.

Sollicitations très Rapides ou très Lentes

On dit que la sollicitation AB est très rapide si la courbe $O\beta_t\beta_t'$ est voisine de la limite $O\beta_0\beta_0'$ et l'on dit qu'elle est très lente si la courbe $O\beta_t\beta_t'$ est voisine de la limite $O\beta_\infty$.

Le paramètre définissant rapide ou lent est $\frac{\Delta\sigma}{\Delta t}$ qui a les dimensions de $\dot{\sigma}$.

Pour un $\Delta\sigma$ donné, «rapide» ou «lent» est défini par la grandeur du temps Δt.

Pour un corps donné et un $\Delta\sigma$ donné, on dira qu'un temps est très petit si la sollicitation est très rapide. Selon les cas, un temps «très petit» pourra aussi bien désigner un $\frac{1}{10.000}$ de seconde que 10.000 siècles.

Mêmes remarques pour un temps très grand.

Élasticité

Considérons le trajet aller-retour $ABBA$ parcouru à la même vitesse à l'aller et au retour (et sans s'arrêter au point B). Si le trajet déformation est le même au retour qu'à l'aller, soit $O\beta_t\beta_tO$, on dit que le trajet est réversible et que la transformation est élastique.

On pose, en principe IX, que les trajets réversibles ne peuvent exister que pour les sollicitations très rapides ou très lentes (les 2 cas limites).

Pour toutes les autres sollicitations aller-retour, le travail des forces au contour est positif, donc négatif celui des forces intérieures.

Ce principe dont l'énoncé ci-dessus est rigoureux peut être exprimé de façon plus intuitive de la façon suivante: les forces de viscosité dissipent de l'énergie. Cette dissipation ne peut être nulle *eventuellement* que dans 2 cas. Dans les expériences très rapides, la dissipation est nulle *faute de temps*; dans les expériences très lentes, la dissipation est nulle *faute de vitesse*.

Domaines Élastiques

Il peut donc *éventuellement* exister un domaine élastique pour les sollicitations très rapides et un domaine élastique pour les sollicitations très lentes. Domaines indentiques ou non.

Dans tous les cas, les définitions de élastique, élasto-plastique, plasticité infinitésimale et plasticité infinie, sont les mêmes qu'au chapitre précédent, mais sont à appliquer deux fois: une fois pour les sollicitations très rapides et une fois pour les très lentes.

Nous parlerons donc d'élasto-plasticité instantanée et d'élasto-plasticité finale.

Elasticité Différée

Un cas particulier de la visco-élasticité est celui de l'élasticité différée. Il est défini comme suit.

Supposons que le trajet-contrainte AB de la fig. 7 soit réversible aussi bien en expérience très rapide que très lente. Dans ce cas, $O\beta_0$ et $O\beta_\infty$ sont élastiques.

Supposons, en outre, que toute la courbe $\beta'_0\,\beta'_t\dots$ soit réduite à un seul point β_∞. C'est cette hypothèse supplémentaire qui définit l'élasticité différée.

Quand elle est satisfaite, l'effet de la viscosité est simplement de retarder la réalisation de la forme finale mais non de la modifier.

Une loi de ce genre semble convenir pour les caoutchoucs et les plastiques.

Mais il est bien entendu qu'un solide pourrait mériter l'appellation visco-élastique sans obéir à l'élasticité différée.

Remarquons que, si l'on veut représenter par un modèle mécanique composé de ressorts et de dash-pots le comportement d'un corps visco-élastique, si compliqué que soit ce modèle, se sera toujours un modèle d'élasticité différée, donc un cas très particulier.

Bien entendu, les corps à élasticité différée sont des corps à forme neutre.

Fluide Visqueux

Sollicitations très Lentes. Plaçons-nous au point contrainte A (fig. 8) et maintenons cette contrainte constante.

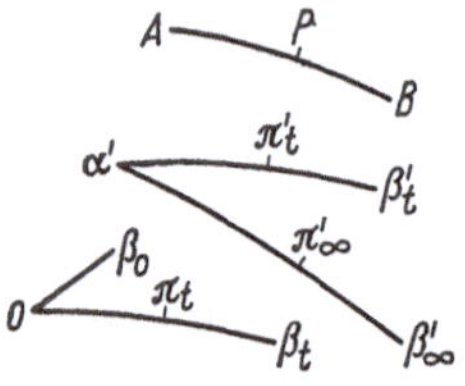

Fig. 8.

Le corps étant un fluide, sa vitesse de déformation $\mathcal{D}$ tend vers une limite.

Attendons suffisamment longtemps pour que cette limite soit atteinte ou presque.

Soit α' le point représentant cette vitesse de déformation.

Considérons maintenant le trajet contrainte $AB = \Delta\sigma$ parcouru dans le temps Δt, c'est-à-dire à la vitesse $\dot\sigma$. Nous pouvons tracer le lieu des points $\mathcal{D}$ correspondants. Au point P correspond le point π'_t et à la courbe AB la courbe $\alpha'\beta'_t$.

Nous admettons que quand Δt tend vers l'∞, c'est-à-dire $\dot\sigma$ vers O, la courbe $\alpha'\beta'_t$ tend vers une limite $\alpha'\beta'\infty$ et le point π'_t vers le point π'_∞. Ce point π'_∞ est la vitesse de déformation limite correspondant au point P.

On dit que la sollicitation est très lente si la courbe $\alpha'\beta'_t$ est voisine de la courbe $\alpha'\beta'\infty$. La lenteur d'une sollicitation est caractérisée par la petitesse de $\dot\sigma$, c'est-à-dire, pour $\Delta\sigma$ donné, par la grandeur de Δt.

Sollicitations très Rapides

Reprenons le cas de la sollicitation du paragraphe précédent. Choisissons comme forme de référence la forme du corps correspondant au point A, et considérons la transformation T figurée par le point β_t correspondant au point P.

Au trajet AB parcouru dans le temps Δt correspond la courbe $O\pi_t\beta_t$. Il est évident que quand Δt tend vers l'∞, la courbe $O\beta_t$ s'allonge vers l'∞.

Quand Δt tend vers O, on admet que la courbe $O\beta_t$ tend vers une limite $O\beta_0$ qui, en particulier, pourra être de longueur nulle, c'est-à-dire se réduire au point O.

On dit que la sollicitation est très rapide si la courbe $O\beta_t$ est voisine de $O\beta_0$ (supposée non nulle).

L'élasticité et — éventuellement — l'élasto-plasticité des fluides ne peut se définir que pour des sollicitations très rapides.

Nous allons donc distinguer 2 cas selon que la courbe $O\beta_0$ se réduit au point O ou non.

Fluide Purement Visqueux

On suppose que la courbe $O\beta_0$ se réduit au point O.

Un saut instantané de contrainte $\Delta\sigma$ n'entraîne pas de déformation instantanée D. Il ne peut être question d'élasticité ni de plasticité. On dit que le fluide est purement visqueux. Sa loi s'exprime alors par une relation entre trajet contrainte et trajet vitesse de déformation.

A priori, cette relation peut être aussi complexe que l'on veut, c'est-à-dire que le trajet $\alpha'\beta_t'$ dépend non seulement du trajet AB, mais encore de la loi des temps sur le trajet AB.

Mais, l'on fait habituellement les hypothèses simplificatrices qui suivent.

Hypothèses Simplificatrices

On suppose d'arbord que $\alpha'\beta_t'$ ne dépend pas de la loi des temps, donc est confondu avec $\alpha'\beta'\infty$.

On suppose, en outre, que le trajet lui-même n'intervient pas et qu'il y a une relation biunivoque entre point contrainte et point vitesse de déformation.

Cette relation fait le pendant de celle existant en Elasticité pure des solides entre contrainte et déformation.

Il existe alors une fonction scalaire du point contrainte, dite fonction de dissipation, et qui représente la puissance dissipée par unité de volume.

Cette fonction peut se rapprocher de la fonction énergie élastique. Toutefois, on notera que son existence n'entraîne pas de relation d'intégrabilité dans les relations contrainte — vitesse de déformation, ce qui donne une plus grande latitude pour la loi purement visqueuse que pour la loi purement élastique.

En revanche, la loi élastique peut être isotrope ou non, tandis que la loi purement visqueuse est en général supposée isotrope. Ce qui signifie que, si l'on transmute la contrainte par une rotation, on transmute aussi la vitesse de déformation par la même rotation.

Ainsi simplifiée, la loi de viscosité d'un fluide se trouve infiniment particulière, comparée à celle du fluide purement visqueux le plus général.

Magré cela, cette loi se trouve infiniment plus générale que celle de la viscosité Newtonnienne.

Fluide Visco-Élastique

L'élasticité d'un fluide ne se définit que pour des sollicitations instantanées. Elle correspond au cas où le trajet $O\beta_0$ de la fig. 8 est réversible. Plaçons-nous dans ce cas.

Forme Neutre Instantanée

Un fluide subissant à l'instant O la contrainte σ et animé de la vitesse de déformation $\mathcal{D}$, si l'on fait brusquement sauter la contrainte de σ à O, prend une nouvelle forme appelée, par définition, forme neutre instantanée. La forme actuelle du fluide à chaque instant peut être considérée comme obtenue par une déformation élastique à partir de la forme neutre instantanée.

Cette forme neutre instantanée varie évidemment d'un instant à l'autre.

Si l'on se donne la loi de déformation élastique, il suffit de se donner la loi de variation de la forme neutre instantanée pour avoir la loi du fluide.

Si l'on remplace à chaque instant la forme réelle du corps par sa forme neutre instantanée, on obtient un corps fictif qui, lui, est purement visqueux et rentre donc dans la classification étudiée aux paragraphes précédents.

1.3 Concept of Path Independence and Material Stability for Soils [1]

By

D. C. Drucker

Abstract

The development is traced of the description of soil behavior in terms of the stress-strain relations of the mathematical theory of plasticity. Structural metals and water-soil systems are compared and contrasted with particular reference to the role of the bound water. Major attention is devoted to experimental information on soil as a work-hardening material, including the appearance of corners in the yield curves. The value and the limitations of isotropic stress-hardening are discussed for radial and for general loading paths in the triaxial test. A perfectly plastic idealization is seen to be unsuitable for soils because the limit line bears no resemblance at all to an envelope of successive yield surfaces. The character of both time-dependent and time-independent behavior of stable material provides a frame of reference for the analysis of the stable and the unstable behavior of soil. As an aside, the design implications of the mechanical instability associated with limit lines and dilatation are described and a possible upper bound or failure theorem is discussed.

Introduction

In the years which have elapsed since the proposal was advanced by DRUCKER, GIBSON and HENKEL [1] in 1955 to treat soil as a work-hardening material, additional experimental data has been gathered and interpreted and a much greater understanding has been achieved of the significance of triaxial and shear tests and of the importance of time effects [2] to [10]. HENKEL in his doctoral thesis [3] of 1958 conclu-ded that much of the available experimental information lay outside

[1] The conclusions presented in this paper were derived in the course of research sponsored by the National Science Foundation under Grant No. GP 1115 to Brown University.

the scope of a useful theory of plasticity. However, rather recently CALLADINE, SCHOFIELD, and ROSCOE and his students have stated in their papers [5] to [9] that "wet" clays can be described remarkably well by a simple isotropic work-hardening idealization. ROSCOE and SCHOFIELD have discussed stress-strain relations for such a soil starting from my postulate of stability of material [11], [12]. Perhaps then, this is an appropriate time to reassess the situation to place current knowledge in perspective, and to look forward to future developments.

The first point to be emphasized is that the inelastic behavior of real materials is enormously complicated. Any explicit description in phenomenological or mathematical terms is bound to be a drastic idealization of actual behavior and should not be expected to be valid over a wide range of conditions. This state of affairs is understood quite well for metals because so much research effort has been expended by so many investigators [13]. Yet even for metals there often is confusion in the literature when it is rediscovered that the simple idealizations such as perfect plasticity, isotropic hardening, and kinematic hardening [14] fall so far short of accurate description of real behavior for general paths of loading. Nevertheless it is generally understood that an author's assumption of perfect plasticity does not imply his ignorance of the fact that metals do work-harden, that his assumption of isotropic hardening does not mean that he is unaware of the Bauschinger effect and allied cross-effects [15] to [17].

Drastic idealizations are valuable not only conceptually but also in practical engineering design. Such diverse physical objects as steel structures fabricated of a material which exhibits large plastic deformation at a lower yield point stress, aluminum structures composed of a material which work-hardens over its entire plastic range, and reinforced concrete structures formed of a composite of steel and a soil-like material, all can be and in many cases are designed on the basis of perfect plasticity.

Some of the confusion which exists results from a difference in objective which is not always stated clearly. An attempt to describe real behavior observed experimentally leads to ever increasing elaboration in detail of a mathematical or physical model as more types of tests are made. The desire to design efficiently and economically leads to the opposite requirement of the simplest of idealizations suitable for the limited purpose in hand. Therefore, for metal structures, perfect plasticity is an excellent design assumption, while very complex stress-strain relations which include work-hardening, BAUSCHINGER and allied cross-effects provide only crude approximations to the actual details of the distribution of stress and strain in a structure subjected to the usual complex pattern of loading history.

Perfect plasticity is an appropriate idealization for a structural metal because it contains the essential features of its behavior: the tangent modulus when loading in the plastic range is small compared with the elastic modulus, and the unloading response is elastic.

Perfect plasticity is not nearly as appropriate for soils. Some of the troubles were discussed at the time the idealization was proposed by DRUCKER and PRAGER [18] and extended and improved upon by SHIELD [19]. The subtle but strong difference between frictional and plastic behavior was brought out in subsequent discussion [21].

In the Sections which follow, these points are described and, hopefully, clarified. Isotropic hardening is discussed, and an effort is made to demonstrate both its value for simple loadings and its limitations for more general paths of loading [20], [22]. Here the essential feature, as in metals, lies in the appreciable irreversible or plastic deformation which one is willing to ignore in any simplification. That an idealization permissible for one purpose may not be suitable for another is exhibited. Observed isotropic hardening [6] to [9], the ROS-COE-SCHOFIELD-WROTH [5] HVORSLEV yield or state surface, the unique failure envelope of HENKEL [3] permit many interpretations. The inability of one elementary explanation to cover another experimental fact is not necessarily an invalidating contradiction.

Essential differences between initial or subsequent yield surfaces and a limiting stress surface are brought out and shown to be of great significance for soils as contrasted with metals. The limiting surface is discussed from both the experimental and the analytical point of view. Reversible paths and more general types of path independence are treated within the framework of the stress-strain relations of plasticity and are related to the information available on the response of water-soil systems.

Time effects which appear over long times during a nominally elastic recovery of volume, and rate effects in testing are discussed in terms of time-dependent stress-strain relations [12]. The corresponding asymptotic states also are considered from the time-independent point of view. Differences which arise in the degree of material stability to be expected are examined.

The concept of stability of material gives the essential features of time-independent plasticity. Its application is broadened here to include some of the special aspects of soils [22] as distinguished from metals. Although the fundamental question of frictional versus plastic behavior is not settled satisfactorily, greater consistency of approach than given in previous discussions can be achieved through the inclusion of chemical potential, now under careful study by PALMER, and by the fact that not all loading paths are permissible or will be followed.

Design implications of the difference in behavior between soils and metals are explored further. The essence of the upper and lower bound theorems of perfect plasticity carries over to work-hardening metals and to metallic structures which are stable. Because of the fundamentally different character of the limit state for a soil as contrasted with a metal, the lower bound theorem or its equivalent in terms of safe loading is lost. With it, perhaps catastrophically, go many of the intuitive ideas on which structural design or slope and foundation design are based. The upper bound theorem also disappears. A modified upper bound theorem of possible practical value for stability problems is described.

Some Relevant Features of the Plastic Behavior of Structural Metals

Structural metals are very complex when examined on the microscale and on the atomistic scale. Nevertheless, under quasi-static loading at ordinary and low temperatures, they may be idealized as time-independent with separable increments of elastic and plastic deformation. This means that in simple tension, for example, there is a single stress-strain curve with elastic unloading sharply differentiated from elastic-plastic loading, Fig. 1a. More generally, at each stage of plastic deformation a yield or loading surface in stress space, Fig. 1b, separates

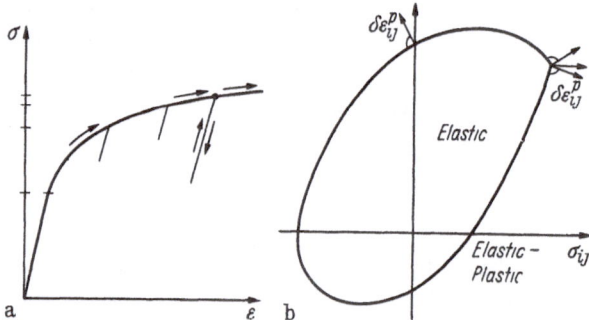

Fig. 1a and b. A yield or loading surface (curve) $f(\sigma_{ij}) = $ const separates elastic from elastic-plastic behavior at each stage of plastic deformation of a metal. Convexity of $f = $ const and normality of the vector representing the increment of plastic strain follow from the postulate of stability of material.

those states of stress which can be reached by purely reversible (elastic) paths from those outside which cannot be reached without producing some additional plastic deformation. The term yield surface is used to emphasize the fact that three or more components of stress may be independent variables. Two dimensional pictures only are drawn, however. The yield surfaces thus are represented by yield curves or actually become yield curves when two independent components of stress are studied.

The concept of stability of material [*11*] then leads directly to the two basic features of the relation between stress and strain in the plastic range. All initial and subsequent yield surfaces are convex. The normal at any point of a current yield surface gives the ratios of the components of the increment of plastic strain produced by further loading from the yield state of stress represented by the point, Fig. 1b. If the yield surface has a corner at the point, or is idealized to have one, the vector representing the increment of plastic strain $\delta\varepsilon_{ij}^p$ must lie between the normals to adjacent points as illustrated. The normality requirement does not give the magnitude of the plastic strain but just determines or restricts the ratios of the components such as $\delta\varepsilon_x^p/\delta\varepsilon_y^p$, $\delta\varepsilon_x^p/\delta\gamma_{xy}^p$, etc.

The familiar simple tension or simple shear tests are particular examples of radial or proportional loading tests, Fig. 2. The components of stress are increased in ratio so that the stress point, whose coordinates represent the state of stress, travels out from the origin on a radial line. Many such tests have been made on a wide variety of

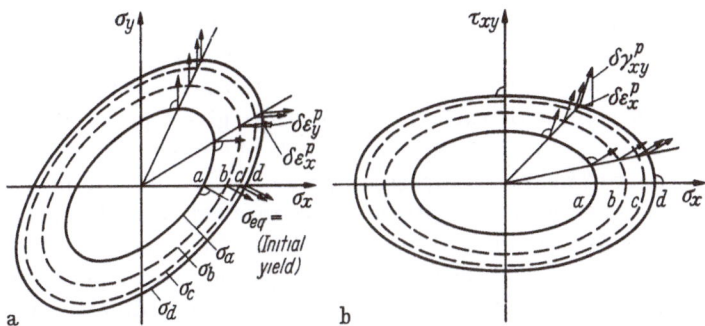

Fig. 2 a and b. Radial or proportional loading for a structural metal and the assumption of isotropic stress-hardening. Successive surfaces are spaced at equal increments of plastic strain indicated on Fig. 1 a.

structural metals [*13*]. When the metal is isotropic to start, the result reported is that the ratios of the increments of plastic (total minus elastic) strain remain constant as the radial stress path is traversed. All the data on one material for all radial loading paths plot on the same stress-strain curve when an appropriate combination of stress components is chosen as an equivalent stress σ_{eq} and a corresponding choice is made for equivalent strain ε_{eq}, or better an equivalent plastic strain ε_{eq}^p. Either the octahedral shearing stress τ_0 or the maximum shearing stress τ_{max} is found to be a reasonable measure of equivalent stress for structural metals.

In general, initial isotropy imposes the formal mathematical requirement for radial loading paths that the equivalent stress be a

function of the three invariants of the stress tensor. One choice is the
set of principal stresses $\sigma_1, \sigma_2, \sigma_3$. A convenient choice is the set J_1, J_2, J_3
where J_1 is the first invariant of the stress tensor

$$J_1 = \sigma_1 + \sigma_2 + \sigma_3 = -3p \tag{1}$$

defined here as three times the average tension or the negative of three
times the average pressure. J_2 is the second invariant of the stress
deviation tensor

$$\begin{aligned} J_2 &= \frac{1}{6} \left[(\sigma_1 - \sigma_2)^2 + (\sigma_2 - \sigma_3)^2 + (\sigma_3 - \sigma_1)^2 \right], \\ &= \frac{3}{2} \tau_0^2 \end{aligned} \tag{2}$$

and J_3 is the third invariant of the stress deviation tensor

$$\cdot J_3 = \frac{1}{27} (2\sigma_1 - \sigma_2 - \sigma_3)(2\sigma_2 - \sigma_3 - \sigma_1)(2\sigma_3 - \sigma_1 - \sigma_2). \tag{3}$$

The initial yield surface then is given by σ_{eq}, or some function of
the invariants $f(J_1, J_2, J_3)$, equal to a constant. The initial plastic strain
increments for continued loading are normal to this surface. To pre-
serve constancy of direction of the normal along radial lines, surfaces
of larger equivalent stress are simply expanded versions of the initial
yield surface as shown in Fig. 2. Concentricity of the surfaces requires
f or σ_{eq} to be a homogeneous function of the principal stress components. These concentric surfaces are drawn to give a pictorial representation of their spacing for equal increments of equivalent plastic strain.

The effect of hydrostatic pressure, $p = -J_1/3$, on structural metals
is negligible until very high pressures are reached. Therefore, the equivalent stress for metals is a function of J_2 (the MISES criterion) and
J_3 alone. The maximum shearing stress can be expressed in terms of
J_2, J_3. Experimental data for radial loading paths ordinarily lie some-
where between this criterion of TRESCA and the one of MISES, usually
favoring the latter.

If now the concentric surfaces of constant equivalent stress are
imagined to be successive yield surfaces established by any path of
loading, radial or not, the assumption is termed isotropic stress-hardening. The current yield surface is determined by the maximum σ_{eq}
reached in the past but is otherwise path independent. Isotropic stress-
hardening satisfies all the mathematical requirements for valid stress-
strain relations and often is employed in the solution of problems when
work-hardening must be given some consideration. However, not only
is there no reason to suppose that so simple an hypothesis has physical
validity, ample experimental evidence exists to show how badly in
disagreement with reality it is. Nevertheless, before looking at the
evidence, it is worth repeating that despite its shortcomings it may well
be the appropriate assumption to make for many problems.

Fig. 3 compares isotropic hardening with the experimental results for initial and subsequent yield surfaces obtained by IVEY [16] and by BERTSCH and FINDLEY [17]. A large body of experimental information of a confirmatory nature exists in addition, especially on the tremendous BAUSCHINGER effect found by IVEY, showing the yield surface to pull away from and not enclose the origin. Also the results agree with intuitive concepts based upon the mechanism of plastic deformation.

Isotropic hardening has a single simple parameter which controls or gives the state of the material. Maximum equivalent stress, or equivalent plastic strain, or plastic work are interchangeable descriptions or labels for the current yield surface of an isotropic stress-hardening theory. Nothing so simple should be expected to, or does in fact, control the state of the material as portrayed by the shape and size of the current yield surface for a real metal. Strong path dependence is found. Hardening is markedly anisotropic under most conditions. The set of successive yield surfaces for one direction of loading in stress space bears little resemblance to that for another direction.

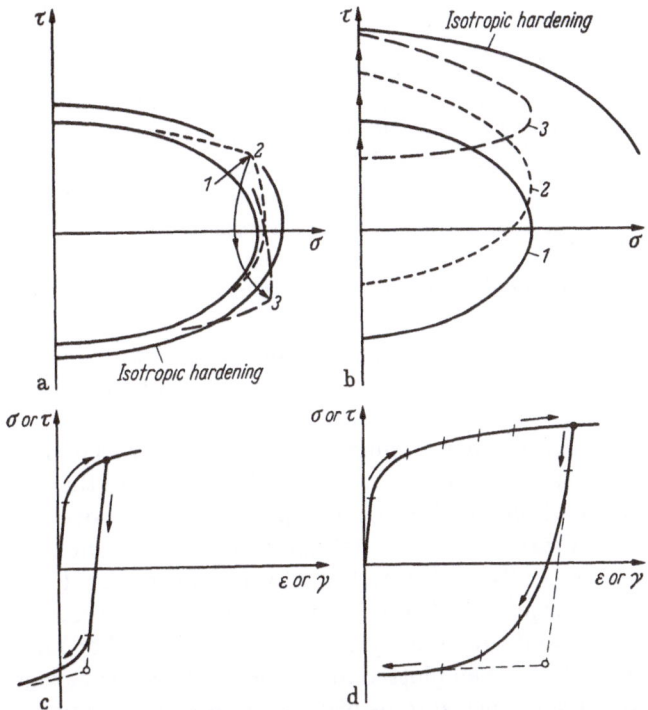

Fig. 3a—d. Comparison of actual yield surfaces for aluminum alloys with isotropic stress-hardening. a) Data of BERTSCH and FINDLEY [17] based on very small offset; b) Data of IVEY [16] similarly based; c), d) Schematic representation of BAUSCHINGER effect and alternate definitions of yield values. Initial values indicated by horizontal lines, large offset values by slant lines, and the TAYLOR-QUINNEY definition by the open circles.

All of these complexities and many more are found in a detailed look at the yield surfaces and at the corresponding stress-strain relations. However, if a very crude approach is taken instead, a less complicated picture emerges for a variety of loading paths which does not differ nearly so greatly from isotropic hardening.

If plastic deformations several times as large as total elastic strains are ignored, then as shown in Fig. 3c, the offset yield strengths for reversed loading do not exhibit the large and real BAUSCHINGER effect which does exist. TAYLOR and QUINNEY adopted a similar point of view for their tension-torsion tests which followed non-radial loading paths but did not include reversal of the sign of the applied stress. As a consequence they obtained reasonable agreement with an isotropic J_2 theory [13]. There is not a great deal of accurate experimental information of this type on a wide range of materials and paths of loading. Yet enough evidence is available to suggest the result shown schematically in Fig. 4 and contained in ILYUSHIN's postulate of delay and isotropy. The yield surface for additional plastic strains of, say, 1%, or 10,000 microinches per inch, will resemble an expanded version of the initial yield surface far more than the translating and shrinking yield surfaces defined by small plastic deformations of the order of 20 microinches per inch as in Figs. 3a, b.

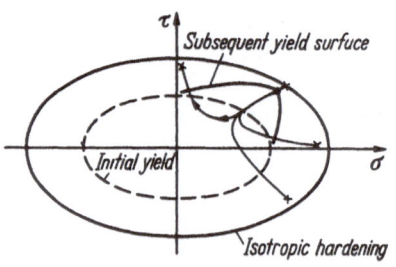

Fig. 4. A crude isotropic hardening would be obtained if plastic strain increments of the order of total elastic strains were ignored (see Fig. 3 c).

Yield surfaces *are* a matter of definition. The choice is not an absolute one, but is determined by the most significant features of the problem to be solved. Both crude and refined measures of yielding have their place. For design purposes, the crude may be more relevant than the refined. Yet it is always well to keep in mind the basic experimental information which has been approximated so drastically.

Far more drastic idealizations than isotropic hardening are permissible mathematically and useful in physical problems. Perfect plasticity, Fig. 5, is especially helpful. Successive surfaces for isotropic hardening, Fig. 2, and successive yield surfaces for radial or for non-radial loading, Figs. 3, 4, when plotted in positions which correspond to equal increments of some measure of plastic strain or plastic work, give the appearance of piling up tangentially against an almost fixed limit surface at which the plastic strain is large. A choice of a fixed yield surface in the middle of the stress range of interest, as indicated on the simple

tension diagram of Fig. 5a, provides a perfectly plastic idealization
of real value for many problems. Because the real yield surfaces come
in closely tangential to this artificial yield surface, the ratios of the
components of the plastic strains given by the normals do correspond
to reality. Also, of course, the stress
level is proper on the average.

Nevertheless, the idealization is
severe and it is necessary to guard
against improper interpretation. All
plastic strains which occur in the
real material, as the state of stress
moves about inside the perfectly
plastic yield surface, are neglected
in this model of a metal. Also, the
possibility of unlimited plastic deformation at constant stress, the
hallmark of the perfectly plastic
idealization, ignores the real work-
hardening of the metal beyond
the arbitrarily chosen yield level.
Therefore the absence of plastic
deformation in the model must be
interpreted as meaning that no
more than small plastic deformation
takes place in the prototype. Infinite strain in the model when limit
loads are reached means large strain
in the prototype. These are limitations and cautions to be observed

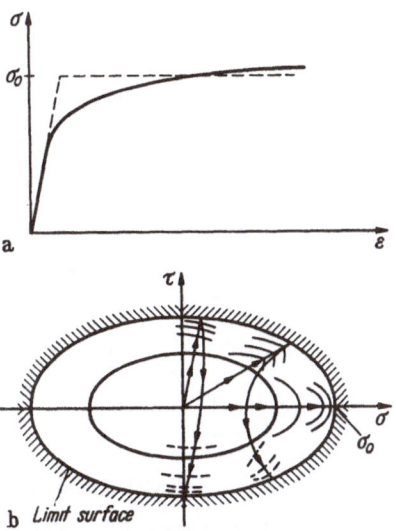

Fig. 5a and b. Perfectly plastic idealization.
The strain level of interest in the problem
determines the choice of σ_0. As illustrated
schematically in b), successive yield surfaces
at equal increments of plastic strain, for
various loading paths, give the impression of
piling up tangentially against the limit surface. Large strains are produced by small
additional loading past this arbitrarily chosen
surface.

but they interpose few conceptual difficulties in the translation of the
results of the useful practical idealization of perfect plasticity to a
structural metal.

Now attention can be shifted to a state of stress of direct relevance
to the study of water-soil systems. Fig. 6 is a schematic plot of triaxial
test results for a structural metal, but with compression taken as positive to correspond to the usual sign convention of soil mechanics. The
triaxial test has two independent stress variables σ_1 and $\sigma_2 = \sigma_3$ which
could be chosen as coordinates. This choice would not be as appropriate
as either of the two sets of coordinates shown. One, σ_1 vs $\sigma_2\sqrt{2}$, represents to scale the plane in principal stress space passing through the
σ_1-axis and bisecting the angle between the $\sigma_1\sigma_2$-plane and the $\sigma_1\sigma_3$-
plane. As $\sigma_2 = \sigma_3$ at all times in the triaxial test, all loading and unloading paths for such tests must lie in this plane. The other set of

coordinate axes p, q, where

$$q = \sigma_1 - \sigma_2 \qquad (4)$$

emphasizes the fact that an all-around cell pressure p and superposed additional axial stress q are an equally appropriate set of independent stress components. As indicated in Fig. 6, the scales for p and for q are not the same and both differ from the scale for σ_1.

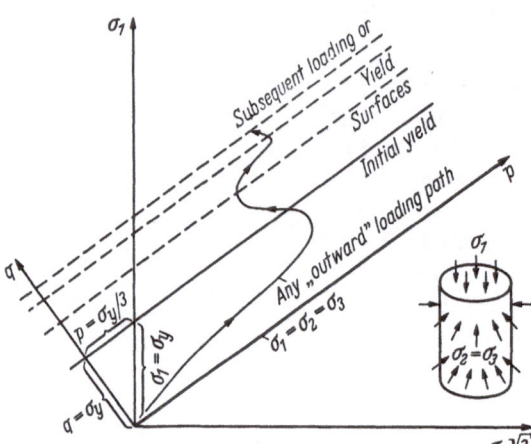

Fig. 6. Triaxial test on metal is equivalent to simple compression.
Note: *Compression* is drawn as *positive*.

$$p = \frac{1}{3}(\sigma_1 + \sigma_2 + \sigma_3) = \frac{1}{3}\sigma_1 + \frac{2}{3}\sigma_2, \qquad q = \sigma_1 - \sigma_2.$$

All scales are linear but unequal as indicated.

No consideration of isotropy has entered into the triaxial test discussion so far. No assumption of isotropy is needed for the choice of axes and the plotting of results in Fig. 6. The salient feature of σ_1 and σ_2 is not that they are principal stresses but that they are a set of independent components of the loading system.

For the record, with compression positive the invariants of the stress tensor and stress deviation tensor are:

$$J_1 = -3p = -(\sigma_1 + 2\sigma_2), \qquad (5)$$

$$J_2 = \frac{1}{3}(\sigma_1 - \sigma_2)^2 = \frac{3}{2}\tau_0^2 = \frac{4}{3}\tau_{\max}^2 = \frac{1}{3}q^2, \qquad (6)$$

$$J_3 = 0. \qquad (7)$$

In this special case, the octahedral shearing stress and the maximum shearing stress are proportional and so provide the same isotropic measure of stress.

Successive yield surfaces appear in Fig. 6 as lines parallel to the p-axis, $\sigma_1 = \sigma_2 = \sigma_3$, because the hydrostatic pressure p has no effect on the plastic deformation of the metal. The plastic strain in a triaxial test on a metal is the same as that produced in a simple compression test with a compressive stress of $\sigma = \sigma_1 - \sigma_2$. Large or small excursions of the loading path in the direction of the positive or negative p-axis cause purely elastic response.

With this lengthy preliminary description of relevant known aspects of the plastic stress-strain relations for metals, water-soil systems can be examined with some perspective.

Isotropic Hardening Idealization for Soils

Consolidation under equal all-around pressure $\sigma_1 = \sigma_2 = \sigma_3$ plays the same role for a water-soil system as simple tension or shear does for a metal. Effective stresses are the significant stresses for soils so that the time of loading must be sufficiently long to permit pore water to escape without pressure build-up, or the pore water pressure must be measured and subtracted from the total state of stress. Although there is an important difference between the soil curve, which has a continuously increasing tangent modulus, and the curve for a metal (Fig. 1a) which exhibits a decreasing tangent modulus as stress is increased in magnitude, both are work-hardening. An increase in stress is needed to produce additional plastic strain. Also, the initial response to a decrease in stress is an elastic recovery.

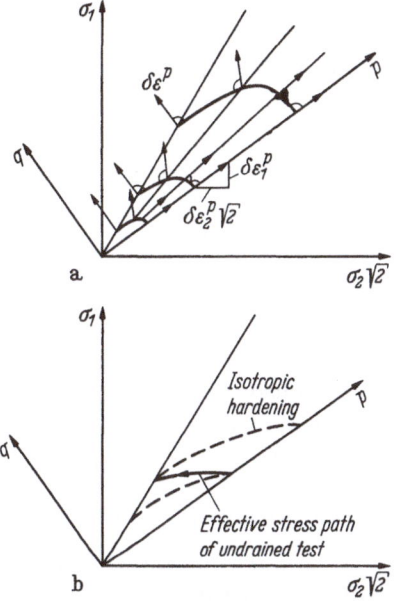

a

b

For soils as for metals it is reasonable to suppose that the results for simple loading extend to all radial or proportional loading paths in the stable domain of the material. Triaxial tests following radial lines, Fig. 7a, have been termed anisotropic consolidation tests by ROSCOE and POOROOSHASB [6] who find the same result as for initially isotropic metals. The ratios of the strain

Fig. 7a and b. Radial loading paths in the stable "wet-clay" region and the assumption of isotropic stress-hardening (see Fig. 2). a) Data of ROSCOE and POOROOSHASB [6] termed anisotropic consolidation; b) Undrained test discussed by CALLADINE [9] and by ROSCOE and SCHOFIELD [8] shown schematically for comparison with another set of isotropic hardening curves [8].

components remain constant as the stresses increase in ratio. It should be emphasized again that this simple result is not an essential ingredient of the theory of plastic stress-strain relations, but it does make for very convenient description [15]. Just as for metals, an isotropic stress-hardening stress-strain relation will fit the data.

The first invariant of the stress tensor J_1 or p must appear prominently in addition to the second invariant of the stress deviation tensor J_2 because appreciable inelastic deformation occurs under hydrostatic pressure. As for metals it is likely that the third invariant also would be needed if data for more elaborate (not triaxial) tests were to be matched with high precision. In the triaxial tests a useful form is obtained with σ_1 and σ_2 or with J_1 and J_2 or with p and q alone because there are only two independent components of applied load.

There is no difficulty in devising a perfectly reasonable test to demonstrate that any isotropic hardening form is badly in error, Fig. 3. Oscillation of q from the one radial limiting line shown (compression test) to the other (extension test) at constant p would be purely elastic according to isotropic theory but undoubtedly would involve appreciable plastic deformation in actuality. This is not the proper point of view at this stage. The first question to be examined is how far isotropic hardening can be pushed before it no longer is useful for soils. As for metals, the answer to this question will depend upon the alternatives available. For soil as for metal, it should be appreciated that hardening is anisotropic and very complex. Yet if the path of loading has only moderate deviations from the outward radial, isotropic hardening should give reasonable answers which differ from reality by strains of the order of elastic strains. Considerable deviation is acceptable in preference to the use of a stress-strain relation so elaborate that practical problems require an inordinate amount of work for their solution. Also, the present state of the art of triaxial or any other testing of soils does not generate excessive faith in the accuracy and reproducibility of results. Pore pressure measurement and equalization are somewhat uncertain, strain measurements offer their difficulties, end effects always are present, and time effects need study.

Rather crude agreement with any theory is cause for rejoicing rather than sorrow. For example, the differences found by CALLADINE [9] in the comparison of undrained tests and radial loading tests are comparable to those for metals. In the undrained test, the total volume change is zero, the plastic contraction balances the elastic expansion. The effective path of loading is far from radial, Fig. 7b, so that deviations from isotropic theory of the order of elastic strains are to be expected. Consequently, the undrained test is likely to place isotropic hardening in a most unfavorable light. The qualitative similarity found over much of the range is extremely encouraging.

Before examining further the shortcomings of the isotropic stress-hardening formulation for general loading paths, it should be noted that in such a theory the state variables may be looked at equivalently as stress and consolidation pressure or stress and plastic change in volume or void ratio. The successive states of the material established by any loading path are identically the same as those established by conventional consolidation under hydrostatic pressure. Isotropic hardening thus is consistent with the ROSCOE-SCHOFIELD-WROTH-HVORSLEV picture [5] and with HENKEL's statement [3] that the con-solidation pressure establishes the state. There is no converse implication however that these concepts require isotropic hardening. Different loading paths, radial or otherwise, need not establish the same states, and in fact will not except in an approximate sense.

For example, the entire domain inside each of the successive isotropic hardening curves is not an elastic or reversible one in actuality. If, Fig. 8a, the all around pressure is reduced to 40% or less of the consolidation pressure and the axial compressive stress σ_1 then is increased, considerable plastic deformation will occur at stresses well below the isotropic hardening curve and the limiting stress. Nevertheless, the agreement between the actual yield curve and a reasonable set of isotropic hardening curves in Fig. 8a seems as good as for metals, Figs. 3, 4.

A real difficulty does arise in connection with the limiting radial line which separates the stable region of behavior from the incipient unstable or the forbidden. The idea of isotropic hardening could be extended beyond

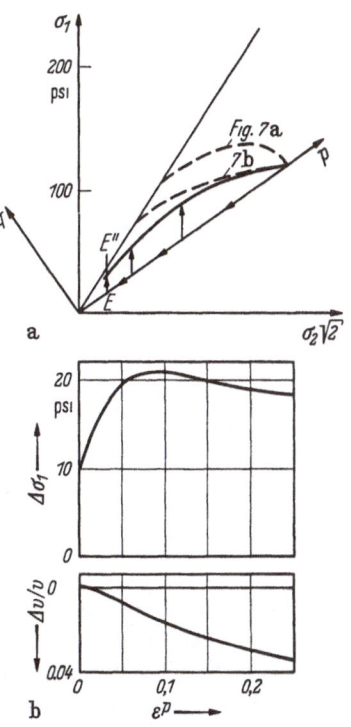

Fig. 8 a and b. Data of HENKEL [1], [3] on remolded silty clay. a) Comparison of Fig. 7 a and Fig. 7 b stress-hardening curves with yield curve [22] found by unloading along $q = 0$ and then adding σ_1 (cf. Fig. 3); b) Loading path $E\,E''$ showing volume *expansion* and drop in stress as plastic deformation proceeds.

the radial line limit by drawing a set of limiting COULOMB lines of the conventional $c-\varphi$ type which also move outward in proportion to the consolidation pressure as shown dotted in Fig. 9. Although overconsolidation does permit reaching these states of stress, the isotropic hardening curves outside the stable zone do not convey an appropriate physical meaning. A radial loading path such as σ_1 alone, or even worse q alone, starting from the origin with zero consolidation is not a permissible path. The soil simply deforms at essentially zero load.

3*

Limitations on the range of validity of a simple idealization are to be expected along with disagreement in detail with experimental data.

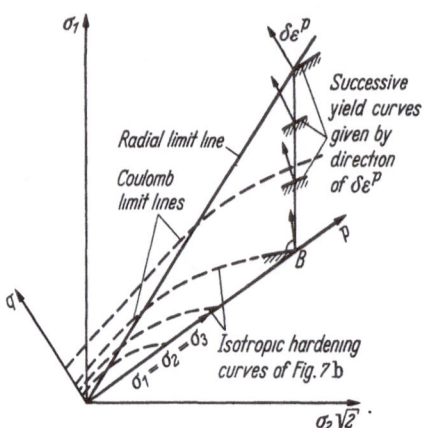

The surprising result indicated in Figs. 7, 8 and 9 is the remarkably good qualitative picture provided by isotropic stress-hardening.

Perfectly Plastic Idealization for Soils

Perfect plasticity is a suitable idealization for metals because successive yield surfaces tend to pile up tangentially against a limit surface as shown in Fig. 5. A soil exhibits neither a piling up nor tangential approach to the limit line. Neither a COULOMB limit line nor the radial limit line is the envelope of successive yield curves but instead is a succession of singular points at which deformation takes place

Fig. 9. Perfect plasticity is not a good idealization of the physical behavior of soil. At the radial limit line, the plastic volume change is zero and the yield curves are parallel to the p-axis. The limit line is not the envelope of the successive yield curves (see Fig. 5) or of successive isotropic hardening curves Fig. 7b. Also, no closing limit curve is appropriate because yield surfaces representing equal increments of consolidation are spaced farther apart as p increases.

with zero volume change (radial line) or small volume change (COULOMB line). The normal to the limit line then is not the limit of the normals to successive yield curves. Therefore the large volume increase predicted by a perfectly plastic idealization based on the COULOMB line is seriously in error. An even larger expansion would be given by the radial line. Volume expansion of smaller magnitude does occur on the two loading paths indicated on Fig. 8 for very highly overconsolidated clay, Fig. 8b.

Another indication of difficulty is the fact that the COULOMB line represents an unstable state of stress Fig. 8b and so is not really within the scope of the mathematical theory of plasticity. Still another is that successive yield curves are spaced farther and farther apart along radial lines instead of closer and closer together as for a metal. The limit approached along radial lines is a locking material, not a perfectly plastic material. This last feature is somewhat worrisome also in the concept of soil as a work-hardening plastic material.

Nevertheless, the combination of perfect plasticity and the tension cut-off proposed [18] to take into account the inability of a soil to carry appreciable tension does lead to a sensible and experimentally correct result for the stable height of a free-standing vertical embankment or ditch. This probably is not due to the perfectly plastic assumption but instead indicates that lack of tensile strength is the governing feature. Any appropriate consistent mechanism of energy dissipation is likely to do as well.

Material Stability and Stress-Strain Relations for Water-Soil Systems

The inelastic volume changes which occur in a saturated soil are in marked contrast to the reversible elastic behavior of structural metals under normal variation of hydrostatic pressure. However, as already discussed, this difference is not a crucial one in itself. Plastic volume changes can be included without difficulty in plastic stress-strain relations. In fact, a stable foamed structural metal could be imagined in which uniformly dispersed microscopic spherical holes occupied an appreciable fraction of the total volume. Irreversible changes in the volume of the holes would occur as the metal was subjected to triaxial tests in the plastic range. Radial paths of loading would give results which resemble those of Fig. 7. Convexity of yield surfaces and normality of increments of plastic strain for all paths of loading would follow from the postulate of stability of material.

If the volume of holes in the foamed metal were very large, as in a soil, the elastic compressibility would be appreciably greater than for a solid metal. As the applied hydrostatic pressure increased and caused plastic deformation, the compressibility would decrease measurably. Assuming local stability of all portions of the foamed metal, the increase in effective elastic moduli would be far less than for a saturated clay undergoing consolidation. In part, however, this is a matter of the scale of stress, in particular the ratio of yield stress to YOUNG's modulus. It does not involve a fundamental qualitative difference between the behavior of the foamed metal and of soil. In both cases, some degree of concavity of the yield surface becomes permissible when the elastic constants change radically with plastic deformation but is not likely to occur. When it appears again, the word "convexity" will be understood to include this limited concavity possibility.

Similarly, the inability of soil to carry appreciable tension can be duplicated in the imagined metal by taking the yield stress and work-hardening in tension as zero or as small in comparison with the corresponding quantities in compression.

In fact, a complete and consistent picture can be obtained which looks just like the soil with two significant and at times overriding omissions. One is the absence of a limiting radial line which cuts through the successive yield surfaces or curves. The other is the absence of an effect equivalent to that produced by the change in water content.

The change in water content introduces a chemical potential which must be taken into account in a complete analysis of the stability of a water-soil system. Transfer of mass as the free water is driven out, or sucked in, does not appear to be the significant aspect. It is the much smaller amount of water, involved in bonds between the particles of soil, whose contribution to the energy of the system must be considered. As soil is consolidated, free water is driven out, but the bound water content increases. Part of the work done on the system is stored as chemical energy. When a soil expands as it shears, the bound water decreases and chemical energy is released.

Bound water is the state variable, not void ratio or total water or free water. A reversible path then is one in which the bound water content does not change. Recovery on partial release of p and subsequent recompression back to the original value of p changes the void ratio and the total water content appreciably. To a first approximation free water only is driven out and taken back and the cycle of unloading and reloading may be considered as reversible or elastic. Of course, with time an expanded soil will lose some of its bound water and a compressed soil will increase its bound water. Therefore in the very strict sense, no reversible paths exist. Experimental evidence does indicate, however, that reversibility is a permissible idealization for extensive changes of p in the triaxial test over customary time scales.

The increase in stored energy during the process of consolidation adds to the stability of the system of water, soil, and the applied forces. Therefore in the terminology of the postulate of stability, [11], [12] the work done by the external agency on the change in displacements it causes will be positive. When the soil plus the loads applied to it are considered as a system, it is not possible to extract energy from the system through any cycle of application and removal of additional load which stays inside the radial limiting line and does not lead to dilatation. As for a metal, "convexity" of yield surfaces or curves follows in the region of stability of water-soil systems along with the normality of the plastic strain increment vector as illustrated in Fig. 1 and Fig. 7a. On the contrary, in the region of dilatation, the release of binding energy can lead to mechanical instability. Overall stability in combined mechanical and chemical terms is likely but there is no obvious requirement of convexity and normality. Nevertheless, the plastic strain increment vectors determined experimentally [22] are normal to the initial yield curve of Fig. 8a, a curve which is convex. Perhaps stability in the mechanical sense does hold below the radial limiting line. The stable, or rising, stress-strain curve for this region, Fig. 8b, is a hopeful sign.

These and related questions are being studied carefully by PALMER and will form the basis of a later paper. The main points to be emphasized here are that in a wide variety of experimental programs and in much of the natural history of water-soil systems in situ, mechanically stable behavior equivalent to that of structural metals is observed and is to be expected. Yet it remains true that water-soil systems exhibit a basic instability of material. Energy can be extracted repeatedly from a system of soil and applied loads p, q by a suitable choice of a stress cycle involving a decrease of p. This can lead to a failure of a type which is not possible in a proper stable plastic system.

A mechanical instability has grave consequences of obvious nature but also of a more subtle type. In large measure, structural design is based intuitively on the concept of stability. It is almost always true that adjustments in configuration required by fabrication can be depended upon to take place safely, peak stresses tend to be smoothed out by plastic action, the materials and the structures do their best to carry the design loads and occasional overloads. These consequences of the lower bound theorem of plastic limit analysis and design, as well as consequences of the upper bound theorem, often are described as obvious because they do contain and express in precise form the intuitive feelings of the designer. Unfortunately, neither they nor their counterparts for work-hardening material do hold in general for a material or for a structure which is mechanically unstable over part of its range of behavior.

A soil mass, or a laboratory specimen, may not cooperate to do its best to carry the applied loads. An equilibrium state of stress may exist which is perfectly safe and yet the soil may progress toward failure. The lower bound theorem of limit analysis does not hold directly or in an extended manner for work-hardening.

Unfortunately also, the basic principle of the upper bound theorem is inapplicable in general because there is no way of computing the energy dissipation from the state of the material and the plastic strain alone. The normal pressure on the plane of sliding, or the value of p, is needed in addition and is not known in the usual statically indeterminate problem. A possible but not very convenient upper bound theorem was proposed previously [22] which philosophically is the converse of the concept of stability of material. The postulate is that *failure* will occur if at all possible and so requires a look at all modes of failure under all equilibrium distributions of stress:

Failure should be assumed to occur if for any pattern of deformation the work done by the applied forces exceeds the least allowable dissipation. The allowable dissipation is to be computed from any equilibrium distribution of normal stresses (or p) on the assumed surface of sliding, or region of plastic deformation.

Corners or Rounded Noses on Yield Curves

If hardening were isotropic, the yield curves of Figs. 7—9 would be symmetric about the p-axis. Hydrostatic pressure produces hydrostatic contraction and the expectation might be that the yield curves established by consolidation would be like the smooth curves of Fig. 7a which intersect the p-axis at right angles. However, these curves were drawn for radial loading and do not necessarily represent actual yield curves. The implication in Figs. 7b, 8a, and 9 is strong that the standard consolidation test does produce a very sharp corner on the p-axis, a corner which moves radially outward with the stress point. The ratios of the plastic strain increments as observed by HENKEL [3], the undrained test in Fig. 7b, and the more recent analyses of the Cambridge group [7], [8] are in agreement here.

Whether the corner is absolutely sharp or is well-rounded, as those of Fig. 3a for an aluminium alloy, is not too relevant here. When a small loading increment of stress is applied with an appreciable component at right angles to the direction of the loading which produced the corner or rounded nose, the vector representing the increment of plastic strain rotates rather abruptly through a large angle. This is shown in Fig. 9 by the left-of-vertical direction of the arrow at B which gives the plastic strain increment as σ_1 is increased at constant σ_2. An abrupt counterclockwise rotation of about 60° from the positive direction of the p-axis occurred as soon as the loading path left the consolidation line. As σ_1 is increased, the very large component of distortion dominates the additional consolidation numerically.

The question of the stability of the consolidation test arises immediately. A continual succession of small perturbations in the equality of σ_1 and σ_2 would appear eventually to induce large inequalities in the plastic strain components ε_1^p and ε_2^p which could not reasonably be expected to average out in each test. All radial loading paths are likely to produce a corner and yet Fig. 7a indicates no deviations in plastic strain ratios as the loading proceeds. A possible but not the most probable answer is that as in aluminum the corner really is a rounded nose and the likely perturbations are too small to be effective. Probably, the main stabilizing influence is the pore water pressure and the appreciable time required for deformation to occur. Inadvertent small fluctuations in load over short periods of time have little if any effect. Also, if for any planned ratio of σ_1 to σ_2 the test set-up permits σ_1 to increase over σ_2 temporarily, the stable reaction of the system is to have ε_1 increase over ε_2 and so reduce σ_1 immediately in most loading systems.

Time-Dependence

Water-soil systems are time-dependent in the laboratory and in the field. Perhaps then the time-dependence is primary and it is never permissible to idealize behavior as time-independent simply by taking sufficient time to reach the asymptotic or equilibrium state. However, the same argument can be raised for structural metals at ordinary temperatures where the answer clearly is that time-independence is a reasonable approximation for the usual rates of loading. Metals at elevated temperature and many polymers at ordinary temperatures are so time-dependent that ignoring time-effects generally is not a permissible idealization.

A viscoelastic or a viscoplastic or a more complex idealization of a soil system can be devised easily. Energy release of elastic elements in an elasto-visco-plastic material is comparable in some sense to the energy release in a soil as bound water becomes free water.

Reasonable requirements of stability for a time dependent system [12] are far weaker than for a time-independent one. Work can be extracted in a cycle of application and removal of external load. It is the rates of strain which are restricted rather than the total increments in strain over a cycle. Unfortunately all would not be well in such a consistent system of stress-strain relations for soil mechanics. Time-dependent systems can be, and often are, mechanically unstable in the large. For example, columns composed of a viscous or a Maxwell material inevitably must buckle given enough time. Instability of soil masses, of foundations, or of retaining walls may well be only a matter of time but this prospect appears neither helpful nor encouraging. For some practical, non-geologic, time scales at least, time-independence of the asymptotic states should be a permissible idealization.

To an appreciable extent, the appropriate asymptotic state does depend upon the time scale chosen. Correspondingly, for continued loading, the proper rate of testing is somewhat arbitrary. Different choices are possible and reasonable, although conceivably they may lead to results as far apart as the two isotropic hardening curves drawn side by side in Fig. 8a.

Concluding Remarks

Much additional work is needed on each of the topics touched on in the paper and the many more which were omitted. The thermodynamic study of stress-strain relations including the chemical potential associated with the bound water and allied topics should prove especially fruitful. Differences between frictional and plastic behavior must be explored further in an effort to broaden the scope of the present descriptions of mechanical properties in mathematical or pictorial form.

Stability of embankments, foundations, and similar configurations or structures should be examined with great care to determine the realm of absolute stability, of stability over ordinary, time, and of stability over geological time. Useful theorems are needed by designers for practical calculations. Appropriate theorems require clarification of the degree of stability to be expected. Closely associated with this point is the question of whether time-dependence is so essential an aspect of water-soil systems that a time-independent idealization has but little significance except in special cases. The rather tentative conclusion is reached that for times of the order of long laboratory experiments, a time-independent approximation will prove useful over a wide range of conditions.

Isotropic hardening has been shown to be a far better approximation to reality than should be expected on the basis of the behavior of metals. It is a drastic idealization which should not be pushed too far. However, an extensive series of tests to examine the yield surfaces which are established by various paths of loading should prove as useful for soils as it has been for metals.

References

[1] DRUCKER, D. C., R. E. GIBSON and D. J. HENKEL: Soil Mechanics and Work-Hardening Theories of Plasticity. Proceedings A.S.C.E. **81**, 1955, Separate 798, Transactions A.S.C.E. **122**, 338—346 (1957).

[2] BISHOP, A. W., and D. J. HENKEL: The Measurement of Soil Properties in the Triaxial Test, London: Arnold 1957.

[3] HENKEL, D. J.: The Correlation Between Deformation, Pore Water Pressure and Strength Characteristics of Saturated Clays. Ph. D. Thesis, Imperial College University of London, 1958. See also Géotechnique **6**, 139—150 (1956), **9**, 119—135 (1959), **10**, 41—54 (1960).

[4] PARRY, R. H. G.: Triaxial Compression and Extension Tests in Remoulded Saturated Clay. Géotechnique **10**, 166—180 (1960).

[5] ROSCOE, K. H., A. N. SCHOFIELD and C. P. WROTH: On the Yielding of Soils. Géotechnique **8**, 22—53 (1958).

[6] ROSCOE, K. H., and H. B. POOROOSHASB: A Theoretical and Experimental Study of Strains in Triaxial Compression Tests on Normally Consolidated Clays. Géotechnique **13**, 12—38 (1963).

[7] ROSCOE, K. H., A. N. SCHOFIELD and A. THURAIRAJAH: Yielding of 'Wet' Clays. Géotechnique **13**, 211—240 (1963).

[8] ROSCOE, K. H., and A. N. SCHOFIELD: Mechanical Behavior of an Idealised 'Wet-Clay', ASTM-NRC (Canada) Symposium on Laboratory Shear Testing of Soils, Ottawa, 1963.

[9] CALLADINE, C. R.: The Yielding of Clay. Géotechnique **13**, 250—255 (1963).

[10] HAYTHORNTHWAITE, R. M.: Stress and Strain in Soils. Proceedings 2nd Symposium on Naval Structural Mechanics (Brown University), Ed. E. H. LEE and P. S. SYMONDS, Pergamon Press, 1960, pp. 185—193.

[11] DRUCKER, D. C.: A More Fundamental Approach to Stress-Strain Relations. Proceedings First U. S. National Congress of Applied Mechanics, A.S.M.E., 1951, pp. 487—491.

[*12*] DRUCKER, D. C.: A Definition of Stable Inelastic Material. J. Applied Mechanics **26**, 101—186 (1959), also see J. Mécanique, June 1964.

[*13*] DRUCKER, D. C.: On the Role of Experiment in the Development of Theory, Proceedings 4th U. S. National Congress of Applied Mechanics, A.S.M.E., 1962, pp. 15—33; Basic Concepts in Plasticity, Chap. 46, Handbook of Engineering Mechanics, Ed. W. FLÜGGE, McGraw-Hill, 1962; Chap. IV in Rheology, Vol. 1, Academic Press, 1956; Stress-Strain Relations in the Plastic Range — A Survey of Theory and Experiment, Brown University Report A-11, S-1 to the Office of Naval Research, 1950, for numerous references.

[*14*] PRAGER, W.: An Introduction to Plasticity, Addison-Wesley, 1959.

[*15*] EDELMAN, F., and D. C. DRUCKER: Some Extensions of Elementary Plasticity Theory. J. Franklin Inst. **251**, 581—605 (1951).

[*16*] IVEY, H. J.: Plastic Stress-Strain Relations and Yield Surfaces for Aluminum Alloys. J. Mechanical Sciences, IME, **3**, 15—31 (1961).

[*17*] BERTSCH, P., and W. N. FINDLEY: An Experimental Plasticity Study of Corners, Normality and Bauschinger Effect on Subsequent Yield Surfaces. Proceedings 4th U.S. National Congress of Applied Mechanics, A.S.M.E., 1962.

[*18*] DRUCKER, D. C., and W. PRAGER: Soil Mechanics and Plastic Analysis or Limit Design. Quart. Applied Mathematics **10**, 157—165 (1952).

[*19*] SHIELD, R. T.: On Coulomb's Law of Failure in Soils. J. Mechanics and Physics of Solids **4**, 10—16 (1955).

[*20*] JENIKE, A. W., and R. T. SHIELD: On the Plastic Flow of Coulomb Solids beyond Original Failure. J. Applied Mechanics **26**, 599—602 (1959).

[*21*] DRUCKER, D. C.: Coulomb Friction, Plasticity, and Limit Loads. J. Applied Mechanics **21**, 71—74 (1954).

[*22*] DRUCKER, D. C.: On Stress-Strain Relations for Soils and Load Carrying Capacity, Proceedings of the 1st International Conference on the Mechanics of Soil- Vehicle Systems, Torino: Minerva Tecnica 1961, pp. 15—23.

Discussion

Question posée par P. ANGLES d'AURIAC: Le «principe» de M. DRUCKER est-il une conséquence des principes de la thermodynamique ou bien un postulat nouveau surajouté aux autres.

Dans cette deuxième hypothèse, à quelle idée *physique* correspond ce nouveau principe.

Réponse de D. C. DRUCKER: In answer to Professor ANGLES d'AURIAC, I should like to make it very clear that my postulate of stability of material is a classification of material. It is not a law of nature or principle of thermodynamics. Both time-dependent and time-independent materials have been treated by considering the system of material plus the forces acting on the material. A time-independent system should not change its state spontaneously. Another aspect of my postulate is that energy cannot be extracted from the system in a cycle of application and removal of *additional* forces. This reduces to the second law of thermodynamics when the system is one of material alone, no forces applied. In general, however, the impossibility of extracting work is an expression of the behavior of most materials but not all. Mechanically unstable materials do exist as shown by the falling portion of the stress-strain curve for an overconsolidated clay, Fig. 8b of the paper. These and allied questions are discussed in References 11, 12, 21, 11 and in the Journal de Mécanique, June 1964.

Question posée par   R. HAYTHORNTHWAITE: You have described certain experimental data which is in conformity with the existence of a plastic potential. Do you consider that the way is now clear for the application of the theory of plasticity to practical situations where such soils are present?

Questions posées par BENT HANSEN: Mr. DRUCKER has explained how the normality condition is valid for a work-hardening soil without creating a contradiction at the point of final failure. Such points are singular, the direction of the individual failure surfaces not coinciding with the curve through the points (the MOHR envelope).

However, in practical engineering one is concerned with states of final failure. In this case one has the entire mass of soil partitioned into failure zones and rigid bodies. In the failure zones one must use the equilibrium conditions, the failure condition which corresponds to the MOHR envelope, and the movement conditions which do not then obey the normality condition (when $\varphi \neq 0$).

This does not matter so much when the zones are statically determined, i. e. they can be determined uniquely by the equilibrium and failure conditions. But in a number of important cases the zones have mixed boundary conditions, being entirely surrounded by rigid bodies.

In such cases it seems that the normality condition cannot be used in practice. Thus, the uniqueness of solutions cannot be proved, and the limiting conditions are also not valid. In this situation there seems to be two possible ways. One is to use more advanced limiting theorems as proposed by de JOSSELIN de JONG, but for some problems it is rather difficult to construct statically admissible stress fields, and if one is not careful they may be much too far on the safe side. The other possibility is to start with the prestressings, supposed to be known, and carry the calculations through the intermediate elastic-plastic states. For soils with internal friction, the final state of failure may in fact depend on the prestressings in the unloaded state.

Réponse de D. C. DRUCKER: The question posed by Professor HAYTHORNTHWAITE has been answered to a considerable extent by Mr. BENT HANSEN. An isotropic hardening stress-strain relation or a much more elaborate form can be used with modern electronic computers to solve problems step by step. Initial conditions may well be important and determine the final failure state because plastic deformation is path-dependent and the radial or COULOMB limit lines, Fig. 9, are not envelopes of successive yield curves. Whether or not such step by step computation up to failure would be worth while depends upon the need to obtain better answers to problems of practice and the faith one has in the generality of the very limited experimental data presently available.

Question posée par A. W. SKEMPTON:

These diagrams illustrate typical stress-strain curves for sands and clays, showing the tendency to approach an unique 'residual' strength at large strains, independent of the original density of the material. It is to be noted

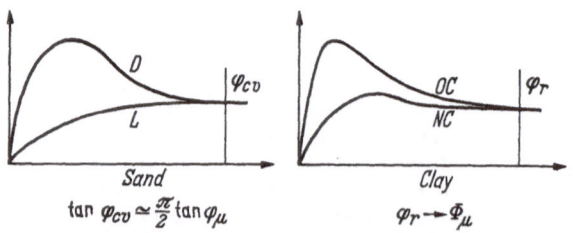

$$\tan \varphi_{cv} \simeq \frac{\pi}{2} \tan \varphi_\mu$$

$$\varphi_r \to \Phi_\mu$$

that whilst loose sand rises continuously towards this limiting strength, normally-consolidated clays generally show a peak, the subsequent decrease in strength being due to particle orientation.

Contribution de K. H. Roscoe. Professor Skempton has suggested that attempts to make analogies between the mechanical behaviour of cohesionless and cohesive media are unwarranted and may be particularly misleading when considering the long term stability of clays. He illustrates this argument by comparing the results of drained standard shear box reversal tests (constant piston load) on a sand and a clay as indicated in a somewhat idealised manner in Figs. 1 and 2 respectively. In the sand tests (Fig. 1) the relationship between the observed mean shear stress (τ) and the cumulative relative displacement of the two halves of the box (x) correspond to ODX_1R_1 for initially dense and OLX_1R_1 for initially loose samples. On the other hand Skempton states that the corresponding curves for over-consolidated clays is OPX_2R_2 in Fig. 2 and for normally consolidated clays is OQX_2R_2. He notes two particular features of dissimilarity between the sands and clays. Firstly the residual strength after great strain τ_{R_1} is the same as the ultimate (or critical state) strength τ_{X_1} for sands but τ_{R_2} is much less than τ_Q or τ_{X_2} for clays. Secondly the shear stress attains a distinct peak value at Q for normally consolidated clays. He implies that the critical state concept is not valid for such clays.

In contrast, I believe that there are many points of similarity in the mechanical behaviour of cohesionless and cohesive media but do not thereby suggest that all cohesionless and cohesive media behave similarly under a given set of conditions. The concept of the critical state was developed for conditions of random packing of isotropic media. The sand (Fig. 1) would appear to be such a medium but the clay (Fig. 2) is obviously not. The probable explanation of the gradual fall in strength of the clay to the low residual value τ_{R_2} is that it corresponds to the gradual attainment of a regular orientation of the clay particles in the shearing zone. If Skempton had compared the behaviour of his clay to that of a cohesionless medium which could have developed anisotropy during shear, he would have observed a very marked measure of similarity. Such a medium can be obtained by taking a pack of uniform steel balls. Loose and dense samples of steel balls in initially random states of packing will, for a given piston load, attain a common critical state in which they are still randomly packed. Typical curves for

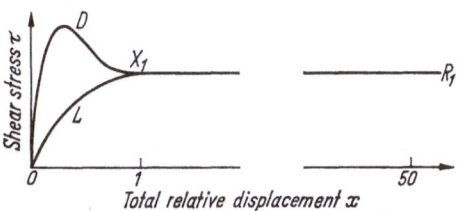

Fig. 1. Form of stress-strain curves for drained shear box reversal tests on dense (D) and loose (L) sands.

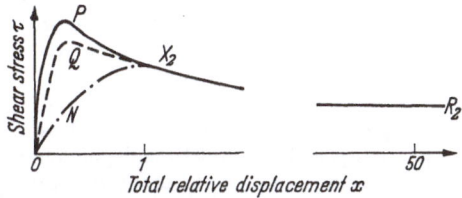

Fig. 2. Form of stress-strain curves for drained direct shear reversal tests on media that can develop particle orientation. OPX_2R_2 for overconsolidated clays or dense steel balls; OQX_2R_2 for normally consolidated clays in the standard shear box; ONX_2R_2 for normally consolidated clays or loose steel balls in the simple shear apparatus.

the dense and loose samples in the simple shear apparatus would be OPX_2 and ONX_2 respectively in Fig. 2. If either loose or dense samples are then sheared to strains of the order of fifty times larger than those required to attain the critical state they develop a regular state of packing. In this process the density of the medium *increases* by 30% while the shear strength (as measured in the simple shear apparatus) *diminishes* by more than 30%. For further details see ROSCOE (1961) or WROTH (1958). I suggest that strains cannot be measured reliably in the standard shear box and that the strain distribution is not uniform (see ROSCOE 1953). This will cause the value of τ_{max} in tests on dense, or over-consolidated, samples to be underestimated and the corresponding strain will also be erroneous. Hence the location of P, and possibly Q, in Fig. 2 is probably incorrect. In any case the magnitudes of the strains are not a relevant parameter in so far as the critical state concept is concerned. I believe that over-consolidated and normally consolidated samples of saturated remoulded clays, which can develop structural anisotropy, behave in a very similar manner to the dense and loose samples of steel balls. Both clay samples will, if under the same normal stress, first attain the critical state and then their strength will gradually diminish as the aniso-tropic packing develops.

Finally I would mention that whereas when yielding in the randomly packed conditions the stresses can properly be related to the voids ratio, in the case of regular packings this is not so. A more relevant parameter than voids ratio would probably be the number of discontinuities in the regularity of the packing.

Réponse de D. C. DRUCKER: Professor SKEMPTON's discussion of the stress-strain behavior of sand and of clay is most interesting as is the contribution by Mr. ROSCOE to the points raised by Professor JOSSELIN DE JONG. Clearly, when two such discerning and careful experimenters are not in agreement on funda-mental issues, much more data must be obtained by a variety of proven experi-mental techniques.

Question posée par TAN TJONG-KIE: 1. The concept of time-independent plasticity needs further study. As far as I know the failure strength is dependent on the way of loading or strain increase. As time-effects are primary factors in clays, it is possible that the yield surface will not be a definite one, but will vary within some limits dependent on the stress or strain paths.

2. Further in common tests the failure-strength is measured, which is larger than the upper-yield-value. When the stresses exceed this upper yield value accelerating flow leading to failure generally occurs, which makes an unique determination of the failure strength more complicated. Therefore I am of opinion thet the yield surface (s) should be related to the upper yield value better than to the customary failure strength.

Réponse de D. C. DRUCKER: Professor TAN TJONG-KIE is entirely correct in emphasizing, as I do, that time-dependence is real and often is governing. How-ever, the same situation exists for an ordinary structural steel. Very rapid loading rates will show a dynamic yield stress twice the static value. Yet for a large range of strain rates of great practical importance, steel can be and is idealized as time-independent. The belief is stated in the paper that this is true for soil as well.

References

ROSCOE, K. H. (1953): "An apparatus for the application of simple shear to soil samples". Proc. 3rd. Int. Conf. Soil Mech. Vol. 1, pp. 186—191.

ROSCOE, K. H. (1961): Contribution to the discussion on "Soil properties and their measurement". Proc. 5th. Int. Conf. Soil Mech. Vol. III, p. 105—107.

WROTH, C. P. (1958): The behaviour of soils and other granular material when subjected to shear. Ph. D. Thesis, Cambridge University.

1.4 On Elastic/Visco-Plastic Soils

By

W. Olszak and **P. Perzyna**

Introduction

The paper consists of two parts. The first part is devoted to a discussion of the constitutive equations describing the dynamic properties of soils. In the second part, we shall consider the problem of propagation of stress waves in elastic/visco-plastic soils.

Many problems of soil mechanics are often treated as problems of plasticity. The soil is replaced by an idealized material which behaves elastically up to some state of stress at which yielding occurs. This useful idea was introduced by D. C. Drucker and W. Prager [5].

Recent experimental results show clearly that soils exhibit rheological effects and are sensitive to the change of the strain rate (see for instance N. A. Alexeev, Ch. A. Rakhmatulin and A. Y. Sagamonian [1]). Thus the feature of the behaviour of soils, especially their dynamic response, is the time dependence of the deformation process.

It is clear that in real soils the anelastic strain rate tensor depends on time stress-history as well as on path-stress history. In inviscid plasticity theory of soils, on the other hand, the plastic strain rate tensor is only path-dependent.

The basic aim of this paper is a description of the rheological and dynamical properties of a class of soils.

We assume the Drucker's postulate of a stable inelastic material and introduce the additional restrictions that soils behaves in the plastic range as path-dependent and time-dependent material[1].

We shall introduce the notion of an elastic/visco-plastic body (see [8], [18], [17], [7], [13], [14]). This enables us to describe the influence of the strain rate on the behaviour of a soil. It is assumed that the soil is purely elastic before the plastic state is reached and becomes elastic/visco-plastic if that limit is exceeded. Owing to this assumption, the original yield condition, which will be called the static yield crite-

---

[1] See a classification scheme for materials given by D. C. Drucker [6].

rion, will not differ from the known yield criteria of the inviscid plasticity theory of soil [5], [11], [12].

1. General Constitutive Equations

Let us recall first the definition of stable inelastic materials with the inclusion of dynamic terms. This postulate was introduced by D. C. Drucker [4], and can be stated in the following way:

The work done by the external energy on the change in displacement produced must be positive or zero.

With reference to rectangular Cartesian coordinates, used throughout the paper, the mathematical expression of this postulate for an isothermal mechanical system leads to the following condition:

$$\int_{t=0}^{t_c} \left\{ \int_V [\sigma_{ij}^{(2)} - \sigma_{ij}^{(1)}][\dot{\varepsilon}_{ij}^{(2)} - \dot{\varepsilon}_{ij}^{(1)}]\, dV \right\} dt$$
$$+ \left\{ \int_V \frac{1}{2}\varrho[\dot{u}_i^{(2)} - \dot{u}_i^{(1)}]\, dV \right\}_{t=0}^{t_c} \geqq 0, \tag{1.1}$$

where the stress tensors $\sigma_{ij}^{(1)}$ and $\sigma_{ij}^{(2)}$ and the corresponding strain rate tensors $\dot{\varepsilon}_{ij}^{(1)}$ and $\dot{\varepsilon}_{ij}^{(2)}$ refer to two distinct mechanical paths. The first integral is extended over a finite interval of time $0 \leqq t \leqq t_c$, the second term is the value of the volume integral at t_c minus the value of divergence of paths at the time $t = 0$. Necessarily, in all cases the velocity $\dot{u}_i^{(2)}$ throughout the volume is the same as $\dot{u}_i^{(1)}$ at $t = 0$, and the second term of (1.1) is positive or zero at all times t_c. Therefore, the inequality

$$\int_{t=0}^{t} \left\{ \int_V [\sigma_{ij}^{(2)} - \sigma_{ij}^{(1)}][\dot{\varepsilon}_{ij}^{(2)} - \dot{\varepsilon}_{ij}^{(1)}]\, dV \right\} dt \geqq 0 \tag{1.2}$$

is much more significant. It is obvious that the inclusion of dynamic terms does not affect the conclusions concerning the stress-strain relations to be drawn by neglecting these terms.

Let us assume that the strain rate tensor can be decomposed into elastic and anelastic parts

$$\dot{\varepsilon}_{ij} = \dot{\varepsilon}_{ij}^e + \dot{\varepsilon}_{ij}^p. \tag{1.3}$$

The anelastic strain rate tensor $\dot{\varepsilon}_{ij}^p$ represents both the viscous and plastic response of the material.

As basic conclusions from the inequality (1.2) and in consequence of Eq. (1.3), we obtain the condition of the convexity of the subsequent dynamic loading surface and of the normality of the anelastic strain rate vector to the yield surface.

We might have introduced these two conditions also in the usual way, without reference to Drucker's postulat.

To describe the work-hardening of soils, let us introduce the statical yield function in the form

$$F\left(\sigma_{ij}, \varepsilon_{ij}^{p}\right) = \frac{f\left(\sigma_{ij}, \varepsilon_{ii}^{p}\right)}{\varkappa} - 1, \qquad (1.4)$$

where the function $f(\sigma_{ij}, \varepsilon_{ij}^{p})$ depends on the state of stress σ_{ij} and of the state of anelastic strain ε_{ij}^{p}, and where

$$\varkappa = \varkappa \left(\int_{0}^{\varepsilon_{ij}^{p}} \sigma_{kl} d\varepsilon_{kl}^{p} \right) \qquad (1.5)$$

is the work-hardening parameter.

We shall assume that the loading surface $F = 0$ considered in the stress space is regular and convex.

An elastic/visco-plastic soil may be described by the following constitutive equations

$$\dot{\varepsilon}_{ij} = \frac{\dot{s}_{ij}}{2\mu} + \frac{1-2\nu}{E} \dot{\sigma}_m \delta_{ij} + \gamma \langle \Phi(F) \rangle \frac{\partial f}{\partial \sigma_{ij}}, \qquad (1.6)$$

where ν, μ and E are elastic constants, γ is the constant of viscosity of soil, $\sigma_m = 1/3\sigma_{kk}$, $s_{ij} = \sigma_{ij} - \sigma_m \delta_{ij}$, and the dot denotes differentiation with respect to time. The symbol $\langle \Phi(F) \rangle$ is defined thus

$$\langle \Phi(F) \rangle = \begin{cases} 0 & \text{for } F \leq 0, \\ \Phi(F) & \text{for } F > 0. \end{cases} \qquad (1.7)$$

The function $\Phi(F)$ should be selected on the basis of experimental results of dynamic properties of soils.

The constitutive Eq. (1.6) are based on the assumption that the anelastic components of the strain rate tensor $\dot{\varepsilon}_{ij}^{p}$ are functions of the difference between the actual state of stress and the state corresponding to the static yield condition. The function of this stress difference $\Phi(F)$ generates anelastic strain rates according to MAXWELL's viscosity law. The elastic part of the strain tensor is considered to be independent of the strain rate.

In view of the assumption that $f(\sigma_{ij}, \varepsilon_{ij}^{p})$ is a function of the three invariants of the stress tensor, we have $\partial f/\partial \sigma_{ii} \neq 0$, and we find easily

$$\dot{\varepsilon}_{ii} = \frac{1-2\nu}{E} \dot{\sigma}_{ii} + \gamma \langle \Phi(F) \rangle \frac{\partial f}{\partial \sigma_{ii}}. \qquad (1.8)$$

The relation (1.8) determines the rate of cubical dilatation of the soil.

To discuss more accurately the constitutive Eq. (1.6), let us consider the anelastic part of the strain rate

$$\dot{\varepsilon}_{ij}^{p} = \gamma \langle \Phi(F) \rangle \frac{\partial f}{\partial \sigma_{ij}}. \qquad (1.9)$$

Squaring both sides of Eq. (1.9) and denoting by $I_2^p = 1/2\dot{\varepsilon}_{ij}^p\dot{\varepsilon}_{ij}^p$ the invariant of the anelastic strain rate tensor, we obtain

$$I_2^p = \frac{1}{2}\gamma^2 \langle \Phi^2(F)\rangle \frac{\partial f}{\partial \sigma_{kl}} \frac{\partial f}{\partial \sigma_{kl}}. \tag{1.10}$$

The Eq. (1.10) leads to the following dynamic yield criterion

$$f(\sigma_{ij}, \varepsilon_{ij}^p) = \varkappa \left\{ 1 + \Phi^{-1}\left[\frac{(I_2^p)^{1/2}}{\gamma}\left(\frac{1}{2}\frac{\partial f}{\partial \sigma_{kl}}\frac{\partial f}{\partial \sigma_{kl}}\right)^{-1/2}\right]\right\}, \tag{1.11}$$

where Φ^{-1} denotes the inverse function of Φ.

The expression (1.11) shows the character of the influence of the strain rate on the yield criterion. At the same time it describes the actual variability of the surface of flow during the process of anelastic deformation. This variability is produced by isotropic and anisotropic work-hardening of the soil and the rheological effects.

2. Special Cases of Constitutive Equations

As a particular case of the constitutive Eq. (1.6), we shall study the elastic/visco-perfectly plastic soil for the following static yield function

$$F = \frac{\alpha J_1 + J_2^{1/2}}{k} - 1, \tag{2.1}$$

where α is a constant describing the dilatation rate of the soil, k denotes the plastic constant and J_1 and J_2 are the first invariant of the stress tensor and the second invariant of the stress deviation, respectively.

In this case the constitutive Eq. (1.6) give

$$\dot{\varepsilon}_{ij} = \frac{\dot{s}_{ij}}{2\mu} + \frac{1-2\nu}{E}\dot{\sigma}_m\delta_{ij} + \gamma\left\langle \Phi\left[\frac{\alpha J_1 + J_2^{1/2}}{k} - 1\right]\right\rangle\left(a\delta_{ij} + \frac{s_{ij}}{2J_2^{1/2}}\right). \tag{2.2}$$

The rate of cubical dilatation takes the following form

$$\dot{\varepsilon}_{ii} = \frac{1-2\nu}{E}\dot{\sigma}_{ii} + 3\alpha\gamma\left\langle \Phi\left[\frac{\alpha J_1 + J_2^{1/2}}{k} - 1\right]\right\rangle. \tag{2.3}$$

Eq. (2.3) shows that the anelastic deformation is accompanied by an increase of volume, if $\alpha \neq 0$. This property is known as dilatancy.

The dynamic yield condition following from (2.2) has the form

$$\alpha J_1 + J_2^{1/2} = k\left\{1 + \Phi^{-1}\left[\frac{(I_2^p)^{1/2}}{\gamma}\left(\frac{3}{2}\alpha^2 + \frac{1}{4}\right)^{-1/2}\right]\right\}. \tag{2.4}$$

It may be interesting to show that we can obtain the constitutive equations of the flow theory as a limit case from the general constitutive equations of elastic/visco-plastic soil (see [14]).

With this assumption let us put that $\gamma \to \infty$. According to (1.4), this means that

$$F = 0 \quad \text{or} \quad f(\sigma_{ij}, \varepsilon_{ij}^p) = \varkappa. \tag{2.5}$$

From (2.5) and by definition of the function $\Phi(F)$ in the limit case $(\gamma \to \infty)$, we find $\Phi(F) \to 0$; then $\gamma \Phi(F) = \Lambda$ is an indeterminate parameter. In this case, we have

$$\dot{\varepsilon}_{ij}^p = \Lambda \frac{\partial f}{\partial \sigma_{ij}}. \tag{2.6}$$

It is possible to determine the parameter Λ by satisfying the static yield criterion (2.5). Since the subsequent static loading surface (2.5) passes through the loading point, and since during loading

$$\dot{F} = \frac{\partial F}{\partial \sigma_{ij}} \dot{\sigma}_{ij} + \frac{\partial F}{\partial \varepsilon_{kl}^p} \dot{\varepsilon}_{kl}^p = 0, \tag{2.7}$$

then by means of Eq. (2.6) we have

$$\Lambda = \left| -\frac{\dfrac{\partial F}{\partial \sigma_{ij}} \dot{\sigma}_{ij}}{\dfrac{\partial F}{\partial \varepsilon_{kl}^p} \dfrac{\partial f}{\partial \sigma_{kl}}} \right|. \tag{2.8}$$

In the special case of elastic/visco-perfectly plastic soils, we obtain from Eqs. (2.2) the known constitutive equation for an elastic-plastic soil (see D. C. Drucker and W. Prager [5]):

$$\dot{\varepsilon}_{ij}^p = \lambda \left[\alpha \delta_{ij} + \frac{s_{ij}}{2 J_2^{1/2}} \right], \tag{2.9}$$

where

$$\lambda = \left[I_2^p \Big/ \left(\frac{3}{2} \alpha^2 + \frac{1}{4} \right) \right]^{1/2}. \tag{2.10}$$

In this limit case the plastic rate of cubical dilatation is expressed by the relation

$$\dot{\varepsilon}_{ii}^p = 3 \alpha \lambda. \tag{2.11}$$

3. The Propagation of Stress Waves in Elastic/Visco-Plastic Soils

As an example of application of the constitutive equations previously established, we shall study the problem of propagation of stress waves in an infinite elastic/visco-perfectly plastic soil, [see the constitutive Eq. (2.2)]. Three wave types will be analysed, namely spherical, cylindrical and plane.

It is assumed that a uniform pressure constituting any function of time $p = p(t)$ is applied to the spherical surface of the cavity of radius r_0, or to the cylindrical surface of radius r_0, or the to plane $r = r_0 = 0$.

It will be shown that the solution of the propagation problem of stress waves of all the three types in an infinite elastic/visco-perfectly plastic soil may be reduced to a unique mathematical problem. When appropriately changing the coefficients of the differential equations obtained, and the boundary values, we obtain the description of

4*

spherical, cylindrical radial and plane waves respectively. This will be done in both elastic and anelastic regions.

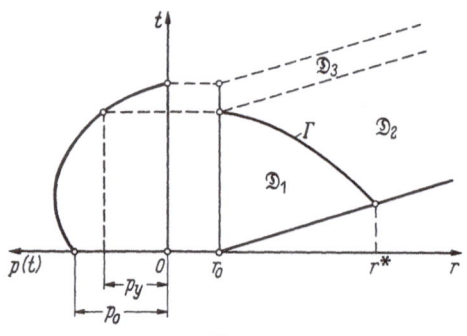

Fig. 1.

Even if we assume that the soil is elastic/visco-perfectly plastic, we have the possibility to introduce many parameters describing the mechanical properties of the material; their appropriate choice enables us to approach in a satisfactory way the properties of real soils.

In each of the cases under consideration, the problem in the range of elastic/visco-plastic deformations ($\mathfrak{D}_1$ in Figs. 1 and 2) is reduced to the solution of a quasi-linear first order hyperbolic system of partial differential equations of the form

$$U_t + A U_r + B = 0, \quad (3.1)$$

where U is the corresponding column vector with the n components, A is an $n \times n$ matrix and B is an n element column vector; A and B depend on r, t and U.

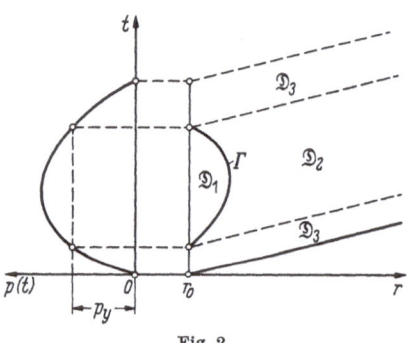

Fig. 2.

The characteristic lines of the system (3.1) have the form

$$r = \text{const}, \quad r = r_0 \pm at + \text{const}, \quad \text{where} \quad a = \sqrt{\frac{4\mu + 3K}{3\varrho}}, \quad (3.2)$$

if ϱ denotes the density of the soil. Depending on the type of wave, the characteristic line $r = \text{const}$ is a single or a double characteristic line.

Along the characteristic lines (3.2), the following conditions must be satisfied:

$$a_{i1} dt + b_{i1} d\sigma_{rr} + c_{i1} d\sigma_{\varphi\varphi} + d_{i1} d\vartheta = 0,$$
$$a_{k2} dr + b_{k2} d\sigma_{rr} + c_{k2} d\sigma_{\varphi\varphi} + d_{k2} d\vartheta = 0,$$
$$(3.3)$$

where $k = 1, 2$ and $i = 1$ if $r = \text{const}$ is a single characteristic line, $i = 1, 2$ if $r = \text{const}$ is a double characteristic line, σ_{rr} and $\sigma_{\varphi\varphi}$ are components of stress tensor and ϑ denotes the velocity. The coefficients in Eqs. (3.3) depend on r, t and U.

If a pressure exceeding the plasticity limit of soil, $p_0 > p_y$, is suddenly applied to the boundary surface, then the characteristic line

$r = r_0 + at$ will be a strong discontinuity or shock wave (see W. PRA-GER [16]). Along a discontinuity, additional conditions must be satis-fied which will be called the conditions of kinematic and dynamic continuity. These conditions have the form (see H. G. HOPKINS [9])

$$\vartheta + a\varepsilon_{rr} = 0,$$
$$\varrho a\vartheta + \sigma_{rr} = 0,$$

(3.4)

respectively.

Since the strain rate $\dot{\varepsilon}_{ij}$ at the shock wave must be regarded as infinity, we obtain the additional conditions from the constitutive equations.

Using the relation which is satisfied along the characteristic line $r = r_0 + at$ and the conditions satisfied along the discontinuity, we obtain on the wave front the same differential equation for each of the three wave types

$$\frac{d\sigma_{rr}}{dr} = -\psi(r, \sigma_{rr}) \quad \text{with the condition} \quad \sigma_{rr}\big|_{\substack{r=r_0 \\ t=0}} = p_0. \quad (3.5)$$

This equation may be reduced to the non-linear Volterra integral equation of the second kind

$$\sigma_{rr} = p_0 - \int_{r_0}^{r} \psi[\xi, \sigma_{rr}(\xi)]d\xi. \quad (3.6)$$

Assuming that the function ψ satisfies the LIPSCHITZ condition, we can obtain the solution of Eqs. (3.6) by means of the recurrence formula

$$\sigma_{rr} = \lim_{n \to \infty} \sigma_{rr}^{(n)}, \quad \text{where} \quad \sigma_{rr}^{(n)} = p_0 - \int_{r_0}^{r} \psi[\xi, \sigma_{rr}^{(n-1)}(\xi)]d\xi. \quad (3.7)$$

The solution given by (3.7) is valid for $r < r^*$, Fig. 1, where r^* satisfies the condition

$$\alpha J_1(r^*) + J_2^{1/2}(r^*) = k. \quad (3.8)$$

The solution on the shock wave for $r > r^*$ has the form

$$\sigma_{rr} = p_0 - \int_{r_0}^{r} \omega(\xi) \sigma_{rr}(\xi)d\xi, \quad (3.9)$$

where ω is a known function of r.

Consider now the solution in the elastic/visco-plastic regions (regions $\mathfrak{D}_1$ at Figs. 1 and 2). The elastic/visco-plastic region $\mathfrak{D}_1$ at Fig. 1 is bounded by the discontinuous wave $r = r_0 + at$ and the straight line $r = r_0$. On the discontinuous wave all components of the vector U are determined by means of the solution given by (3.7), while the quantity σ_{rr} is known on the line $r = r_0$ from the boundary condition.

Using the procedure of finite differences, we obtain by (3.3) in every point of the characteristic net, a system of algebraic equations with respect to all components of the vector U.

The same procedure can be applied to the elastic/visco-plastic region in the case shown at Fig. 2. In this case we known the components of the vector U at the point $t = t_1$ on the line $r = r_0$ from the elastic solution.

The boundary Γ of the elastic/visco-plastic region (see Figs. 1 and 2) can be obtained by an approximate method using the condition

$$\alpha J_1(r,t) + J_2^{1/2}(r,t) = k. \tag{3.10}$$

In elastic regions, the problem can be reduced to the solution of the hyperbolic second order differential equation

$$\frac{\partial^2 u}{\partial \xi \, \partial \eta} = \frac{\Delta}{\xi - \eta}\left(\frac{\partial u}{\partial \xi} - \frac{\partial u}{\partial \eta}\right) + \frac{\delta}{(\xi - \eta)^2}\, u, \tag{3.11}$$

where u is the radial component of displacement vector and ξ and η are new coordinates along the characteristic lines, Δ and δ being constants.

The considered elastic regions in the $\xi\eta$-plane are shown in Fig. 3. The equation is valid in the region $\mathfrak{D}_2$. This region is bounded by the characteristic line $\eta = 0$ and the curve $\xi = \tau(\eta)$. On the characteristic line $\eta = 0$ we have $u = 0$, whereas on the curve $\xi = \tau(\eta)$ by (3.10) we obtain the condition

$$\frac{\partial u}{\partial \xi} = \Psi\left(\eta, u, \frac{\partial u}{\partial \eta}\right). \tag{3.12}$$

Thus the problem in the elastic region was reduced to the non linear generalized Picard problem (see G. Majcher [10]). Using the Riemann function, this problem can be solved by means of the interaction method.

4. Detailed Discussion of the Spherical Waves

With reference to the spherical coordinates (r, φ, θ), we have

$$U = \begin{pmatrix} \vartheta \\ \sigma_{rr} \\ \sigma_{\varphi\varphi} \end{pmatrix}, \qquad A \begin{pmatrix} 0 & -\dfrac{1}{\varrho} & 0 \\ -\left(K + \dfrac{4}{3}\mu\right) & 0 & 0 \\ -\left(K + \dfrac{2}{3}\mu\right) & 0 & 0 \end{pmatrix},$$

$$B = \begin{pmatrix} -\dfrac{2}{\varrho}\dfrac{\sigma_{rr} - \sigma_{\varphi\varphi}}{r} \\ \left(-2K + \dfrac{4}{3}\mu\right)\dfrac{\vartheta}{r} + \gamma\langle\Phi(F)\rangle\left(3\alpha K + \dfrac{2\mu}{\sqrt{3}}\right) \\ -\left(2K + \dfrac{2}{3}\mu\right)\dfrac{\vartheta}{r} + \gamma\langle\Phi(F)\rangle\left(3\alpha K - \dfrac{\mu}{\sqrt{3}}\right) \end{pmatrix}, \tag{4.1}$$

where

$$F = \left[\left(\alpha + \frac{1}{\sqrt{3}}\right)\sigma_{rr} + \left(2\alpha - \frac{1}{\sqrt{3}}\right)\sigma_{\varphi\varphi}\right]\Big/ k - 1. \qquad (4.2)$$

The coefficients of the characteristic set (3.3) have the following values

$$a_{k1} = 3\mu K \left[2\frac{\vartheta}{r} + \gamma\langle\Phi(F)\rangle\left(\frac{1}{\sqrt{3}} - 2\alpha\right)\right],$$

$$b_{k1} = K - \frac{2}{3}\mu, \qquad c_{k1} = -\left(K + \frac{4}{3}\mu\right), \qquad (4.3)$$

$$d_{k1} = 0, \qquad k = 1;$$

$$a_{i2} = \left[\mp a\left(2K + \frac{2}{3}\mu\right) + \frac{6\mu K}{\varrho a}\right]\frac{\vartheta}{r} + \gamma\langle\Phi(F)\rangle$$

$$\times \left[\pm a\left(3\alpha K + \frac{3\mu K}{\varrho a}\right)\left(\frac{1}{\sqrt{3}} - 2\alpha\right) + \left(K + \frac{4}{3}\mu\right)\frac{2}{\varrho}\frac{\sigma_{rr} - \sigma_{\varphi\varphi}}{r}\right],$$

$$b_{i2} = \frac{1}{\varrho}\left(K - \frac{2}{3}\mu\right), \qquad c_{i2} = 0, \qquad (4.4)$$

$$d_{i2} = \mp\left(K - \frac{2}{3}\mu\right)a, \qquad i = 1, 2.$$

In the spherical case, the additional condition at the shock wave has the form

$$\sigma_{\varphi\varphi} = \sigma_{rr}\left(1 - \frac{2}{\varrho a^2}\right), \qquad (4.5)$$

and the function $\psi(r, \sigma_{rr})$ is

$$\psi(r, \sigma_{rr}) = \frac{1}{r}\sigma_{rr} + \frac{3\alpha K + \frac{2}{\sqrt{3}}\mu}{2a}\gamma\langle\Phi(F)\rangle, \qquad (4.6)$$

where

$$F = \frac{\left(\alpha + \frac{1}{\sqrt{3}}\right) + \left(2\alpha - \frac{1}{\sqrt{3}}\right)\left(1 - \frac{2\mu}{\varrho a^2}\right)}{k}\sigma_{rr} - 1.$$

The solution on the spherical shock wave for $r > r^*$ has the closed form

$$\sigma_{rr} = \frac{1}{R}\frac{r^*}{r}, \qquad \vartheta = -\frac{1}{\varrho a R}\frac{r^*}{r}, \qquad \sigma_{\varphi\varphi} = \sigma_{\theta\theta} = \left(1 - \frac{2\mu}{\varrho a^2}\right)\frac{1}{R}\frac{r^*}{r}, \qquad (4.7)$$

where $R = $ const.

In the elastic region, the Eq. (3.11) is

$$\frac{\partial^2 u}{\partial\xi\,\partial\eta} = \frac{1}{\xi - \eta}\left(\frac{\partial u}{\partial\xi} - \frac{\partial u}{\partial\eta}\right) - \frac{2}{(\xi - \eta)^2}u \qquad (4.8)$$

and the condition (3.12) takes now the form

$$h_1\left(\frac{\partial u}{\partial\xi} - \frac{\partial u}{\partial\eta}\right) + h_2 u = h_3, \qquad (4.9)$$

where h_1, h_2 and h_3 are constants.

Hence for spherical waves the problem in the elastic region is reduced to the linear generalized PICARD problem[1] (see G. MAJCHER [10]) the solution of which may be expressed as follows

$$u(\xi, \eta) = \int_0^\eta \mathcal{R}(\xi, \eta; \xi_0, z) \, \Omega(z) \, dz, \qquad (4.10)$$

where $\mathcal{R}(\xi, \eta; s, z)$ denotes the RIEMANN function for the Eq. (4.8) and $\xi_0 = $ const. The function $\Omega(z)$ should be determined so that the solution (4.10) satisfies the condition (4.9). The RIEMANN function for the Eq. (4.8) can be expressed by the hypergeometric function (see R. COURANT [3] and E. T. COPSON [2]).

References

[1] Алексеев, Н. А., Х. А. Рахматулин и А. Я. Сагомонян: Об основ- ных уравнениях динамики грунта. Журнал прикладной механики и технической физики 2, 147 (1963).

[2] Copson, E. T.: On the Riemann-Green function. Arch. Rat. Mech. Anal. 1, 324 (1958).

[3] Courant, R., and D. Hilbert: Methods of mathematical physics, Vol. 2: Partial Differential Equations, by R. Courant, New York/London: Inter- science Publishers, 1962.

[4] Drucker, D. C.: A definition of stable inelastic material. J. Appl. Mech. 26, 101 (1959).

[5] Drucker, D. C., and W. Prager: Soil mechanics and plastic analysis or limit design. Quart. Appl. Math. 10, 157 (1952).

[6] Drucker, D. C.: Some remarks on fracture and flow. Technical Report, Brown University 1963.

[7] Freudenthal, A. M.: The mathematical theories of the inelastic con- tinuum. Handbuch der Physik, Vol. VI, Berlin/Göttingen/Heidelberg: Springer 1958.

[8] Hohenemser, K., and W. Prager: Über die Ansätze der Mechanik iso- troper Kontinua. ZAMM 12, 216 (1932).

[9] Hopkins, H. G.: Dynamic expension of spherical cavities in metals. Pro- gress in Solid Mechanics, 1, 83 (1960).

[10] Majcher, G.: Sur un problème mixte pour l'équation du type hyperbolique. Ann. Polon. Math. 2, 121 (1958).

[11] Olszak, W.: On some basic aspects of the theory of non-homogeneous loose and cohesive media. Arch. Mech. Stos. 5, 751 (1956).

[12] Olszak, W., Z. Mróz and P. Perzyna: Recent trends in the development of the theory of plasticity, Oxford/Warszawa: Pergamon 1963.

[13] Perzyna, P.: The constitutive equations for rate sensitive plastic materials. Quart. Appl. Math. 20, 321 (1963).

[14] Perzyna, P.: The constitutive equations for work-hardening and rate sen- sitive plastic materials. Proc. Vibr. Probl. 4, 281 (1963).

[15] Perzyna, P.: On the propagation of stress waves in a rate sensitive plastic medium. ZAMP 14, 241 (1963).

---

[1] For detailed analysis see Ref. [15].

[*16*] PRAGER, W.: Discontinuous fields of plastic stress and flow. Proceedings of the Second U.S. National Congress of Applied Mechanics, 1954, p. 21.

[*17*] PRAGER, W.: Introduction to Mechanics of Continua, Boston: Ginn 1961.

[*18*] Соколовский, В. В.: Распространение упруго-вязко-пластических волн в стержниях. Докл. Ан СССР **60**, № 5. 775—778 (1948).

Discussion

Question posée par R. V. WHITMAN: The problem of a footing on a clay undoubtedly involves time-dependent creep following yielding of some portion of the clay mass. How you attempted to apply your theory to this problem? If not, what mathematical methods might possibly permit such an application?

Réponse de W. OLSZAK: The constitutive relations proposed should be used in the usual way for solving specified boundary-value problems. These problems may, of course, be rather complex. I do not think there is any general method of solving them. However, particular problems may effectively be solved, sometimes even in a closed form.

In our paper, we showed how the proposed constitutive equations may be applied to problems of propagation of different kinds of waves: spherical, cylindral and plane ones, the first case (spherical waves) having been discussed in detail.

1.5 Conditions de Stabilité et Postulat de Drucker

Par

J. Mandel

L'objet de cette communication est de discuter les conditions de stabilité mécanique d'un élément du sol et notamment le postulat de DRUCKER. On montre que le postulat de DRUCKER est une condition suffisante de stabilité, mais qu'il n'est pas une condition nécessaire lorsqu'il y a des frottements intérieurs du type de COULOMB. On propose une condition nécessaire, découlant du fait qu'un matériau stable doit pouvoir propager une perturbation sous forme d'ondes réelles.

1. Préliminaires

Nous rappelons le postulat de DRUCKER et diverses conséquences immédiates [1].

Selon ce postulat, un agent extérieur superposé aux forces qui existent dans l'état d'équilibre actuel doit, pour que l'équilibre soit stable, effectuer un travail non négatif durant son application et aussi dans un cycle complet d'application et d'enlèvement de cet agent. On en déduit les deux inégalités:

$$(\sigma_{ij} - \sigma_{ij}^*)\, \dot{\varepsilon}_{ij}^p \geq 0, \tag{1}$$

$$\dot{\sigma}_{ij}\dot{\varepsilon}_{ij}^p \geq 0 \text{ [1,2]}. \tag{2}$$

σ_{ij} désigne les contraintes dans un état situé à la frontière d'écoulement caractérisée par l'équation:

$$f(\sigma_{ij}) = 0. \tag{3}$$

$\dot{\varepsilon}_{ij}^p$ est la vitesse de déformation plastique correspondante,
σ_{ij}^* est un état de contraintes admissible, c'est-à-dire tel que:

$$f(\sigma_{ij}^*) \leqq 0. \tag{4}$$

De (1) résulte que, dans l'espace de coordonnées σ_{ij} la surface de charge (3) est convexe et qu'en un point ordinaire de cette surface,

---

[1] Égalité dans le cas d'un corps parfaitement plastique.
[2] On adopte dans toute la suite la conventions de l'indice muet.

le vecteur $\dot{\varepsilon}^p$ est dirigé suivant la normale extérieure à la surface [ce qu'indique également l'inégalité (2)]. Cette vitesse devant s'annuler en même temps que $\dfrac{\partial f}{\partial \sigma_{hk}} \dot{\sigma}_{hk}$, on a :

$$\dot{\varepsilon}_{ij}^p = g \, \frac{\partial f}{\partial \sigma_{ij}} \, \frac{\partial f}{\partial \sigma_{hk}} \, \dot{\sigma}_{hk}\ {}^1 \qquad g > 0 \tag{5}$$

soit :

$$\dot{\varepsilon}_{ij}^p = P_{ij,hk} \, \dot{\sigma}_{hk}$$

avec :

$$P_{ij,hk} = P_{hk,ij}.$$

Pour la vitesse de déformation élastique, on a d'autre part :

$$\dot{\varepsilon}_{ij}^e = \Lambda_{ij,hk} \, \dot{\sigma}_{hk}$$

avec

$$\Lambda_{ij,hk} = \Lambda_{hk,ij}$$

d'où, pour la vitesse de déformation totale :

$$\dot{\varepsilon}_{ij} = (P_{ij,hk} + \Lambda_{ij,hk}) \, \dot{\sigma}_{ij}$$

et en inversant ces relations :

$$\dot{\sigma}_{ij} = A_{ij,hk} \, \dot{\varepsilon}_{hk} \tag{6}$$

toujours avec la même symétrie :

$$A_{ij,hk} = A_{hk,ij}. \tag{7}$$

De plus, d'après les propriétés du potentiel élastique :

$$\dot{\sigma}_{ij} \dot{\varepsilon}_{ij}^e \geq 0$$

d'où :

$$\dot{\sigma}_{ij} \dot{\varepsilon}_{ij} \geq 0 \quad \text{ou} \quad A_{ij,hk} \dot{\varepsilon}_{ij} \dot{\varepsilon}_{hk} \geq 0 \ {}^2 \tag{8}$$

2. Le Postulat de Drucker est Condition Suffisante de Stabilité

Le postulat de Drucker est condition suffisante de stabilité de l'équilibre d'un volume matériel V, lorsque les conditions à la surface S

---

[1] En principe on doit introduire dans cette formule non pas la vitesse de contraintes $\dot{\sigma}_{hk}$ par rapport à des axes fixes, mais la vitesse de contraintes $\dfrac{D\sigma_{hk}}{Dt}$ par rapport à des axes animés de la vitesse de rotation ω_{ij} de l'élément matériel. On a :

$$\frac{D\sigma_{ij}}{Dt} = \dot{\sigma}_{ij} - \omega_{ik} \sigma_{kj} - \omega_{jk} \sigma_{ki}.$$

Mais nous supposons la vitesse de rotation ω et les contraintes σ suffisamment faibles pour que $\dfrac{D\sigma_{ij}}{Dt}$ puisse être remplacé par $\dot{\sigma}_{ij}$.

[2] Égalité dans le cas d'un corps parfaitement plastique, qui peut subir des déformations sous contrainte constante.

sont: contraintes données $\vec{T}_0$ sur une partie S_T, déplacements donnés $\vec{\xi}_0$ sur le reste S_ξ de cette surface. Les déplacements sont supposés infiniment petits.

En effet l'application au volume V du théorème de l'énergie cinétique donne, π désignant la puissance des forces s'exerçant sur S et dans V, E_c l'énergie cinétique:

$$\pi = \int\limits_V \sigma_{ij}\dot{\varepsilon}_{ij}dV + \frac{dE_c}{dt}. \tag{9}$$

Or:

$$\pi = \int\limits_{S_T} T_i^0 \dot{\xi}_i dS + \int\limits_V X_i^0 \dot{\xi}_i dV.$$

X_i^0 désignant les coordonnées de la force de masse par unité de volume.

Puisque $\dot{\xi}_i = 0$ sur S_ξ l'intégrale de surface peut être étendue à la surface totale S et transformée en intégrale de volume. On obtient ainsi, σ_{ij}^0 désignant les contraintes dans l'état d'équilibre:

$$\pi = \int\limits_V \sigma_{ij}^0 \dot{\varepsilon}_{ij}dV$$

et l'équation (9) devient:

$$\int\limits_V (\sigma_{ij} - \sigma_{ij}^0)\,\dot{\varepsilon}_{ij}dV + \frac{dE_c}{dt} = 0.$$

Décomposons $\dot{\varepsilon}_{ij}$ en $\dot{\varepsilon}_{ij}^e$ et $\dot{\varepsilon}_{ij}^p$ et posons:

$$\Phi = \frac{1}{2}\int\limits_V \Lambda_{ij,hk}(\sigma_{ij} - \sigma_{ij}^0)(\sigma_{hk} - \sigma_{hk}^0)\,dV.$$

On a:

$$\int\limits_V (\sigma_{ij} - \sigma_{ij}^0)\,\dot{\varepsilon}_{ij}^e dV = \frac{d\Phi}{dt}$$

et d'après le postulat (1):

$$\int\limits_V (\sigma_{ij} - \sigma_{ij}^0)\,\dot{\varepsilon}_{ij}^p dV \leq 0$$

d'où:

$$\frac{d}{dt}(\Phi + E_c) \leqq 0. \tag{10}$$

Φ étant minimum strict dans la position d'équilibre, la stabilité est assurée. Toutefois il peut y avoir exception dans le cas d'un corps parfaitement plastique[1], parce que dans ce cas le minimum de Φ, comme fonctionnelle des déplacements, peut n'être pas strict (le corps pouvant subir des déformations sous contraintes constantes).

---

[1] En d'autres termes le postulat est condition suffisante si les inégalités (1) et (2) sont strictes, ce qui exclut le corps parfaitement plastique.

3. Le Postulat de Drucker n'est pas Condition Nécessaire de Stabilité, lorsqu'il existe des Frottements Intérieurs du Type de Coulomb

Il suffit, pour le montrer, de considérer le modèle de la fig. 1. Les ressorts a et b rendent compte des déformations dans le domaine d'élasticité, mais ne joueront en fait aucun rôle dans la question. Le patin correspond au frottement de COULOMB, le ressort G à un écrouissage dû à un blocage d'énergie élasti-

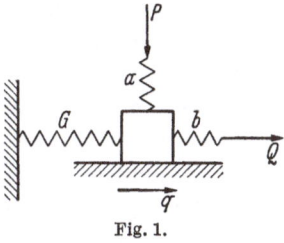

Fig. 1.

que. Soit P, Q la force extérieure appliquée, R la tension du ressort G, q le glissement du patin, c'est à dire la partie irréversible du déplacement ($q = \varepsilon^p$). Plaçons nous à l'équilibre limite, le patin étant sur le point de se déplacer vers la droite ($Q - R = fP$, f désignant le coefficient de frottement). Si nous diminuons P de $-dP$ et simultanément Q de $-dQ$ tel que $-dQ < f(-dP)$, le patin glisse vers la droite: $dq > 0$ et puisque $dQ < 0$, *l'inégalité* (2) *est violée*: $dQ\,dq < 0$. Cependant, grâce à la présence du ressort G, la stabilité est assurée.

Etudions un peu plus ce modèle. Il s'agit d'un corps élastoplastique avec écrouissage, dont le domaine d'élasticité, défini par:

$$- fP \leq Q - R \leq fP$$

est limité dans le plan P, Q par les deux demi-droites D et D' (fig. 2). La position de cette «courbe de charge» dépend de la valeur actuelle R de la tension du ressort. Si le point σ de coordonnées P, Q est sur la droite D et $\vec{\sigma}$ dirigé vers l'extérieur du domaine d'élasticité, le déplacement plastique qui se produit est représenté par un vecteur paralléle à l'axe $0Q$: *la règle de normalité est violée*.

Soit σX la parallèle à $0P$ menée par le point σ. Si l'on choisit le point σ^* correspondant à un état admissible dans l'angle $X\sigma D$, *l'inégalité* (1) *est violée*. Si le vecteur $\vec{\sigma}$ est contenu dans l'angle opposé à $X\sigma D$, *l'inégalité* (2) *est violée*.

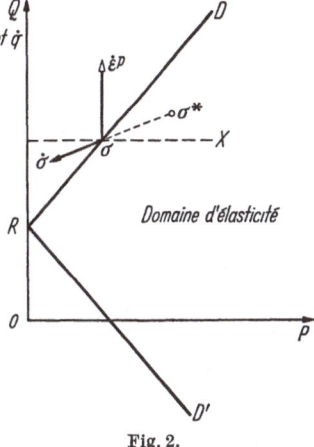

Fig. 2.

4. Frottement de Coulomb et Résistance de Schmid (ou de cohésion)

Il est manifeste dans l'exemple précédent, que c'est l'intervention de la contrainte normale dans la loi de frottement de COULOMB qui met en échec le postulat de Drucker. Si la résistance au glissement du patin était indépendante de la force normale P, comme cela a lieu pour les glissements à l'intérieur des cristaux métalliques (loi de Schmid), le domaine d'élasticité défini par:

$$-s \leqq Q - R \leqq s$$

s désignant le seuil du patin, indépendant de P, serait limité par deux droites DD' parallèles à l'axe $0P$. La règle de normalité et les inégalités (1) et (2) seraient alors vérifiées.

Dans le cas d'un milieu où la déformation plastique s'effectue par glissements obéissant à la loi de Schmid, l'inégalité (1), qui constitue le principe du travail maximal de Hill, a été démontrée par cet auteur [2] de la manière suivante. En négligeant les déformations élastiques[1], le théorème des travaux virtuels appliqué à un élément de volume V, donne:

$$V (\sigma_{ij} - \sigma_{ij}^*) \dot{\varepsilon}_{ij}^p = \varSigma (\tau - \tau^*) \dot{\gamma}, \tag{11}$$

σ_{ij} désigne le tenseur de contraintes réel s'exerçant sur l'élément, $\dot{\varepsilon}_{ij}^p$ la vitesse de déformation plastique correspondante, $\dot{\gamma}$ les vitesses de glissement et τ les cissions sur les plans de glissement (dans la direction du glissement) correspondant à σ_{ij}. σ_{ij}^* est un tenseur de contraintes admissible ($f(\sigma_{ij}^*) \leq 0$), auquel on peut par conséquent faire correspondre un équilibre des cissions τ^* qui respectent les lois du frottement sur les différentes facettes de glissement possibles.

Ceci étant, si les glissements obéissent à la loi de Schmid, τ^* ne peut pas dépasser τ sur les plans de glissements actifs ($\dot{\gamma} > 0$). Par conséquent $\varSigma (\tau - \tau^*) \dot{\gamma} \geq 0$. L'inégalité (1) en découle. On en déduit ensuite, pour un point ordinaire de la surface de charge, la règle de normalité, puis l'inégalité (2).

Mais si les glissements obéissent à la loi de Coulomb, ou plus généralement à une loi de la forme $\tau = f(\nu)$ où f est une fonction croissante de la pression normale ν sur le plan de glissement, τ^* pourra dépasser τ sur les plans actifs, pourvu que ν^* soit plus grand que ν. L'inégalité (1) cesse d'être vérifiée.

Dans le cas des argiles pas trop consolidées, les particules solides ne sont pas en contact et le frottement de Coulomb n'intervient pas. Il n'est pas impossible que dans ce cas (le processus irréversible, rupture de liaisons électrostatiques entre les grains, s'apparentant quelque peu avec la rupture de liaison entre atomes d'un cristal) l'inégalité (1) soit vérifiée. Mais il n'en est certainement pas ainsi pour les sables et la majorité des sols. Les grains solides sont alors en con-

---

[1] On peut montrer que la relation (11) reste vraie, sans qu'il soit nécessaire de négliger les déformations élastiques, pourvu que l'élasticité soit linéaire.

tact. Les actions de contact entre grains obéissant à la loi de Coulomb, le principe du travail maximal et le postulat de Drucker ne s'appliquent pas.

5. Une Condition Nécessaire de Stabilité

Si nous n'utilisons plus le postulat de Drucker ou le principe du travail maximal de Hill, la relation (5) est remplacée [3] par:

$$\dot{\varepsilon}_{ij}^{p} = g\,\varphi_{ij}\frac{df}{d\sigma_{hk}}\,\dot{\sigma}_{hk} \qquad (12)$$

$f(\sigma_{ij}) = 0$ désignant toujours la frontière d'écoulement actuelle, φ_{ij} un tenseur symétrique maintenant différent de $\dfrac{\partial f}{\partial \sigma_{ij}}$.

On en déduit encore:

$$\dot{\sigma}_{ij} = A_{ij,hk}\,\dot{\varepsilon}_{hk} \qquad (13)$$

mais la matrice A ne possède plus la symétrie (7) entre ij et hk.

Une petite perturbation par rapport à un état d'équilibre stable se propage par ondes. α_i désignant les cosinus directeurs de la normale à une onde S, Ω la célérité de S par rapport à la matière dont ϱ est la masse volumique, $[\vec{\gamma}]$ la discontinuité d'accélération de part et d'autre de S, B_{ik} la matrice:

$$B_{ik} = A_{ij,hk}\,\alpha_j\alpha_h \qquad (14)$$

nous avons établi en [3] que l'on doit avoir:

$$B_{ik}[\gamma_k] = \varrho\,\Omega^2[\gamma_i]\,{}^{1}. \qquad (15)$$

Autrement dit le vecteur $[\vec{\gamma}]$ est un vecteur propre de la matrice B_{ik} et $\varrho\,\Omega^2$ est la valeur propre correspondante.

Ceci étant, si les 3 valeurs propres de la matrice B sont réelles positives, nous pouvons décomposer une perturbation imposée en un point du corps en 3 vecteurs propres réels qui se propagent dans la matière. Si 2 seulement des valeurs propres sont réelles positives, la troisième étant réelle négative, une des composantes de la perturbation ne peut pas se propager. Cela signifie qu'il y a localisation de la déformation (formation d'une surface de glissement ou d'une fissure). On sait par exemple que pour une tige plastique la célérité des ondes longitudinales est $\sqrt{\dfrac{M}{\varrho}}$, M étant le module d'écrouissage. Pour $M \leq 0$ il n'y a plus propagation mais formation de lignes d'Hartman-Lüders ou de surfaces de glissement. Ceci correspond à une instabilité. On doit noter que cette instabilité peut être interdite par la présence de parois fixes qui bloquent les glissements. Dans tout ce qui suit, nous

---

[1] Si les accélérations sont continues jusqu'à l'ordre $p-1$, l'accélération p ième étant discontinue, l'èquation (15) s'applique à $[\vec{\gamma}^{(p)}]$.

supposons donc les contraintes à la surface imposées mais les déplacements libres.

Pour démontrer l'instabilité, considérons un feuillet de normale $\vec{\alpha}$ limité par les plans $X = -h$ et $X = +h$ ($X = \alpha_i x_i$). Dans l'état d'équilibre les contraintes σ_{ij}^0 sont uniformes dans le feuillet. Les deux faces étant soumises à une contrainte fixe, le feuillet peut prendre des mouvements par ondes planes, de la forme:

$$\vec{\xi} = \vec{\lambda}\varphi(X, t) \tag{16}$$

$\vec{\lambda}$: vecteur constant, $\vec{\xi}$: déplacement.

Dans le cas de « charge » les variations de contraintes sont:

$$\sigma_{ij} - \sigma_{ij}^0 = A_{ij,hk}\,\varepsilon_{hk} = A_{ij,hk}\lambda_k\alpha_h\varphi'_X(X, t) \tag{17}$$

et les équations de la dynamique donnent alors:

$$B_{ik}\lambda_k\varphi''_{X^2} = \varrho\,\lambda_i\varphi''_{t^2} {}^1 \tag{18}$$

$\vec{\lambda}$ doit donc être un vecteur propre de la matrice B.

Supposons alors que la valeur propre correspondante soit réelle négative $-\varrho K^2$.

Les équations (18) se réduisent à: $K^2\varphi''_{X^2} + \varphi''_{t^2} = 0$.

Un mouvement satisfaisant aux conditions imposées s'obtient en prenant par exemple:

$$\varphi = A\,Sh\,mkt\,\sin mX, \qquad m = \frac{\pi}{2h}$$

φ ne restant pas infiniment petit quand t augmente, il y a instabilité.

Il résulte de là qu'une condition nécessaire de stabilité sous contraintes imposées est que les trois valeurs propres de la matrice B définie par (17) soient réelles positives, quel que soit $\vec{\alpha}$.

On vérifie que si la condition de DRUCKER (2) (suffisante) est satisfaite, il en est de même de la condition nécessaire précédente. En effet dans ce cas, grâce à la symétrie (7) de la matrice A, la matrice B est symétrique. Ses 3 valeurs propres sont donc réelles. De plus elles sont non négatives, car:

$$B_{ik}x_i x_k = A_{ij,hk}(x_i\alpha_j)(x_k\alpha_h) \geq 0 {}^2$$

d'après (8).

La condition nécessaire précédente n'est probablement pas suffisante, mais elle permet, comme on va le voir, de préciser les limites du domaine de stabilité d'un matériau.

---

[1] On retrouve l'équation (15) en prenant $\varphi(X, t) = \varphi(X - \Omega t)$.

[2] Égalité dans le cas d'un corps parfaitement plastique, en accord avec le fait que, dans ce cas (№ 2 in fine), la condition (2) peut n'être pas suffisante.

6. Caractéristiques Statiques et Stabilité

La condition nécessaire précédente revient à dire que *les surfaces caractéristiques des équations aux dérivées partielles de l'équilibre ne doivent pas être réelles*. (En effet les ondes étant les caractéristiques des équations du mouvement, les caractéristiques statiques sont des ondes de célérité nulle.) En d'autres termes l'apparition de caractéristiques réelles au cours d'un chargement progressif est, pour les éléments qu'elles atteignent, le signe de l'instabilité.

Dans le cas du potentiel plastique, c'est à dire lorsque la condition (2) est vérifiée, nous avons montré en [3] que des caractéristiques réelles ne peuvent exister que si les deux conditions suivantes sont réunies:

1. le corps est parfaitement plastique (en accord avec la note de la fin du No. 5),

2. l'état de contraintes est tel que l'une des 3 valeurs principales du tenseur $\frac{\partial f}{\partial \sigma_{ij}}$ soit nulle et les deux autres de signes contraires.

Le tenseur vitesse de déformation est alors plan et il existe deux facettes caractéristiques réelles perpendiculaires à son plan: ce sont les caractéristiques réelles que l'on rencontre dans les problèmes classiques de déformation plane[1]. Elles sont éliminées par le moindre écrouissage.

Considérons maintenant le cas où il n'y a pas potentiel plastique [relations (12)]. En supposant pour simplifier le milieu isotrope et les directions principales des contraintes et des vitesses de déformation en coïncidence et en désignant par φ_1, φ_2, φ_3 les valeurs principales du tenseur φ_{ij}, par f_1, f_2, f_3 celles du tenseur $\frac{\partial f}{\partial \sigma_{ij}}$, le cône des normales aux facettes caractéristiques a pour équation [3] par rapport aux axes principaux:

$$0 = \frac{1-\nu}{2\mu g}(\alpha_1^2 + \alpha_2^2 + \alpha_3^2)^2 + [f_2\varphi_2 + f_3\varphi_3 + \nu(f_2\varphi_3 + f_3\varphi_2)]\,\alpha_1^4$$
$$+ [2f_3\varphi_3 + f_1\varphi_2 + f_2\varphi_1 + \nu(f_1 + f_2)\,\varphi_3 + \nu(\varphi_1 + \varphi_2)f_3]\,\alpha_1^2\alpha_2^2 + \cdots,$$

ν désignant le rapport de POISSON, μ le deuxième coefficient de LAMÉ.

Pour que ce cône n'ait pas de génératrices réelles, il faut que l'on ait:

$$\frac{1-\nu}{2\mu g} > \frac{[f_1\varphi_2 - f_2\varphi_1 + \nu\varphi_3(f_1 - f_2) + \nu f_3(\varphi_1 - \varphi_2)]^2}{4(f_2 - f_1)(\varphi_2 - \varphi_1)} - (1-\nu^2)\,f_3\varphi_3 \quad (19)$$

et deux autres conditions analogues se déduisant de celle-ci par permutation circulaire des indices 1, 2, 3.

---

[1] Les caractéristiques réelles dans les problèmes dits de *contrainte* plane résultent de l'approximation consentie dans ces problèmes. On sait qu'en fait cette approximation viole les conditions de compatibilité tridimensionnelles.

On voit que pour un corps parfaitement plastique ($g = + \infty$), il y a toujours instabilité dans le cas où la déformation est plane ($\varphi_3 = 0$).

Dans le cas d'un critère de plasticité de la forme de Mohr:

$$f \equiv \sigma_1 - H(\sigma_2) \quad \text{avec} \quad \sigma_1 < \sigma_3 < \sigma_2$$

(les contraintes positives étant les pressions); on a:

$$f_1 = 1, \quad f_2 = - H'(\sigma_2) = -j, \quad f_3 = 0, \quad 0 \leq j \leq 1.$$

En admettant:

$$\varphi_1 = 1, \quad \varphi_2 = -k, \quad \varphi_3 = 0, \quad 0 \leq k \leq 1$$

les conditions (19) se réduisent à la suivante (la plus sévère des trois):

$$\frac{M}{\mu}(1 - \nu) > \frac{(j - k)^2}{2(1 + j)(1 + k)}$$

où l'on a posé: $g^{-1} = M$ (module d'écrouissage).

Au cours d'un chargement progressif, tant que le rapport $\dfrac{M}{\mu}(1 - \nu)$ dépasse la valeur ci-dessus, la condition nécessaire de stabilité du No. 5 est remplie. Elle cesse de l'être quand $\dfrac{M}{\mu}(1 - \nu)$ atteint cette valeur. A ce moment apparaissent deux facettes caractéristiques réelles (passant par la direction de la contrainte principale intermédiaire σ_3, symétriques par rapport aux deux autres directions principales) et des surfaces de glissement commencent à se développer.

Notons à ce propos qu'un élément, qui, individuellement, serait instable sous les contraintes qui lui sont imposées si ses déformations étaient libres, peut parfaitement être stable dans un massif. En fait c'est ce qui se produit dans le cas des déformations plastiques contenues. Ceci autorise l'existence de caractéristiques réelles dans un domaine limité. Demander, suivant le point de vue de M. Drucker [1], qu'un élément d'une structure stable soit lui-même stable sous les contraintes qui lui sont appliquées (supposées maintenues fixes), est une exigence excessive, qui peut n'être pas strictement respectée.

7. Conclusions

1. Le postulat de Drucker[1] est une condition suffisante, mais n'est pas une condition nécessaire pour la stabilité d'un élément du sol.

2. Nous avons donné une condition nécessaire: l'absence de caractéristiques réelles pour l'élément en question.

3. Mais la stabilité individuelle d'un élément n'est pas une condition nécessaire pour la stabilité d'un massif. Ceci autorise l'existence de caractéristiques réelles dans un domaine limité.

---

[1] Au sens strict, cf. note No. 2.

References

[1] DRUCKER, D. C.: A more fundamental approach to stress-strain relations. Proc. First U.S. National Congress of Applied Mechanics, Am. Soc. Mech. Engrs 1951, p. 487—491.

[2] HILL, R.: The Mechanics of quasi-static plastic deformation in metals. Surveys in Mechanics 1956, p. 12; voir aussi J. F. W. BISHOP et R. HILL, Phil. Mag. **42**, 414 (1951).

[3] MANDEL, J.: Propagation des surfaces de discontinuité dans un milieu élastoplastique. Int. Symp. On Stress Waves in Anelastic Solids, Brown University 1963, Berlin/Göttingen/Heidelberg: Springer 1964.

[4] HILL, R.: Acceleration Waves in Solids. J. Mech. Phys. Solids. **10**, 1—16 (1962).

Discussion

Questions posées par P. PERZYNA:
1. Would you comment why in the first part of your consideration (postulate of stable plastic material) did you neglect the term which represents the kinetic energy. On the other hand in the second part of your consideration (that is for the condition of stability) you introduced that term.

2. Do you think that the condition of reality of the waves in plastic material is equivalent to the condition of the unique solution of basic boundary value problem?

Réponse de J. MANDEL:
1. En fait, la première question ne concerne pas mon exposé. Je me suis borné à rappeler les inégalités de base proposées par M. DRUCKER. J'observe toutefois qu'il n'y a pas lieu d'introduire l'énergie cinétique dans l'établissement des inégalités de DRUCKER, à partir de son postulat selon lequel un agent extérieur perturbateur ne peut extraire de l'énergie. Il doit être entendu que l'agent extérieur est appliqué à une vitesse infiniment lente, de telle manière que l'équilibre soit maintenu et aucune énergie cinétique n'apparaisse. Bien entendu, dans la discussion de la stabilité, qui est une question de dynamique, il doit être tenu compte de l'énergie cinétique.

2. Je pense en effet que la réalité des vitesses des ondes et l'unicité de la solution sont liées. A titre d'exemple je citerai le cas du flambage d'une tige. Lorsque la charge imposée atteint la valeur critique, d'une part l'unicité de la solution disparaît, d'autre part la fréquence fondamentale de vibration s'annule, ce qui correspond à la nullité de la célérité des ondes. Ainsi les deux choses paraissent liées. Cependant je ne peux pas énoncer de proposition générale concernant cette relation.

Question posée par D. C. DRUCKER: Professor MANDEL has given an elegant presentation of the meaning and consequences of my postulate of stability for time-independent material and systems. As he points out, the postulate is a sufficient condition for stability and ensures uniqueness in dynamic as well as static problems. These and related questions were discussed by me in a series of papers (1951, 1956, 1959) and by Professor KOITER in his penetrating review paper in Progress in Solid Mechanics. The distinction between stability in the large (labelled as 1 by Professor MANDEL) and in the small (labelled as 2) is impotant. Professor MANDEL's treatment of wave propagation and the friction example are based on stability in the small. Professor RIVLIN and Professor TRUESDELL some 3 and 2 years ago analyzed waves in a medium under finite strain from a similar point of view.

Although my postulate is not a necessary one for stability I do not know of
any systems which are stable in any sense against more than infinitesimal disturbance and yet violate the postulate of stability in the small stable frictional
systems do not obey stability in the large but this is another matter. Stable soil
systems do appear to be stable in the small in my sense. They are not stable in
the large and this is why the bound theorems break down.

My question is whether Professor Mandel knows of any "unstable-stable"
system which would illustrate the lack of necessity of my postulate and for
which a solution is known. Such examples should exist.

Réponse de J. Mandel: En réponse à la question de M. Drucker, voici
un autre exemple de système stable n'obéissant pas à son postulat. Un curseur
pesant, mobile avec frottement d'angle φ sur une courbe de plan vertical, est placé

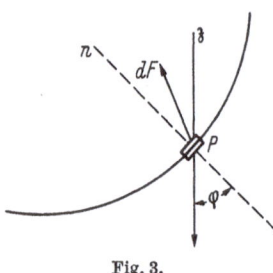

Fig. 3.

dans la position d'équilibre limite où la tangente
à la courbe fait l'angle φ avec le plan horizontal.
Si on lui applique une force dF contenue dans
l'angle nPz (figure 3), il se déplace vers le bas:
le travail de dF est donc négatif. Tous les systèmes
comportant des patins à friction mettant en jeu
le frottement de Coulomb donnent des résultats
analogues.

M. Drucker introduit la distinction entre la
stabilité dans l'infiniment petit et la stabilité
"dans le large", c'est-à-dire pour des déplacements
finis. Mais cette distinction ne joue pas pour le
modèle que j'ai présenté, ni pour le modèle ci-dessus: ces modèles sont stables
vis à vis des variations non infiniment petites.

1.6 Lower Bound Collapse Theorem and Lack of Normality of Strainrate to Yield Surface for Soils

By

G. de Josselin de Jong

In soil mechanics practice there is a need for a lower bound collapse theorem, which permits an analysis with a result on the safe side. The usual analysis of slip surfaces may give unsafe results for a purely cohesive soil, since it is based upon a kinematically admissable collapse system and therefore constitutes an upper bound. It is therefore necessary to investigate a great number of slip surfaces and the smallest load is an approximation to the actual load which will produce collapse, but it is never known how much the computed load exceeds the actual one.

Upper bound theorems for a material possessing COULOMB friction have been treated by DRUCKER (1954, 1961), but it is still necessary to establish a lower bound theorem. Indeed a lower bound theorem would seem to be of more practical value since it would lead to a result on the safe side. Unfortunately the virtual work proofs of lower bound theorems break down if the material does not obey the postulate of DRUCKER: that additional loads cannot extract useful net energy from the body and any system of initial stresses.

Now in soils there are two possible ways of extracting work, since soils in general are friction systems. The first possibility was mentioned by DRUCKER [1954] and is obtained by changing the isotropic stress in the body with internal friction. The second way to extract work is a consequence of the possible deviation angle between the principal directions of strain rate and stress tensors. This can be shown by considering the extreme case of deviation corresponding to the sliding of the upperleft block in Fig. 1 along a slip surface at $\left(45° - \frac{1}{2}\varphi\right)$ to the direction of the major principal stress. The slip occurs under constant volume conditions. Initially the stress state is represented by the points AA in the stress diagram of Fig. 2, lying just inside the limit circle. The additional forces are the stresses AB which bring

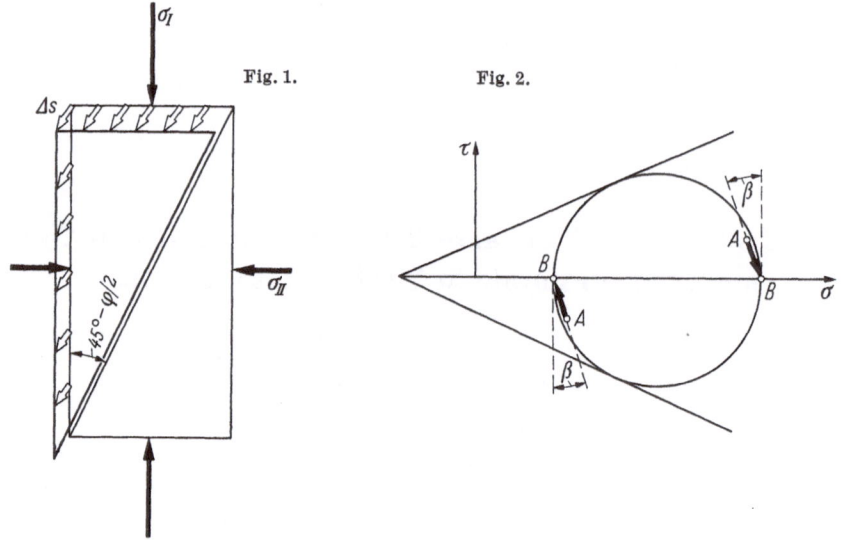

Fig. 1.

Fig. 2.

Fig. 3.

Fig. 4.

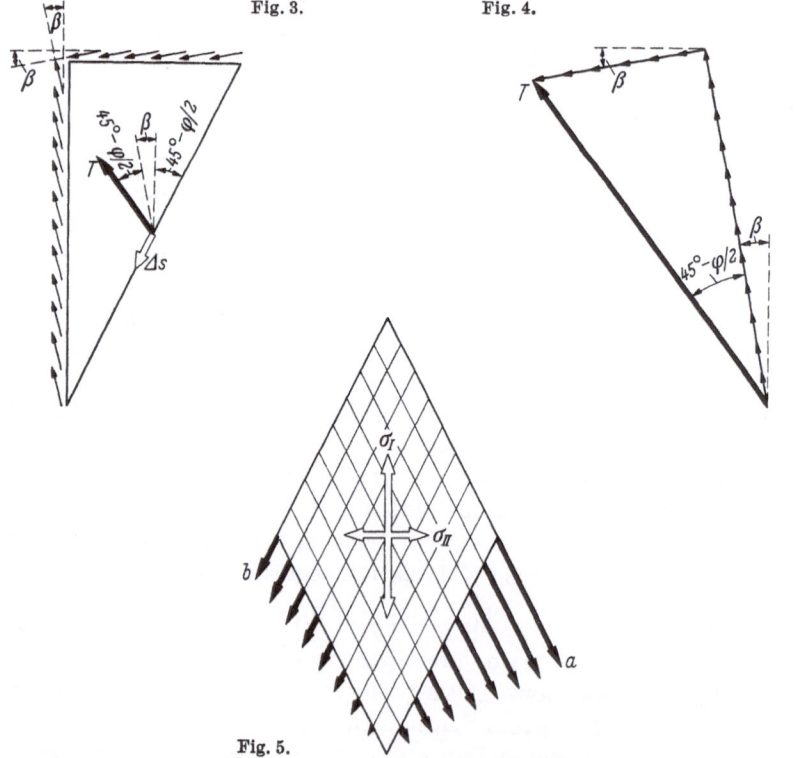

Fig. 5.

the system to a failure condition at BB. Let us consider the case when the vectors AB make an angle β with the τ-axis. The angle β can be made as small as we please by letting A approach B.

The additional loads on the moving upper left block then consist of stresses uniformly distributed along the vertical and horizontal faces and acting at an inclination β to these faces, Fig. 3. The resultant T of the additional forces on the upper left block is shown in Fig. 4 to make an angle of $\left(45° - \frac{1}{2}\varphi + \beta \right)$ with the vertical.

Under the influence of the existing stresses the block slides in a direction, at $\left(45° - \frac{1}{2}\varphi \right)$ downwards. If the displacement of the block is ΔS in that direction, then the work done by the body and the system of initial stresses on the added stress resultant T is equal to ΔS times the component of T in the direction opposite to ΔS. The work is therefore.

$$\Delta S \cdot T \cos (90° - \varphi + \beta) = \Delta S \cdot T \sin (\varphi - \beta).$$

This is positive if β is smaller than φ, thus positive work can be extracted.

Work can be extracted from a yielding system if the plastic strain rate tensor plotted as a vector in the corresponding generalised stress space is not normal to the yield surface.

In order to show the lack of normality in the case of soil explicitly, it is convenient to consider a stack of parallel cylinders which form a two dimensional analogy of a grain system with internal friction. Then the generalised stresses are the 4 stresses σ_x, σ_y, τ_{xy}, τ_{yx}, and the generalised stress space is therefore 4-dimensional. Fortunately τ_{xy} is equal to τ_{yx} and only the diagonal of length $\tau \sqrt{2}$ is a relevant coordinate. Therefore the generalised stress space can be reduced to the 3-dimensional space of Fig. 6, with coordinates σ_x, σ_y, $\tau \sqrt{2}$.

Let the material obey a COULOMB friction law, such that the yield criterion is:

$$(\sigma_x - \sigma_y)^2 + 4\tau^2 = [\sin \varphi (\sigma_x + \sigma_y + 2c \cot g \, \varphi)]^2. \tag{1}$$

To obtain a simpler expression for the yield surface, the coordinates are changed in the orthogonal system p, q, t according to

$$p = \frac{1}{2} \sqrt{2} (\sigma_x - \sigma_y),$$

$$q = \frac{1}{2} \sqrt{2} (\sigma_x + \sigma_y + 2c \cot g \, \varphi),$$

$$t = \tau \sqrt{2}$$

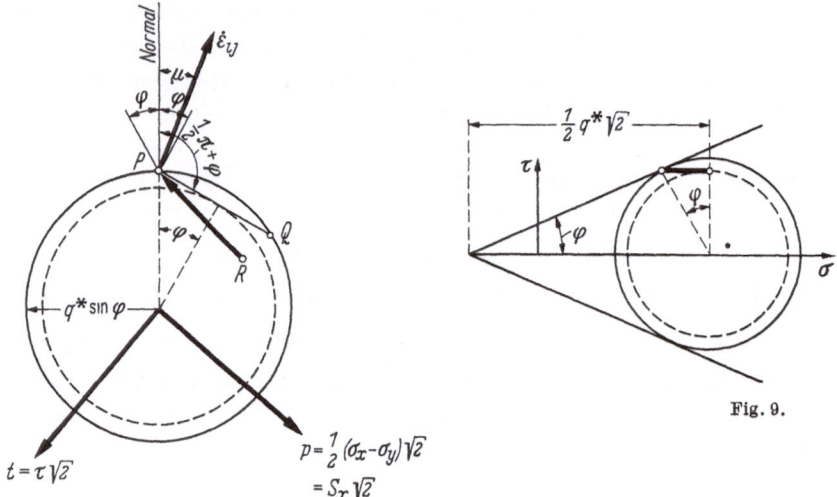

Fig. 8.

Fig. 9.

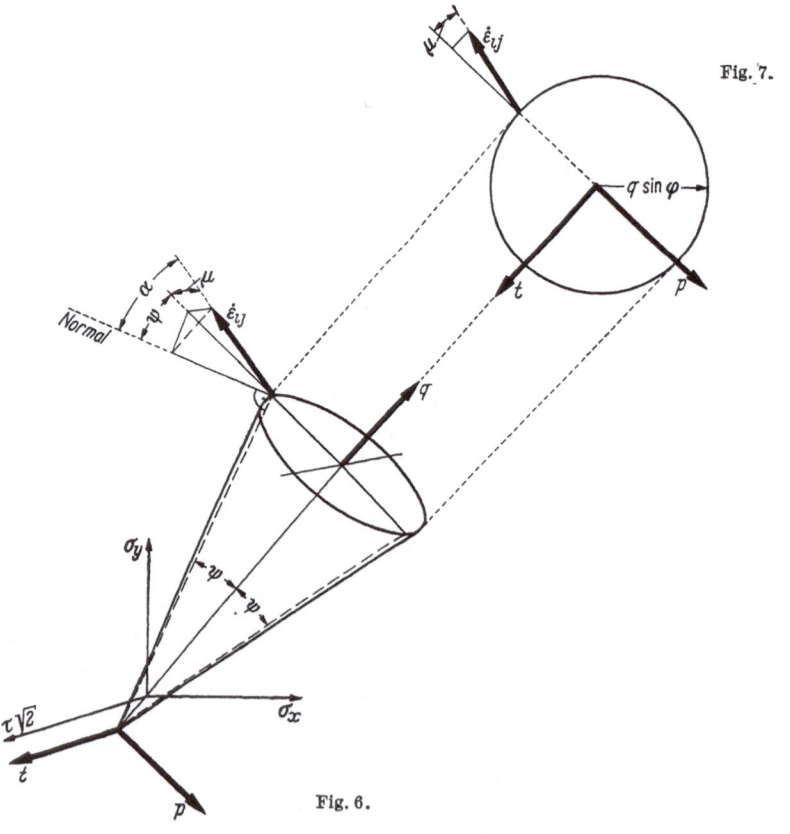

Fig. 7.

Fig. 6.

Then q is the bisectrix of σ_x and σ_y, and p is a coordinate in the σ_x, σ_y-plane perpendicular to q. In these coordinates the yield criterion is

$$2p^2 + 2t^2 = 2q^2 \sin^2 \varphi. \tag{2}$$

This shows that the yield surface is a cone with q as axis and which intersects the planes for $q = $ constant by a circle with radius $q \sin \varphi$. The angle ψ is then related to φ by

$$\tan \psi = \sin \varphi. \tag{3}$$

If the rod material is assumed to behave as the mechanical model proposed by the author (1958, 1959) plastic shear strain rates consist of volume conserving slip in the directions at $\left(45° - \dfrac{1}{2}\varphi\right)$ with the major principal stress. The two conjugate shear strain rates need not be equal. If they are a and b respectively as shown in Fig. 5, then the deviation angle μ between principal directions of strain rate tensor is given by

$$\tan \mu = \frac{a - b}{a + b} \tan \varphi. \tag{4}$$

Since a and b can only be positive, this relation implies

$$-\varphi \leq \mu \leq \varphi. \tag{5}$$

It can be shown by a straightforward but somewhat tedious computation that the deviation angle α between the strain rate vector and the normal to the yield surface is then given by:

$$\cos \alpha = \cos \psi \cos \mu. \tag{6}$$

Since the sliding motion is considered to take place at constant volume the strain rate vector $\dot{\varepsilon}_{ij}$, plotted in a coordinate system corresponding to the generalised stresses, lies in the $q = $ constant plane. This plane makes an angle ψ with the normal to the yield surface as shown in Fig. 6. In order that the angle α between $\dot{\varepsilon}_{ij}$ and the normal obeys (6) it is necessary that $\dot{\varepsilon}_{ij}$ is not normal to the circle in the $q = $ constant plane of Fig. 7, but makes an angle μ with the radius of that circle.

According to the first collapse theorem a body is capable of supporting the external loads in any loading program, if it is possible to find a safe statically admissable stress distribution $\sigma_{ij}^{*(s)}$. A stress distribution is called statically admissable if it obeys the equilibrium conditions inside the body, if it satisfies boundary conditions on the part of the boundary where surface tractions are given and if a yield inequality is nowhere violated. For perfectly plastic materials the yield inequality simply requires that $\sigma_{ij}^{*(s)}$ lies inside the yield surface. This requirement is clearly necessary and is also sufficient because convexity of the yield surface and normality of the strain rate vector to that surface ensure that the real collapse stress state σ_{ij} is such that the quantity

$$[\sigma_{ij} - \sigma_{ij}^{*(s)}]\,\mathring{\varepsilon}_{ij}$$

is always positive. The proof of the first collapse theorem follows then by use of virtual work considerations [for a comprehensive description of this theorem and related matter see f.i. KOITER (1960)].

Since there is not always normality in the case of soils the yield inequality condition has to be modified. The modification necessary to take care of the angle μ is only small if by some other means it is possible to prove that q cannot decrease below a certain value q^*.

If the mechanical model of Fig. 5 is applicable, Eqs. (4) and (5) say that the absolute value of μ cannot exceed φ. Now let P represent a real collapse stress state $\sigma_{ij}^{(P)}$, then P lies on the circle with radius $q^* \sin \varphi$ in the plane $q = q^*$, Fig. 8. All stress states $\sigma_{ij}^{(R)}$ represented by a point R lying below PQ, the line at an angle $\left(\frac{1}{2}\pi + \varphi\right)$ to the normal in P, may be called statically admissable with respect to P, because the angle, between any line PR and the vector $\mathring{\varepsilon}_{ij}$ for $\mu = \varphi$, will be larger than $\frac{1}{2}\pi$. Therefore the quantity

$$[\sigma_{ij}^{(P)} - \sigma_{ij}^{(R)}]\,\mathring{\varepsilon}_{ij}$$

will always be positive for $\mu = \varphi$, and clearly this result is generally valid in the interval $0 \le \mu \le \varphi$.

Since the actual collapse stress will be everywhere on the circle, the statically admissable stress state $\sigma_{ij}^{*(s)}$ is limited by all lines PQ drawn from all points of the circumference. This means that the stress states are limited by the dotted circle in Fig. 8, with a radius of length $q^* \sin \varphi \cos \varphi$.

Since the coordinates p and t actually are $\sqrt{2}$ times the deviator-stresses s_x, τ_{xy}, the requirement of the dotted circle can be represented in the usual MOHR-diagram of Fig. 9 by the dotted circle whose radius is equal to the shear stress at the tangent point of MOHR-circle and COULOMB envelope line. This means that a safe statically admissable stress state is limited by the dotted circle (Fig. 9) which is equivalent to reducing the angle of shearing resistance to a value φ^* given by:

$$\sin \varphi^* = \sin \varphi \cos \varphi.$$

Although by this modification of the definition for a statically admissable stress state, the difficulties created by the uncertainty about the deviation angle between principal directions of stress tensor and strain rate tensor are circumvented, it must be emphasized that this only applies if by other means it is established that q cannot decrease below the value q^*. The region limiting the statically admissable stress states is therefore given by a circular cylinder starting

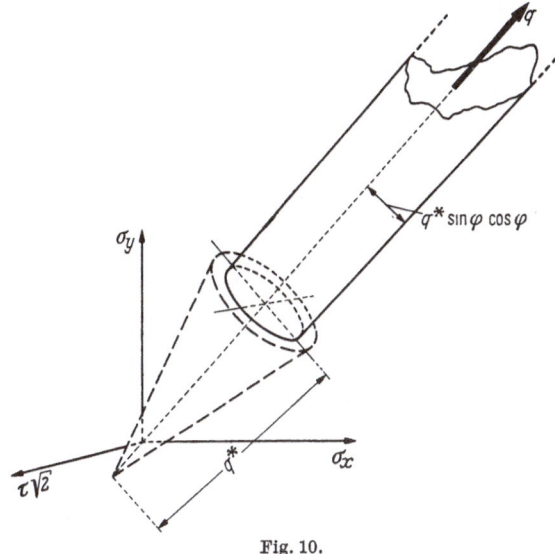

Fig. 10.

on the base of the cone with height q^* and running up to infinity with a radius $q^* \sin \varphi \cos \varphi$.

Literature

DRUCKER, D. C.: Coulomb Friction, Plasticity and Limit Loads. Appl. Mech. **21**, №. 1, 71—74 (1954).

DRUCKER, D. C.: On Stress-Strain Relations for Soils and Load Carrying Capacity. Proc. 1st Int. Conf. Mech. of Soil Vehicle Systems, Turin, 1961.

DE JOSSELIN DE JONG, G.: Indefinitness in Kinematics for Friction Materials. Proc. Conf. Brussels on Earth Pressure Problems, 1958, Vol. I, Brussels 1958, pp. 55—70.

DE JOSSELIN DE JONG, G.: Statics and Kinematics in the Failable Zone of a Granular Material. Doctors Thesis Delft, 1959.

KOITER, W. T.: General Theorems for Elastic-Plastic Solids. Progress in Solid Mechanics, Vol. I, 1960, pp. 165—221.

Discussion

Contribution de K. H. ROSCOE: I would like to question the universal application of Professor DE JONG's statement that the normality condition does not apply to soils. The following remarks are very tentative since I have not had an opportunity to make a proper study of DE JONG's proposals. It does however seem that he is considering soil to be a non-dilatant material possessing constant cohesion and constant internal friction and he is concerned only with states of failure of such a medium. I wish to make two observations regarding these assumptions. Firstly soil is a dilatant medium and as it dilates the apparent cohesion and internal friction will change. Secondly the Mohr-Coulomb envelope is not a true yield surface for soils. If yield is defined as permanent irrecoverable deformation then soils yield, and of course dilate, at stress levels well below those required to satisfy the Mohr-Coulomb criterion of failure.

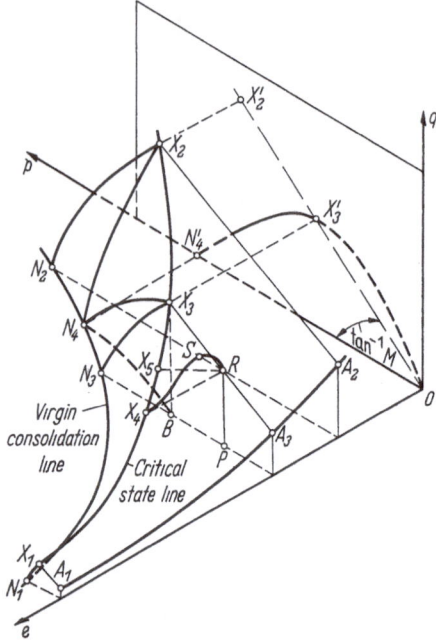

Fig. 1. Isometric view of idealised (p, q, e) yield surface for clays.

The position can be made clearer by referring to Fig. 1 which represents our concepts of the yield surface, obtained from triaxial tests on samples of a saturated remoulded clay, in p, q, e space; where $p = (\sigma_1' + 2\sigma_3')$, $q = (\sigma_1' - \sigma_3')$, e is the voids ratio and σ_1' and σ_3' are the major and minor principal effective compressive stresses respectively. In Fig. 1 the curve $N_1 N_2$ is the isotropic virgin consolidation curve and $X_1 X_2$ is the critical state line. The projection of the critical state line on the (p, q) plane is the straight line $O X_2'$. When a sample reaches a state corresponding to a point on the curve $X_1 X_2$ it will continue to distort in shear without further dilation and without change of stress.

The (p, q, e) yield surface for virgin and lightly over-consolidated clays is represented by the curved surface $N_1 N_2 X_2 X_1$ and in my paper to this symposium I have endeavoured to show that there is some experimental justification for such a surface. Its precise shape is open to some doubt as discussed by Roscoe, Schofield and Thurairajah (1963), and Roscoe and Schofield (1963). Typical (p, q, e) state paths for undrained tests on normally consolidated samples are represented by curves $N_1 X_1$, $N_2 X_2$ and $N_3 X_3$, while a typical path for a drained test is $N_4 X_2$. It is important to notice that whenever a sample is at a state corresponding to a point on the surface $N_1 N_2 X_2 X_1$, and the deviator stress is increasing, it will be yielding. Consider for example a sample initially at state N_4. If it traverses any state path on the yield surface within the sector $N_2 N_4 X_3$ it will work harden as it yields but it will not fail until the critical state is attained. If the state change corresponds to the path $N_4 X_3$ which lies vertically above the elastic swelling curve $N_4 B$ then the sample will yield and not work harden. The relevant plastic potential curve is then $N_4' X_3'$. We have called the curve $N_4 X_3$ an elastic limit curve. As a sample work hardens the relevant plastic potential curve continuously grows in size but remains geometrically similar to curve $N_4' X_3'$. We have proposed that the form of the plastic potential curves is governed by the equation $q = M p \log_e \dfrac{p_0}{p}$ where p_0 is the initial consolidation pressure, and M is as shown in Fig. 1.

Let us now consider more heavily over-consolidated clays. The experimental data that is available for such clays is much less reliable than for lightly over-consolidated clays, hence the following remarks are extremely tentative. We suggest that the (p, q, e) yield surface for undrained tests is $A_1 A_2 X_2 X_1$ in Fig. 1. Consider an over-consolidated sample initially in a state represented by the point P. If it is subjected to an undrained test it will follow a state path which may be idealised by the path PRX_3 in Fig. 1. During the portion PR the sample behaves virtually elastically but it begins to yield at R and continues to yield and work harden until it reaches the peak deviator stress, as well as the critical state, at X_3. If the sample was allowed to dilate during a test then present evidence suggests that the state path comes above the undrained surface. For example an ideal representation of a $p =$ constant test is given by the path $PRSX_4$. In such a test yield begins at R but the sample continues to work harden over the range RS and attains the peak deviator stress at S. The sample then becomes unstable and subsequent successive states correspond to SX_4. I suggest that some path above a line such as RX_5 may be found in which this unstable portion is not present. For such a test the deviator stress would never diminish as the state changed from P to X_5. Hence as a sample, of initial state P, traverses any state path between RX_5 and RX_3, it will continually work harden until it attains the critical state when it fails. It is possible that a family of plastic potential curves of the type shown by OX_3' apply during all the work hardening processes undergone by over-consolidated samples. The curve OX_3' may have the same equation as $N_4' X_3'$, but adequate exerimental evidence is not available to be able to see how such plastic potentials relate to the yield surfaces for anything other than lightly over-consolidated clays. We have a little indirect evidence on the heavily overcon-solidated or "dense" side from simple shear tests on steel balls. This medium appears, during any work hardening process, to have plastic potential curves of the type shown in Fig. 2. The equation of these curves is $\tau = M \sigma \log_e \frac{\sigma_0}{\sigma}$, where τ is the maximum shear stress and σ the mean normal stress under conditions of plane

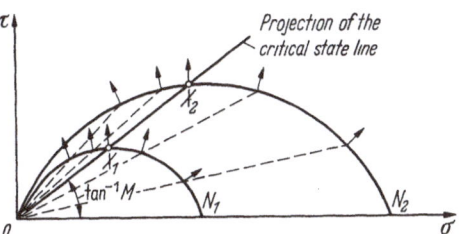

Fig. 2. Plastic potential curves for steel balls.

strain. This equation follows directly from the application of the normality condition to the boundary energy equation which was discussed by POOROOSHASB and ROSCOE (1961) for steel balls. Further work is still required to connect these potential curves with the observed yield surfaces.

Finally I would like to make the point that far too much effort has been made in soil mechanics to study failure conditions. Engineers design, and hope their structures operate, at much lower stress levels. This is the region of yielding that should be studied in detail. The MOHR-COULOMB envelope may, or may not, be shown to be valid for the failure of soils but it is not a yield surface in the true sense of the word since the yielding of a sample cannot be related to a movement on the envelope. With such a theory yield does not occur until failure takes place.

References

POOROOSHASB, H. B., and H. H. ROSCOE (1961): The correlation of the results of shear tests with varying degrees of dilatation. Proc. 5th. Int. Conf. Soil Mech. Vol. 1, pp. 297—304.

ROSCOE, K. H., and A. N. SCHOFIELD (1963): Mechanical behaviour of an idealised "wet-clay". Proc. European Conf. Soil Mech., Wiesbaden, October 1963, Vol. 1, pp. 47—54.

ROSCOE, K. H., A. N. SCHOFIELD and A. THURAIRAJAH (1963): Yielding of clays in state wetter than critical. Géotechnique 13, No. 3, 211—240.

Réponse de G. DE JOSSELIN DE JONG: It was not my intention to say that for soils there never is normality, but that normality is not necessary. In the cases studied by M. ROSCOE normality may have been observed, but these are special cases, which are not representative for the situation in general.

That M. ROSCOE did not observe the deviation of the principal directions of stress and strain rate tensors, is due to the fact, that the stress coordinates p and q in his diagrams are not the complete set of generalised stresses. The samples were 3-dimensional, so the testresults require a representation in a 9 dimensional stress space. Since shearstresses on perpendicular faces are equal the amount of dimensions can be reduced to 6. The system I talked about this morning, is 2-dimensional and so there are 4 generalised stresses, from which τ_{xy} is τ_{yx}, reducing the system to 3 stress coordinates.

Since M. ROSCOE only considers the stress combinations p and q, his graphs correspond in a way to the σ_x, σ_y plane which intersects the cone enclosed by the yield surface along the axis. The deviation of the principal directions of the tensors is only visible in the plane perpendicular to the axis.

Cf. aussi, p. 46, la citation de D. C. DRUCKER.

1.7 Plasticity and Creep of Cohesive Medium

By

S. S. Vyalov

Due to their internal friction, cohesive soils have different resistance to tension and compression, which should be taken into consideration when examining their rheological equation of state. The existing theories of plasticity and creep are based on the conception that the change in the shape of a material depends only upon the stress deviator, while the change in its volume depends only upon the hydrostatic pressure, that is

$$\sigma_i = \varphi(\varepsilon_i) \quad \text{and} \quad \sigma = \varphi^*(\varepsilon), \tag{1}$$

where σ_i and ε_i — intensity of tangent stresses and intensity of shearing strain, while σ and ε — average normal (hydrostatic) pressure and average (volumetric) deformation.

The rheological equation of state is obtained by introducing the time factor t into the relationship (1); the method of introducing t depends on the adopted creep theory; the simplest variant may be written as follows:

$$\sigma_i = \varphi(\varepsilon_i) F(t) \quad \text{and} \quad \sigma = \varphi^*(\varepsilon) F^*(t), \tag{2}$$

where $F(t)$ — time function.

The form of functions φ and F is determined from tests in a simple stressed state (extension or compression, pure shear). The above statements, however can not be applied to materials with different resistance to extension and compression, for example to cohesive soils. It has been shown by soviet scientists in 1940, that the change of the shape and volume of such soils depends both upon σ_i and σ, that is

$$\sigma_i = \varphi(\varepsilon_i, \sigma) \quad \text{and} \quad \sigma = \varphi^*(\varepsilon, \sigma_i). \tag{3}$$

Upon examination of the surface in coordinates $\sigma_i - \varepsilon_i - \sigma$ the following form of the first of these equations has been acquired by author

$$\sigma_i = \varphi_1(\varepsilon_i) + \varphi_2(\varepsilon_i)\phi(\sigma), \tag{4}$$

where the first term characterizes resistance to pure shear, while the second term characterizes the increase of this resistance due to the effect of the hydrostatic pressure σ. In the general case this equation will have an integral form. A similar equation is given for volumetric deformation.

The time factor is taken in account by simultaneous solution of the Eq. (2) and the conventional creep equation. As a result, the rheological equation of state for a cohesive medium is obtained, which establishes the relationship between the intensity of tangent stresses, the intensity of shear strain (or their rates), hydrostatic pressure and time. As a result the following equation must be written instead of Eq. (2)

$$\sigma_i = \varphi_1(\varepsilon_i) F_1(t) + \varphi_2(\varepsilon_i) \phi(\sigma) F_2(t). \tag{5}$$

For the general case, this equation is expressed by the integral form of VOLTERRA-BOLTZMAN. Different kinds of functions included in the Eq. (5) and various particular variants of these equations have been examined. With some assumptions proved by the experiments, it may be assumed that

$$\varphi_1(\varepsilon_i) = \tau = A_0 \varepsilon_i^m, \quad \varphi_2(\varepsilon_i) = \tan \psi = B_0 \varepsilon_i^n,$$

$$F_1(t) = \frac{A(t)}{A_0}, \quad F_2(t) = \frac{B(t)}{B_0}, \quad \phi(\sigma) = \sigma,$$

where τ — pure shear, while ψ — deflection angle on an octahedral platform.

Then Eq. (5) may be written as follows:

$$\sigma_i = A(t) \varepsilon_i^m + B(t) \sigma \varepsilon_i^n. \tag{6}$$

The forms of the functions $A(t)$ and $B(t)$ characterize the law of development of deformations in time (creep law). In particular, it may be assumed that

$$A(t) = \frac{A_0 \xi}{\xi + A_0 t^\lambda} \quad \text{or} \quad A(t) = \frac{A_0 \xi}{\xi + b \ln (1 + \xi t)}.$$

Parameters m and n and the forms of the creep functions $A(t)$ and $B(t)$ are determined from the test for triaxial compression or torsion and compression under the conditions of creep. With $m = n$ the equation is significantly simplified:

$$\sigma_i = A(t) \varepsilon_i^m \left[1 + \frac{\sigma}{H(t)} \right], \tag{7}$$

where

$$H(t) = \frac{\tan \psi}{\tau} = \frac{A(t)}{B(t)}.$$

Similar equations are obtained when the rate of deformation $\frac{d\varepsilon_i}{dt}$ is considered instead of the deformation ε_i itself. Such equations have been considered for NEWTONian, non-NEWTONian and BINGHAM flows.

The equation of limit equilibrium taking into consideration the time factor can be obtained from Eq. (6) for the limit values $\tau = \tau_s$ and $\psi = \psi_s$, corresponding to the moment of soil distruction

$$\sigma_i = \tau_s(t) + \sigma \tan \psi_s(t) = \tau_s(t)\left[1 + \frac{\sigma}{H_s(t)}\right]. \tag{8}$$

The change of $\tau_s(t)$ and $H_s(t)$ with time is determined by the equation of stress-rupture strength, for instance

$$\tau_s(t) = \frac{\beta}{\ln\dfrac{t+1}{\alpha}}. \tag{9}$$

The creep and limit equilibrium of a frozen soil cylindrical guard, made for the purpose of sinking mine shafts by the method of artificial freezing, have been considered as an example. The example shows that the inclusion of the effect of the hydrostatic pressure σ increases the supporting power of the soil.

1.8 An Application of the Random Walk Argument to the Mechanics of Granular Media

By

Jerzy Litwiniszyn

Strata movements in Nature due to Man's mining activity and similarly laboratory investigations of dry sand models show that the mathematical theory based on the notion of continuum mechanics does not describe the motion in question adequately.

Since a few years the Cracow Centre develops theoretical and experimental investigations of the granular medium considered as a set (or collection) of individual particles [1].

The theory of motion of such a collection of particles which is being indentified with a granular medium has been founded on a system of postulates resulting in the SMOLUCHOWSKI-KOLMOGOROV equation associated usually with Markovian stochastic processes [2].

An equation derived from a random walk argument given in this paper is a generalization of S. GOLDSTEIN's approach [3]. A *hyperbolic* equation is obtained in general and as a particular case the parabolic equation of MARKOV processes is seen to follow.

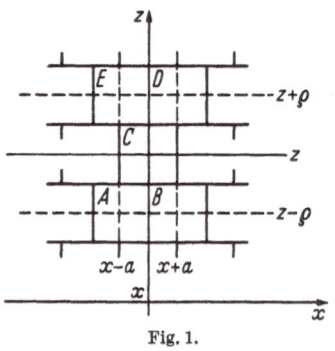

Fig. 1.

Let us imagine a set of identical particles distributed in a twodimenstional system of cages as shown on Fig. 1. This cage system is referred to a Cartesian coordinate system (x, z), the z-axis being directe vertically upwards. The width of each cage is $2a$, its height being ϱ. The cage coordinates are defined as those of the point of intersection of the diagonals of the cage rectangle. The material particles are subjected to the action of a gravity force directed vertically downwards and parallel to the z-axis. This force excludes any upward motion of the particles. The particles are assumed to remain in the (x, z) plane.

We assume further that the particle displacements occur by successive steps or "jumps".

A single step from cage C (Fig. 1) may be performed to one of the cages immediately below, thus from C to A or from C to B. This step takes place during a finite time interval, say τ. Displacements of particles in a "jump" from C to A or from C to B are accompanied by corresponding displacements of "cavities" from A to C or from B to C. Thus, removal of a particle from cage A results in a cavity moving from A to C if the particle is displaced from C to A.

We assume that the appearance of a cavity in cage C is due to *one* out of two possible events, namely displacement of a cavity from either cage A or cage B to cage C immediately above.

Let p denote the probability that the cavity will continue to move left (or resp. right) when it moved previously to the left (or resp. right) in its displacement from a lower cage to an upper one.

The quantity:

$$q = 1 - p \tag{1}$$

denotes the probability that the wandering cavity will change its direction. Thus, the probability that a particle will move elong the path $A \rightarrow C \rightarrow D$ is p whereas the path $A \rightarrow C \rightarrow E$ has the probability $q = 1 - p$.

We introduce further the quantity:

$$c = p - q \tag{2}$$

which will further be referred to as the coefficient of correlation. The case when $c \neq 0$ is interpreted as a kind of "memory", a particle or cavity "remembering" its previous step.

When $c = 0$, we have $p = q = \frac{1}{2}$ and the particle in its next step moves in the right or left direction with equal probability, independently of the direction of motion in the previous step.

Let us denote by $P = P(x, z, t)$ the probability that a cavity will be produced at the (x, z) cage at time t. Assuming that the cavity may move upwards only, the following difference equation is obtained for the probability $P(x, z, t)$ after some manipulations [4]:

$$P(x, z + \varrho, t + \tau) = p\left[P(x - a, z, t) + P(x + a, z, t)\right]$$

$$- c\,P(x, z - \varrho, t - \tau), \tag{3}$$

where a, ϱ, τ denote the magnitude of the "jumps" of the coordinates x, z and the time t.

6*

By a limiting process such that

$$\lim_{\substack{\varrho \to 0 \\ \tau \to 0}} \frac{\varrho}{\tau} = A > 0; \qquad \lim_{\substack{a \to 0 \\ \varrho \to 0 \\ \tau \to 0}} \frac{a^2}{\varrho \tau} = B > 0; \qquad \lim_{\substack{c \to 1 \\ \tau \to 0}} \frac{1 - c}{\tau} = D > 0 \qquad (4)$$

the difference Eq. (3) becomes a linear hyperbolic equation:

$$\frac{\partial^2 P}{\partial z^2} + \frac{2}{A} \frac{\partial^2 P}{\partial z \, \partial t} + \frac{1}{A^2} \frac{\partial^2 P}{\partial t^2} - \frac{B}{A} \frac{\partial^2 P}{\partial x^2} + \frac{D}{A} \frac{\partial P}{\partial z} + \frac{D}{A^2} \frac{\partial P}{\partial t} = 0. \qquad (5)$$

A solution $P = P(x, z, t)$ may, for instance, be interpreted as follows:

If a single particle is removed from the cage at (x_0, z_0) at time t_0, a cavity appears in this cage. This cavity propagates upwards i.e. to the cages at $z > z_0$, as time increases $(t > t_0)$. The probability that the wandering cavity for $t > t_0$ reaches the cage at (x, z) is $P = P(x, z, t)$.

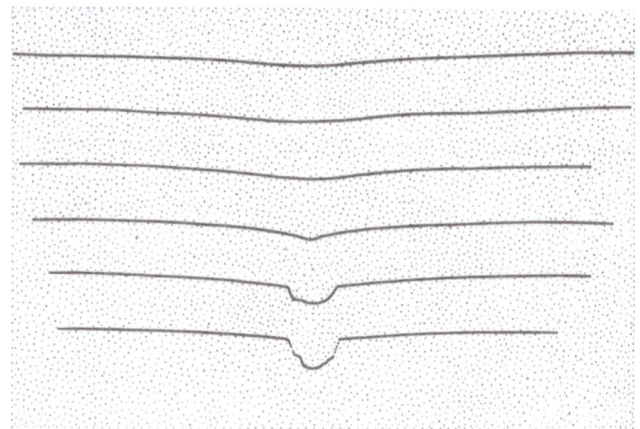

Fig. 2.

When a sufficiently large number of particles is removed from cage $|x_0, z_0|$ at time t_0 a corresponding number of cavities is formed in the region at $z > z_0$.

The solution $P = P(x, z, t)$ is informative on the ratio of the number of particles which moved through the cage at (x, z) during the time interval $[t_0, t]$ to the number of particles which were removed from the cage (x_0, z_0) at time t_0.

The above said may be imagined further in the following way. The magnitude $P = P(x, z, t)$ defines a quantity proportional to the volume of the granular medium which is sifted through a unit area (x, z) of a horizontal surface during the interval $[t_0, t]$.

This process of sifting produces a "trough"-shaped displacement profile at every horizon z. The ordinate of the profile can be *measured*.

A trough pattern obtained in a dry sand experiment by an out-pour of sand through a bottom slot is shown in photo 2. This pattern corresponds to a certain state which is no more variable in time, i.e. for which $\frac{\partial P}{\partial t} = 0$.

The process of trough formation in time is not easily accessible to observation and no appropriate equipment was available. There-fore a deteiled investigation was limited to the case when finally.

The process of trough formation was thus terminated and finally formed troughs were measured only.

The Eq. (5) becomes in this case

$$\frac{\partial^2 P(x,z)}{\partial z^2} + \frac{D}{A}\frac{\partial P(x,z)}{\partial z} = \frac{B}{A}\frac{\partial^2 P(x,z)}{\partial x^2}. \tag{6}$$

Solution of this equation were compared with measured troughs. As may be seen from the photo 2, there are two kinds of troughs produced by an out sand through a narrow slot in the bottom of a chamber. A few troughs in the lower part of the sand mass show marked dis-continuities which are symmetrical with regards to the central vertical axis passing through the bottom slot. These discontinuities disappear in the upper part of the sand mass.

As aforesaid the troughs produced by an outpour bottom (or initial conditions) of this kind can be written as follows:

(a) $P(x,0) = \gamma\,\delta(x),$

(b) $\left(\dfrac{\partial P(x,z)}{\partial z}\right)_{z=0} = 0.$ (7)

The symbol $\delta = \delta(x)$ denotes the "DIRAC function", γ is the sand volume poured out. This quantity can be easily measured. The initial conditions (7a, b) allow to obtain a solution of the hyperbolic Eq. (6) in the form:

$$P(x,z) = \gamma\,\frac{D}{2}\,(A\,B)^{-1/2}\exp\left(-\frac{zD}{2A}\right)\left[I_0(y) + \frac{zD}{2A}\frac{I_1(y)}{y}\right]$$

$$\text{for} \quad |x| < \left(\frac{B}{A}\right)^{1/2} z \tag{8}$$

$$P(x,z) \equiv 0 \quad \text{for} \quad |x| > \left(\frac{B}{A}\right)^{1/2} z,$$

where:

$$y = \frac{\left(\frac{B}{A}z^2 - x^2\right)^{1/2}}{2\sqrt{A\,B}},$$

further I_0 and I_1 are the BESSEL functions of the second kind. Intro-ducing the dimensionless variables:

$$\xi = \frac{xD}{2\sqrt{A\,B}}; \qquad \eta = \frac{zD}{2A}$$

we obtain from (8)

$$P = P(x, z) = \overline{P}(\xi, \eta).$$

The behaviour of this function is shown on Fig. 3 where $\overline{P} = \overline{P}(\xi, \eta)$ was plotted for a few constant η values, i.e. constant z values.

The characteristics of Eq. (6) are given by the two equations:

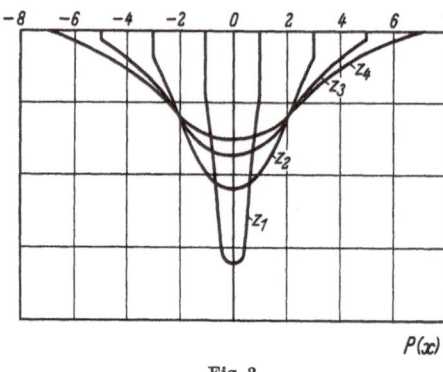

Fig. 3.

$$x = \pm \sqrt{\frac{B}{A}}\, z.$$

Discontinuities of the solution (8) propagate along these characteristics.

The solution plotted in Fig. 3 may be associated with the troughs shown in photo 2 especially at the lower horizons. The troughs in this region approach more and more the shape of the "Dirac function" which is an abstraction of an outpour of sand through a narrow slot. A qualitative resemblance of the trough shapes observed in the lower part of the sand mass with the theoretical solution is clearly visible.

The geometrical loci of the discontinuities may be interpreted as two characteristics of the Eq. (6) issuing from the point where the "DIRAC" bottom conditions are given.

The above interpretation of the characteristics is reminiscent of the interpretation of characteristics as surfaces of slip observed in continuous media.

It seems worth noting that probabilistic considerations lead to a similar interpretation.

When $c = 0$ is put in the initial difference Eq. (3) i.e. it is assumed that the correlation coefficient is zero we obtain $p = q = \frac{1}{2}$. Particles possess no memory, however short, in this latter case. For $\frac{\partial P}{\partial t} = 0$ we have

$$P(x, z + \varrho) = \frac{1}{2}\left[P(x - a, z) + P(x + a, z)\right]. \qquad (9)$$

After a limiting process given by

$$\lim_{\substack{a \to 0 \\ \varrho \to 0}} \frac{a^2}{\varrho} = K \qquad (10)$$

we obtain a parabolic equation:

$$\frac{\partial P(x, z)}{\partial z} = B \frac{\partial^2 P(x, z)}{\partial x^2}, \quad \text{where} \quad B = \frac{K}{2}. \tag{11}$$

The solution of this equation for initial conditions (7a) assumes the form:

$$P(x, z) = \gamma (4 \pi B z)^{-1/2} \exp \left[-\frac{x^2}{4 B z} \right]. \tag{12}$$

It is known that the solutions of the parabolic Eq. (11) possess the property of smoothin out of discontinuities of initial conditions given for $z = 0$. The troughs formed in the lower part of the sand mass are not smoothed out but a propagating and finally decaying discontinuity is clearly visible. These discontinuities propagate along the characteristics of the hyperbolic Eq. (6).

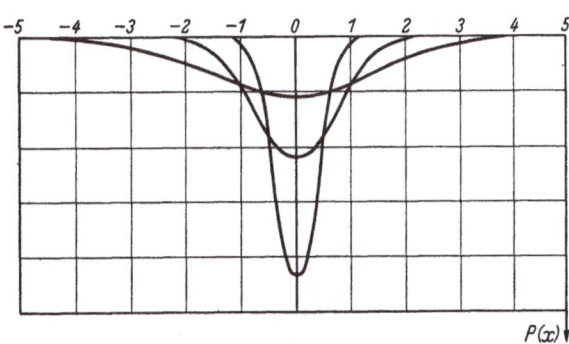

Fig. 4.

The $P = P(x, z)$ curves assume a GAUSSIAN shape for fixed z values (cf. Fig. 4).

Troughs due the initial conditions (7a) measured at the upper horizons (i.e. for large z values) are shown in functional scale in Fig. 5.

Thus the trough formation process at higher horizons for large positive z-values is explained with a sufficient accuracy by the Eq. (11).

The transition from discontinuous troughs at lower horizons to the continuous ones at higher levels is still a problem demanding further investigations.

It seems that this effect might be explained by introducing a non-linear (or quasi-linear) hyperbolic equation where the functional coefficient of the $\frac{\partial^2 P}{\partial z^2}$ term [in Eq. (6)] depends on $\frac{\partial P}{\partial x}$. For $\frac{\partial P}{\partial x} \to 0$ this coefficient should tend to zero. The hyperbolic equation in this case tends to its parabolic counterpart.

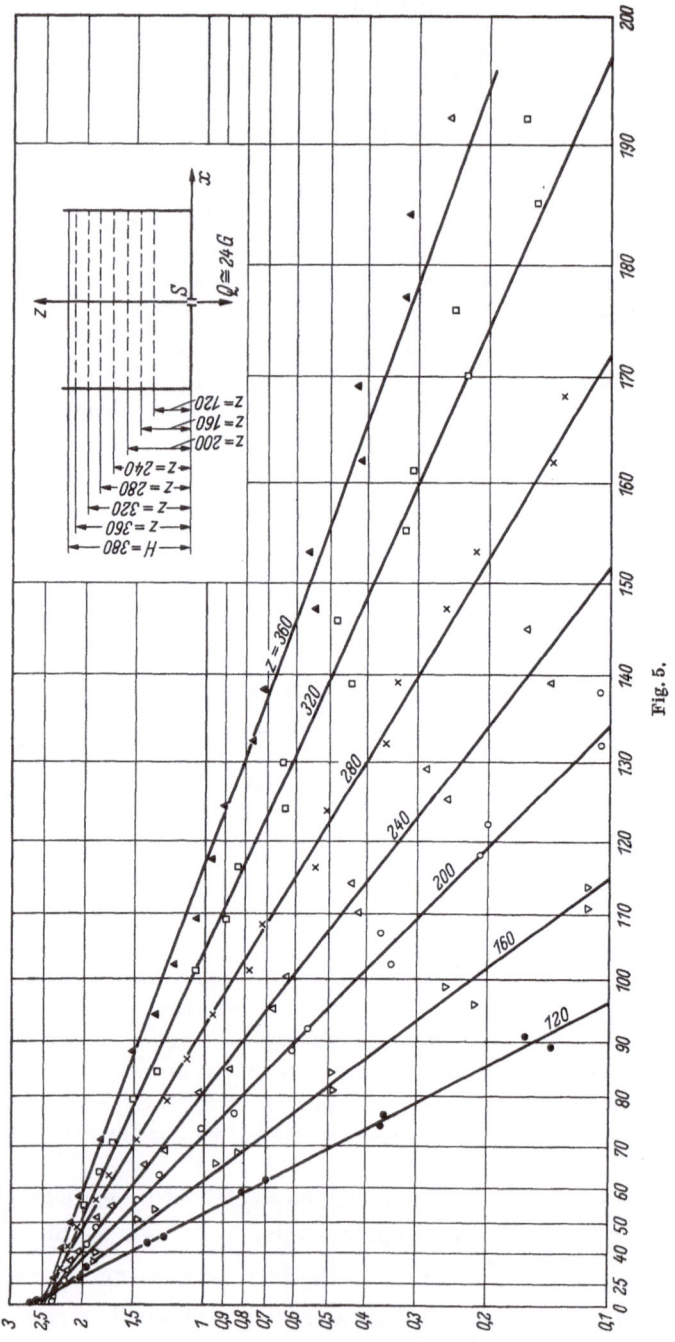

Fig. 5.

References

[1] LITWINISZYN, J.: Die Mechanik diskontinuierlicher Medien und ihre Anwendung in der Felsmechanik. — Felsmechanik und Ingenieurgeologie. Rock Mechanics and Engineering Geology, Vol. I/3—4, 1963.

[2] KOLMOGOROV, A.: Über die analytischen Methoden der Wahrscheinlichkeitsrechnung. Math. Ann. B. 104, 1931.

[3] GOLDSTEIN, S.: On Diffusion by Discontinuous Movements, and on the Telegraph Equation. The Quarterly Journal of Mechanics and Applied Mathematics, Vol. IV, Part 2, June 1951.

[4] LITWINISZYN, J.: The Model of a Random Walk of Particles Adapted to Researches on Problems of Mechanics of Loose Media. II. Bull. Acad. Polon. Sci. Sér., Sci. techn. Vol. XII, No. 5, 1964.

1.9 Rheological and Mechanical Models of Saturated Clay Particle Systems[1]

By

E. C. W. A. Geuze

I. Introduction

Rheological models have been used to characterize the observed stress-strain-time behavior of clays. In this manner model parameters and numerical values of coefficients were obtained from experiments involving specific test conditions in terms of stress-time sequences and stress and/or displacement control at the boundaries of the specimen. In some cases the expulsion of liquid at the boundaries of the specimen was prevented, in other cases the expulsion of liquid could occur at certain boundaries without this restriction.

The evidence obtained in this manner and with a view towards establishing the rheological characteristics of the bulk material has however remained limited. Nevertheless, a few authors have used this limited evidence as a basis for prediction of rheological behavior of clay systems under more general environmental and loading conditions.

This procedure, however commendable for instructional reasons, is subject to a number of restrictions stemming from the limited extent of experimental proof of the validity of the models used. This situation would however improve with time, as more experimental evidence and record data would become available.

Far more serious objections should be raised against the basis of the model technique; the superposition of components representing the various mechanisms inherent to the deformation of clay systems. The present use of models is restricted to a two-dimensional arrangement of components (springs, dashpots, friction elements, etc.). These elements appear either in parallel or in series. Addition of a third dimension is considered a possibility to include space-oriented variables. This does however not alter the fact that system behavior predicted

---

[1] In the text the word "clay" will usually replace the more elaborate description of a system composed of clay mineral particles and an electrolyte solution.

by the model is based on the principle of superposition of stresses and strains, which is strictly valid only when these are small magnitudes. The arbitrary nature of this definition does not invalidate the principle. Our interest lies primarily with large-strain behavior, resulting from the characteristic of flow.

In the present paper an attempt will be made to remove some of the limitations inherent to the rheological model by presenting, what the author has named, a mechanical model.

The adjective "mechanical" has been chosen to emphasize the fact that its principle is based upon the mechanism of forces and displacements of clay particle structures[1].

II. Mechanical Model of a Uniformly Flocculated Clay Structure

The geometrical characteristics of a flocculated clay particle structure are determined by the particle shape and the nature of the particle contacts.

To facilitate the treatment of the problem it is assumed that:

(a) the particle structure is two-dimensional and of unit thickness,

(b) the particles are platelets of small, uniform thickness and comparitively large lengths,

(c) the particle contacts are edge-to-face.

The particles are thus orientated with respect to each other at varying contact angles (Fig. 1).

It is further assumed that:

(d) particle orientations are uniformly distributed through the system.

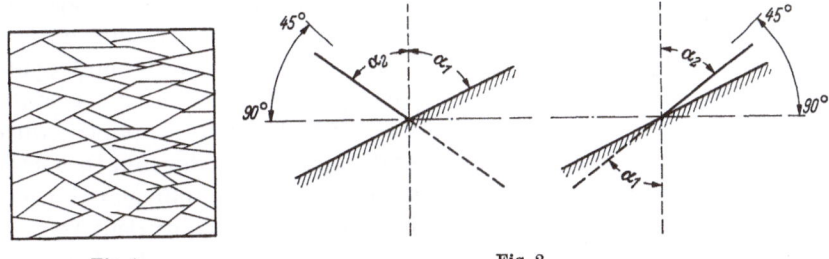

Fig. 1.   Fig. 2.

Uniformity of distribution would require a floc of infinite dimensions, if the distribution is a continuous function of the orientation angle. Hence:

---

[1] The term "structure" will repeatedly be used in its geometrical sense. The adjective "structural" relates quantities such as forces, displacements, etc. to the geometrical entity.

(e) a representative floc from a uniformly distributed system requires a finite number of particle orientations in order to be of finite dimensions.

It is finally assumed that:

(f) orientations are limited by a maximum and a minimum value of the angle enclosed with the axis of symmetry. As a result particles appear in two combinations of directions (Fig. 2).

1. Forces between Platelets

The forces between the platelets can be divided in two main groups:

1. Forces resulting from the energy of interaction between the atoms of the platelets, the ions adsorbed on the platelets and the ions of the void liquid.

2. Forces resulting from static or dynamic forces at the boundaries of the clay system or from intrinsic forces.

Force system 1, can again be divided in two groups: the attractive and the repulsive forces. Their magnitudes depend on:

(a) the physico-chemical properties of the platelets and the electrolyte,

(b) the mutual orientation of the platelets.

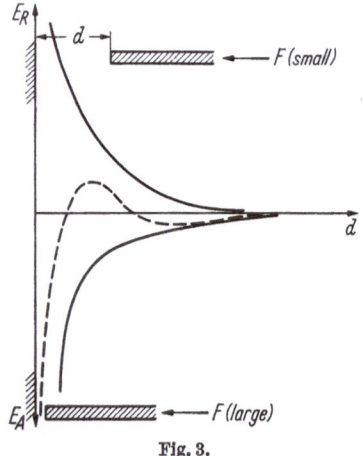

Fig. 3.

An extensive literature is available on subject a. Though many gaps still exist in this area of knowledge, special attention has already been given to forces acting between platelets in face-to-face (parallel) positions. Theoretical results on the balance between attractive and repulsive forces have been checked against experimental results obtained with parallel platelets subjected to a mechanical force perpendicular to their direction. Forces, equivalent to normal pressures of hundreds of kg/cm² were required to approach one platelet to a distance of 10 Å from its face and pressures of thousands of kg/cm² to a distance of 3 Å (Fig. 3).

It is the author's contention that these results are not only indicative of the existence of very strong repulsive energies at these short distances, but that sufficient evidence points at a similar increase of the viscosity of the electrolyte over comparable distances (Fig. 4).

Comparitively little information is available on the balance between the attractive and repulsive forces of a platelet in edge-to-face contact with another platelet.

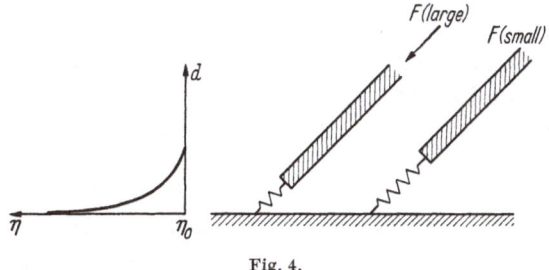

Fig. 4.

It can however be shown, that the total repulsive force on the platelet decreases as the contact angle approaches 90°, whence it reaches its minimum value. As a result the attractive force gains in importance and the edge comes closer to the face (Fig. 4). This argument has been used to advocate a cubic platelet arrangement as the logical consequence of edge-to-face attractive forces. A rotation of one platelet with respect to the other implies a decrease of the angle of contact and a net increase of the repulsive forces on the one platelet. This effect tends to separate the edge from the face.

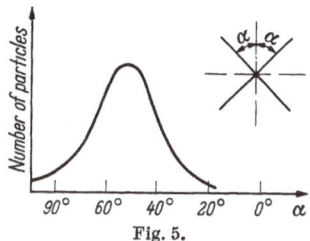

Fig. 5.

2. The Nature of the Edge-to-Face "Hinge"

Though some controversy still exists concerning the attraction force between edge and face, due to the VAN DER WAALS attraction, it seems that an expression:

$$F_a = \text{constant} \times \text{area} \times \text{distance}^{(-n)}$$

with: $\qquad n = 3$ to 4

is the most probable relationship.

From a mechanical viewpoint F_a is represented by a non-linear tensile spring attached to the edge and the face of a pair of platelets. Assuming that the face is a smooth plane with uniform physico-chemical properties the spring may be shifted to any location on the face without change of its tensile properties. Its rotation around the edge will affect the spring constant to a minor degree.

The shifting of the hinge across the face occurs by a slipping motion. This motion will usually occur in concurrence with a rotation of the one particle with respect to the other as a result of structural displacements. This motion will be named "hinge-slip".

The resistance to the motion is significant as it establishes one of the basic mechanisms in the changes of particle configuration due to structural displacements. The forces opposing the motion are of viscous nature. Their magnitude depends on the distance of the edge from the face and on the contact angle (Fig. 6).

The strong progressive increase of the viscosity near the surface of the platelet is significant for two reasons.

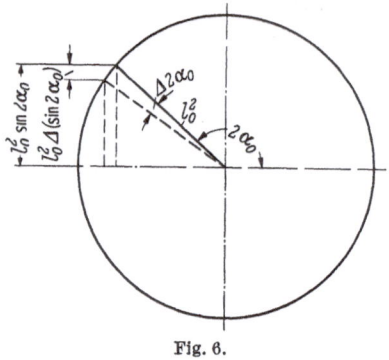

Fig. 6.

1. It provides the particle with a coating of higher viscosity than that of the void electrolyte.

2. It enables penetration of particle edges at large contact angles (approaching 90°) under the action of short range atomic forces and of structural compressive forces into a layer of increasing rigidity, providing for a part rigid and part viscous fixation.

These combined effects allow for the design of a mechanical model as proposed in paragraph II; in first approximation as a two-dimensional structure of rigid lineal elements in edge-to-face contacts.

3. Particle Forces by Displacement of a Symmetrical Model

Conditions II, d) and e) imply that all directions of the particles in a structural unit occur symmetrically with respect to the coordinate axis. The sizes (lengths) of the particles may vary between certain limits; we will however exclude the possibility of arrangements of triangular shape and limit ourselves to units of quadrangular shape. When a network of such units is compressed in the direction of one coordinate axis, expansion takes place in the direction of the other coordinate axis. When the particle orientations occur between maximum and minimum limits of the angle enclosed with the vertical axis of symmetry, α_{max} and α_{min}, according to the condition II, f the resulting displacement ratio in both directions depends on the path of the angle vs. particle number distribution curve. Because of the assumed uniformity of the particle arrangement, the resulting structural unit is a parallelogram with sides oriented at $\alpha_{res} = \alpha_0$ and with lengths $l_{res} = l_0$. The displacement ratio depends on the magnitude of α_0. The change in volume of the resulting unit depends on both the angle α_0 and the side-length l_0. The displacement ratio

$$m = \frac{h}{v} = -\cot \alpha_0,$$

where:

$$h = \text{lateral displacement,}$$
$$v = \text{vertical displacement.}$$

When $\alpha_0 > 45°$, $m < 1$, the structure is contracting,
when $\alpha_0 = 45°$, $m = 1$, the structure is isotropic,
when $\alpha_0 < 45°$, $m > 1$, the structure is dilatant.

In a structure with α varying between $0°$ and $90°$, the resulting ratio depends on the numerical contribution of the structural units with orientations $\geqq 45°$ (Fig. 5).

We will assume a structural system with $\alpha_0 \geqq 45°$. Deformation by compression in one coordinate direction then involves a *contraction* of the structural system.

4. The Function of the Void Liquid

Since we assumed the platelets to be rigid structural units, the function of the void liquid is to transmit hydrostatic stresses to the structure, which occur as a consequence of the contraction of the system. To facilitate the treatment of the problem we postulate an incompressible liquid. In a saturated system this postulate is a reasonable approximation of realistic conditions, which depends however primarily on the rigidity of the hinge-springs[1].

The volume decrease of the resultant structural unit is:

$$\Delta A_1 = l_0^2 \cdot \Delta(\sin 2\alpha_0) \qquad (\text{width} = \text{unity})$$

when the resultant angle α_0 increases by $\Delta\alpha_0$. The volume increase by the extension of the hinge-springs is:

$$\Delta A_2 = (l_0 + \Delta d)^2 \sin 2\alpha_0,$$

where Δd is the extension of the springs.

From the condition of incompressibility of the liquid follows:

$$\Delta A_1 = \Delta A_2.$$

Hence:

$$\frac{\Delta d}{l_0} = \sqrt{\cot 2\alpha_0 \cdot \Delta(2\alpha_0)}$$

where l_0 and $\cot 2\alpha_0$ represent the geometrical parameters of the model. For a model with known parametrical values, the relationship between Δd and $\Delta(2\alpha_0)$ is a cubic parabola.

---

[1] The two-dimensional model does not provide for an outlet of the void liquid in a direction perpendicular to the model plane.

5. The Strength of the Hinge

According to part. II, 2 the strength of the hinge is expressed by the attraction force:

$$F = - B\frac{\delta \cdot l}{d^4},$$

where:

δ thickness of the platelet $= 10^{-7}$ cm,
l = unit width of the platelet $= 1$ cm,
B coefficient of attraction $= 10^{-19}$ erg $\cdot$ cm,
d = distance from edge to face.

We will assume an initial distance of 10 Å. The initial strength of the hinge then is:

$F_0 = - 10^{-26} \cdot 10^{28}$ dynes $= - 100$ dynes ≈ 0.1 gram. (When the width of the particle would be 1000 Å, F_0 equals $- 10^{-3}$ dynes. Evidence from literature indicates an order of magnitude of 10^{-4} dynes.)

The magnitude of F decreases progressively with the distance d. With d increased to 20 Å the force decreases to 6.25% of its initial value.

The author's tests on the strength of clay in pure shear indicates a value of about 10^{-3} radians of shearing deformation at the limit of strength. From this result would follow a hinge displacement in the order of 5—10 Å.

It is evident that compressive forces acting on the contact points in the directions of the platelets would increase the initial strength of the system.

It follows from the considerations as given above that the shear displacements involve a volume contraction of the structural units (for angles of orientation larger than 45°). The void liquid however prevents this contraction. The rigidity of the particle system then depends to a large extent on the magnitude of the attraction energy at the edge-to-face contacts.

The shear displacements cause a loss of strength of the particle system, which largely depends on the orientation and size of the structural units. When a certain vertical displacement is imposed on the system, those units with high values of α_0 (tending towards vertical positions) and composed of large-size platelets will be ruptured at the contact points first. As the vertical displacement increases more and more units with lower values of α_0 and those composed of small-size units will fail one after another.

This conclusion is confirmed by the results of the author's tests, which shows a decreasing rigidity of the structure at increasing shearing displacements, though it maintains the property of recovery of the strains at the release of the stress.

6. The Flow Range

Beyond the limit of structural strength the system can no longer be considered as a structure in the accepted sense of the word, since the hinges have mostly lost their function of cross-linking the particles. The initial geometrical arrangement of the structure however has been maintained as a result of the small magnitude of the rotation.

Further displacements in one coordinate direction force the particles to move with respect to another by sliding of the edges along the faces. This motion is opposed by viscous forces as substantiated in part II, 2.

The author's experiments have proved this point by series of results showing linear viscosity relationships over the range of flow and this uniqueness has been shown to be valid over a wide range of stress deviator time paths.

Since the structural system is no longer composed of coherent units, liquid migration can in principle take place from one point of the system to another. The resistance to the displacements is now governed primarily by the nature of the sliding resistance and the direction and magnitude of the edge-to-face forces.

III. Flow in Compression

It has been postulated that strain-time effects in compression by consolidation of clays could be explained in terms of a superposition of the effects of the hydrostatic and the deviator components of stress. This postulation has been made with particular reference to the so-called secondary time effect in consolidation.

Without further proof the preceding considerations apply to the performance of the mechanical model in a hydrostatic state of stress, when expulsion of the void liquid is permitted at one or both boundaries. The force system is then identical with the geometry of the model. Rupture of the bonds occurs commensurate with the magnitude and direction of the individual forces at the edge-to-face contacts. At large angles of intersection the particles will tend to penetrate into the viscous layer and become more strongly fixed, whereas at small angles the slip of the hinges will prevail. This process will continue to the measure of liquid expulsion.

The process of consolidation of the model therefore is inherent with the viscous slip motion at edge-to face contacts.

IV. Conclusions

Rheological models of saturated clays can be used to predict their behavior only when the strains (or stresses) are limited to small quantities.

A mechanical model of a two-dimensional, uniformly flocculated clay is used to demonstrate that:

(a) The limit of structural strength is due to the rupture of tensile bonds at the edge-to-face contacts if and when the expulsion of the void liquid is prevented at the boundaries of the system.

(b) Within this limit the system behaves as a material with non-linear rigidity (decreasing at increasing displacements).

(c) Beyond this limit tensile bonds become mostly inoperative and particle displacements take place in continuous edge-to-face sliding of a viscous nature.

(d) The same procedure applies to a system in a state of hydrostatic stress, providing for expulsion of void liquid at the boundary (or boundaries). Here the rupture of the bonds takes place without preceding volume increase of the voids and subsequent separation of the edges from the faces.

(e) The rupture takes place first of all at those contacts where the angle of intersection is small and proceeds at contacts points of more favourable particle positions (closer to a rectangular juxtapostion).

(f) A combination of the hydrostatic component and the deviator component of stress in consolidation can not be considered for the superposition of the strains. Failure of particle bonds occur both in elastic and viscous ranges for different reasons at different points of contact, depending primarily on the orientations of the particles and their sizes.

(g) The strains resulting from the displacements of the particles can therefore not be obtained by superposition of the displacements due to each of the components of stress separately.

1.10 Flow and Stress Relaxation of Clays

By

Sakuro Murayama and Toru Shibata

Introduction

This is a part of the theoretical studies on the rheological properties of clays, especially on the behaviours in flow, stress relaxation and secondary compression of clays which accompany no fracture in clay skeleton.

In this paper, authors intend to revise and improve some unsatisfactory assumptions and treatments used in their previous papers (1958, 1959, 1961) on the same subject and to give more rational explanation on the mechanical model of clay skeleton proposed in the previous papers. The improved points in this paper are laid mainly on the clarification of the mechanism of clay behaviours by considering statistically the micrometric structure of clay skeleton and on the deduction of theories concerning the rheological macrocharacters of clay. The formulae thus obtained well agree with the results of various experiments for verification.

Study in this paper is founded on the authors' previous papers above stated, but their revised parts in the theoretical considerations are mainly performed by the first author and their experimental studies are developed mainly by the second author.

I. Fundamental Studies on the Rheological Properties of Clay Skeleton

In this chapter, general theories on the rheological properties of clay skeleton are obtained through following procedures.

1. Conditions and Assumptions Applied on the Solution

(a) **Structure of Clay Skeleton.** It is assumed that the structure of the clay skeleton is composed of a heap of micrometric clay segments (i. e. mineral particles which move as units) in a card-house structure, and between the segments connecting in edge to face contact, there lies thin layer of adsorbed water which binds up the segments. Segments in the skeleton are assumed to bind themselves by interparticle force and friction between segments.

If the maximum shearing stress τ caused by the deviatoric stress σ is applied on the clay skeleton, the average magnitude of the force f produced on individual segment is given by

$$f = \tau/\delta N \quad \text{or} \quad f = \sigma/2\delta N, \tag{1^1}$$

where N is the number of segments per unit area of clay and is assumed as a constant unless the clay skeleton is subjected to failure or consolidation, and δ is a coefficient to adjust the deviation of the direction of the force produced on the individual segment from that of the applied maximum shearing stress.

(b) **Types of Joint.** The connecting joint of segments is classified in accordance with its mechanical behaviour caused by the force f into two types, i. e. the elastic joint and the visco-elastic joint as tentatively designated.

(α) *The Elastic Joint.* An example of this type of joint is illustrated schematically in Fig. 1a. At the elastic joint, no relative sliding between segments is expected though variation is produced in the intercepting angle between the segments by the force f applied on the segment.

When the force f displaces the segment relatively from its equilibrium position by v_1 in the direction of f, recovering force proportional to v_1, i.e. $e \cdot v_1$ (e: coefficient of proportion, v_1: angular displacement) is generated on the segment in the reverse direction of v_1 due to the unbalance of the repulsive force and the attractive force existing between segments. It may be acceptable that the coefficient e is assumed as a constant on the average for any group of segments, provided the number of segments in the group is large enough. As e is assumed as a constant the recovering force behaves as like as an elastic reaction. Then the coefficient e is designated tentatively as the elastic constant of segment. The equilibrium relation of forces on the elastic joint is given by

$$f = e \cdot v_1. \tag{2}$$

---

[1] Eq. (1) is obtained as follows. Let $(f)_i$ be a force on an individual segment and $(\delta)_i$ a cosine of the angle between the direction of $(f)_i$ and that of τ.

$$f = \sum_1^N (f)_i/N,$$

$$\tau = \sum_1^N \{(\delta)_i \cdot (f)_i\} \equiv \delta \cdot N \cdot f,$$

$$\therefore \ f = \tau/(\delta \cdot N). \tag{1}$$

Besides such elastic property of clay segment as above stated, its elasticity may be also due to flexure of thin plate-like clay segments and variation in thickness of water film between segments.

The relations of forces due to these causes are assumed to be included in Eq. (2).

(β) *The Visco-Elastic Joint.* An example of this type of joint is shown in Fig. 1b. As the segment of this type of joint slides on the surface of adjacent segment holding the adsorbed water around its joint, the applied force on the segment f is supported by the same elastic resistance as above stated, the frictional resistance between segments f_r and the viscous resistance of adsorbed water at the joint f_2 simultaneously. As the joint of this type behaves plasto-visco-elastically, it is designated as visco-elastic joint abbreviately.

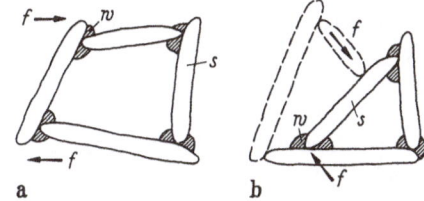

Fig. 1. Schematic diagrams of a) elastic joint and b) visco-elastic joint. *s* Segment; *w* Adsorbed water around the joint of segment; *f* Force on a segment produced by shearing stress on clay.

While all elastic joints are mobilized simultaneously, visco-elastic joints do not always slide simultaneously, and only the joint whose applied force exceeds its frictional resistance is able to slide actually. As for an individual visco-elastic joints, applied force on a segment is denoted by $(f)_i$, its viscous resistance by $(f_2)_i$, its frictional resistance by $(f_r)_i$, angular displacement of segment in direction of $(f)_i$ from its equilibrium position by $(v_2)_i$ and magnitudes of $(v_2)_i$ at initial time of load application by $(v_r)_i$, — where ( )$_i$ means the magnitude for each individual segment or joint. Above described symbol in the bracket without ( )$_i$ means the average magnitude among total visco-elastic joints including sliding and fixed joints[1].

As magnitudes of $(f)_i$, $(f_r)_i$ and $(ev_r)_i$ vary individually for each joint, subsequent behaviour of a joint can be devided into following 3 cases.

(i) When

$$(f)_i < (e \cdot v_r + f_r)_i \tag{3}$$

or

$$(f)_i > (e \cdot v_r - f_r)_i, \tag{4}$$

no displacement at the joint succeeds, i. e. $dv_2/dt = 0$.

---

[1] The average angular displacement v_1 and v_2 may be represented by

$$v_1 = \Delta\lambda_1/\lambda, \qquad v_2 = \Delta\lambda_2/\lambda.$$

Where $\Delta\lambda_1$ and $\Delta\lambda_2$: average linear displacements of the elastic joint and the total visco-elastic joint respectively, λ: average distance between the connecting joint of segment.

(ii) When

$$(f)_i > (e \cdot v_r + f_r)_i, \tag{5}$$

normal flow displacement in the direction of f is produced at the joint, i. e. $dv_2/dt > 0$, and the equilibrium relation of forces is expressed as follows:

$$(f)_i = (e \cdot v_2 + f_r + f_2)_i. \tag{6}$$

(iii) When

$$(f)_i < (e \cdot v_r - f_r)_i, \tag{7}$$

recovery flow displacement in the reverse direction of f is expected at the joint, i. e. $dv_2/dt < 0$, and the equilibrium relation of forces is given by

$$(f)_i = (e \cdot v_2 - f_r + f_2)_i. \tag{8}$$

(γ) *Number of Joints.* Among N joints per unit area of clay, the number of elastic joints and that of visco-elastic joints are denoted by $N_1 (= \alpha_1 \cdot N)$ and $N_2 (= \alpha_2 \cdot N)$ respectively, and α_1 and α_2 are assumed to be invariable irrespective of the intensity of statical deviatoric stress σ. Hence, the followings are obtained.

$$\left. \begin{array}{l} \alpha_1 + \alpha_2 = 1, \\ N_1 = \alpha_1 \cdot N, \qquad N_2 = \alpha_2 \cdot N. \end{array} \right\} \tag{9}$$

If the probability of the sliding segment or the ratio of sliding joints to total visco-elastic joints is denoted by P, actual number of sliding joints per unit area of clay is given by $P \cdot N_2$.

(c) **Probability of Sliding Segment.** The frequency distribution of the applied force on a segment $(f)_i$ (rewritten by x_1) is assumed to be expressed by a GAUSSIAN distribution function $f(x_1)$ or $N(m_1, \varrho_1^2)$ whose mean value is $f(= m_1)$ and standard deviation is ϱ_1. Similarly, the distribution of $(f_r)_i$ (rewritten by x_2) is assumed by a GAUSSIAN function $\varphi(x_2)$ or or $N(m_2, \varrho_2^2)$. Where $m_2 = f_r$.

$$\left. \begin{array}{l} f(x_1) = \dfrac{1}{\sqrt{2\pi} \cdot \varrho_1} \cdot \exp\left\{ -\dfrac{(x_1 - m_1)^2}{2\varrho_1^2} \right\}, \\[2mm] \equiv N(m_1, \varrho_1^2), \\[2mm] \varphi(x_2) = N(m_2, \varrho_2^2). \end{array} \right\} \tag{10}$$

Eq. (10) is shown in Fig. 2a. Standard deviations ϱ_1 and ϱ_2 may be dependent on structure of clay skeleton.

In this paper, the normal flow of clay skeleton which is subjected to no residual strain will be treated. In this case, sliding condition of a segment is obtained by the substitution of $v_r = 0$ into Eq. (5) as follows,

$$x_2 - x_1 \equiv y < 0. \tag{11}$$

As x_1 and x_2 are independent variables, the frequency distribution function of $(x_2 - x_1)$ (or y) can be obtained through the statistical theorem as follows.

$$p(y) = \frac{1}{\sqrt{2\pi} \cdot \varrho} \exp\left\{-\frac{(y-m)^2}{2\varrho^2}\right\} \equiv N(m, \varrho^2),$$

where

$$m = m_2 - m_1, \quad \varrho = \sqrt{\varrho_1^2 + \varrho_2^2}.$$

$$(12)$$

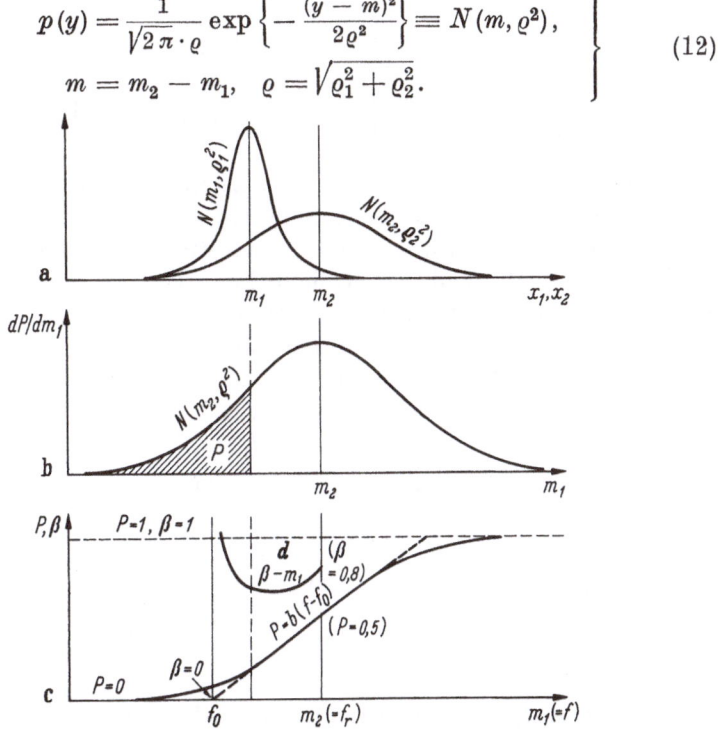

Fig. 2. a) Frequency distribution curves of applied stress on segment and its frictional resistance; b) Frequency distribution curve of mobilized segment; c) Cumulative distribution curve of mobilized segment; d) Relation between β and m_1.

Therefore, the probability of sliding segment P is obtained by following integration of the above equation within the region of $y < 0$.

$$P = \int_{-\infty}^{0} p(y) \cdot dy. \tag{13}$$

By the substitution of $(y - m)/\varrho = \theta$, P is transformed as

$$P = \int_{-\infty}^{-(m/\varrho)} \frac{1}{\sqrt{2\pi}} \exp\left(-\frac{\theta^2}{2}\right) \cdot d\theta. \tag{14}$$

As m and θ are independent variables, dP/dm_1 can be obtained from Eq. (14) as follows.

$$\frac{dP}{dm_1} = \frac{1}{\sqrt{2\pi}}\left[\exp\left(-\frac{\theta^2}{2}\right)\right]_{\theta=-m/\varrho} \cdot \frac{d}{dm_1}\left(-\frac{m}{\varrho}\right)$$

$$= N(m_2, \varrho^2). \tag{15}$$

Hence, dP/dm_1 is represented by the GAUSSIAN distribution curves whose mean value is m_2 and standard deviation is ϱ as shown in Fig. 2b. Accordingly, P is given by Eq. (16) and is shown in Fig. 2c.

$$\left. \begin{aligned} P &= \int_{-\infty}^{m_1} N(m_2, \varrho^2)\, dm_1 \\ &= \int_{-\infty}^{(m_1/\varrho)} N(m_2/\varrho, 1)\, d\left(\frac{m_1}{\varrho}\right). \end{aligned} \right\} \tag{16}$$

We approximate this cumulative distribution curve by the straight line which contacts with the cumulative curve at the middle point of the curve ($m_1 = m_2$, $P = 0.5$), neglecting fairly large deviation existing near both ends of straight line. The tangent of the slope angle of the straight line is given by dP/dm_1 at $m_1 = m_2$. If the abscissa of the point where the straight line intersects with axis of $P = 0$ is denoted by f_0, equation of the approximated straight line can be express-ed by P_a as

$$\left. \begin{aligned} P_a &= b(m_1 - f_0) = b(f - f_0), \\ b &= [dP/dm_1]_{m_1 = m_2} = 1/(\sqrt{2\pi}, \varrho), \\ f_0 &= f_r - 1/(2b) = m_2 - 1/(2b). \end{aligned} \right\} \tag{17}$$

Another representation of P_a is given by

$$P_a = 0.5 + b \cdot m = 0.5 + (m/\varrho)/\sqrt{2\pi}, \tag{18}$$

where $\qquad m = m_2 - m_1.$

Where f_0 is designated as the lower yield value of segment because f_0 is assumed as the apparent internal resistance below which no sliding of segment or no flow of clay exists.

Total sum of excess magnitudes of applied forces on sliding seg-ments than their frictional resistances is given by $\int_{-\infty}^{0} - y \cdot p(y) \cdot dy$ and this solution can be obtained through the integral calculation as follows.

$$\left. \begin{aligned} \int_{-\infty}^{0} - y \cdot p(y)\, dy &= \varrho \cdot g(\theta_{y=0}) - m \cdot F(\theta_{y=0}) \\ &= \varrho \cdot g\left(-\frac{m}{\varrho}\right) - m \cdot F\left(-\frac{m}{\varrho}\right), \\ F(\theta_{y=0}) = F(-m/\varrho) &= \int_{-\infty}^{-m/\varrho} \frac{1}{\sqrt{2\pi}} \exp\left(-\frac{\theta^2}{2}\right) d\theta, \\ g(\theta_{y=0}) = g(-m/\varrho) &= \left[\frac{1}{\sqrt{2\pi}} \exp\left(-\frac{\theta^2}{2}\right)\right]_{\theta = -m/\varrho}. \end{aligned} \right\} \tag{19}$$

where

For convenience' sake, applying a parameter β, we put as follows.

$$g(-m/\varrho) - (m/\varrho) \cdot F(-m/\varrho) = \beta \cdot [P_a]_{at(-m/\varrho)}. \qquad (20)$$

As numerical values of function g and F in Eq. (19) and that of P_a in Eq. (18) can be calculated by the numerical table of GAUSSIAN function and by simple calculation, relation between β and (m/ϱ) can be obtained easily and thus obtained result is shown in Fig. 2d. From this figure, β may be assumed approximately as a constant $(0.64 \sim 0.82 \doteqdot 0.73)$ within the range of $-1.0 < (-m/p) < 0$. Therefore, from Eqs. (18), (19) and (20), following approximation may be obtained

$$\int_{-\infty}^{0} -y \cdot p(y) dy \doteqdot \beta_2(f - f_0), \qquad (21)$$

where

$$\beta_2 = \varrho \cdot \beta \cdot b = \beta/\sqrt{2\pi} \doteqdot 0.29.$$

Since the sum of the excess magnitudes of applied forces on sliding segments than their frictional resistances should be equal to the elastic resistance $e v_2$ and viscous resistance f_2 of the visco-elastic segment, following relation is obtained.

$$\beta_2(f - f_0) = e v_2 + f_2. \qquad (22)$$

Average displacement among sliding segments v_{2s} (suffix s means a magnitude belonging to sliding segment) and their average viscous resistance f_{2s} can be represented by

$$\left. \begin{array}{l} v_{2s} = v_2 N_2/(P_a \cdot N_2) = v_2/P_a, \\ f_{2s} = f_2 N_2/(P_a \cdot N_2) = f_2/P_a. \end{array} \right\} \qquad (23)$$

By the substitution of Eqs. (17) and (22) into Eq. (23), Eq. (23) is transformed:

$$\left. \begin{array}{l} v_{2s} = \dfrac{v_2}{(b/\beta_2) f_{20}}, \quad f_{2s} = \dfrac{f_2}{(b/\beta_2) f_{20}}, \\ f_{20} = [f_2]_{v_2=0} = \beta_2(f - f_0). \end{array} \right\} \qquad (24)$$

Where f_{20} is the viscous resistance at initial time when stress is applied.

(d) Viscous Resistance Against Sliding of Joint. It is assumed that the adsorbed water in a certain effective zone around the connecting joint exhibits viscous resistance against sliding of the joint. The viscosity of the clay skeleton will be derived from the viscosity of adsorbed water in the effective zone above mentioned by applying EYRING's "hole theory".

A shearing force applied on each water molecule in the effective water zone is expressed by $f_{2s}/(n \cdot A_w)$, where f_{2s} is the shearing force on each sliding segment or each effective water zone, A_w is the sectional area of the effective water zone parallel to the direction of f_{2s} and is assumed to be equal for any effective zone, and n is the number of water molecules per unit area of each effective water zone and is assumed also to be equal for any effective zone. According to Eyring's theory (1941), the flow process of polymeric material is produced by exchange of position between particles and their neighbouring holes i. e. by jump of particles.

The probability of getting this jump per unit time or the jump frequency J in the direction of the applied force is obtained by the difference between the frequency of forward jump J_+ and that of backward jump J_- and is calculated by statistical mechanics as follows.

$$J = J_+ + J_- = \frac{\varkappa T}{h} \exp\left\{\frac{-(E_0 - \alpha \cdot f_{2s})}{\varkappa T}\right\} - \frac{\varkappa T}{h} \exp\left\{\frac{-(E_0 + \alpha \cdot f_{2s})}{\varkappa T}\right\}$$

$$= \frac{2\varkappa T}{h} \exp\left(\frac{-E_0}{\varkappa T}\right) \cdot \sinh\left(\frac{\alpha \cdot f_{2s}}{\varkappa T}\right),$$

where $\alpha = (\varLambda/2) \cdot (1/n \cdot A_w)$, and $\varLambda$: the average distance projected in the direction of force f_{2s} between equilibrium position of two molecules, T: the absolute temperature of the water, E_0: the free energy of activation of water molecule for the jump at $T°\mathrm{K}$, $\varkappa$: Boltzmann's constant $(1.3808 \cdot 10^{-16}$ erg. $\cdot$ degree$^{-1})$ and h: Planck's constant $(6.626 \cdot 10^{-27}$ erg. $\cdot$ sec$)$. Above equation may be written in abbreviated form:

$$\left.\begin{aligned} J &= 2A \cdot \sinh(B \cdot f_{2s}), \\ A &= \frac{\varkappa T}{h} \exp\left(\frac{-E_0}{\varkappa T}\right), \quad B = \frac{\alpha}{\varkappa T}, \\ \alpha &= \varLambda/(2 n A_w). \end{aligned}\right\} \qquad (25)$$

If there are m holes in series per one effective water zone in the direction of f_{2s}, the rate of shearing displacement of the connecting joint of a sliding segment dv_{2s}/dt should be

$$\lambda \frac{dv_{2s}}{dt} = m \cdot \varLambda \cdot J = 2m \cdot \varLambda \cdot A \cdot \sinh(B \cdot f_{2s}), \qquad (26)$$

where λ: average distance between the connecting joints of the segment or the average mutual distance between the segments.

Substituting Eqs. (24) and (25) into Eq. (26), we get

$$\frac{dv_2}{dt} = 2 \cdot \frac{m \cdot \varLambda \cdot A \cdot b}{\lambda \beta_2} \cdot f_{20} \cdot \sinh\left(\frac{B\beta_2}{b} \cdot \frac{f_2}{f_{20}}\right). \qquad (27)$$

(e) Relation between Displacement of Segment and Macro-Deformation of Clay Skeleton. The segments are assumed to be independent of each other i. e. the applying force for any segment is unaffected by the displacement which may be occurring in other segments. Furthermore, a displacement of a segment of any type v_i (v_1 or v_2) is assumed to contribute an increment $c \cdot \delta \cdot v_i$ (c: constant) to the overall shearing strain of clay skeleton γ.

Among the overall shearing strain γ, if the shearing strain of clay skeleton contributed by the elastic joints is denoted by γ_1 and that by the visco-elastic joints by γ_2, equations are given by

$$\left.\begin{array}{l} \gamma = \gamma_1 + \gamma_2, \\ \gamma_1 = c \cdot \delta \cdot N_1 \cdot v_1, \quad \gamma_2 = c \cdot \delta \cdot N_2 \cdot v_2. \end{array}\right\} \quad (28)$$

Where, every shearing strain (γ, γ_1 and γ_2) is measured from the state of no residual shearing strain in the direction of the maximum shearing stress τ.

Substituting Eq. (9) into (28), we get

$$\left.\begin{array}{l} \gamma = \gamma_1 + \gamma_2, \\ \gamma_1 = c \cdot \delta \cdot \alpha_1 \cdot N \cdot v_1, \quad \gamma_2 = c \cdot \delta \cdot \alpha_2 \cdot N \cdot v_2. \end{array}\right\} \quad (29)$$

(α) *For the Elastic Joint.* Substituting Eqs. (1) and (29) into left and right hand side of Eq. (2) respectively, we get

$$\left.\begin{array}{l} \tau = G_1 \cdot \gamma_1, \\ G_1 = e/(c \cdot \alpha_1). \end{array}\right\} \quad (30)$$

where

Since G_1 is a proportional coefficient relating τ and γ_1, it should be the apparent shear modulus of clay skeleton by which clay skeleton represents instantaneous elasticity.

(β) *For the Visco-Elastic Joint.* If $d\gamma_2/dt > 0$, the relation between γ_2 and τ is calculated as follows by the substitution of Eqs. (1), (24) and (29) into Eq. (27).

$$\left.\begin{array}{l} \dfrac{d\gamma_2}{dt} = 2 \cdot A_2 \cdot \tau_{20} \cdot \sinh\left(B_2 \dfrac{\tau_2}{\tau_{20}}\right), \\[2mm] \text{where} \\ A_2 = m \cdot \left(\dfrac{\Lambda}{\lambda}\right) A \cdot \alpha_2 \cdot b \cdot c = \alpha_2 \cdot b \cdot c \cdot m \cdot \left(\dfrac{\Lambda}{\lambda}\right)\left(\dfrac{\varkappa T}{h}\right) \cdot \exp\left(\dfrac{-E_0}{\varkappa T}\right), \\[2mm] B_2 = \beta_2 \cdot B/b = (\beta_2 \Lambda/2 n A_w b \varkappa)/T \equiv B_{20}/T, \\[2mm] G_2 = e/(\beta_2 \cdot c \cdot \alpha_2), \quad \tau_0 = \delta N f_0, \\[2mm] \tau_2 = \delta N f_2/\beta_2 = \tau - \tau_0 - G_2 \gamma_2, \\[2mm] \tau_{20} = \delta N f_{20}/\beta_2 = \tau - \tau_0 > 0. \end{array}\right\} \quad (31)$$

Where A_2 and B_2 are coefficients dependent on soil structure and temperature, G_2 is the apparent shear modulus of clay skeleton by which clay skeleton represents visco-elasticity, τ_0 is the apparent internal friction of clay which is designated as the lower yield value of clay skeleton, τ_2 is the apparent viscous resistance of clay skeleton and τ_{20} is τ_2 at initial time when external stress is applied.

(γ) *Mechanical Model of Clay Skeleton.* As represented by the Ist equation in Eq. (29), the shearing strain of clay skeleton consists of the pure elastic strain γ_1 expressed by Eq. (30) and the plasto-visco-elastic strain γ_2 whose equation is expressed by Eq. (31). Above mentioned Eqs. (29), (30) and (31) are the simultaneous equations which represent

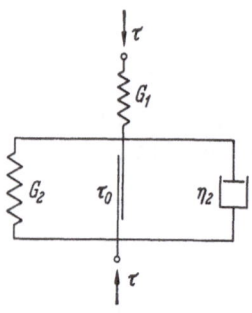

the behaviour of clay skeleton. Therefore the simultaneous equations can be easily analogized with the behaviour of the mechanical model which consists of an independent HOOKEan spring G_1 connected in series with a modified KELVIN or VOIGT element as shown in Fig. 3. The latter element is composed of a HOOKEan spring G_2, a slider τ_0 and a dasphot whose coefficient η_2 is obtained from Eq. (31) as follows.

Fig. 3. Mechanical model for clay skeleton.

$$\eta_2 = \frac{\tau_2}{d\gamma_2/dt} = \frac{\tau_2}{2A_2 \cdot \tau_{20} \cdot \sinh(B_2\tau_2/\tau_{20})}. \quad (32)$$

As represented by Eq. (32), this viscosity of clay is a non-NEWTONIAN viscosity or a kind of structural viscosity.

As for the simultaneous equations or the mechanical model, as above described, they are deduced under the condition where any deformation of clay skeleton accompanies neither fracture nor consolidation. The critical maximum shearing stress (or deviatoric stress) below which no fracture in clay skeleton is produced is defined as the upper yield value of clay and denoted by τ_u (or σ_u). The upper yield value, therefore, should be the true critical strength of clay under constant loading. Hence, the condition for no fracture in clay skeleton is given by

$$\tau < \tau_u \quad (\text{or} \quad \sigma < \sigma_u). \quad (33)$$

The critical maximum stress below which no consolidation of clay proceeds is, as well known, designated as the preconsolidation stress. The behaviour of clay under the stress exceeding the preconsolidation stress will be discussed in Chapter V.

2. Approximate Equation Representing Behaviour of Clay Skeleton

(a) **Equation under Shearing Stress.** As described in 1 (e) (β), τ_2 is equal to τ_{20} at initial time ($t = 0$), but it decreases with time and finally ($t \to \infty$) approaches to zero, i. e. $\tau_{20} \geq \tau_2 \geq 0$.

As far as τ_2 is within the range expressed by

$$B_2 \geq B_2 \cdot \tau_2/\tau_{20} \geq 2, \quad (34)$$

the following approximation holds within error of 2%:

$$2\sinh(B_2\tau_2/\tau_{20}) \doteqdot \exp(B_2\tau_2/\tau_{20}). \tag{35}$$

Hence, as far as the condition expressed by Eqs. (33) and (34) are valid, behaviour of clay skeleton or that of the mechanical model (shown in Fig. 3) can be represented approximately by the following simultaneous equations which consist of Eqs. (29) and (30) and Eq. (31) transformed by Eq. (35).

If $\qquad\qquad \tau < \tau_u \quad \text{and} \quad B_2 \geq B_2\tau_2/\tau_{20} \geq 2, \tag{36-1}$

$$\left.\begin{aligned} \gamma &= \gamma_1 + \gamma_2, \quad \gamma_1 = \tau/G_1, \\ \frac{d\gamma_2}{dt} &= A_2 \cdot \tau_{20} \cdot \exp(B_2\tau_2/\tau_{20}), \end{aligned}\right\} \tag{36-2}$$

where $\qquad\qquad \tau_2 = \tau - \tau_0 - G_2\gamma_2, \quad \tau_{20} = \tau - \tau_0. \tag{36-3}$

The lower yield value τ_0 has been obtained from f_0. As described in 1 (c), f_0 is the apparent value below which no probability of sliding segment exists. However, according to the theoretical calculation expressed by Eq. (16), even if applied force f is less than f_0, there exist still a little probability of sliding segment. Therefore, as far as the clay has no residual shearing strain, applied stress τ less than τ_0 has the probability to produce a feeble flow in clay skeleton.

(b) Equation under Uni-Axial Compression. On the uni-axial compression, there are following wellknown relations.

$$\left.\begin{aligned} \varepsilon_{\mathrm{I}} &= \sigma/E, \quad \varepsilon_{\mathrm{II}} = \varepsilon_{\mathrm{III}} = -\nu\sigma/E, \\ \gamma &= \varepsilon_{\mathrm{I}} - \varepsilon_{\mathrm{III}} = (1+\nu)\cdot\varepsilon_{\mathrm{I}}, \quad \sigma = 2\tau, \\ \gamma &= \tau/G, \quad E = 2(1+\nu)G. \end{aligned}\right\} \tag{37}$$

Where, σ: uni-axial compressive stress or deviatoric stress, $\varepsilon_{\mathrm{I}}(\equiv\varepsilon)$: axial principal strain, $\varepsilon_{\mathrm{II}}, \varepsilon_{\mathrm{III}}$: lateral principal strains, τ: maximum shearing stress, γ: maximum shearing strain, E: YOUNG's modulus, ν: POISSON's ratio and G: shear modulus.

If the above relations are valid for the elastic relation in the independent spring element of the mechanical model, its axial strain $\varepsilon_{\mathrm{I}.1}(\equiv\varepsilon_1)$ (suffix 1 is used for the independent spring element) is given by

$$\left.\begin{aligned} \varepsilon_1 &= \sigma/E_1 = \gamma_1/(1+\nu), \\ E_1 &= 2(1+\nu)G_1. \end{aligned}\right\} \tag{38}$$

where

Similarly, the axial strain $\varepsilon_{\mathrm{I}.2}(\equiv\varepsilon_2)$ of the spring element in the modified VOIGT model (suffix 2) is given by

$$\left.\begin{aligned} \varepsilon_2 &= \gamma_2/(1+\nu), \\ E_2 &= 2(1+\nu)G_2. \end{aligned}\right\} \tag{39}$$

Various shearing stresses are written as follows:

upper yield value: $2\tau_u = \sigma_u$,

lower yield value: $2\tau_0 = \sigma_0$,   (40)

$$2\tau_2 = \sigma_2, \quad 2\tau_{20} = \sigma_{20}.$$

By the substitution of Eqs. (37)—(40) into Eq. (36), the approximate equations for the rheological behaviour under uni-axial compression are obtained as follows.

If $\sigma < \sigma_u$ and $B_2 \geq B_2\sigma_2/\sigma_{20} \geq 2$,   (41—1)

$$\varepsilon = \varepsilon_1 + \varepsilon_2, \quad \varepsilon_1 = \sigma/E_1,$$
$$\frac{d\varepsilon_2}{dt} = A_{2\nu} \cdot \sigma_{20} \cdot \exp\left[B_2 \frac{\sigma_2}{\sigma_{20}}\right],$$ (41—2)

where

$$\sigma_2 = \sigma - \sigma_0 - E_2\varepsilon_2, \quad \sigma_{20} = \sigma - \sigma_0,$$
$$A_{2\nu} = A_2/\{2(1 + \nu)\}. \qquad \cdot$$ (41—3)

Therefore, the mechanical model under uni-axial compression is obtained by replacing τ, G_1, G_2, τ_0 and η_2 shown in Fig. 3 with σ, E_1, E_2, σ_0 and $\eta_{2\nu}$. Where

$$\eta_{2\nu} = \sigma_2/(d\varepsilon_2/dt) = 2(1 + \nu)\,\eta_2.$$ (42)

II. Flow of Clay Skeleton

3. General Equation for Normal Flow

General equation for normal flow of the clay skeleton which is subjected to no residual strain can be obtained by the solution of Eq. (41) under the condition of σ = constant.

From Eq. (41) [the 2nd equation of Eq. (41—1), Eqs. (41—2) and (41—3)] the validity limit of ε is given by

$$\frac{\sigma}{E_1} + \frac{\sigma - \sigma_0}{B_2 E_2}(B_2 - 2) \geq \varepsilon \geq \frac{\sigma}{E_1}.$$ (43)

By the substitution of

$$u = B_2\sigma_2/\sigma_{20}(= B_2(\sigma - \sigma_0 - E_2\varepsilon_2)/(\sigma - \sigma_0))$$ (44)

into the 3rd equation of Eq. (41—2), we obtain

$$- du/dt = A_{2\nu} \cdot B_2 \cdot E_2 \cdot \exp(u).$$

Integrating

$$- \int_{u_0}^{u} \exp(-u)\, du = A_{2\nu} \cdot B_2 \cdot E_2 \cdot t,$$ (45)

where

$$u_0 = [u]_{t=0} = B_2.$$

If B_2 is large and $\exp(-B_2)$ is negligibly small, Eq. (45) becomes approximately as

$$\exp(-u) = A_{2\nu} \cdot B_2 \cdot E_2 \cdot t, \qquad (46)$$

from which

$$\varepsilon_2 = \frac{\sigma - \sigma_0}{E_2} + \frac{\sigma - \sigma_0}{B_2 E_2} \log(A_{2\nu} \cdot B_2 \cdot E_2 \cdot t). \qquad (47)$$

Substituting the above into Eq. (41—2), ε is given by

$$\varepsilon = \frac{\sigma}{E_1} + \frac{\sigma - \sigma_0}{E_2} + \frac{\sigma - \sigma_0}{B_2 E_2} \log(A_{2\nu} \cdot B_2 \cdot E_2 \cdot t). \qquad (48)$$

If ε is beyonds the limit expressed by Eq. (43) and at $t \to \infty$, σ_2 becomes zero, and accordingly ε can be obtained as follows.

$$\varepsilon_{t\to\infty} = \varepsilon_1 + (\varepsilon_2)_{t\to\infty} = \frac{\sigma}{E_1} + \frac{\sigma - \sigma_0}{E_2}. \qquad (49)$$

Eqs. (48) and (49) show that the flow strain of clay ε increases proportionally with the logarithm of time t within the limiting value of ε expressed by Eq. (43) and then ε approaches to a final finite value of $\varepsilon_{t\to\infty}$ as shown schematically in Fig. 4.

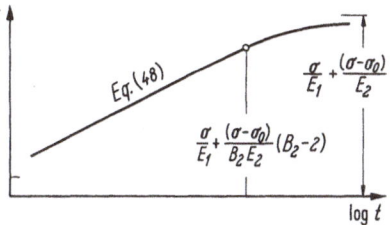

Fig. 4. Schematic diagram representing flow strain vs. elapsed time relation.

4. Flow Tests under Uni-Axial Compression

The flow tests under uni-axial compression were performed on undisturbed clay under undrained condition (1958). Each fresh specimen which had never received any testing load was used for each loading. A typical result of strain-time diagram at temperature of 22.4 °C is shown in Fig. 5, in which flow strain ε increases directly proportional to $\log_{10} t$ within the stress range up to the upper yield value of $\sigma_u = 0.45$ kg/cm². The curves whose applied stresses are larger than 0.45 kg/cm² rise concave upwards and suggest the occurence of failure in future.

(a) Viscosity Diagrams. By the differentiation of Eq. (48), we get

$$\frac{d\varepsilon}{dt} = \frac{\sigma - \sigma_0}{B_2 E_2} \cdot \frac{1}{t}. \qquad (50)$$

For verification of the above equation, the viscosity diagrams at various elapsed times as illustrated in Fig. 6 are constructed with the data obtained from Fig. 5. This figure shows that the relations are linear within the stress range up to $\sigma = 0.45$ kg/cm², whereas they rise concave upward beyound this limit. Moreover all linear parts of these lines concentrate at a point on the σ-axis. The abscissa of this point

shows the lower yield value σ_0, below which no flow deformation seems
to take place. The stresses corresponding to the upper ends of linear
parts are equal for every line and this stress gives the upper yield value

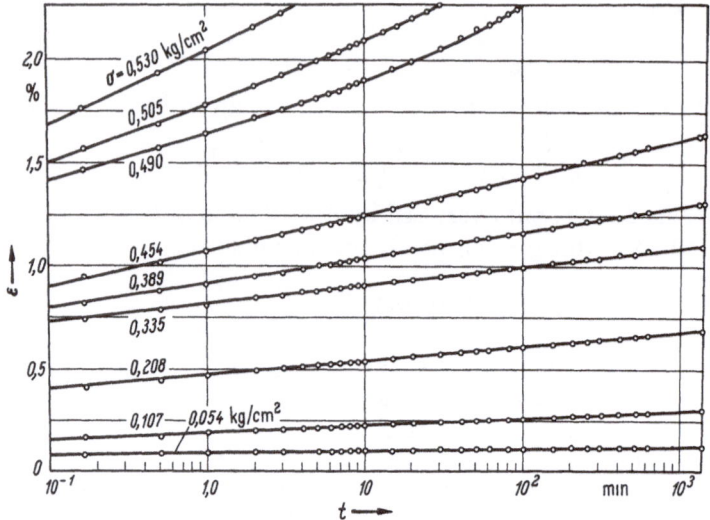

Fig. 5. Curves representing flow strain ε vs. time t relation.

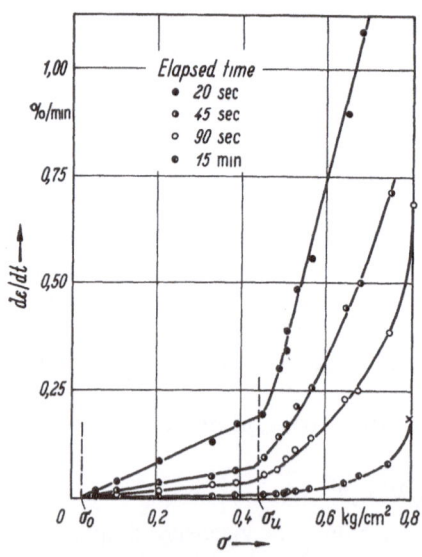

Fig. 6. Viscosity diagrams.

σ_u, below which no failure
occurs. From Fig. 6 it is obtained as $\sigma_0 = 0.025$ kg/cm²
and $\sigma_u = 0.45$ kg/cm². Furthermore, the value of $B_2 E_2$
can be computed from the
slopes of lines lying between σ_0
and σ_u: $(B_2 E_2 = 688$ kg/cm²$)$.

(b) Stress-Strain Characters.
E. C. W. A. GEUZE and T. K.
TAN (1953) found experimen-
taly that the ratio of stress
and strain can be determined
uniquely as the time function
$\phi(t)$. Such time function can
be obtained from Eq. (48) as
follows by neglecting small
stress of σ_0:

$$\frac{\varepsilon}{\sigma} \left(\equiv \phi(t)\right) = \frac{1}{E_1} + \frac{1}{E_2} + \frac{1}{B_2 E_2} \log\left(A_{2\nu} \cdot B_2 \cdot E_2 \cdot t\right)$$

$$= a' + b' \log t \quad (a', b'; \text{consts.}).$$

(51)

An example of the time function curve is shown in Fig. 7 which is constructed from the curves for $\sigma < \sigma_u$ in Fig. 5.

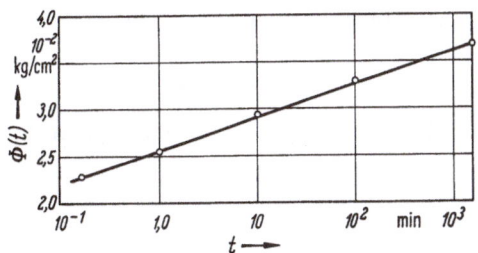

Fig. 7. Variation of $\Phi(t)$ with time constructed from Fig. 5.

5. Thermal Effect on Flow

Substituting B_2 in Eq. (31) into the differential of Eq. (48), we obtain

$$\frac{d\varepsilon}{d\log t} = \frac{\sigma - \sigma_0}{B_2 E_2} = \frac{\sigma - \sigma_0}{B_{20} E_2} \cdot T, \tag{52}$$

where $B_{20} = \beta_2 \Lambda/(2 n A_w b \varkappa)$ (constant).

Eq. (52) shows that $d\varepsilon/d \log t$ and temperature T are directly proportional if the applied stress σ is constant. But the results of compression flow tests performed under the temperature $5\,°C - 40\,°C$ (1961) show that $d\varepsilon/d \log t$ is directly proportional to temperature $(T - T_i)$, (where T_i: freezing temperature of water), as illustrated in Fig. 8. Therefore, the result gives the following empirical equation instead of Eq. (52):

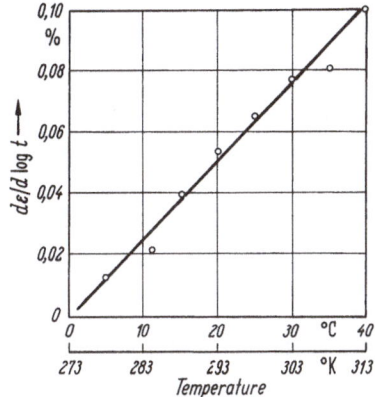

Fig. 8. Thermal effect on the rate of flow.

$$\frac{d\varepsilon}{d\log t} = (\text{const}) \cdot (\sigma - \sigma_0) \cdot (T - T_i). \tag{53}$$

III. Stress Relaxation of Clay Skeleton

6. General Equation for Stress Relaxation

The relaxation of stress in clay skeleton under a constant axial deformation, i. e. the relation between axial stress σ and time t can be analysed through the solution of Eq. (41) under the following condition:

$$\varepsilon = \varepsilon_1 + \varepsilon_2 = \varepsilon_0 (\text{const}), \tag{54}$$

where, ε_0 is the initial constant strain. Substituting the 2nd equation of Eq. (41—2) into the above equation, we obtain

$$\varepsilon_2 = \varepsilon_0 - \varepsilon_1 = \varepsilon_0 - \sigma/E_1, \tag{55}$$

$$\frac{d\varepsilon_2}{dt} = -\frac{1}{E_1}\frac{d\sigma}{dt}. \tag{56}$$

Stress in clay skeleton σ at $t = 0$ is obtained from Eq. (55) as follows, because ε_2 is zero at $t = 0$.

$$\sigma_{t=0} = E_1\varepsilon_0. \tag{57}$$

Substituting Eq. (55) into σ_2 in Eq. (41—3), σ_2 becomes as

$$\sigma_2 = \{(E_1 + E_2)\,\sigma - E_1\sigma_0 - E_1E_2\varepsilon_0\}/E_1. \tag{58}$$

Substituting Eq. (57) into σ_{20} in Eq. (41—3), σ_{20} becomes as

$$\sigma_{20} = E_1\varepsilon_0 - \sigma_0. \tag{59}$$

By the substitution of Eqs. (58) and (59) into the 2nd equation of Eq. (41—1), the validity condition for σ is transformed as

$$\varepsilon_0 E_1 > \sigma \geq \frac{E_1 E_2}{E_1 + E_2}\left\{\varepsilon_0\left(1 + \frac{2E_1}{B_2 E_2}\right) + \frac{\sigma_0}{E_2}\left(1 - \frac{2}{B_2}\right)\right\}. \tag{60}$$

Fundamental equation of stress relaxation is obtained by the substitution of Eqs. (56), (58) and (59) into the 3rd equation of Eq. (41—2) as follows.

$$-\frac{d\sigma}{dt} = A_{2\nu}\cdot E_1(E_1\varepsilon_0 - \sigma_0)\cdot\exp\left[B_2\frac{\sigma_2}{\sigma_{20}}\right]. \tag{61}$$

By the substitution of

$$u = B_2\frac{\sigma_2}{\sigma_{20}} = \frac{B_2\{\sigma(E_1 + E_2) - \sigma_0 E_1 - E_1 E_2\varepsilon_0\}}{E_1(E_1\varepsilon_0 - \sigma_0)} \tag{62}$$

into Eq. (61) and integrating, we obtain

$$\left.\begin{array}{c}-\displaystyle\int_{u_0}^{u}\exp(-u)\,du = Rt, \\[2mm] u_0 = [u]_{t=0} = B_2, \\[2mm] R = A_{2\nu}\cdot B_2\cdot(E_1 + E_2).\end{array}\right\} \tag{63}$$

where

If B_2 is large and $\exp(-B_2)$ is negligibly small, the solution of Eq. (63) is approximately given by

$$\exp(-u) = Rt, \tag{64}$$

from which

$$\sigma = \frac{E_1 E_2}{E_1 + E_2}\left\{\left(\varepsilon_0 + \frac{\sigma_0}{E_2}\right) - \left(\varepsilon_0 - \frac{\sigma_0}{E_1}\right)\frac{E_1}{B_2 E_2}\log(Rt)\right\}. \tag{65}$$

If the lower yield value σ_0 is negligibly small compared with $\varepsilon_0 E_1$ and $\varepsilon_0 E_2$, Eq. (65) may be written approximately as

$$\sigma = \frac{E_1 E_2}{E_1 + E_2} \left\{ 1 - \frac{E_1}{B_2 E_2} \log (Rt) \right\} \varepsilon_0. \tag{66}$$

If σ is beyond the limit expressed by Eq. (60) and at $t \to \infty$, σ is obtained through substituting $\sigma_2 = 0$ and $\sigma_0 \doteqdot 0$ into Eq. (58) as follows:

$$\sigma_{t \to \infty} = \frac{E_1 E_2}{E_1 + E_2} \varepsilon_0. \tag{67}$$

It becomes clear from Eqs. (66) and (67) that the stress σ decreases linearly with the logarithm of time until σ reaches the limiting value expressed by Eq. (60) and finally it relaxes to a finite value of $\sigma_{t \to \infty}$.

As expressed by the 1st equation of Eq. (41—1), above described relation is valid only when axial stress σ is less than σ_u. Therefore, critical initial strain ε_{0c} beyond which the instantaneous axial stress caused by initial strain exceeds σ_u is given by

$$\varepsilon_{0c} = \sigma_u / E_1. \tag{68}$$

7. Stress Relaxation Tests

(a) Stress Relaxation Tests under Uni-Axial Strain. An example of the result of stress relaxation test under uni-axial strain obtained by the plastometer (1961) is plotted in Fig. 9 (for sample No. I). This shows the same behaviour as predicted in Eq. (66). From Eq. (66) the rate of stress relaxation $(d\sigma/d \log t)$ is given by

$$-\frac{d\sigma}{d \log t} = \frac{E_1^2}{B_2 (E_1 + E_2)} \varepsilon_0. \tag{69}$$

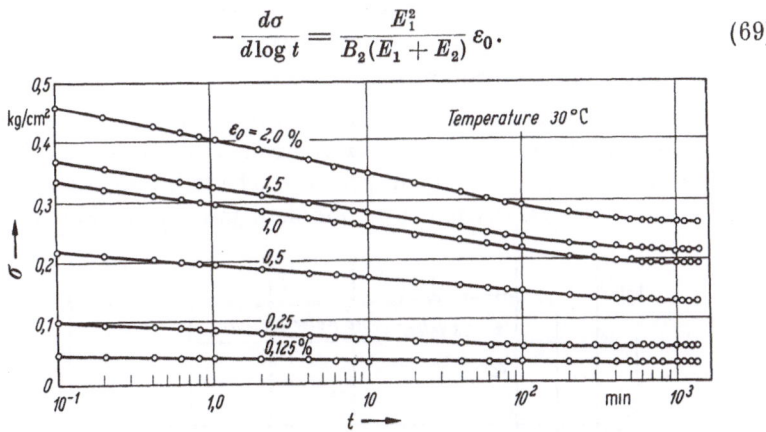

Fig. 9. Curves representing stress relaxation vs. elapsed time relation.

In Fig. 10, the rate of stress relaxation $(d\sigma/d \log t)$ obtained from Fig. 9 is plotted against the initial strain ε_0. As the experimental data lie on a straight line whose inflection point coincides with the critical

8*

initial strain obtained by Eq. (68), it can be verified that Eqs. (68) and (69) are valid and $E_1^2/\{B_2[E_1 + E_2]\}$ is constant within the range of $\varepsilon_0 < \varepsilon_{0c}$.

From measured values of $\sigma_{t=0}$ and $\sigma_{t\to\infty}$ in these tests, the elastic moduli of clay skeleton E_1 and E_2 can be computed by Eqs. (57) and (67).

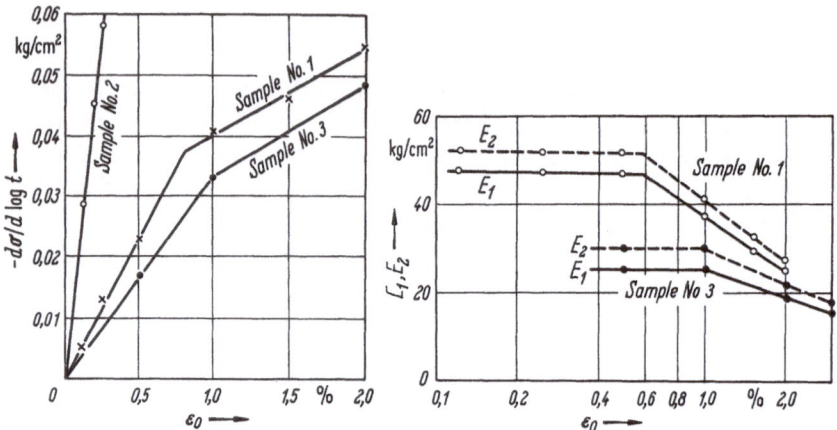

Fig. 10. Relation of rate of stress relaxation and initial strain constructed from Fig. 9.

Fig. 11. Elastic moduli related to initially applied constant strain ε_0.

The elastic moduli thus computed are shown in Fig. 11. From these figures, it is observed that E_1 and E_2 are kept constant within the limit of ε_{0c}, but beyond this limit they decrease with ε_0. This suggests that the fracture in clay skeleton proceeds with the applied strain exceeding the critical initial strain of ε_{0c}.

(b) **Stress Relaxation Test under Tri-Axial Compression.** The stress relaxation tests under a constant axial strain were performed in a tri-axial cell with a constant ambient pressure. The following is an example. Prior to the stress relaxation tests, ambient pressure σ_3 was increased

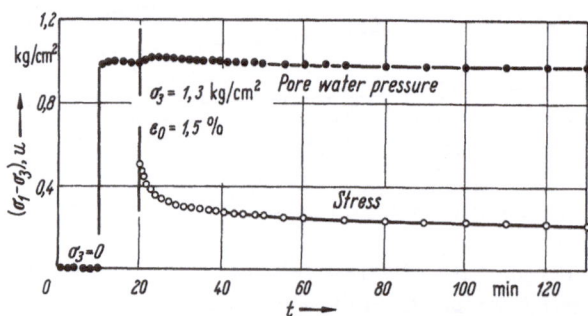

Fig. 12. Variations in pore water pressure and axial stress during stress relaxation test.

suddenly by $1.0 \, kg/cm^2$. The excess pore water pressure u in the undisturbed clay specimen rised almost at the same time by the same magnitude or $1.0 \, kg/cm^2$ as shown in Fig. 12. After the pore water pressure
reached equilibrium state, constant axial strain ($\varepsilon_0 = 1.5\%$) was given
and the deviatoric stress ($\sigma_1 - \sigma_3$) and the pore water pressure u were
measured with time. If ($\sigma_1 - \sigma_3$) — t relation in Fig. 12 is replotted on a
semi-logarithmic paper, ($\sigma_1 - \sigma_3$) can be found to decrease linearly
with $\log t$ and finally approach to a finite value as similar as Fig. 9.

(c) **Effect of Ambient Pressure on Stress Relaxation.** A series of
stress relaxation tests with the same initial strain (in this case $\varepsilon_0 = 1.0\%$) were performed under various ambient pressure in a triaxial

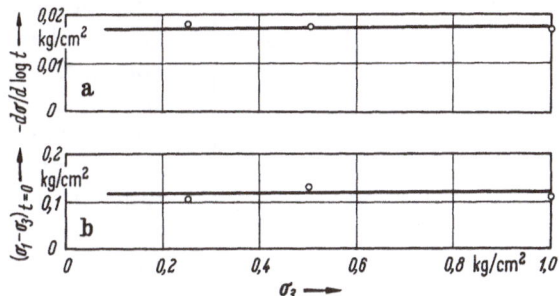

Fig. 13. Influence of ambient pressure on a) rate of stress relaxation and b) instantaneous stress.

apparatus. Results of these tests are summarized in Fig. 13, in which
(a) shows the relation between $d(\sigma_1 - \sigma_3)/d \log t$ and σ_3 ($\sigma = \sigma_1 - \sigma_3$, σ_3:
ambient pressure) and (b) represents the relation between $(\sigma_1 - \sigma_3)_{t=0}$ and
σ_3. As represented in these figures, it is recognized that the rate of
stress relaxation and instantaneous stress due to the application of
initial strain are independent of the ambient pressure.

IV. Upper Yield Value of Clays

8. A Proposed Method to Measure the Upper Yield Value

Generally the upper yield value can be determined graphically with
the viscosity curve as stated in 4. But that procedure requires much
time and many fresh specimens cut from the same sample. Instead
of the above stated measuring procedure, the authors (1958) proposed
a new method using the stress-controlled compression test which can
be performed quickly with one clay specimen.

In this test, the compressive stress is applied by equal stress increment at uniform time interval stepwise and compressive strain is
measured at equal time lapse from each beginning of the step of stress.

It was showed that thus obtained stress-strain relationships can be represented approximately by a straight line on the logarithmic paper as far as the applied stress is less than the upper yield value. Therefore, the upper yield value of clays can be measured as the stress corresponding to the first inflection or bend point of stress-strain curve on a logarithmic paper from which linear relation invalidates.

On this measuring method following were observed experimentally.

1. Though the failure strength of clay is affected by the rate of stress increment in the stress-controlled compression test, the upper yield value obtained with this test is independent of the rate of stress increment in the test. Fig. 14 shows the relations: the each stress corresponding to the first inflection point of the curve obtained by each stress-controlled test of different rate of stress increment ($\alpha = 5 \times 10^{-3}$, 1×10^{-2} or 2.5×10^{-2} kg/cm^2/min) on the same clay is equal to 0.33 kg/cm^2.

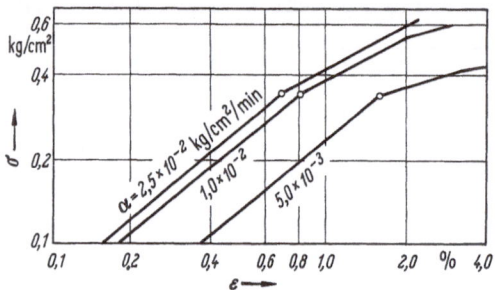

Fig. 14. Influence of rate of stress increment on upper yield value.

2. The upper yield value obtained by the proposed method for various clay samples are well coincident with those obtained by their viscosity curves as same as shown in Fig. 6.

3. The upper yield value can be determined uniquely by the stress-controlled undrained triaxial compression test independently of the ambient pressure. In these tests, applied stresses were restrained not to exceed the pre-consolidation pressure of the clay. Similarly the yield bearing capacity of various type of foundation in clayey ground can be determined by the load corresponding to the first inflection point of the load-settlement curve plotted on the logarithmic paper, which is obtained by the load-controlled test whose load is added in equal increment at uniform time interval.

9. Effect of Water Content on the Upper Yield Value and Safety Factor for Soil Structure

The failure compressive strengths (σ_s) and the upper yield values (σ_u) of clay samples consolidated under various consolidation pressures (σ_m) were measured by stress-controlled triaxial compression tests. σ_s and σ_u are expressed by deviatoric stresses and σ_m by mean principal stress. In spite of the difference in the kind of stress, if σ_s, σ_u and σ_m are plotted against water content w on the same semi-logarithmic paper with the same scale of stress, these relations are represented by parallel straight lines as shown in Fig. 15. But $w - \log_{10}\sigma_u$ line lies always lower side of $w - \log_{10}\sigma_s$ line.

Therefore, following relation may be obtained:

$$\sigma_m = A_m \exp\left(-2.3w/C_c\right), \tag{70}$$

$$\sigma_s = A_s \exp\left(-2.3w/C_c\right), \tag{71}$$

$$\sigma_u = A_u \exp\left(-2.3w/C_c\right), \tag{72}$$

where A_m, A_s, A_u and C_c (compression index) are constants.

As an example, from Fig. 15 (σ_m/σ_u) and (σ_s/σ_u) are measured to be 1/0.5 and 1/0.72 respectivelly. Accordingly we get

$$\frac{\sigma_m}{\sigma_u} = \frac{A_m}{A_u} \div \frac{1}{0.5} = 2.0, \tag{73}$$

$$\frac{\sigma_s}{\sigma_u} = \frac{A_s}{A_u} \div \frac{1}{0.72} = 1.4, \quad A_s > A_u. \tag{74}$$

Since the upper yield value of clay is always smaller than its failure strength, bearing capacity of foundation which is based on such failure strength of clay lies always in dangerous side as compared with the true critical strength based on the upper yield value below which the load is supported permanently.

Therefore, on the stability investigation based on the failure strength of clay measured by the routine testing method, it should be necessary to introduce at least the minimum safety

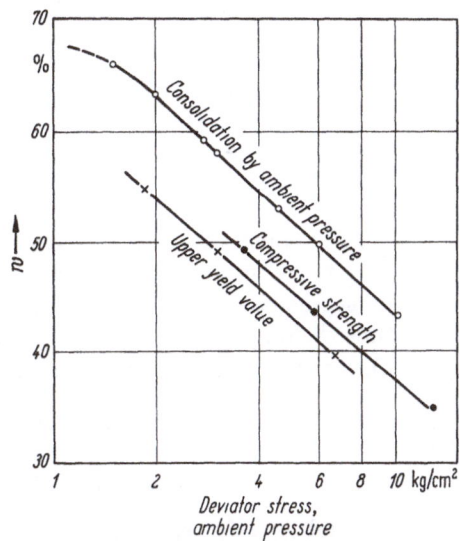

Fig. 15. Relation among consolidation pressure, compressive strength and upper yield value.

factor of (σ_s/σ_u) for permanent constant loading. This may be a meaning of the safety factor for stability calculation.

V. Effect of Consolidation on the Rheological Properties of Clay Skeleton

10. Effect of Consolidation on the Upper Yield Value

As wellknown, clay skeleton is consolidated only in the case when the applied mean principal stress σ_m exceeds the preconsolidation mean principal stress σ_{mc} of the clay, and no recovery of consolidation follows even if the applied stress is removed. This phenomenon is supposed as the clay segments are settled through the consolidation at a new stable orientation of minimum potential which is created by disorder in the electric double layer of adsorbed water accompanied with the nearness of segments and the flow of pore water during consolidation. Therefore, it can be said that the average mutual distance between clay segments in equilibrium is reduced by consolidation.

Although the relation of forces existing between clay segments and the physico-chemical influence on the mechanical character of clay are not well clarified yet, some analysises will be tried to obtain the influence of consolidation on the upper yield value and the elasticity by introducing SALAS and SERRATOSA's research based on DEBYE's theory.

According to SALAS and SERRATOSA's research (1953), the compressive force of clay particle in equilibrium under the application of this force is equal to the force of repulsion by the electric charge between two particles (in this paper, the particles are considered as segments) laid at a mutual distance λ. As the force of electric repulsion is proportional to $\exp(-K\lambda)$, according to DEBYE's theory, the relation between the compressive force of clay particle f_m and λ can be represented by

$$(f_m)_i = K_c \cdot \exp\{-K \cdot (\lambda)_i\}, \tag{75}$$

where K, K_c: constants, $(f_m)_i$: individual compressive force of segment caused by mean principal stress σ_m, $(\lambda)_i$: individual mutual distance between segments of clay consolidated by σ_m and $(\ )_i$ means the magnitude belonging to each individual segment. Corresponding to Eq. (75), average compressive force may be expressed by

where
$$\left.\begin{aligned} f_m &= K_c \cdot \exp(-K \cdot \lambda), \\ \lambda &= \sum_{}^{N} (\lambda)_i / N, \end{aligned}\right\} \tag{76}$$

where f_m and λ: effective compressive force of segment and average mutual distance between segments of the clay consolidated by σ_m, N: number of segments per unit area of clay.

As the consolidation mean principal stress σ_m is the resultant of the compressive forces of various direction on segments in unit area,

σ_m may be represented as follows

$$\sigma_m = \sum_{i}^{N} \{(f_m)_i \cdot (\delta_m)_i\}$$
$$= \delta_m \cdot N \cdot f_m, \tag{77}$$

where $(\delta_m)_i$: coefficient to adjust the deviation of the direction of $(f_m)_i$ from the principal axis, δ_m: similar coefficient corresponding to f_m. From Eqs. (76) and (77) we get

$$\sigma_m = \delta_m \cdot N \cdot K_c \cdot \exp(-K\lambda). \tag{78}$$

Refering Eq. (70) with Eq. (78), we get

$$\left.\begin{array}{l} \delta_m N K_c = A_m, \\ K\lambda = 2.3w/C_c. \end{array}\right\} \tag{79}$$

Hence, Eqs. (70) and (72) are transformed as

$$\left.\begin{array}{l} \sigma_m = A_m \cdot \exp(-K\lambda), \\ \sigma_u = A_u \cdot \exp(-K\lambda), \end{array}\right\} \tag{80}$$

where σ_u: upper yield value of the clay which is consolidated by σ_m. Similarly, for the clay which is consolidated by σ_{mc}, its upper yield value σ_{uc} and its average mutual distance λ_c — adding suffix c — can be represented as

$$\left.\begin{array}{l} \sigma_{mc} = A_m \cdot \exp(-K \cdot \lambda_c), \\ \sigma_{uc} = A_u \cdot \exp(-K \cdot \lambda_c). \end{array}\right\} \tag{81}$$

Therefore, it can be said as follows. So far as the applied stress on the clay is kept below its preconsolidation stress σ_{mc}, the upper yield value of the clay σ_{uc} remains constant. But if this clay is consolidated by σ_m exceeding σ_{mc}, σ_{uc} increases to σ_u and these relation is obtained from Eqs. (80) and (81) as follows

$$\frac{\sigma_u}{\sigma_{uc}} = \frac{\sigma_m}{\sigma_{mc}} = \exp\{-K(\lambda - \lambda_c)\}. \tag{82}$$

11. Effect of Consolidation on the Elasticity of Clay

Elasticity of clay skeleton under a shearing stress which is represented as the spring elements of mechanical model has been deduced in Chap. I as the accumulation of repulsive force on each segment against the variation in the intercepting angle between adjacent segments. Since the magnitude of the repulsive force between segments depends on their mutual position and orientation, the elastic constant of segment e or spring constants of the mechanical model (G_1, G_2) or (E_1, E_2) can be assumed as functions of the mutual distance between the segments λ.

Besides, it is said by Salas and Serratosa that following two kinds of forces are existing and balancing between segments of saturated clay oriented in equilibrium, i. e. the one is Van der Waal's force of attractive which is proportional to λ^{-6} and the other is the force produced by electric repulsion which is proportional to $\exp(-K\lambda)$. Therefore, the equilibrium relation between these forces is given by

$$W\lambda^{-6} = K_s \cdot \exp\left(-K\lambda\right)(\equiv f_s), \qquad (83)$$

where W, K_s: constants.

If an external force f which is caused by the applied shearing stress is applied on a segment, λ decreases by $d\lambda$ and reaches a new balancing position at the mutual distance of $(\lambda - d\lambda)$. Hence,

$$f + W(\lambda - d\lambda)^{-6} = K_s \cdot \exp\left\{-K(\lambda - d\lambda)\right\}.$$

By Taylor's expansion,

$$f + W\lambda^{-6} - \frac{d}{d\lambda}(W\lambda^{-6})\,d\lambda \doteqdot K_s \cdot \exp\left\{-K\lambda\right\}$$

$$- \frac{d}{d\lambda}\left\{K_s \exp\left(-K\lambda\right)\right\}d\lambda.$$

Substituting Eq. (83) into the above,

$$f + 6W\lambda^{-6}\frac{d\lambda}{\lambda} = K_s \cdot K \cdot \lambda \cdot \exp\left(-K\lambda\right)\frac{d\lambda}{\lambda}. \qquad (84)$$

From the definition, the elastic constant for a segment e can be represented by

$$e = f/(d\lambda/\lambda). \qquad (85)$$

From Eqs. (83) — (85), e is expressed as follows:

$$e = \left\{K_s \exp\left(-K\lambda\right)\right\}(K\lambda - 6).$$

Above relation is transformed by substituting Eq. (80) in it as follows:

$$\frac{e}{\sigma_m} = \frac{K_s}{A_m}(K\lambda - 6). \qquad (86)$$

Similarly, for the clay consolidated by σ_{mc}, its elastic constant e_c and its average mutual distance λ_c — adding suffix c — is written by

$$\frac{e_c}{\sigma_{mc}} = \frac{K_s}{A_m}(K\lambda_c - 6). \qquad (87)$$

From Eqs. (86), (87) and (82), we get

$$\frac{e}{\sigma_m} = \frac{e_c}{\sigma_{mc}} + \frac{K_s}{A_m}\cdot K(\lambda - \lambda_c) = \frac{e_c}{\sigma_{mc}} - \frac{K_s}{A_m}\log\left(\frac{\sigma_m}{\sigma_{mc}}\right)$$

from which

$$e = \left\{\frac{\sigma_m}{\sigma_{mc}} - \frac{K_s}{e_c}\frac{\sigma_m}{A_m}\log\left(\frac{\sigma_m}{\sigma_{mc}}\right)\right\}e_c.$$

By the substitution of Eqs. (80) and (83) into the above, we get

$$e = \left\{ \frac{\sigma_m}{\sigma_{mc}} - \left(\frac{f_s}{e_c} \right) \cdot \log \left(\frac{\sigma_m}{\sigma_{mc}} \right) \right\} e_c. \tag{88}$$

As for the right hand side of Eq. (88), the term (f_s/e_c) is temporarily assumed to be sufficiently small compared with unity. Furthermore, if σ_m is so large as $\log(\sigma_m/\sigma_{mc})$ of the 2nd term becomes negligibly small compared with (σ_m/σ_{mc}) of the 1st term, Eq. (88) can be written approximately by

$$e = (\sigma_m/\sigma_{mc}) \, e_c. \tag{89}$$

Above assumption used to obtain Eq. (89) from Eq. (88) will be checked experimentally in 12(c) in this Chapter.

As represented by Eq. (89), elasticity of clay e increases with applied stress σ_m, but this relation is valid only when σ_m exceeds the preconsolidation stress of the clay σ_{mc}. Therefore, so far as σ_m is less than σ_{mc}, elasticity of clay e remains unchanged $(= e_c)$. By the substitution of Eq. (89) into Eqs. (30) and (31) or Eqs. (38) and (39), relations of spring constants of the mechanical model of clay skeleton are obtained as follows. If $\sigma_m > \sigma_{mc}$

$$\frac{G_1}{G_{1c}} = \frac{G_2}{G_{2c}} = \frac{E_1}{E_{1c}} = \frac{E_2}{E_{2c}} = \frac{e}{e_c} = \frac{\sigma_m}{\sigma_{mc}}, \tag{90}$$

where (G_{1c}, G_{2c}) or (E_{1c}, E_{2c}) are shear moduli or YOUNG's moduli of the clay which has been consolidaed by σ_{mc} respectively. From Eqs. (83) and (90), it is known that the upper yield value, the elasticity and the preconsolidation stress are proportional to each other at any consolidated state.

12. Secondary Compression

(a) Secondary Compression in Oedometer. In the case of ordinary oedometer test, consolidation stress is usually measured in the vertical pressure (σ_I) applied on the clay specimen instead of the mean principal stress (σ_m). In order to obtain σ_m from σ_I, it is necessary to know the lateral pressure $K_0 \sigma_I$ produced on the consolidation ring, where K_0 is the coefficient of earth pressure at the state of consolidated-equilibrium. According to A. W. BISHOP (1958), value of K_0 which was measured with triaxial compression test under the condition of zero lateral strain seemed to be kept almost constant independent of the increment of the vertical stress, but to vary with kinds of clays. It was also reported that value of K_0 may be influenced by the activity of the clay and by whether it is in the remoulded or undisturbed state.

If σ_I and $K_0 \sigma_I$ are given, the consolidation mean principal stress (σ_m) and deviatoric stress (σ) at the state of consolidated-equilibrium are expressed as

$$\sigma_m = \sigma_I + 2K_0\sigma_I = (1 + 2K_0)\,\sigma_I,$$

$$\sigma = \sigma_I - K_0\sigma_I = (1 - K_0)\,\sigma_I(\equiv \mu\sigma_I).$$

(91)

Therefore, if a clay whose preconsolidation stress is expressed by vertical stress σ_{Ic} is consolidated by vertical stress σ_I, their relation becomes as

$$\sigma_I/\sigma_{Ic} = \sigma_m/\sigma_{mc}.$$

(92)

From Eqs. (90) and (92), we get

If $\sigma_I > \sigma_{Ic}$,

$$E_1/E_{1c} = E_2/E_{2c} = \sigma_I/\sigma_{Ic}.$$

(93)

If $\sigma_I \leq \sigma_{Ic}$,

$$E_1 = E_{1c}, \qquad E_2 = E_{2c}.$$

(94)

Where E_{1c}, E_{2c}: Young's moduli of the clay whose vertical preconsolidation stress is σ_{1c}, E_1, E_2: Young's moduli of the clay in consolidation ring under the vertical stress of σ_I.

Generally secondary compression is said to be a deformation succeeding to the end of primary consolidation. The secondary compression is a kind of flow of such newly consolidated clay under a deviatoric stress. Therefore, substituting σ of Eq. (91) into Eqs. (43), (48) and (49) in Chapter II and neglecting σ_0 because of its relatively small value, the equation of secondary compression due to vertical stress σ_I can be obtained as follows.

As far as

$$\varepsilon < \mu \cdot \sigma_I \left\{ \frac{1}{E_1} + (B_2 - 2)\,\frac{1}{B_2 E_2} \right\},$$

$$\varepsilon = \mu \cdot \sigma_I \left\{ \frac{1}{E_1} + \frac{1}{E_2} + \frac{1}{B_2 E_2} \log\,(A_{2v} B_2 E_2 t) \right\}.$$

If ε is beyond the above expressed limit and at $t \to \infty$,

$$\varepsilon_{t \to \infty} = \mu \cdot \sigma_I \left\{ \frac{1}{E_1} + \frac{1}{E_2} \right\}.$$

(95)

(α) *The Secondary Compression of Normally Consolidated Clay.* The secondary compression or the flow succeeding to the consolidation under the vertical stress σ_I higher than σ_{Ic} can be expressed by the equation which is obtained by substituting Eq. (93) into the 2nd equation of Eq. (95) as follows.

When $\sigma_I > \sigma_{Ic}$,

$$\varepsilon = \mu \cdot \sigma_I \left\{ \left(\frac{1}{E_{1c}} + \frac{1}{E_{2c}} \right) \frac{\sigma_{Ic}}{\sigma_I} + \frac{1}{B_2 E_{2c}} \cdot \frac{\sigma_{Ic}}{\sigma_I} \log (A_{2\nu} B_2 E_2 t) \right\}$$

or $\quad \varepsilon = \mu \cdot \sigma_{Ic} \left\{ \frac{1}{E_{1c}} + \frac{1}{E_{2c}} + \frac{1}{B_2 E_{2c}} \log (A_{2\nu} B_2 E_2 t) \right\}.$

From the above, the rate of secondary compression is given by $\quad$ (96)

$$\frac{d\varepsilon}{d\log t} = \frac{\mu}{B_2 E_{2c}} \sigma_{Ic}.$$

Therefore, the rate of secondary compression succeeding to the consolidation under the stress σ_I remains invariable independent of σ_I, so far as σ_I is larger than σ_{Ic}.

(β) *Secondary Compression of Over-Consolidated Clay.* In this case, applied vertical stress σ_I is lower than the preconsolidation stress σ_{Ic} of the clay. Therefore, the rate of secondary compression is obtained as follows by substituting Eq. (94) into the 2nd equation of Eq. (95) and by differentiating it.

When $\sigma_I \leq \sigma_{Ic}$,

$$\frac{d\varepsilon}{d\log t} = \frac{\mu}{B_2 E_{2c}} \cdot \sigma_I. \tag{97}$$

Therefore, the rate of secondary compression of over-consolidated clay increases in direct proportion to σ_I within the range of $\sigma_I \leq \sigma_{Ic}$.

(γ) *Tests on Secondary Compression in Oedometer.* In order to verify the result above obtained ordinary oedometer tests under various constant vertical stresses were performed on undisturbed saturated clay specimens whose preconsolidation stresses were known. Each specimen was set in the consolidation ring and was soaked in water without applying any stress prior to the oedometer test. Rates of secondary compression obtained through these tests are plotted against the vertical stress σ_I as shown in Fig. 16.

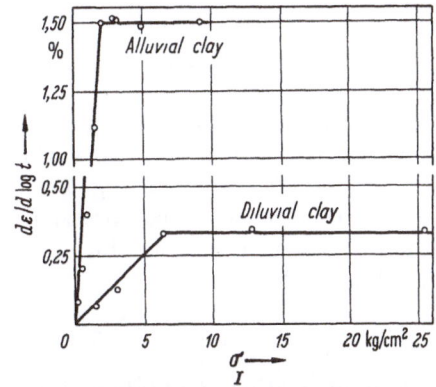

Fig. 16. Relation between applied stress and rate of secondary compression in oedometer. Each inflection point coinside with preconsolidation stress.

In this figure, it can be seen that each line (one is the line of alluvial clay in Osaka, the other is that of diluvial clay in that place) passes through the origin and rises in a straight line until a certain point, then deviates to be horizontal. The each abscissa corresponding to the inflection point coinsides with each preconsolidation stress σ_{Ic}. These results well agree with the theory expressed by Eqs. (96) and (97) which are deduced from Eq. (89). Therefore, the approximation used to obtain Eq. (89) from Eq. (88) may be acceptable.

As shown in these test results, there lies clear difference in the rate of secondary compression according to the history of consolidation.

(b) Effect of Stress Condition on Secondary Compression

(α) *Effect of Coefficient of Earth Pressure on Secondary Compression.* The ratio of axial stress and ambient pressure on the triaxial compression test is termed the coefficient of earth pressure K, viz. $K = \sigma_3/\sigma_1$.

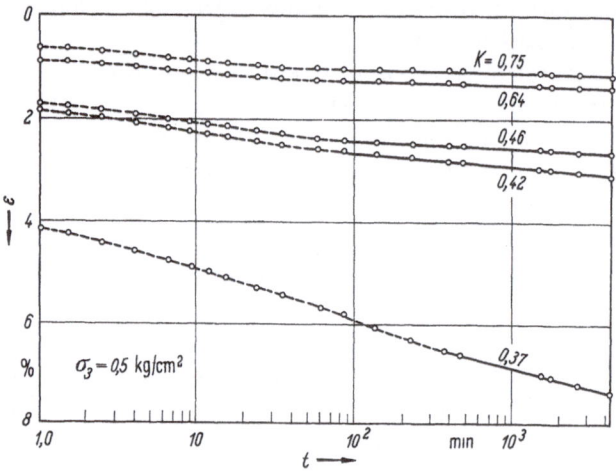

Fig. 17. Strain-time diagram in anisotropic consolidation test under constant σ_3 and various K.

Long term anisotropic consolidation tests were performed under a constant σ_3 of 0.5 kg/cm² and various K in a triaxial apparatus (1962). The results are shown in Fig. 17. In the figures, parts of thick lines represent the secondary compressions. Neglecting small value of σ_0, the rate of secondary compression can be expressed by Eq. (52) in Chapter II as follows

$$\frac{d\varepsilon}{d\log t} = \frac{(\sigma_1 - \sigma_3)}{B_2 E_2} = \frac{\sigma_3}{B_2 E_2}\left(\frac{1}{K} - 1\right). \qquad (98)$$

Values of $d\varepsilon/d\log_{10} t$ obtained from Fig. 17 are plotted against K as shown in Fig. 18. These plots lie in well agreement on the curve obtained by the substitution of $\sigma_3 = 0.5$ kg/cm² and $B_2 E_2 = 280$ kg/cm² into Eq. (98). Dotted part of the curve in Fig. 18 represents the imaginary region where flow failure occurs due to the deviatoric stress exceeding the upper yield value of the clay.

As the rate of flow is in direct proportion to the deviatoric stress as expressed by Eq. (98), considering from a theoretical stand point,

any application of all-round uniform pressure accompanying no deviatoric stress must cause no flow of clay. But unexpected exceptional
results or a feeble flow of clay were often observed when all-round
uniform pressure ($K = 1$) was applied
on undisturbed clay. These phenomena
seem to be due to the residual deviatoric strain existing in the clay prior
to the application of the all-round
pressure.

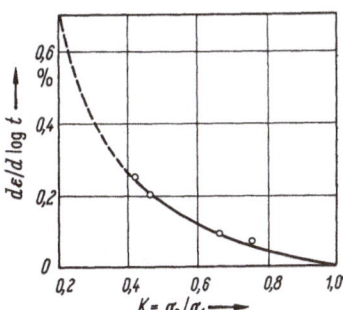

Fig. 18. Relation between rate of secondary compression and coefficient of earth
pressure constructed from Fig. 17.

(β) *Effect of Mean Principal Stress
on the Secondary Compression.* Long
term consolidation tests were performed in a triaxial apparatus under a
constant deviatoric stress viz. $(\sigma_1 - \sigma_3)$
$= 0.2$ kg/cm² but various K. The re-
sults of strain — time relation are
shown in Fig. 19. From this figure, relation between $d\varepsilon/\delta \log_{10} t$ and σ_3
is plotted in Fig. 20, in which $\delta\varepsilon/\delta \log_{10} t$ is found to be almost constant
independent of σ_3.

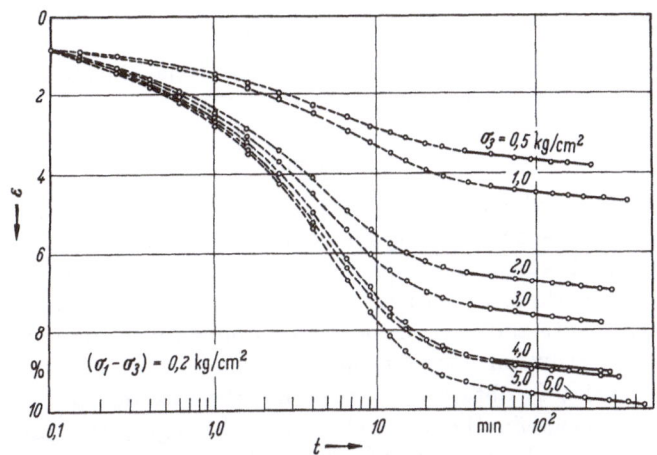

Fig. 19. Strain-time diagram in anisotropic consolidation test under constant deviatoric stress
and various K.

Summarizing the results above stated in (α) and (β), therefore, the
rate of secondary compression may be assumed to be in direct proportion to the deviatoric stress but independent of the mean principal
stress, if there exists no residual deviatoric strain prior to the application of stress.

(c) **Thermal Effect on Secondary Compression.** Theoretical relation
between the rate of secondary compression and temperature can be ob-

tained as the same procedure as described in Chap. II, 5. From
Eqs. (31) and (96), for normally consolidated clay,

$$\frac{d\varepsilon}{d\log t} = \frac{\mu}{B_{20} E_{2c}} \cdot \sigma_{Ic} \cdot T.$$ (99)

From Eqs. (31) and (97), for over-consolidated clay,

$$\frac{d\varepsilon}{d\log t} = \frac{\mu}{B_{20} E_{2c}} \cdot \sigma_{I} \cdot T.$$ (100)

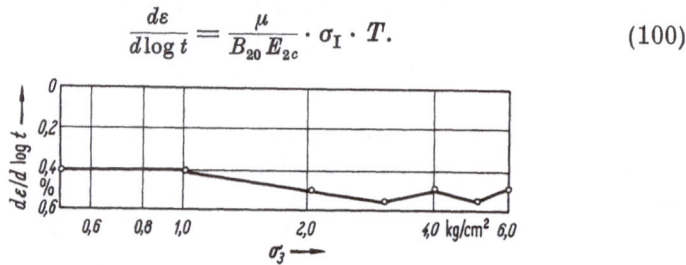

Fig. 20. Influence of σ_3 on rate of secondary compression.
Eq. (99) indicates the linear relationship between the rate of secondary commpression
$d\varepsilon/d \log t$ and temperature T.

But according to the experimental results on normally consolidated
clay, $d\varepsilon/d\log_{10} t$ does not increase directly proportional to T but $(T - T_i)$
(where T_i: freezing temperature of pore water), as expressed by Eq. (99)
and as also found in the experimented result by Morita (1955).
Therefore, the rate of secondary compression of clay increases in direct
proportion to the temperature $(T - T_i)$.

Conclusion

The mechanical model which represents the rheological behaviour
of clay skeleton accompanying no fracture is obtained by some hypotheses. Thus obtained formulae well agree with the results of experiments. Among these hypotheses, distribution of the internal friction
between segments and that of individual applied force on segment are
assumed by Gaussian distribution functions as a first attempt. The
viscosity of clay is obtained by applying the rate process on the adsorbed water between the contacting part of segments. Influence of
consolidation on the upper yield value and the elasticity is obtained by
refering to Salas and Serratosa's research based on Debye's theory.
Applications of these hypotheses may leave room for further investigation, but this study has been performed hoping to get any clue for
the consideration on this problem.

References

Bishop, A. W. (1958): Test Requirements for Measuring the Coefficient of
    Earth Pressure at Rest. Proc. Brussels Conf. 58 on Earth Pressure Problems,
    Vol. 1, 1958, pp. 2—14.

GEUZE, E. C. W. A., and T. K. TAN (1953): The Mechanical Behaviour of Clays. Proc. 2nd Int. Cong. Rheol., 1953, pp. 247—259.

GLASSTONE, S., K. J. LAIDLER and H. EYRING (1941): The Theory of Rate Process, New York, 1941, p. 477.

MORITA, N., and S. SAKATA (1955): Thermal Effect on the Consolidation. Abstract for 11th Conf. of J.S.C.E. p. 70 (in Japanese).

MURAYAMA, S., and T. SHIBATA (1958): On the Rheological Characters of Clay, Part 1. Disaster Prevention Research Inst. Kyoto Univ. Bulletin No. 26.

MURAYAMA, S., and T. SHIBATA (1959): On the Secondary Consolidation of Clay. Proc. 2nd Japan Cong. on Testing Materials, 1959, pp. 178—181.

MURAYAMA, S., and T. SHIBATA (1961): Rheological Properties of Clays. Proc. 5th Int. Conf. S.M.F.E. Vol. 1, pp. 269—273.

SALAS, J. A. J., and J. M. SARRATOSA (1953): Compressibility of Clays. Proc. 3rd Int. Conf. S.M.F.E. Vol. 1, pp. 192—198.

SHIBATA, T. (1962): On the Consolidation of Clays. Disaster Prevention Research Inst. Kyoto Univ. Annual report No. 5.A, 1962, pp. 102—112 (in Japanese).

1.11 The Theoretical Research on the Stress Relaxation of Clay by Thermo-Dynamics and Statistical Mechanics

By

Hiroshi Fujimoto

Synopsis

In order to introduce the thermo-dynamics and the statistical mechanics to the system of soil mechanics, in this paper the author has established the original assumptions on a sectional pattern of clay network structure, on a mechanism of deformation of clay based upon its sectional pattern, and on a relation between the energy-elasticity and the entropy-elasticity with reference to a rheological theory of high-polymer materials.

Consequently, the author has ascertained a possibility on the above introduction that has been thought as it was principally difficult, and has been able to explain molecular or particular theoretically the rheological behaviours of clay.

Prologue

With development of soil engineering, when we stand on a view point that the soil is also one of the engineering materials such as highpolymer or metal, and wish to explain unificatively the complicated mechanical behaviours of soil, it is methodologically necessary that we will must consider an introduction of thermo-dynamics and statistical mechanics based upon a molecular or particular theoretical investigation.

Because the thermo-dynamical analysis is confined to considerations of reversible process of deformation only and the deformations of soil are irreversible in general, therefore, in this case we cannot avoid facing a difficulty to explain the phenomenon. Already this difficulty was indicated by Dr. T. MOGAMI [1], but at present there is no likely a successed example for the theoretical study on the aforsaid theme. There are, merely, several investigations [2]—[5], [7] which partially adopted the molecular or particular theoretical consideration. However, there is kept behinded a possibility of further study on a expression of fundamental relationship between the stress strain and time, it should be deduced by analytical method of thermo-dynamics and statistical mechanics on the basis of a more realistic structural conception of soil skeleton.

1. Imaginary Network Structure and Mechanism of Deformation of Clay

Hitherto, many conceptions for the skeletal structure of clay were proposed by some investigators. The author has synthetically reexamined those structural conceptions, and has adopted the skeletal structure proposed by T. K. Tan [2], which is accepted so reasonable for the clay mineralogically. Namely, the structure of clay under an equilibrium state is supposed to be isotropic and homogeneous three dimensional "card-house" structure shown in Fig. 1. The clay particles under such state are connected each other with the bonding forces through an adsorbed double water layer around the particles. The bonding forces, of course, may be due to COULOMB's attraction force between the positively charged edges and the negatively charged flat sides of particles, VAN DER WAAL's forces, bonding energies by cations, and dipole hydrogen bridges.

Fig. 1. Schematic picture of clay network by T. K. TAN, 1957.

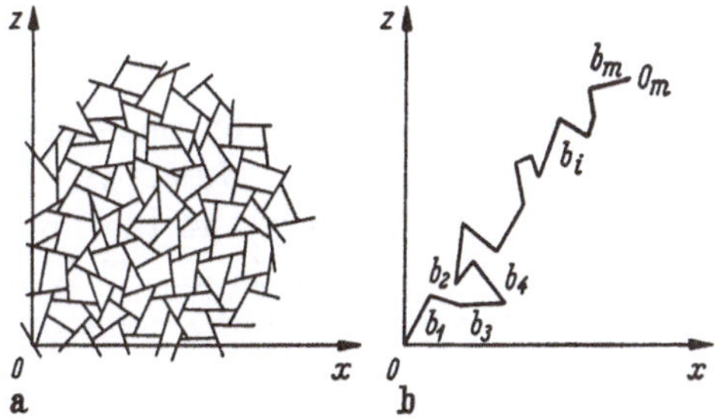

Fig. 2a and b. Graphic pattern of an imaginary cross section and chain of clay network.

Now, T. K. TAN entitled the schematic picture shown in Fig. 1 the "clay network", but we cannot mathematically express a formal distribution of the skeletal structure merely by this network structure. Therefore, the author has presented a pattern shown in Fig. 2a which was a cross section of the clay network in Fig. 1, and by the author this pattern was called anew the "imaginary network" of clay. In Fig. 2b, a broken line turned at random was named the "imaginary chain" of the clay network, further a turning point of the imaginary chain (i.e., a mutual contact point between the particles) was denominated the "unit mechanism of deformation" of clay network by the author. The author has presumed N the number of such unit mechanisms of deformation existed in a unit area of cross section perpendicular to the direction of stress, and M in series per unit length in the direction of stress, respectively, and the edges or corners of the particles of more than two are not contact.

9*

Thus, acting an external force on the clay of such structure, at the moment, a reversible instantaneous elastic deformation occurs, which is not accompanying a change of configuration of the imaginary network. In this process, a micro shear deformation resulted on leaping or slipping of the particles will occur when an accumulated energy at the unit mechanism of deformation by the effective stress component excesses the bonding energy of particle at unit mechanism. This deformation is an irreversible one because that brings a change of configuration of the network. A visible deformation is consist of the instantaneous elastic deformation and the irreversible deformation accumulated in whole body of sample. In this process a part of the stress component acted on the unit mechanism was dissipated as a thermo-energy in consequence of cohesive resistance of double water layer around the clay particles. Consequently, once slipped or leaped particles renew form the unit mechanism of deformation at an another point by the bonding energy. In a word, at a steady state of deformation such as the creep or the process of stress relaxation, the particles are repeating the above slipping or leaping and rebonding, and the number of unit mechanism of deformation is not change.

The author interprete this phenomenon that the restraining internal resistance proposed by S. MURAYAMA and T. SHIBATA [3] is represented the above bond energy at the unit mechanism accumulated in a whole body of sample. And also T. K. TAN [2] has explained this relationship as the lower yielding value of clay.

2. Introduction of Thermo-Dynamics and Statistical Mechanics into the Elastic Zone of Soil

The thermo-dynamics and the statistical mechanics for the elasticity are not principally applicable to the soil, because the deformations of soil are almost irreversible. If we think that, however, the deformation process of soil was as explained in the precedings, the thermo-dynamical expression will be able to apply for the stage of instantaneous micro elastic deformation.

Now, when an external force F acted on a certain material, let us that the elastic deformation of the material was Δl in the direction of F, and a variation of thermal quantity in the sample was ΔQ. Then the external force F is represented by following equation as the sum of the energy-elasticity $(\partial U/\partial l)_{T=\text{const}}$ and the entropy-elasticity $T(\partial S/\partial l)_{T=\text{const}} = -T(\partial F/\partial T)_{l=\text{const}}$,

$$F = \left(\frac{\partial H}{\partial l}\right)_{T=\text{const}} = \left(\frac{\partial U}{\partial l}\right)_{T=\text{const}} - T\left(\frac{\partial S}{\partial l}\right)_{T=\text{const}}, \tag{1}$$

where U is an internal energy of the material, S is an entropy, T is an absolute temperature, $H = U - TS$ is HELMHOLTZ's free energy and l is an axial length of the sample.

Thus, as a rudimental case for the present, we assume that the $F-T$ curve of soil be a straight line so as to simplify the analysis shown in Fig. 3, and represents F in only term of the entropy-elasticity as equation (2),

$$F = - (1 + \alpha)\, T \left(\frac{\partial S}{\partial l}\right)_{T=\text{const}} = - \beta\, T \left(\frac{\partial S}{\partial l}\right)_{T=\text{const}}, \qquad (2)$$

where α and β are the proportional constants as follows,

$$\alpha = \left(\frac{\partial U}{\partial l}\right)_{T=\text{const}} \bigg/ - T\left(\frac{\partial S}{\partial l}\right)_{T=\text{const}}, \quad \beta = 1 + \alpha. \qquad (3)$$

If we are, therefore, able to determine a probability function P that represents a state of formal distribution of clay network, we can describe statistical mechanically the instantaneous elasticity of soil, by applying the fundamental principle in the statistical mechanics;

$$S = k \ln P, \qquad (4)$$

where k is BOLTZMAN's constant.

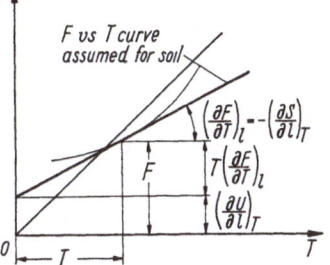

Fig. 3. F vs. T curve, to explain a relation between energy elasticity and entropy elasticity.

3. Determination of Probability Function that Represents the Formal Distribution of the Clay Network

For the determination of the distribution probability of clay network, if we leave out of consideration the "substantial volume exclusion" of clay particles, we can adopt the "random flight problem" in the theory of probability for the state of two dimensional distribution of the imaginary network of clay shown in Fig. 2.

That is, in this problem, let $p_{(x,z)}$ a probability density, when we take the origin O on any unit mechanism of deformation in Fig. 2 (b), the probability that another unit mechanism O_m is apart in distance r which is represented by $r^2 = x^2 + z^2$ in range $(r, r + dr)$ from O, in other words, the probability that O_m is apart in distance r which is stretched for m segments of the broken line turned at random in an elemental area $(dx\,dz)$ is $p_{(x,z)}\,dx\,dz$. Then, since a length of each segment is not equal, the author adopted K. FUSHIMI's [6] equation under mentioned (5) which is an enlarged one of Kluyver for a circular symmetry distribution of segments.

$$p_{(x,z)} = \frac{m^2}{2\pi\varrho^2} \exp\left[-\frac{(x^2 + z^2)}{2\varrho^2}\right], \tag{5}$$

where $\varrho^2 = m\frac{\overline{r}^2}{2}$, $\overline{r}^2/2 = x_i^2 + z_i^2$.

Consequently, the distribution probability P of the clay network in any cross sectional area A is

$$P = \iint\limits_{A} \frac{m^2}{2\pi\varrho^2} \exp\left[-\frac{(x^2 + z^2)}{2\varrho^2}\right] dx\, dz. \tag{6}$$

4. Representation by Thermo-Dynamics and Statistical Mechanics of Instantaneous Elastic Deformation and Stress of Clay

We assume that the each components (x, z) of original distance for the both end of the imaginary chain become $(\zeta x, \xi z)$ when an external force acts on and sample instantaneously deforms.

Then, the number of imaginary chains that each components of original distance for the both end (x, z) are

$$dN = N p_{(x,z)}\, dx\, dz = N \frac{m^2}{2\pi\varrho^2} \exp\cdot\left[-\frac{x^2 + z^2}{2\varrho^2}\right] dx\, dz,$$

$$= N \frac{m^2}{\pi} \Phi^2 \exp\cdot\left[-\Phi(x^2 + z^2)\right] dx\, dz, \tag{7}$$

$$\Phi^2 = 1/2\varrho^2.$$

Therefore, the components of distance for the both ends of the imaginary chains dN become $(\zeta x, \xi z)$, which is due to the instantaneous elastic deformation of the sample.

Hence a deformed chain has the probability $p_{(\zeta x, \xi z)}$ substituted (x, z) for $(\zeta x, \xi z)$, an entropy change s caused by the deformation of a chain is

$$\begin{aligned} s &= k\left[\ln p_{(\zeta x, \xi z)} - \ln p_{(x,z)}\right], \\ &= k\ln\left\{\frac{m^2}{\pi}\Phi^2 \exp\left[-\Phi^2(\zeta x^2 + \xi z^2)\right]\right\} \\ &\quad - k\ln\left\{\frac{m^2}{\pi}\Phi^2 \exp\left[-\Phi^2(x^2 + z^2)\right]\right\}, \\ &= -k\Phi^2\left[(\zeta^2 - 1)x^2 + (\xi^2 - 1)z^2\right]. \end{aligned} \tag{8}$$

Since the number of imaginary chains have the entropy change represented by Eq. (8) are dN, the sum of entropy change S of all chains is

$$S = \int s \, dN$$

$$S = - Nk \frac{m^2}{\pi} \Phi^2 \int\limits_{-\infty}^{+\infty}\!\!\!\int \Phi^2 [(\zeta^2 - 1) x^2 + (\xi^2 - 1) z^2]$$

$$\times \exp\left[- \Phi^2 (x^2 + z^2)\right] dx \, dz \tag{9}$$

$$= - \frac{Nk m^2}{2} (\zeta^2 + \xi^2).$$

Thus, assuming that a volume change of the sample was due to the deformation is nothing and is isotropic, the following equation are organized,

$$\zeta \cdot \iota \cdot \xi = 1, \qquad \zeta = \iota, \qquad \zeta^2 = \iota^2 = 1/\xi, \tag{10}$$

where ι is a strain for y-axial direction of sample.

Consequently, the Eq. (9) becomes

$$S = - \frac{1}{2} Nk m^2 \left(\xi^2 + \frac{1}{\xi}\right), \tag{11}$$

then

$$\frac{\partial S}{\partial \xi} = - Nk m^2 \left(\xi - \frac{1}{2\xi^2}\right). \tag{12}$$

While as a result of instantaneous elastic deformation an original length l_0 in z-axial direction becomes $l = \xi l_0$ (Fig. 4). Now, assumed that we have a sample whose volume is V ($=$ constant) and an area corresponds to l_0 is V/l_0. Therefore a stress σ_z due to F is

$$\sigma_z = \frac{F}{V/l_0}, \qquad \text{or} \qquad F = \sigma_z \frac{V}{l_0}. \tag{13}$$

Accordingly, substituting Eq. (13) into Eq. (2);

$$\sigma_z = - \frac{\beta l_0}{V} T \left(\frac{\partial S}{\partial l}\right). \tag{14}$$

And substituting $l = \xi l_0$ into Eq. (14);

$$\sigma_z = - \frac{\beta l_0}{V} T \left[\frac{\partial S}{\partial(\xi l_0)}\right] = - \frac{\beta}{V} T \left(\frac{\partial S}{\partial \xi}\right). \tag{15}$$

Moreover substituting Eq. (12) for Eq. (15);

$$\sigma_z = \frac{\beta m^2 Nk T}{V} \left(\xi - \frac{1}{2\xi^2}\right). \tag{16}$$

In Eq. (16), considering that $1/(2\xi^2)$ is negligibly small in comparison with ξ;

$$\sigma_z = \frac{\beta Nk m^2 T}{V} \xi = E_0 \xi, \tag{17}$$

$$E_0 = \frac{\beta Nk m^2 T}{V}. \tag{18}$$

5. Deduction of Rheology Equation

In order to enlarge and apply the Eqs. (17) and (18) into an irreversible deformation zone, it will be a reasonable method that we obtain a

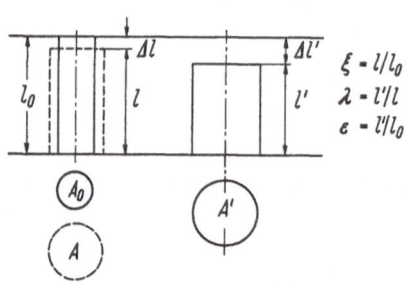

functional relationship between an macro deformation ε as a visible phenomenon and an accumulation λ of micro irreversible deformation which is due to the slipping of clay particles in addition to a conceptional instantaneous elastic deformation ξ, and to replace ξ in the Eq. (17) with ε and λ.

Fig. 4. Relation between instantaneous elastic deformation and plastic deformation of soil.

Namely, in Fig. 4, let define that $\varepsilon = l'/l_0$ and $\xi = l/l_0$, then since λ is represented by $\lambda = l'/l = \varepsilon/\xi$,

$$\ln \varepsilon = \ln \xi + \ln \lambda, \quad \text{or} \quad \bar{\varepsilon} = \bar{\xi} + \bar{\lambda}, \tag{19}$$

where $\ln \lambda$ represent's a so-called rheological deformation of sample. Therefore we define $\ln \lambda$ such as Eq. (20);

$$\ln \lambda = \bar{\lambda} = \varkappa t, \tag{20}$$

where $\varkappa$ is a strain rate of sample under a certain condition.

Eq. (19) consequently becomes,

$$\ln \xi = \ln \varepsilon - \varkappa t. \tag{21}$$

While we can rewrite Eq. (17) as follows,

$$\ln \sigma_z = \ln E_0 + \ln \xi. \tag{22}$$

Substituting Eq. (21), therefore, into Eq. (22);

$$\ln \sigma_z = \ln E_0 + \ln \varepsilon - \varkappa t, \tag{23}$$

i.e.,
$$\sigma_z = E_0 \varepsilon e^{-\varkappa t}. \tag{24}$$

Since E_0 is constant, differentiating Eq. (23) we obtain Eq. (25) and (26)

$$\frac{1}{\sigma_z} \frac{d\sigma_z}{dt} = \frac{1}{\varepsilon} \frac{d\varepsilon}{dt} - \varkappa, \tag{25}$$

$$\frac{d\varepsilon}{dt} = \frac{\varepsilon}{\sigma_z} \frac{d\sigma_z}{dt} + \frac{\varepsilon}{\sigma_z} \varkappa \sigma_z. \tag{26}$$

Hereupon, let us to define that $(\varepsilon/\sigma_z) = 1/G_{(t)} = J_{(t)}$ a "compliance" τ and η are a "relaxation time" and a "coefficient of apperent viscosity" respectively as $(\varepsilon/\sigma_z)\varkappa = (\varepsilon/\sigma_z)/\tau = 1/\eta$, the Eq. (26) is of the same type as Maxwell's equation.

$G_{(t)} = \sigma_z/\varepsilon$ is a variable value with time, but in this case, if we define that $G_{(t)}$ at time $t = 0$ is $G_{(t=0)} = (\sigma_{z(t=0)}/\varepsilon_{(t=0)}) = (\sigma_{z(t=0)}/\xi)$, since σ_z at $t = 0$ is $\sigma_z = \xi E_0$ from the Eq. (17) $G_{(t=0)}$ equals to E_0. Consequently, the Eq. (26) will be written as follows,

$$\frac{d\varepsilon}{dt} = \frac{1}{E_0}\frac{d\sigma_z}{dt} + \frac{1}{\eta}\sigma_z. \tag{27}$$

However, by Eq. (27), it is not yet enough to explain the complicated rheological behaviours of clay. And so, the author has tried to deduce a structural viscosity of the clay by applying EYRING's viscosity theory to the double water layer around the particles.

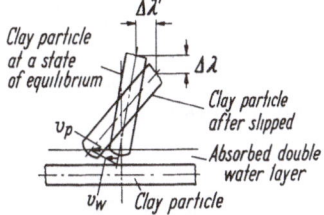

Fig. 5.   Schematic representation of slipping of clay particle at unit mechanism of deformation.

In Fig. 5, let us suppose that the edge of clay particle at the unit mechanism has been slipped for V_p in unit time in the direction of stress component σ_z/N by which was greater than the bonding energy at the unit mechanism, and as a result of the slipping a rate of shear V_w has occurred at any point of double water layer. Then according to EYRING's theory the rate of shear strain V_w is represented by following equation.

$$V_w = 2c\,n_w\varkappa_0 \sin h\left(\frac{c\,c_1 c_2 \sigma_z}{2NkT}\right), \tag{28}$$

where c is a molecular distance of equilibrium position in the direction to the shear strain of double water layer, $c_1 \cdot c_2$ is an effective area which is occupied by a molecule of the double layer, n_w is number of unit mechanism of double layer in series per unit length in the direction of shear strain of double layer, h is PLANCK's constant, $\varkappa_0$ is a constant of strain rate which is equal to $(kT/h)\exp(-U_w/kT)$, and U_w is a potential barrier for transformation of a molecule of double water layer.

Furthermore, in Fig. 5, taking it for granted that there is a proportional corresponding relation between V_w and the micro shear strain $\Delta\lambda$ (or $\Delta\lambda'$) by slipping of clay particle at unit mechanism of deformation Then, since the number of unit mechanism of deformation in series per unit length in direction of σ_z are M, if we define c_3 as a proportional constant between V_w and $\Delta\lambda$, the apperent strain rate $\varkappa$ of sample is represented as follows,

$$\varkappa = 2c\,c_3 n_w M \varkappa_0 \sin h\left(\frac{c\,c_1 c_2 \sigma_z}{2NkT}\right)\bigg/\varepsilon_0,$$

$$= BM\varkappa_0 \sin h\left(D\,\frac{\sigma_z}{N}\right)\bigg/\varepsilon_0, \tag{29}$$

$$B = 2c\,c_3 n_w, \qquad D = \frac{c\,c_1 c_2}{2kT}. \tag{30}$$

In the above definition of $\varkappa$, the author assumed that τ was proportional to the initial strain ε_0.

The coefficient of apperent viscosity η of sample is obtained from the following relation. That is, from 2nd term in the right side of Eqs. (26) and (27),

$$\varkappa = \frac{\ln \lambda}{t} = \frac{1}{\tau} = \frac{G}{\eta},$$

hence

$$\eta = \frac{\sigma_z}{\dfrac{d \ln \lambda}{dt}} = \frac{\sigma_z}{\varkappa \varepsilon_0} = \frac{\sigma_z}{B M \varkappa_0 \sin h \left(D \dfrac{\sigma_z}{N} \right)}. \tag{31}$$

Therefore, substituting Eq. (31) into Eq. (27), the following equation is obtained:

$$\frac{d\varepsilon}{dt} = \frac{1}{E_0} \frac{d\sigma_z}{dt} + B M \varkappa_0 \sin h \left(D \frac{\sigma_z}{N} \right). \tag{32}$$

The Eq. (32) is of the same type as EYRING-TOBOLSKY's molecular rheology equation, but it is differents from EYRING-TOBOLSKY's equation as the 2nd term in right side contain M and N. This is the fundamental rheology equation for the clay obtained by the author.

6. Solution of the Fundamental Rheology Equation and its Application to the Model Analysis

Since the phenomenon of stress relaxation occurs under the condition that the initial strain ε_0 is constant, integrating the Eq. (32) on the condition of $(d\varepsilon/dt) = 0$,

$$\tan h \left(\frac{D}{2N} \sigma_{z(t)} \right) = \tan h \left(\frac{D}{2N} \sigma_{z(t=0)} \right) \exp \left(- \frac{B D M \varkappa_0 E_0}{N} t \right)$$

considering the condition of $\sigma_{z(t=0)} > \sigma_{z(t)} > 1/(2D/N)$ approximately;

$$\sigma_{z(t)} = - \left(\frac{N}{D} \right) \ln \left(\frac{B D M \varkappa_0}{2N} E_0 \right) - \left(\frac{N}{D} \right) \ln t \tag{33}$$

or

$$\sigma_{z(t)} = - C_1 \ln (C_2 E_0) - C_1 \ln t, \tag{33'}$$

$$C_1 = N/D, \quad C_2 = (B D M E_0 \varkappa_0)/2N.$$

The relaxation modulus $G_{(t)} = \sigma_{z(t)}/\varepsilon_0$ is

$$G_{(t)} = - \frac{C_1}{\varepsilon_0} \ln (C_2 E_0) - \frac{C_1}{\varepsilon_0} \ln t, \tag{34}$$

$$= - C_3 \ln (C_2 E_0) - C_3 \ln t. \tag{34'}$$

To apply this theory to the model (Fig. 6) proposed by S. MU-RAYAMA and T. SHIBATA [3], let us correspond to the restraining resistance σ_0 proposed by them the resistance of bonding energy to the slipping of clay particles at the unit mechanism of deformation in a unit area of cross section perpendicular to the direction of σ_z. Further considering an activation of unit mechanism, the number of unit mechanism M and N are as follows,

$$M = a(\sigma_z - \sigma_0),$$
$$N = \dot{b}(\sigma_z - \sigma_0).$$

(35)

And, establishing the following relations, the model analysis is possible.

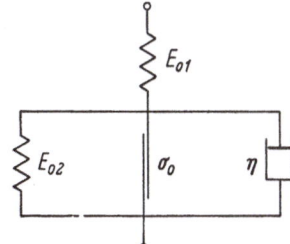

Fig. 6. Rheological model for clay by S. MURAYAMA, 1956.

$$\left.\begin{aligned}\ln \varepsilon &= \ln \xi + \ln \lambda, \\ \bar{\varepsilon} &= \bar{\xi} + \bar{\lambda},\end{aligned}\right\}$$

(36)

$$\sigma_z = \bar{\xi} E_{01},$$

(37)

$$\sigma_z = \bar{\lambda} E_{02} + \frac{N}{D} \sin h^{-1}\left(\frac{1}{A_1 M}\frac{d\bar{\lambda}}{dt}\right) + \sigma_0,$$

(38)

$$A_1 = B\varkappa_0.$$

For the stress relaxation, since

$$\bar{\varepsilon} = \bar{\xi} + \bar{\lambda} = \bar{\varepsilon}_0 = \text{constant},$$

i.e.,

$$\frac{d\bar{\varepsilon}}{dt} = \frac{\cdot 1}{E_{01}}\frac{d\sigma_{z(t)}}{dt} + \frac{d\bar{\lambda}}{dt} = 0.$$

(39)

According to the method that S. MURAYAMA and T. SHIBATA [5] already had been analized, from Eq. (38),

$$\left.\begin{aligned}\frac{d\bar{\lambda}}{dt} &= A_2(\sigma_z - \sigma_0)\sin h\left\{\frac{B_2[(\sigma_z - \sigma_0) - \bar{\lambda}E_{02}]}{(\sigma_z - \sigma_0)}\right\}, \\ A_2 &= A_1 a, \qquad B_2 = D/b.\end{aligned}\right\}$$

(40)

Applying the condition of

$$0 < \bar{\lambda} < \frac{\sigma_z - \sigma_0}{2 B_2 E_{02}}(2 B_2 - 1),$$

(41)

approximately

$$\frac{d\bar{\lambda}}{dt} \doteqdot \frac{A_2}{2} = (\sigma_z - \sigma_0)\exp\left\{\frac{B_2[(\sigma_z - \sigma_0) - \bar{\lambda}E_{02}]}{(\sigma_z - \sigma_0)}\right\}$$

(42)

and since $\bar{\lambda} = \bar{\varepsilon}_0 - \dfrac{\sigma_z}{E_{01}}$,

$$\frac{d\bar{\lambda}}{dt} = \frac{A_2}{2}(\sigma_z - \sigma_0)\exp\left[B_2\left(1 + \frac{E_{02}}{E_{01}}\right) - \frac{B_2 E_{02}}{(\sigma_z - \sigma_0)}\left(\bar{\varepsilon}_0 - \frac{\sigma_0}{E_{01}}\right)\right].$$

(43)

Substituting Eq. (43) into Eq. (39);

$$\frac{d\varepsilon}{dt} = \frac{1}{E_{01}} \frac{d\sigma_z}{dt} + \frac{A_2}{2} (\sigma_z - \sigma_0) \exp\left[B_2\left(1 + \frac{E_{02}}{E_{01}}\right) - \frac{B_2 E_{02}}{(\sigma_z - \sigma_0)}\left(\bar{\varepsilon}_0 - \frac{\sigma_0}{E_0}\right)\right] = 0. \tag{44}$$

To solve Eq. (44), establishing following relations;

$$C = \frac{A_2}{2} E_{01} \exp\left[B_2\left(1 + \frac{E_{02}}{E_{01}}\right)\right], \tag{45}$$

$$f = \frac{B_2 E_{02}}{\sigma_z - \sigma_0}\left(\bar{\varepsilon}_0 - \frac{\sigma_0}{E_{01}}\right) \tag{46}$$

and transform σ_z to f, then;

$$\frac{1}{f} \exp(f) \, df = C \, dt. \tag{47}$$

Integrating Eq. (47) from $f_0 = B_2 \cdot E_{02}/E_{01}$ (f_0 is the value of f at $t = 0$) to f,

$$\int_{f_0}^{f} \frac{1}{f} \exp(f) \, df = Ct. \tag{48}$$

Consequently, if we can obtain the rheology constants A_2, B_2, E_{01}, and E_{02} in the Eq. (48), the relationship between σ_z and t is obtained by numerical analysis.

And σ_z at $t \to \infty$ is

$$\sigma_{z(t \to \infty)} = \frac{E_{01} E_{02}}{E_{01} + E_{02}} \bar{\varepsilon}_0. \tag{49}$$

Addendum

The author in this theoretical investigation has tried that the introduction of thermo-dynamics and statistical mechanics to the system of soil mechanics. As a result of the study the author has deduced a fundamental rheology equation for the clay, which was applicable to the both of a molecular or particular theoretical investigation and a phenomenological model analysis.

There were, however, a few differences in the results of analyses by the both methods, i.e., especially, in the former method the stress at $t = \infty$ was 0, while in the latter method it became

$$(E_{01} E_{02}) \bar{\varepsilon}_0 / (E_{01} + E_{02}).$$

The examination about this point the author will report in the future.

The author wishes to express his many thanks to I. UCHIDA, Prof. of Kyushu Univ., who gave to the author his kind advices and encouragements, and also the author presents the same acknowledgements to M. TAKAYANAGI and M. MIKAMI, Prof. of Kyushu Univ.

References

[1] MOGAMI, T.: An Approach to the System of the Soil Mechanics. J. of the J.S.C.E. **36**, No. 6, 16—20 (1951).

[2] TAN, T. K.: Discussion to the Soil Properties and Their Measurement. Proc., 4th Int. Conf. S.M.F.E., Vol. 3, pp. 87—89, 1958.

[3] MURAYAMA, S., and T. SHIBATA: On the Rheological Characters of Clay. Transaction of the J.S.C.E., No. 40, 1956.

[4] GOLDSTEIN, M. N., V. A. MISUMSKY and L. S. LAPIDUS: The Theory of Probability and Statistics in Relation to the Rheology of Soils. Proc., 5th Int. Conf. S.M.F.E., Vol. 1, pp. 123—126, 1961.

[5] MURAYAMA, S., and T. SHIBATA: On the Stress Relaxation of Clay, Transaction of the J.S.C.E., No. 74, pp. 54—58 (1961).

[6] FUSHIMI, K.: The Theory of Probability and Statistics, (Kawade P. C.) 1948, pp. 211—216.

[7] UCHIDA, I., and H. FUJIMOTO: The Theoretical Research on the Stress Relaxation of Clay. J. of the S.M.S., Japan, "Special Issue on Rheology", **12**, No. 116, 276 (1963).

1.12 Flux Spécifique et Vitesse Effective de la Loi de Darcy — Analyse Critique

Par

S. Irmay et **D. Zaslavsky**

1. Définition d'un Milieu Poreux

Chacun sait ce que c'est qu'un milieu poreux, mais il n'est pas facile de le définir. Une définition possible d'un milieu poreux avec un fluide s'infiltrant à travers lui, requiert que:

(a) Le milieu renferme des substances *hétérogènes*, soit à petite échelle (sable), lorsque chaque portion est une phase homogène; soit à grande échelle (motte de terre), lorsque chaque portion est hétérogène.

(b) Le milieu consiste d'une portion plus ou moins solide ou stable et bien distribuée, appelée *matériau poreux*, que ce soit sous forme de grains meubles (sable) ou à structure cimentée (argile cohérente).

(c) Le milieu est divisible en *sous-systèmes*; ce sont des portions assez larges pour avoir des caractéristiques moyennes statistiquement significatives; et assez petites pour permettre la continuité en moyenne et la différentiation orientée. Si $d =$ le diamètre équivalent des grains, $M =$ surface spécifique entre phases ou portions, $\varDelta x =$ la grandeur d'un sous-système, on a:

$$\varDelta x > 50 \text{ à } 100d, \quad \text{soit} \quad 300 \text{ à } 600/M. \tag{1}$$

$\varDelta x$ doit être tel que les déviations d'une propriété par rapport à la moyenne en deux points quelconques situés à une distance excédant $\varDelta x$, soient statistiquement indépendantes. La grandeur d'un sous-sytème peut varier de quelques mm en limon à des km en une région géographique. Lorsque le système est très irrégulier, on peut le remplacer par un système continu régulier équivalent aux propriétés prises en moyenne dans le sous-système.

(d) Le sous-système doit être observé pendant des *intervalles de temps* $\varDelta t$ assez larges pour aplanir les fluctuations locales temporelles de la vitesse et de la pression du fluide en mouvement; et assez faibles pour permettre la continuité en moyenne et la différentiation par rapport au temps.

Les propriétés prises ainsi en moyenne sur un sous-système et sur un intervalle de temps, sont rapportées au centre de gravité de chaque sous-système. Le mouvement continu de centre à centre produit des propriétés variant d'une façon continue, bien aplanies et différentiables en chaque direction et dans le temps.

2. Mouvement de Particules Fluides

Une *particule fluide* est une portion de fluide composée de nombreuses molécules, mais plus petite qu'un pore. Comme la particule se meut à travers les interstices tortueux, elle peut se scinder et se rescinder, échangeant des molécules avec les particules avoisinantes. La position moyenne d'une particule à l'instant Δt est définie par le centre de gravité de toutes les molécules qui composaient un temps Δt auparavant la particule primaire. Le mouvement d'une particle fluide est le mouvement de son centre de gravité. Le mouvement de nombreuses particules fluides s'appelle *écoulement en masse*.

Ceci nous rappelle le *principe ergodique*, d'après lequel les moyennes dans l'espace et dans le temps sont interchangeables. Si le principe pouvait s'appliquer aux milieux poreux, le mouvement permanent de nombreuses particules, pris en moyenne dans un sous-système et Δt, serait approximativement le même que le mouvement moyen d'une particule pendant des temps prolongés. L'observation montre que la moyenne spatiale dans un sous-système produit des données bien aplanies dans le temps.

3. Flux Spécifique

Considérons une surface de contrôle en forme d'une boîte rectangulaire de volume $\Delta U = \Delta x \Delta y \Delta z = \Delta A \Delta z$, renfermant un sous-système formé d'un milieu poreux saturé de liquide. La face $\Delta A = \Delta y \Delta z$, normale au vecteur-unité 1_x, a une porosité superficielle n et une aire $\Delta A' = n' \Delta A$ inoccupée par les grains solides. Cette aire est partiellement occupée par de l'eau *stagnante* (*ineffective*) qui est stationnaire ou contribue peu à l'écoulement. Le reste est l'aire $\Delta A'_e = n'_e \Delta A$ de *porosité effective superficielle* $n'_e < n'$. Le flux massique total de fluide sortant à travers cette aire en un instant donné, est le produit scalaire: $1_x \cdot \int\limits_{(\Delta A'_e)} \varrho V dA'$, où V est le vecteur vitesse d'une particule fluide en dA'.

S'il n'y avait pas de milieu poreux, le liquide s'écoulant à travers ΔA à une vitesse de vecteur q', donnerait le même flux massique total, si :

$$q' = \int_{(\Delta A_e)} V \, dA' / \Delta A \tag{2}$$

q' est le *flux spécifique superficiel*.

La valeur moyenne de q' en Δx est :

$$q'' = \int_{(\Delta x)} q' \, dx / \Delta x = \int_{(\Delta U_e)} V \, dU' / \Delta U \tag{3}$$

et est égale à la moyenne de V dans tous les pores du sous-système, dont le volume est $\Delta U_e = n_e \Delta U$. Ici n_e est la *porosité effective* du milieu poreux. Par le théorème de la moyenne :

$$q'' = V'' \cdot \Delta U_e / \Delta U = n_e V''. \tag{4}$$

V'' est une vitesse intermédiaire dans le sous-système (boîte).

Comme V, V'', q', q'' fluctuent dans le temps, on prend la valeur moyenne de (4) en un temps Δt :

$$q = \int_{(\Delta t)} q'' \, dt / \Delta t = \int_{(\Delta t)} dt \int_{(\Delta U_e)} V \, dU / \Delta U' \, \Delta t = n_e \int_{(\Delta t)} V'' \, dt / \Delta t = n_e V(m) \tag{5}$$

q est le *vecteur flux spécifique* de la loi de Darcy :

$$q = KJ. \tag{6}$$

$V(m)$ est la *vitesse effective (vraie)* de Dupuit (1863), qui est une certaine vitesse moyenne dans le sens de l'écoulement turbulent moyen de Boussinesq (1877). $J = - \operatorname{grad} \bar{\varphi}$ est le *vecteur gradient hydraulique* pris en moyenne sur ΔU et Δt. La charge piézométrique $\bar{\varphi} = z + \bar{p}/\gamma$ et la pression $\bar{p}$, sont les valeurs de φ et p prises en moyenne sur ΔU et Δt.

q et $V(m)$ se rapportent au centre de gravité du sous-système en milieux stables au repos, non pas à une section plane, comme il est admis parfois.

La distinction entre q et $V(m)$ est essentielle. En écoulement descendant d'un réservoir dont le niveau d'eau baisse de ΔH pendant Δt, le débit spécifique est $q = \Delta H / \Delta t$. En même temps temps le front mouillé avance verticalement en moyenne de $V(m) \Delta t = q \Delta t / n_e = \Delta H / n_e > \Delta H$.

En écoulement compressible, (5) est à remplacer par le *vecteur flux massique spécifique* :

$$q_m = \int_{(\Delta t)} dt \int_{(\Delta U_e)} \varrho \, V \, dU' / \Delta U \, \Delta t. \tag{7}$$

En écoulement non-saturé on obtient des résultats semblables, lorsqu'on remplace la porosité volumique n par la *concentration volumi-*

que c en liquide, et lorsqu'on remplace n_e par la *concentration effective* en liquide $c - (n - n_e)$.

Le milieu poreux avec le liquide s'écoulant à travers lui, en écoulement aplani (pris en moyenne), est remplacé par un continuum fluide imaginaire de vitesse q, dont l'écoulement obéit à la loi de continuité et à la loi macroscopique (6) de DARCY.

Bibliographie

BOUSSINESQ, M. J.: Essai sur la théorie des eaux courantes. En: Acad. Sci. **23,** No. 1, p. 1—666 (1877).

DUPUIT, J.: Etudes théoriques et pratiques sur le mouvement des eaux dans les canaux découverts et à travers les terrains perméables, 2me éd., Paris: Dunod 1863, 364 p.

1.13 A Theoretical Consideration on a Behaviour of Sand

By

Sakuro Murayama

Introduction

Generally, interparticle forces existing in saturated or dry sand mass consist of compressive forces caused by external stress including own weight and frictional forces between particles. Therefore, the character of the interparticle force of sand mass seems to be less complicated than that of clay, but the mechanical behaviour of sand may be rather more complicated than that of clay. Sometimes, sand behaves as if it is even a plasto-visco-elastic matter.

This is a study on the behaviour of sand, especially on the relation between shearing strain and applied stress for the sand at elastic state, as one of the simplest case among the various behaviours. The relation is solved considering the mobilizing mechanism of sand particles.

1. Stress-Strain Relation of Sand at Elastic State

Strictly speaking, elastic state means the state where HOOKE's law is valid, and accordingly stress-strain relation at the loading process coincide with that at unloading process. But in the case of sandy soil, such strict relations as above stated can not be expected. Therefore, in this study, the elastic state of sand means the state where no residual strain is left after applied load is removed.

Such elastic state can be obtained with the sandy specimen in the triaxial apparatus. One of the examples is shown in Fig. 1. This figure shows a relation of axial strain (ε) and deviatoric stress ($\sigma_1 - \sigma_3$) obtained by repetitional loading under drained condition whose maximum deviatoric stress was $4.5\ \mathrm{kg/cm^2}$ but mean principal stress σ_m (in this case, $\sigma_m = 4.0\ \mathrm{kg/cm^2}$) was kept always constant. Where σ_m is $(\sigma_1 + 2\sigma_3)/3$. Each loading curve and unloading one makes a hysteresis loop. As the applied stress is repeated, straight line part in the loading curve becomes longer, and the residual strain at the removal of the load becomes smaller. Finally, at the 23rd repetitional loop in this test, stress-strain loop was fixed at the same path, and sand specimen reached the so-called elastic state. As shown in Fig. 1, every straight line of the loading curve is parallel to each other. The straight line fixed on the final loop is represented by AB. The magnitude of YOUNG's modulus of the

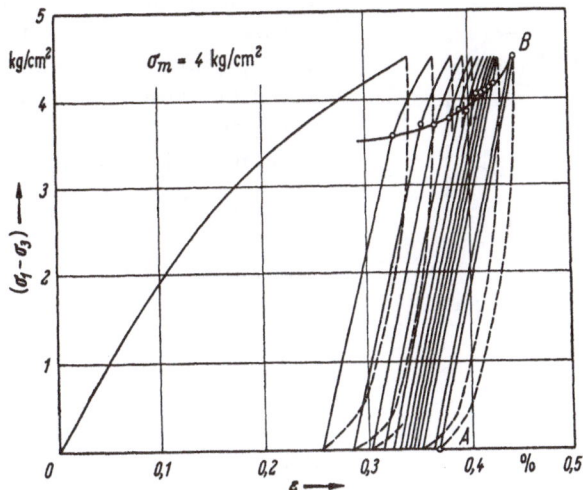

Fig. 1. Relation between axial strain and deviatoric stress under repetitional compression test with constant mean principal stress σ_m.

sandy soil can be obtained from the inclination of the straight line AB. But this value is exceedingly smaller than the modulus of sand grain (i.e. sandy soil: 10^3 kg/cm²-order, sand grain: 10^6 kg/cm²-order).

Therefore, such elasticity of sandy soil should be another elasticity of sand grain. In this study, the elastic property on the shear deformation is explained by the reversible movement of mobilized sand particles against their mutual friction.

2. Forces Acting on Sand Grains

If sand grains in sand mass contact each other at finite contacting area, relation of pore water pressure of saturated sand under undrained condition may be expressed as follows.

$$\sigma = \sigma' + (1 - A_c)\, u,$$

where σ: external stress, σ': effective stress, u: pore water pressure, A_c: mutual contacting area of sand grains per unit area.

But as wellknown, following relation is certified empiricaly with permissible precision:

$$\sigma = \sigma' + u.$$

10*

Therefore, it can be assumed that A_c is negligibly small and sand grains contact each other at points.

· The shearing deformation of the sand under a triaxial compression is caused along the plane where (τ/σ) is maximum as shown in Fig. 2.

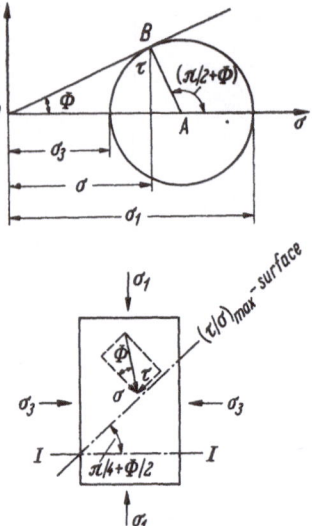

This sliding plane is represented in the figure by a chain line which inclines at $(\pi/4 + \phi/2)$ from the major principal plane. From Mohr's diagram, angle ϕ can be obtained as a function of σ_1 and σ_3 as expressed by Eq. (1):

$$\phi = \tan^{-1}\left(\frac{\tau}{\sigma}\right)_{max} = \sin^{-1}\left(\frac{\sigma_1 - \sigma_3}{\sigma_1 + \sigma_3}\right), \quad (1)$$

where τ: shearing stress, σ: normal stress, ϕ: mobilized internal frictional angle of sand mass which is placed under σ_1 and σ_3, σ_1, σ_3: major and minor principal stress respectively.

Fig. 2. Mobilizing plane where (τ/σ) is maximum under triaxial compression.

But this sliding plane is a common sliding plane along which numerous mobilized sand grains assemble as a group. While individual sand grains situate along this common sliding plane, they mobilize or slide along their individual contacting surfaces of adjacent grains in their own directions. In Fig. 3, sand grains along the common sliding plane are illustrated. If we mark a certain sand grain among them, as shown in the figure, it settles on its adjacent grains and contacts with them at points A and B below the $(\tau/\sigma)_{max}$-plane. In the above side of $(\tau/\sigma)_{max}$-plane, the marked grain is acted by the external force through the contacting points existing above this plane, and resultant of those external forces is expressed by $(f_r)_i$ in the figure. Where $(\ )_i$ means the magnitude for each individual grain.

The intercepting angle between $(f_r)_i$ and the normal to the common sliding plane is denoted by $(\beta)_i$. With incrasing value of τ, angle $(\beta)_i$ becomes larger and accordingly the contact pressure at B becomes smaller, and finally the marked sand grain begins to slide along the surface of adjacent grain touching only at the point A. Let the slope angle of the individual sliding surface against $(\tau/\sigma)_{max}$-plane be $(\theta)_i$, then the sliding condition of the marked grain can be expressed by

$$(\beta)_i - \delta > (\theta)_i \quad \text{or} \quad (\theta)_i - (\beta)_i + \delta = y < 0, \quad (2)$$

where δ is the frictional angle between the grain surfaces and is assumed as a constant for every grain on the average. As the magnitudes of

$\{(\beta)_i - \delta\}$ and $(\theta)_i$ vary at every grain, each may be subjected to a certain frequency distribution function.

Let β and θ be the arithmetic means of $(\beta)_i$ and $(\theta)_i$ respectively, then β and θ can be represented as

$$\beta = \sum_{}^{N} (\beta)_i/N, \qquad \theta = \sum_{}^{N} (\theta)_i/N, \tag{3}$$

where N is the number of grains per unit area of sandy soil.

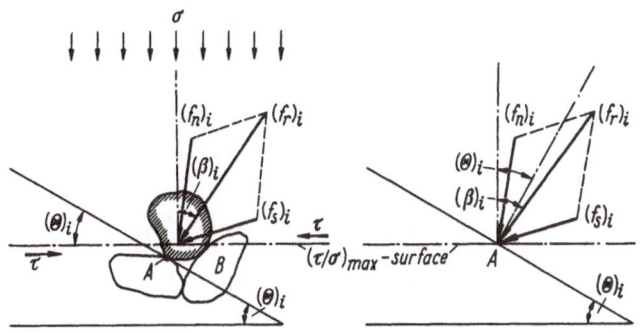

Fig. 3. Forces acting on a sand grain moving along the surface of adjacent grain.

Force $(f_r)_i$ can be analysed into forces $(f_n)_i$ and $(f_s)_i$ as shown in Fig. 3. Where $(f_n)_i$ is the force on the grain caused by the normal stress σ and $(f_s)_i$ by the shearing stress τ. Though the magnitudes and directions of $(f_n)_i$ and $(f_s)_i$ may be random, the average forces f_n and f_s which are expressed by Eqs. (4) and (5) are equal to σ/N and τ/N and directions of f_n and f_s coincide with those of σ and τ respectively.

$$f_n = \sum_{}^{N} (f_n)_i \cdot \cos[(f_n)_i \frown \sigma]/N = \sigma/N, \tag{4}$$

$$f_s = \sum_{}^{N} (f_s)_i \cdot \cos[(f_s)_i \frown \tau]/N = \tau/N. \tag{5}$$

As β can be assumed to be equal to $\tan^{-1}(f_s/f_n)$, β is expressed as

$$\beta = \tan^{-1}\left(\frac{f_s}{f_n}\right) = \tan^{-1}\left(\frac{\tau}{\sigma}\right)_{\max} = \phi. \tag{6}$$

If the applied force is removed after the grain reached its final position, the grain returns from the final position to its initial position as far as the following condition is satisfied.

$$(\beta)_i + \delta < (\theta)_i.$$

Such reciprocating motion of grains may cause the hysteresis loop on the stress-strain diagram at elastic state under repetitional loading.

3. Relation between Shearing Strain and Shearing Stress

The frequency distribution of the angle $\{(\beta)_i - \delta\}$ (rewritten by x_1) and that of the slope angle $(\theta)_i$ (rewritten by x_2) are assumed to be represented by following GAUSSIAN distribution functions whose mean values are m_1, m_2 and standard deviations are ϱ_1, ϱ_2 respectively:

$$f\{(\beta)_i - \delta\} = f(x_1) = \frac{1}{\sqrt{2\pi}\,\varrho_1} \exp\left\{-\frac{(x_1-m_1)^2}{2\varrho_1^2}\right\}$$
$$\equiv N\{(\beta-\delta), \varrho_1^2\} = N(m_1, \varrho_1^2),$$
$$\tag{7}$$

$$\varphi\{(\theta)_i\} = \varphi(x_2) = \frac{1}{\sqrt{2\pi}\,\varrho_2} \exp\left\{-\frac{(x_2-m_2)^2}{2\varrho_2^2}\right\}$$
$$\equiv N(\theta, \varrho_2^2) = N(m_2, \varrho_2^2),$$
$$\tag{8}$$

where

$$x_1 = (\beta)_i - \delta, \qquad x_2 = (\theta)_i,$$
$$m_1 \equiv \beta - \delta, \qquad m_2 \equiv \theta.$$

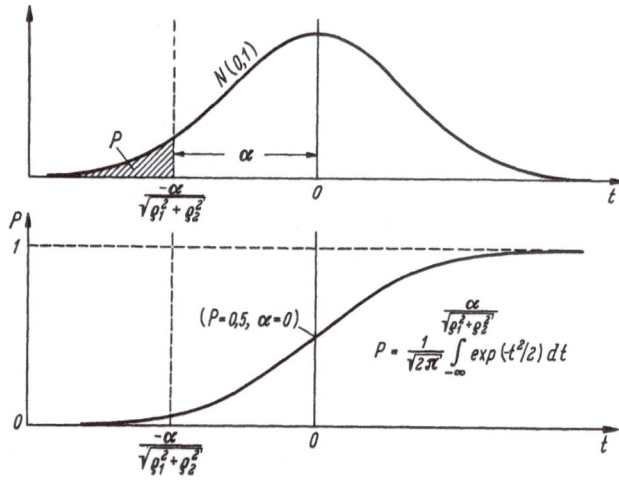

Fig. 4. GAUSSIAN distribution curve (above) and its cumulative distribution curve (below) which is expressed by Eq. (11).

The frequency distribution function of y expressed by Eq. (2) can be obtained through the statistical theorem as follows.

$$p(y) = \frac{1}{\sqrt{2\pi}\,\varrho} \exp\left\{-\frac{(y-\alpha)^2}{2\varrho^2}\right\},$$

where

$$\varrho = \sqrt{\varrho_1^2 + \varrho_2^2}, \qquad \alpha = \theta - \beta + \delta.$$
$$\tag{9}$$

Therefore, the probability of the sliding grain P is obtained by the integration of Eq. (9) within the region of $y < 0$.

$$P = \int_{-\infty}^{0} p(y) \cdot dy. \tag{10}$$

By the substitution of $(y - \alpha)/\varrho = t$, P is transformed as

$$P = \int_{-\infty}^{-(\alpha/\varrho)} \frac{1}{\sqrt{2\pi}} \exp\left(-\frac{t^2}{2}\right) dt \equiv F\left(-\frac{\alpha}{\varrho}\right). \tag{11}$$

Thus obtained P is the cumulative distribution function of GAUSSIAN distribution function whose standard deviation is unity. This GAUSSIAN function and its cumulative one are shown in Fig. 4.

If P_0 expresses the value of P at $\beta = 0$ or the probability of sliding grain under the allround uniform pressure, increment of the probability due to the application of deviatoric stress can be expressed by $(P - P_0)$.

In the sand mass at elastic state, every mobilized grain is assumed to slide a certain finite displacement λ on the average and then to stop there. Moreover, every angular displacement (λ/d) (where d: average interparticle distance of sand grain) is assumed to contribute by $(c\lambda/d)$ (where c: const) to the overall shearing strain along the common sliding plane γ_β (Fig. 5). Therefore, γ_β is given by

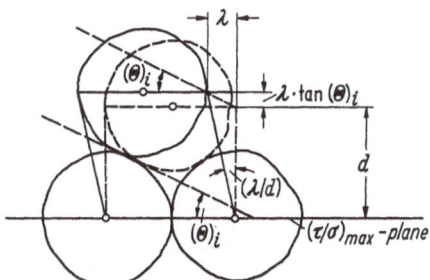

Fig. 5. Relative position of a grain before and after displacement of grain.

$$\gamma_\beta = c\lambda(P - P_0)N/d = A(P - P_0),$$

where

$$P - P_0 = F\left(-\frac{\alpha}{\varrho}\right) - F\left(-\frac{\alpha_0}{\varrho}\right), \qquad A = c\lambda N/d, \tag{12}$$

$$\alpha = \theta - \beta + \delta, \qquad \alpha_0 = \theta + \delta.$$

Using the relation of statics, from γ_β, the maximum shearing strain γ can be obtained as follows:

$$\gamma = \gamma_\beta/\cos\beta = A(P - P_0)/\cos\beta. \tag{13}$$

4. Approximate Solution of Stress-Strain Relation

In order to approximate the cumulative curve expressed by Eq. (11), a sinusoidal curve shown in Fig. 6 and expressed by following equation is adopted in this case.

$$P \equiv F\left(-\frac{\alpha}{\varrho}\right) \doteq \frac{1}{2} + \frac{1}{2}\sin\left(-\frac{\alpha}{2}\right) \equiv P_a. \tag{14}$$

Hence,

$$P - P_0 = \frac{1}{2} \left\{ \sin\left(-\frac{\alpha}{2}\right) - \sin\left(-\frac{\alpha_0}{2}\right) \right\},$$

$$= \frac{1}{2} \left\{ \sin\frac{\beta}{2} \cdot \cos\frac{\theta + \delta}{2} + 2\sin\frac{\theta + \delta}{2} \cdot \sin^2\frac{\beta}{4} \right\}.$$

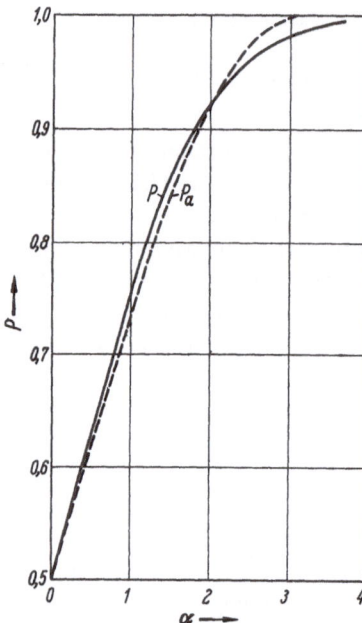

Fig. 6. Proposed sinusoidal curve to approximate the cumulative distribution curve.

Neglecting the 2nd term of the right hand side because of its relatively small value of $\sin^2(\beta/4)$, we get

$$P - P_0 = \frac{1}{2} \cos\frac{\theta + \delta}{2} \cdot \sin\frac{\beta}{2}. \quad (15)$$

Substituting Eq. (15) into Eq. (13), we obtain

$$\gamma = A(P - P_0)/\cos\beta$$

$$= \frac{A}{2} \cdot \cos\frac{\theta + \delta}{2} \cdot \sin\frac{\beta}{2} \Big/ \cos\beta$$

$$= \frac{A}{4} \cdot \cos\frac{\theta + \delta}{2} \cdot \tan\beta \Big/ \cos\frac{\beta}{2}.$$

If angle β is small, $\cos(\beta/2)$ is approximately equal to unity. Hence,

$$\gamma \doteqdot \frac{A}{4} \cdot \cos\frac{\theta + \delta}{2} \cdot \tan\beta. \quad (16)$$

On the other hand, the normal stress on the octahedral plane (σ_{oct}) and the shearing stress on it (τ_{oct}) are represented as

$$\left. \begin{array}{l} \sigma_{oct} = \sigma_m = (\sigma_1 + \sigma_2 + \sigma_3)/3, \\ \tau_{oct} = \sqrt{(\sigma_1 - \sigma_2)^2 + (\sigma_2 - \sigma_3)^2 + (\sigma_3 - \sigma_1)^2}/3, \end{array} \right\} \quad (17)$$

where σ_1, σ_2 and σ_3: principal stresses, σ_m: mean principal stress.

As $\sigma_2 = \sigma_3$ in the triaxial test, σ_m and τ_{oct} become as follows.

$$\left. \begin{array}{l} \sigma_m = (\sigma_1 + 2\sigma_3)/3 = (1 + 2K)\sigma_1/3, \\ \tau_{oct} = \sqrt{2}(\sigma_1 - \sigma_3)/3 = \sqrt{2}(1 - K)\sigma_1/3, \end{array} \right\} \quad (18)$$

where K: ratio of σ_3/σ_1.

Besides, from Eq. (6) $\tan\beta$ can be calculated as

$$\tan\beta = \tan\phi = \frac{\sin\phi}{\sqrt{1 - \sin^2\phi}} = \frac{1 - K}{2\sqrt{K}}. \quad (19)$$

Substituting Eq. (19) into Eq. (18), ratio (τ_{oct}/σ_m) is obtained as

$$\frac{\tau_{oct}}{\sigma_m} = \frac{\sqrt{2}(1 - K)}{1 + 2K} = \frac{2\sqrt{2}\sqrt{K}}{1 + 2K} \tan\beta. \quad (20)$$

As far as K lies between 1 and 0.22 (i.e. $1.0 > K > 0.22$), numerical value of $2\sqrt{2}\sqrt{K}/(1+2K)$ is almost equal to unity within the error of 9%. Therefore above equation may be written as

$$\frac{\tau_{oct}}{\sigma_m} \doteqdot \tan \beta. \tag{21}$$

Substituting Eq. (21) into Eq. (16), we obtain

$$\gamma = \frac{A}{4} \cdot \cos \frac{\theta + \delta}{2} \cdot \frac{\tau_{oct}}{\sigma_m}. \tag{22}$$

According to Eq. (22), maximum shearing strain increases proportionally with (τ_{oct}/σ_m). Therefore, the shear modulus of sandy soil seems to be in direct proportion to σ_m. As θ is the slope angle of sliding surface and δ is the frictional coefficient of the grain surface, $(\theta + \delta)$ represents the internal resistance against shearing deformation of sand. Since the stronger the sand grain interlocks, the larger the angle θ becomes, θ increases as the sand is more compacted. As value of $\cos\{(\theta + \delta)/2\}$ decreases with increasing of $(\theta + \delta)$, shearing strain also decreases with increasing of internal resistance of $(\theta + \delta)$.

5. Effect of Shearing Strain on Dilatancy

When a sand grain moves along the adjacent grain surface, it deviates from $(\tau/\sigma)_{max}$-plane by angle $(\theta)_i$ as shown in Fig. 5. Therefore, the interparticle distance is expanded at the rate of $\lambda \cdot \tan(\theta)_i/d$ for one sliding grain in the direction perpendicular to $(\tau/\sigma)_{max}$-plane, and this elementary expansion is assumed to contribute an increment $c(\lambda \cdot \tan(\theta)_i/d)$ to the overall expansive strain of sandy soil perpendicular to $(\tau/\sigma)_{max}$-plane ε_n. If c is assumed as the same contribution factor with that for γ_β, ε_n can be written by

$$\varepsilon_n = (c\lambda/d) \sum_{NP_0}^{NP} \tan(\theta)_i = \{c\lambda N(P - P_0)/d\} \cdot \tan\theta_e,$$

$$\varepsilon_n = \gamma_\beta \cdot \tan\theta_e, \tag{23}$$

where

$$\left.\begin{array}{l} \tan\theta_e = (P \cdot \tan\theta_m - P_0\tan\theta_{m0})/(P - P_0), \\[2mm] \tan\theta_m = \sum_{1}^{NP} \tan(\theta)_i/(NP), \quad \tan\theta_{m0} = \sum_{1}^{NP_0} \tan(\theta)_i/(NP_0). \end{array}\right\} \tag{24}$$

If value of $(\theta)_i$ is small, $\tan\theta_m$ is obtained approximately as the tangent of the mean value of angle $(\theta)_i$ as follows.

$$\theta_m = \frac{1}{P} \int_{-\infty}^{0} (\theta)_i \cdot p(y) \cdot dy. \tag{25}$$

As for Eq. (25), this can be solved as follows. Now we put as

$$t_1 = \frac{(x_1 - m_1)}{\varrho_1}, \qquad t_2 = \frac{(x_2 - m_2)}{\varrho_2}. \tag{26}$$

Substituting Eq. (26) into Eqs. (2), (7) and (8), they become as

$$t_2 = \left(\frac{\varrho_1}{\varrho_2}\right) t_1 - \frac{\alpha - y}{\varrho_2}, \tag{27}$$

$$f(x_1) \cdot dx_1 = \frac{1}{\sqrt{2\pi}} \cdot \exp\left(-\frac{t_1^2}{2}\right) dt_1 \equiv z_1, \tag{28}$$

$$\varphi(x_2) \cdot dx_2 = \frac{1}{\sqrt{2\pi}} \cdot \exp\left(-\frac{t_2^2}{2}\right) dt_2 \equiv z_2. \tag{29}$$

Taking rectangular co-ordinate axes of t_1, t_2 and z, Eqs. (28) and (29) are plotted in reference to these 3 axes. Since $z_1 = z_2$ for each equation, z- surface or the probability surface becomes axially symmetrical. Therefore, any equivalent probability line projected on $(t_1 \cdot t_2)$-plane becomes a circle with the center at the origin as shown in Fig. 7. The intersection line of the probability surface and the vertical plane containing the line expressed by Eq. (27) gives a symmetrical curve having a maximum point. The co-ordinates of the projected point on $(t_1 \cdot t_2)$-plane of this maximum point are denoted by (t_{10}, t_{20}). Since the projected

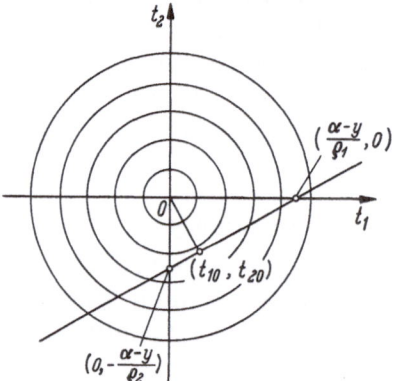

Fig. 7. Equivalent lines of probability surface projected on the plane of $z = 0$.

point (t_{10}, t_{20}) lies on the straight line which is perpendicular to the line expressed by Eq. (27) and passing through the origin, the equation of this straight line through the origin is expressed by

$$t_2 = - (\varrho_2/\varrho_1) \cdot t_1 \tag{30}$$

Values of t_{10}, t_{20} can be obtained as the co-ordinates of the intersection point of the lines expressed by Eqs. (27) and (30).

$$\left.\begin{array}{ll} t_{10} = \varrho_1(\alpha - y)/\varrho^2, & t_{20} = -\varrho_2(\alpha - y)/\varrho^2, \\[2mm] \alpha = m_2 - m_1 = \theta - \beta + \delta, & \varrho^2 = \varrho_1^2 + \varrho_2^2. \end{array}\right\} \tag{31}$$

where

The value of t_{10} (or t_{20}) gives the mode (the most probable value) of t_1 (or t_2) which satisfies the condition expressed by Eq. (27) under a certain value of y.

Let x_{10} and x_{20} be the modes or the average values of x_1-group and x_2-group which are in relation of $x_1 - x_2 = y$ and correspond to t_{10}, t_{20}, then they can be obtained by the substitution of Eq. (31) into Eq. (26) as follows.

$$
\left.\begin{aligned}
x_{10} &= (m_1\varrho_2^2 + m_2\varrho_1^2 - y\varrho_1^2)/(\varrho_1^2 + \varrho_2^2), \\
x_{20} &= (m_1\varrho_2^2 + m_2\varrho_1^2 + y\varrho_2^2)/(\varrho_1^2 + \varrho_2^2).
\end{aligned}\right\} \tag{32}
$$

As these relation are valid for any value of y, average values of all x_1-group and x_2-group which satisfy the condition of $y < 0$ and denoted by $x_{1m} = (\beta_m - \delta)$ and $x_{2m} = \theta_m$ can be obtained by replacing y in Eq. (32) by y_m as follows.

$$
\left.\begin{aligned}
\beta_m - \delta &= (m_1\varrho_2^2 + m_2\varrho_1^2 - y_m\varrho_1^2)/(\varrho_1^2 + \varrho_2^2), \\
\theta_m &= (m_1\varrho_2^2 + m_2\varrho_1^2 + y_m\varrho_2^2)/(\varrho_1^2 + \varrho_2^2),
\end{aligned}\right\} \tag{33}
$$

where y_m is the mean value of y, and y_m can be calculated by following integral:

$$
y_m = \frac{1}{P} \int_{-\infty}^{0} y \cdot p(y) \cdot dy. \tag{34}
$$

Substituting t into Eq. (34), it can be solved as follows.

$$
\left.\begin{aligned}
y_m &= \alpha - \varrho \cdot g\left(-\frac{\alpha}{\varrho}\right)\Big/F\left(-\frac{\alpha}{\varrho}\right), \\
g\left(-\frac{\alpha}{\varrho}\right) &= \left[\frac{1}{\sqrt{2\pi}}\exp\left(-\frac{t^2}{2}\right)\right]_{t=(-\alpha/\varrho)}.
\end{aligned}\right\} \tag{35}
$$

where

6. Triaxial Compression Test under Constant Mean Principal Stress

In order to find the deformation character of sand, drained triaxial compression test on saturated sand of almost same void ratio were performed. Used sand was standard sand for cement test whose particle size was distributed between 0.075 and 0.4 mm in diameter and its uniformity coefficient was 1.75. Every specimen was uniformly compacted denser than its critical void ratio. The initial void ratio of each specimen e which is obtained just after the loading of σ_m is listed in Fig. 9. The applied stress was controlled so as to keep the mean principal stress σ_m always constant while the deviatoric stress $(\sigma_1 - \sigma_2)$ was increased at a constant rate.

Fig. 8 is the obtained relation between applied deviatoric stress and volumetric strain $(\Delta V/V)$ accompanied with shearing deformation due to dilatancy. As shown in Fig. 8 separate curves are obtained for respective constant σ_m.

But if we rewrite these figures according to the co-ordinate axis of $(\sigma_1 - \sigma_3)/\sigma_m$ instead of $(\sigma_1 - \sigma_3)$ as illustrated in Fig. 9, all points rewritten from Fig. 8 lie on a single curve. Therefore, volumetric strain $\Delta V/V$ due to dilatancy may be expressed by following relation:

$$\frac{\Delta V}{V} = D\frac{\sigma_1 - \sigma_3}{\sigma_m},$$

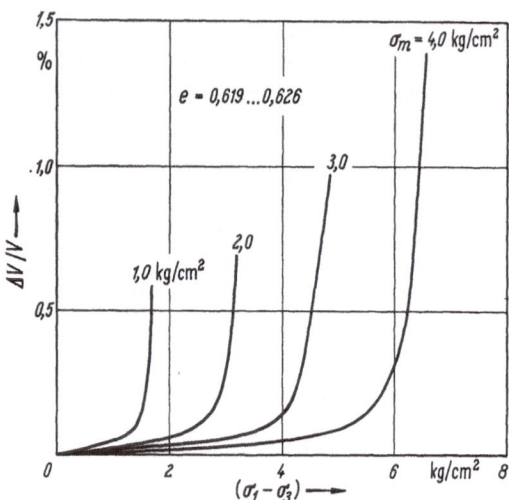

Fig. 8. Volumetric strain due to deviatoric stress.

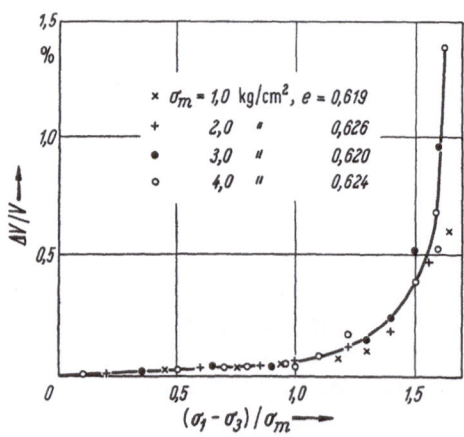

Fig. 9. Relation between volumetric strain owing to dilatancy and stress ratio $(\sigma_1 - \sigma_3)/\sigma_m$.

where D is a coefficient dependent on dilatancy and initial void ratio. Furthermore, D seems to be a constant within a certain limit of $(\sigma_1 - \sigma_3)/\sigma_m$, but beyond this limit $.D$ increases acceleratively with $(\sigma_1 - \sigma_3)/\sigma_m$.

The relation between volumetric strain and shearing strain γ obtained on these tests are plotted as a straight line on a logarithmic paper as shown in Fig. 10. Since the inclination of the straight line is almost equal to 45°, the relation of $\Delta V/V$ and γ can be represented as follows.

$$\gamma = a(\Delta V/V)^b \qquad (a: \text{const}, \quad b \doteqdot 1).$$

According to the above experimentally obtained results, following equation may be obtained:

$$\gamma = a \cdot D \frac{\sigma_1 - \sigma_3}{\sigma_m}.$$

Within the region where D remains constant, above relation well coincides with theoretically obtained relation of Eq. (22).

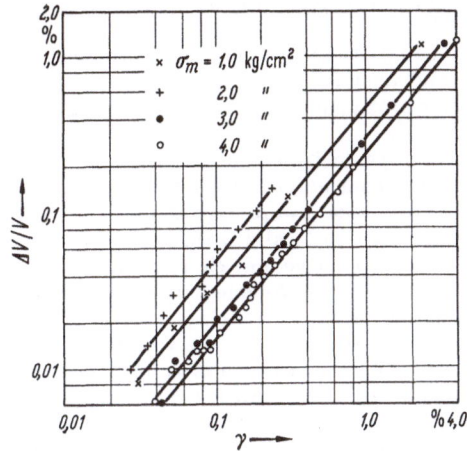

Fig. 10. Relation between volumetric strain owing to dilatancy and shearing strain γ with parameter of σ_m.

Acknowledgements

The author wishes to express his deep appreciation to Dr. M. KADOYA for his usefull suggestion on the statistical study and to Messrs. K. IWAI and N. YAGI for their sincere co-operations in experimental study.

Discussion

Question posée par R. V. WHITMAN: I wish to comment upon the physical mechanisms causing the effect of strain-rate upon the ultimate shear resistance of clays and sands during undrained shear.

The ultimate shear resistance τ_f of a soil is given by the failure law:

$$\tau_f = c + \sigma'_f \tan \phi = c + (\sigma_f - u_f) \tan \phi,$$

where σ_f = total normal stress at failure,
u_f = pore pressure at failure,
σ'_f = effective normal stress at failure,
c = cohesion in terms of effective stress,
ϕ = friction in terms of effective stress.

Thus, if τ_f is a function of the rate of loading, then either the strength parameters c and ϕ or the effective stress σ'_f (or both) must be functions of the rate of loading. The source of the rate effect can thus be determined by measuring the pore pressures generated as soil is strained at different rates under a given σ_f.

In all tests with reliable measurements of pore pressure, it has been found that u_f is a function of strain-rate, and that this variation of u_f (and hence σ'_f) with strain-rate accounts almost fully for the variation of τ_f with strain-rate. That is, the strength paramters c and ϕ are essentially independent of strain-rate. Such measurements have been made in saturated sands with times-to-failure as short as 5 milliseconds (WHITMAN and HEALY, 1962; HEALY, 1962) and in clays with failure times as short as several seconds (HEALY, 1963b; RICHARDSON and WHITMAN, 1964.)

The following arguments explain why σ'_f (and hence u_f) vary with strain-rate during undrained shear.

1. If a system of soil particles is deforming continuously, then particles must be moving around and over adjacent particles. Generally speaking, there is a tendency for expansion of the particulate system in order to minimize geometrical interference.

2. If this particulate system is to deform continuously at some given constant volume, then there must be a normal stress between particles (effective stress) which counteracts the tendency toward expansion. There will, at any instant, be both expansion and contraction locally within different clusters of particles, so that the overall effect is zero net volume change. The magnitude of the effective stress required to maintain a given volume during shear depends upon the degree of geometrical interference between particles.

3. The degree of geometrical interference within a cluster is not a constant, but rather continuously changes with time as the particles shift positions. During slow shear deformation, any one particle moves only when the local particle arrangement permits motion with a minimum of local expansion. During rapid shear deformations, however, a particle may be forced to move at a time when the local particle arrangement requires somewhat greater local expansion.

4. Thus, a greater effective stress will be required to maintain a given overall volume during rapid continuous shear than during slow continuous shear.

The foregoing arguments have been proven by special torsional ring shear tests in which the volume of dry sand during continuous shear has been measured at various rates of shear (HEALY, 1963a).

References

HEALY, K. A. (1962): Triaxial Tests upon Fine Silty Sand. M. I. T. Research Report 63—4 to U. S. Army Engineer Waterways Experiment Station.

HEALY, K. A. (1963a): The Dependence of Dilation in Sand on Rate of Shear Strain. ScD thesis, Department of Civil Engineering, M. I. T.; also M. I. T. Research Report 63—19 to U. S. Army Engineer Waterways Experiment Station.

HEALY, K. A. (1963b): Undrained Strength of Saturated Clayey Silt, M. I. T. Research Report 63—19 to U. S. Army Engineer Waterways Experiment Station.

RICHARDSON, A. M., and R. V. WHITMAN (1964): Effect of Strain-Rate upon Undrained Shear Resistance of a Saturated Remolded Fat Clay. Geotechnique, Vol. 12, pp. 310—324.

WHITMAN, R. V., and K. A. HEALY (1961): Shear Strength of Sands during Rapid Loadings. Proc. American Society of Civil Engineers, Vol. 88, SM 2, pp. 99—132.

Réponse de S. MURAYAMA: The effect of strain rate on the strength of clays may be caused by two reasons. The one is the hydraulic lag in the permeation of porewater and the other is in the character of soil skeleton.

As for a phenomenon which relates to the latter effect, characteristic curves of flow failure under various deviatoric stresses can be illustrated.

If any constant stress exceeding the upper yield value is applied on clay skeleton, the clay can support the stress flowing for a certain duration but it fails finally. If such stress is designated as the long-term strength, the long-term strength does not indicate a definite strength but means various stresses having various failure durations.

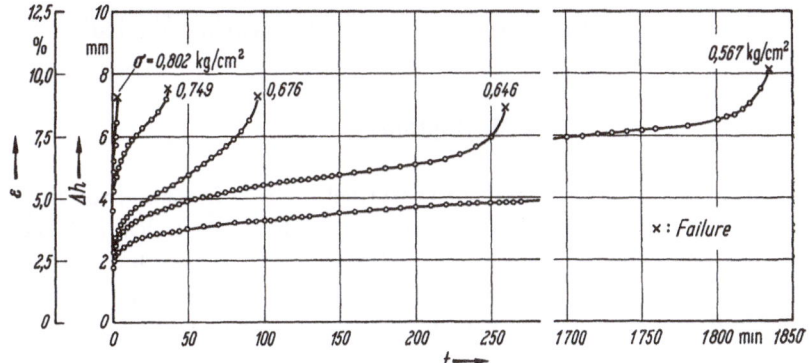

Fig. 1. Characteristic curves of the flow failure under various stresses.

About these characters, I would like to introduce an example of result in flow test performed with undisturbed clay (MURAYAMA and SHIBATA, 1961). Fig. 1 shows the flow curves and those points where failure took place obtained by unconfined compression tests under various constant stress of σ exceeding the upper yield value. These σ are shown as parameters in the figures.

The tested specimens had a cylindrical form with a hight of 8.0 cm. and a diameter of 3.5 cm, and had the lower yield value of $\sigma_0 = 0.025$ kg/cm², the upper yield value of $\sigma_u = 0.45$ kg/cm² and water content of 65% and tests were performed at the temperature of 10 °C.

In this figure, it can be observed that the higher the intensity of the long-term strength or the applied stress, the shorter the elapsed time until the flow failure occurs.

Reference

MURAYAMA, S., and T. SHIBATA (1961): Rheological Properties of Clays. Proc. 5th Int. Conf. S. M. F. E. Vol. 1, p. 272.

1.14 The Role of Friction in Granular Media

By

R. M. Haythornthwaite

It is widely assumed that both interlocking and frictional resistance contribute to the strength of aggregates of loose particles, but there have been few systematic studies. Most that there are have started with the assumption, manifestly untrue, that principal directions of stress and strain necessarily coincide. In this paper, certain regular two-dimensional packings are examined on the basis of a more general theory that discards this assumption, and the results are used to draw tentative conclusions concerning the behavior under plane strain conditions of random aggregates of particles which are macroscopically isotropic.

An analysis is developed for a regular arrangement of uniform cylinders in a partly dilated state. All the cylinders are supposed to be loaded equally when the stress field is uniform, and the loads at the interfaces then become determinate in the case where motion occurs by rolling on two opposite contact points of a given cylinder and sliding on the other two. A consideration of the equilibrium of an individual cylinder leads to the following relation between the principal stresses:

$$\frac{\sigma_2 - \sigma_1}{\sigma_2 + \sigma_1} = \frac{\cos(2\alpha \pm \phi_\mu)}{\cos(2\theta \pm \phi_\mu)}, \tag{1}$$

where $\sigma_1 > \sigma_2$ (compression positive), α is the larger angle subtended between rows of cylinders (a measure of porosity), θ the angle subtended between the principal directions of stress and strain, and ϕ_μ the angle of sliding friction between the cylinders. The positive signs apply when α is decreasing. The expression (1) reduces to that used previously by Rowe (Proc. Roy. Soc. A, 269, 500—527 (1962)) in the particular case $\theta = 0$, i.e. when the principal directions of stress and strain coincide.

The uses of the above result in the construction of a rational theory for random granular media under conditions of plane strain are discussed. One possible model for the random medium is a group of numerous small patches of regular packings at various arbitrary orientations. Neglecting any effects due to irregularity in the packing at the interfaces between the patches, the relationship between the principal stresses can be computed on the basis that all orientations of the packing are equally likely. This is equivalent to assuming that the local directions of principal stress in each microscopic patch do not deviate from the macroscopic average directions, although the magnitudes can vary according to the orientation of the patch. The following expression is obtained when $30° \geq \phi_\mu \geq 0$:

$$\frac{\sigma_2 - \sigma_1}{\sigma_2 + \sigma_1} = \frac{1}{\pi}\left[\cos(2\alpha + \phi_\mu)\ln\frac{\tan\left[\dfrac{3\pi}{4} - \alpha - \dfrac{\phi_\mu}{2}\right]}{\tan\left[\dfrac{\pi}{4} + \dfrac{\phi_\mu}{2}\right]}\right.$$
$$\left. - \cos(2\alpha - \phi_\mu)\ln\frac{\tan\left[\dfrac{\pi}{4}A + \alpha - \dfrac{\phi_\mu}{2}\right]}{\tan\left[\dfrac{\pi}{4} - \dfrac{\phi_\mu}{2}\right]} - 2B\phi_\mu\right],$$

(2)[1]

where $A = 3$, $B = 0$ when $\alpha > \dfrac{\pi}{4} + \dfrac{\phi_\mu}{2}$ and $A = 1$, $B = 1$ when $\alpha < \dfrac{\pi}{4} + \dfrac{\phi_\mu}{2}$.

This result can be used to study the stress-dilation characteristics of media with various statistical distributions in the closeness of the packing.

The special case of the random granular mass at the critical void ratio is considered in detail. Continuing deformation occurs at constant volume overall, so that the particles must be rolling over one another, with the local angle α varying continuously over the range of possible values. If the medium does not develop directional properties in the course of the deformation, all values of α will be equally likely and integration of Eq. (2) leads to an estimate of ϕ_{cv}, the angle of internal friction at the critical void ratio. Values of ϕ_{cv} obtained by numerical integration are given in Table 1. These computed values can be represented by the formula

$$\tan\phi_{cv} = \frac{19}{13}\sin\phi_\mu$$

(3)

with a maximum error in ϕ_{cv} of $0.6°$ in the range tabulated.

Table 1

ϕ_μ	0	2	6	10	14	18	22	26	30
ϕ_{cv}	0	3.5	8.7	14.9	19.6	24.1	28.4	32.6	36.5

[1] In subsequent work, it has been found that Eq. (2) is but one of several possible alternative forms which cannot be distinguished without introducing further assumptions. In terms of the approximate formula, Eq. (3), these alternatives give values for the coefficient which lie between the value stated and 22/17. Details will appear elsewhere.

Eq. (3) is most likely to be valid for cases of plane strain because it is based on the analysis of a bed of cylinders which of necessity deforms in plane strain. Nevertheless it might prove to be of use in other cases such as axial symmetry. Experimental evidence in support of this is given below.

The theory for ϕ_{cv} may be compared with previous theories by CAQUOT (Équilibre des massifs pulvérulents à frottement interne, Paris, Gauthier Villars 1934 and 1948) and BISHOP (Géotechnique, **4**, 43 (1954). CAQUOT's theory, based on an assumption that sliding occurs on all the tangent planes of a spherical surface, leads to the formula

$$\tan \phi_{cv} = \frac{\pi}{2} \tan \phi_\mu. \tag{4}$$

BISHOP has presented the following approximate expressions to describe the results of an analysis stated to be based on energy considerations.

For triaxial compression, $\sigma_2 = \sigma_3 < \sigma_1$,

$$\sin \phi_{cv} = \frac{15 \tan \phi_\mu}{10 + 3 \tan \phi_\mu}. \tag{5}$$

For plane strain, $\sigma_2 = (\sigma_1 + \sigma_3)/2$,

$$\sin \phi_{cv} = \frac{3}{2} \tan \phi_\mu. \tag{6}$$

Formulae (4) and (5) agree closely with the present theory when ϕ_μ is less than 10°. At larger values they diverge, the values of ϕ_{cv} at $\phi_\mu = 30°$ being about 17% higher. Formula (6) predicts much higher values of ϕ_{cv} throughout.

A comparison of the theoretical values of ϕ_{cv} (in degrees). with data from triaxial tests described by ROWE (loc. cit.) is given in Table 2.

Table 2

Material	Theories			Tests
	CAQUOT	BISHOP	Table 1	
Medium-fine sand, $\phi_\mu = 26°$	37.5	39.6	32.6	31.5
Glass ballotini, $\phi_\mu = 17°$	25.6	24.8	23.0	25

The tests would seem to favor the new theory slightly, but this scanty data cannot be considered as in any way conclusive.

This work was supported by the Land Locomotion Laboratory, United States Army Ordnance Arsenal, Detroit, under Contract DA-20-018-AMC-0980 T with The University of Michigan.

Discussion

Contribution de P. HABIB: Il est nécessaire de souligner que la mesure de φ_μ est extrêmement délicate en particulier pour les sables naturels et plus particulièrement pour les sables fins. Le frottement physique est en effet défini par le glissement de deux surfaces identiques. Il n'est pas possible dans la plupart des cas de faire frotter un ou trois grains de sable sur un plan identique et en particulier sur un plan du même minéral: le résultat obtenu dépend du polissage du plan et on connaît trop l'influence de l'état de surface sur la valeur du coefficient de frottement pour que cette méthode puisse être employée sans précaution.

En plus un sable naturel est souvent composé de plusieurs minéraux: quartz, felspath, calcaires. Les frottements deux à deux sont peut-être bien définis, la moyenne ne l'est pas.

La recherche de la relation entre φ_μ et φ_{cv} doit donc se faire avec des matériaux artificiels.

Deux méthodes peuvent être employées: on peut utiliser des grains de formes quelconques d'un matériau dont on peut définir l'état de surface avec précision (sable métallique, sable de verre broyé etc...) de façon à faire une mesure du frottement φ_μ entre un plan et des grains avec de bonnes garanties de fidélité. On peut utiliser aussi des grains d'un matériau moins bien connu mais de forme précise. Par exemple des sphères. On peut alors imaginer des dispositifs engendrant le glissement d'un nombre limité de grains et permettant une mesure acceptable du coefficient de frottement physique.

L'utilisation de grains sphériques présente un autre avantage: la détermination de φ_{cv} relativement simple si la granulométrie n'est pas trop uniforme; avec un aggrégat de sphères on constate en effet que la variation de la densité sèche est faible entre les compacités maximale et minimale. La courbe ci dessous a été déterminée avec cette méthode. Le frottement physique φ_μ a été mesuré directement sur les grains; des films lubrifiants divers ont été utilisés pour obtenir les valeurs les plus basses; leur présence introduisait dans certain cas une légère cohésion (quelques dizaines de g/cm²). L'angle φ_{cv} a été déterminé par des essais triaxiaux avec des pressions latérales allant jusqu'à cinq bars. L'indice des vides critique était de l'ordre de $e = 0{,}56$. Les valeurs les plus élevées ont été obtenues à sec

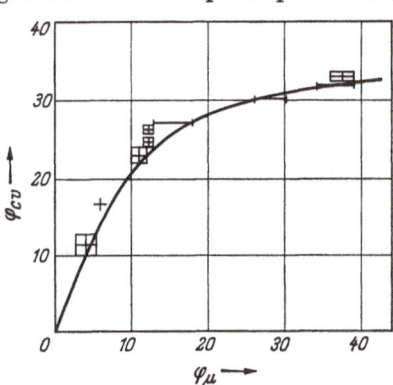

Relation entre l'angle de frottement physique φ_μ et l'angle de frottement interne φ_{cv}.

avec des produits divers comme par exemple des sphères d'alumine pour lesquelles $\varphi_\mu = 38°$.

On constate que la courbe expérimentale est loin d'être rectiligne; il est difficile d'atteindre des valeurs de φ_{cv} supérieures à 32° et ceci montre vraisemblablement l'influence de la rotation des grains telle que M. HAYTHORNTHWAITE l'a précisée dans sa communication.

1.15 Contributions to the Investigations of Granular Systems

Á. Kézdi

1. Introduction

Every investigation on the behavior of granular systems used to start with the examination of ideal packings. The individual elements of the grain assemblies are mostly taken as spheres. This is an idealization of nature, of course, an abstraction, and it is understood that the results of such an investigation are not quite ready for direct application in practice. However, the outcome of the mental work on idealized systems may be a better understanding of nature, the realization of some important facts and the establishment of some, at least qualitative statements. The concept of grain assemblies as macromeritic liquids, introduced by Prof. WINTERKORN, throws a new light on the investigations of packings of equal spheres; the laws of physics, with respect to liquids, can be applied to grain assemblies, thus giving a new line to the research. There are many results available to prove the usefulness of this concept.

Comparing the characteristics of different states of matter, WINTERKORN lists different properties for the different conditions of state

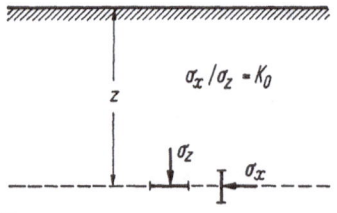

Fig. 1. State of stress in the infinite half space.

including the grain assemblies as a separate state along with the solid, liquid and gaseous states (WINTERKORN, 1953). To the tabulation of the properties we may add that sands and gravels may be listed under the heading "liquids" as well. The variation of the coefficient of lateral pressure K_0 (known as coefficient of earth pressure at rest in Soil Mechanics; see Fig. 1) shows clearly the fields for every condition of state. If a solid body displays a very great cohesion ($c \rightarrow \infty$), then K_0 tends to zero. With decreasing binding forces between the elementary particles (caused by an increase of temperature, by vibration or electric effects) the K_0 value increases and reaches the value $K_0 = 0.4-0.5$, which is characteristic for grain assemblies. Proceeding toward still grater values, we have the viscous liquids, where the internal shearing resistance is much smaller and we arrive at $K_0 = 1$, a value for ideal liquids with zero internal friction. In the case of gases we have $K_0 > 1$, due to the movements of the atomic particles.

The increase of the value K_0 can be achieved by transmitting a certain amount of energy to the system. This may be either a mechanical energy forcing particles out of the solid, or a vibration energy, decreasing the number of contacts between the particles, thus reducing the inner resistance in a transient manner, or a thermal energy which

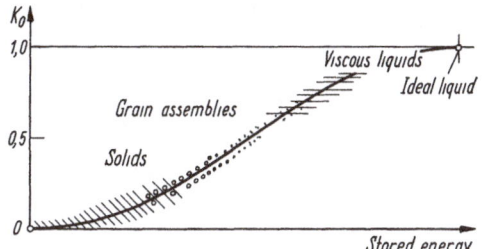

Fig. 2. Relation between conditions of state and the pressure coefficient.

enlarges the distance between the particles and increases their speed, thus turning the solid body to a liquid, the liquid to a gas. By applying a common yardstick for the different kinds of energy, it will be possible to plot the K_0-values as a function of the stored energy in the system thus giving the range for the different conditions of state (Fig. 2).

2. Some Properties of Packings of Uniform Spheres

The first systematic treatment of packings of uniform spheres was given by SLICHTER (1899). He established the different arrangements of the spheres and gave formulas to calculate their density. Before citing some of his results, it is necessary to fix a few assumptions. These are:

(a) The spheres are ideally rigid, undeformable.

(b) The system of spheres is of infinite extent.

(c) There are no adhesive forces between the spheres; frictional forces will be mobilized by movements only.

The smallest volume (V_0), necessary to arrange the units of the packings is given by the well-known hexagonal arrangement, where every sphere has twelve neighbours (Fig. 3). This state may be taken as the solid state of the packing: its shearing resistance equals infinity, because every shearing deformation has to be accompanied by an increase of the volume, this is, however, with respect to the assumption under (b) impossible. If the volume of the packing is greater than V_0, then, with increasing value of $(V - V_0)$, the disorder in the arrangement of the particles will be greater and greater. We may assume, that the free volume, i.e. $V - V_0$, is given by individual pores or "holes", the existence of which does not exclude the existence of a

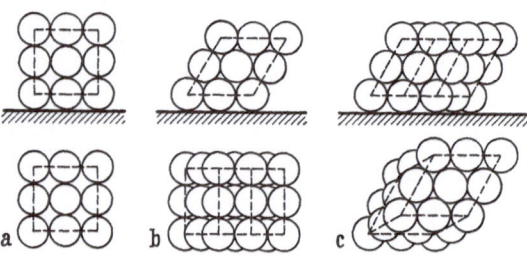

Fig. 3a—c. Regular packings of spheres. a) Cubic; b) orthorhombic; c) rhombohedral packing.

crystal lattice but, on the contrary, implies it. The lattice structure disappears only if the volume of the holes is greater than a certain *"amorphization volume"*. In this case, the system behaves like an ideal fluid, the void ratio in this state represents the melting point of the system. As we can see, the void ratio plays a role comparable to that of the temperature; starting from the densest state we arrive at the melting point with a uniform increase of the volume. In the melting state, the system has a cubic arrangement; the porosity amounts to $n = 0.476$.

Now we transform the cubic system (Fig. 4) to an orthorhombic and to a rhombohedral one. In the first case, every sphere of a given layer glides on a sphere of the lower layer; the direction of the movement being parallel to the straight line connecting the centers of the given row of spheres. To describe the character of variations, it is necessary to consider *eight* spheres. The centers of these spheres form a "unit cell". This unit cell is a cube at the start; it becomes a rhombohedron if we carry out the movement described above. The amount of movement can be given by the variation of the orientation angle ($90° \geq \alpha \geq 60°$; see Fig. 4). At $\alpha = 60°$, we have the orthorhombic system.

The porosity of the system, in terms of α is given by

$$n = 1 - \frac{\pi}{6 \sin \alpha}. \tag{1}$$

The volume of the unit cell varies as

$$V = \sin \alpha.$$

The same relation applies to the height of the unit cell. The variation of V and n is given on Fig. 5. From the definition of n we have

$$n = \frac{V - V_s}{V}$$

where V_s is the volume of solids in the cell, i.e.

$$V_s = V(1-n) = \frac{\pi}{6}.$$

The volume of the spheres in the unit cell remains constant during the movement.

The case described above may be considered as the case of *plane deformation*, the unit cell being deformed in one direction only.

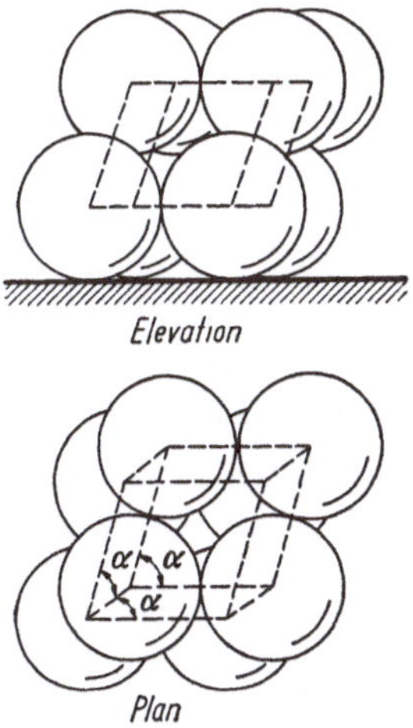

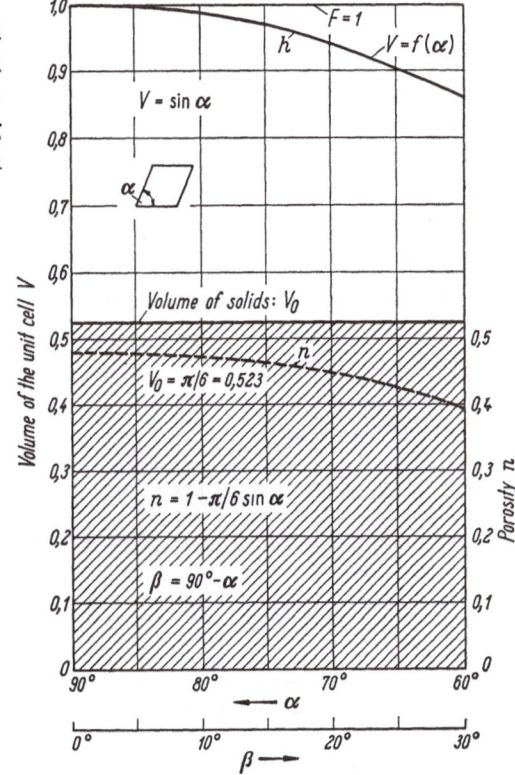

Fig. 4. Rhombohedral packing.

Fig. 5. Variation of V, F and h of the unit cell as functions of the angle of orientation (Plane deformation).

In the second basic case, we have to apply to the system several types of movements. Besides the movement described before, there will be a movement in the direction normal to the first, and the rows of spheres glide along each other. The movement will be carried out uniformly, i.e. the deformation and the compression of the unit cell occurs at a uniform rate; the angle of orientation, that is the angle between two edges of the same side on the unit cell is the same for every two edges. The starting configuration (cubic system) and the "end product" (rhombohedral system) have been given in Fig. 4.

With this assumption in mind, we are able to describe the movement with the help of the angle α; the volume of the unit cell and the porosity can be given as functions of α.

The formula for the porosity has been given by SLICHTER as early as 1889:

$$n = 1 - \frac{\pi}{6(1 - \cos \alpha)\sqrt{1 + 2\cos\alpha}}. \tag{2}$$

The variation of the volume of the unit cell:

$$V = (1 - \cos\alpha)\sqrt{1 + 1\cos\alpha} \tag{3}$$

and the volume of solids

$$V_s = (1 - n)V = \pi/6 = \text{const} \tag{4}$$

as before. The void ratio is given, therefore, by the simple expression

$$e = \frac{n}{1 - n} = \frac{6V}{\pi} - 1. \tag{5}$$

The variation of the height of the unit cell is given by

$$h = D\frac{(1 - \cos\alpha)\sqrt{1 + 2\cos\alpha}}{\sin\alpha}. \tag{6}$$

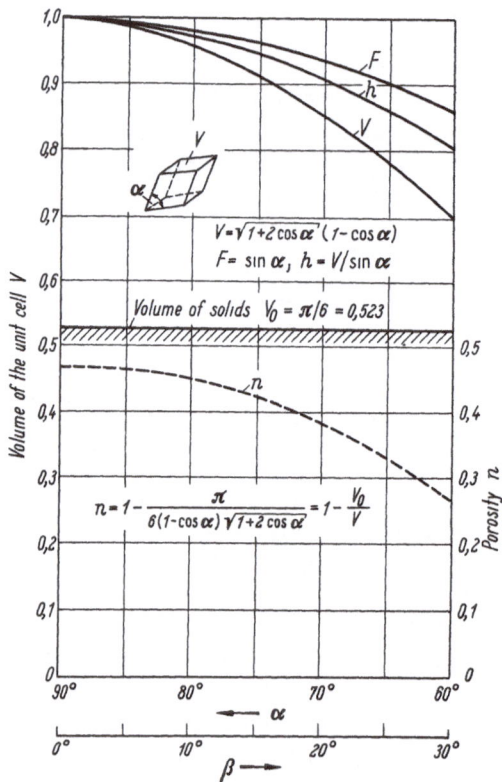

Fig. 6. Variation of V, F and h while forming the rhombohedral packing.

Fig. 6 shows the variation of V, F and h (for $D = 1$) in terms of α.

The term "*coordination number*" gives the number of spheres which are in direct contact with any given shere. This coordination number equals to 6 in the case of the cubic packing and it becomes 12 in the rhombohedral packing. However, it does not vary continuosly during the movement; it takes the final value in the last moment only after performing the described movement. SMITH (1929), with the intention of applying the results derived for uniform spheres to particle systems, assumed that the actual system may for statistical purposes be treated as composed of separate clusters of rhombohedral or cubic arrangements, these being present in such a proportion as to give the observed porosity of the assembly. This consideration leads to the following expression for the average coordination number N in terms of the porosity:

$$N = 26.4858 - \frac{10.7262}{1 - n}. \qquad (7)$$

The curve representing this is shown in Fig. 7, after SMITH. It agrees well with the observed experimental values. Considering now the relationship between the angle of orientation and the porosity, we are able to construct the curve $N = f(\alpha)$, i.e. the relationship between the angle of orientation and the coordination number (Fig. 7b). We get an interesting relation if we plot N as a function of the relative volume of the unit cell. It is a straight line that represents, most likely, the assumption made by SMITH in another form. The equation of the straight line is:

$$N = 6 + 6(2 + \sqrt{2})(1 - V)$$
$$= 6(4.41 - 3.41\,V) \qquad (8)$$

Another possibility to relate coordination number and porosity consists of plotting the values given in Table 1 for different packings constructed afte FILEP (1936). For porosities and the relative volumes of the unit cell, we get the curves given in Fig. 8. The plot consists, of course, of soi-

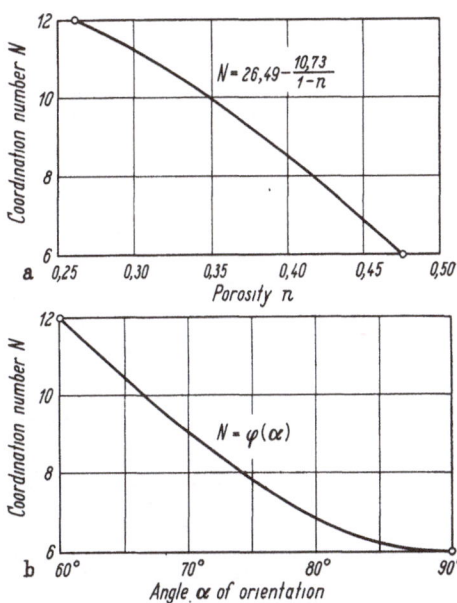

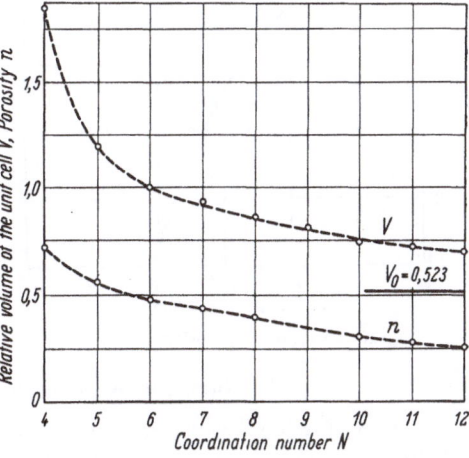

Fig. 7a and b. Coordination number according to the assumption of SMITH.

Fig. 8. Coordination number and relative volume.

lated points, for the integers N; the connecting dotted line is given only to show the trend of variation. It is interesting to show that Fig. 9, giving $N = f(\alpha)$, and constructed on the base of the curves on Fig. 8 does not differ much from the data given on Fig. 7b.

Table 1. *Porosity of Different Regular Packings*

Coordination number	Porosity	Relative volume
4	0.718	1.853
5	0.558	1.195
6	0.476	1.000
7	0.439	0.932
8	0.395	0.864
9	0.352	0.807
10	0.302	0.750
11	0.281	0.728
12	0.259	0.707

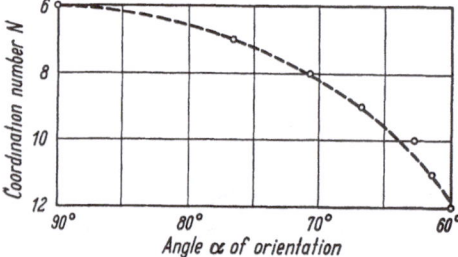

Fig. 9. Coordination numbers for the packings o Table 1.

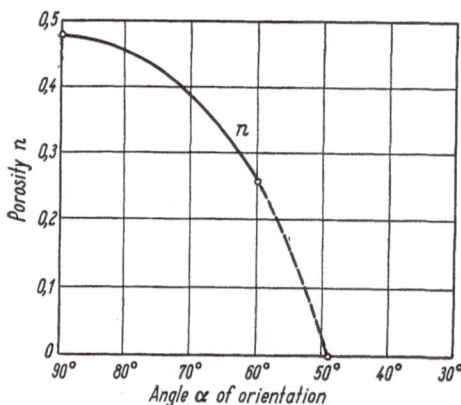

Fig. 10. Variation of porosity with α.

We considered, up to now, the range $90° \geq \alpha \geq 60°$ for the variation of the angle of orientation; $\alpha = 60°$ represents the densest packing, with the coordination number $N = 12$. Extending the interpretation of α beyond 60°, we get decreasing values of n; the spheres intersect each other. The rate of decrease is considerable, because the volumes of the intersecting parts have to be taken twice into consideration. If we arrive at a point, where this volume equals the volume of the remaining voids, we get the theoretical value $n = 0$ for the porosity. This occurs at $\alpha = 49°$; here the volume of the unit cell is equal to $V = 0.523 = \pi/6$. The variation of n with α for the range $90° \geq \alpha \geq 49°$ is given in Fig. 10. The part $60° > \alpha \geq 49°$ of the curve may be used to determine the angle of orientation for packings of non-uniform spheres, having a porosity $n < 25\%$. This angle may be taken as a characteristic of the substitute packing of uniform spheres.

2. Shearing Resistance of Packings

In the following, we try to determine the stresses that are necessary to bring the cubic system into the rhombohedral system. First of all, we can state that the shearing resistance of the densest packing ($N = 12$), *filling the entire space*, will be infinitely great, thus forming a closed system. An increase of the volume cannot take place, even if the shearing stresses increase to infinity. The porosity of a certain arrangement in the infinite space can decrease only as the effect of shearing stresses, any combination of stresses causes a *tendency of densification*.

As displayed by Fig. 5 and 6, the unit cell suffers a compression and a distortion during the movements that make a rhombohedral system ($N = 12$) out of a cubic one ($N = 6$). The decrease of volume can be achieved by the application of a uniform all-around pressure, the decrease of the angle of orientation may be caused by uniform shear (Fig. 11). Let us determine the relation between volume change and shear strain and the respective stresses, making use of the general laws related to liquids.

We assume — as a first approximation — a *linear relationship* between volume change and hydrostatic pressure. The volume in consideration — the volume of the unit cell — is occupied partly by voids and partly by sphere parts; for the present purpose, we assume a substitute liquid filling the unit cell; then, the volume change from $\alpha = 90°$ to $\alpha = 60°$, is given by

$$\frac{V - V_0}{V_0} = \frac{\Delta V}{V_0} = \frac{\sigma}{C}$$

and

$$\sigma = \frac{C}{V_0}(V - V_0) = C_1(V - V_0), \qquad (9)$$

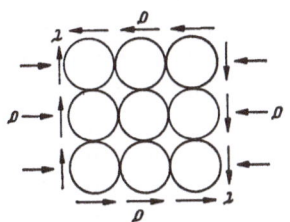

Fig. 11. Packing of spheres loaded by hydrostatic and shearing stresses.

where C is the bulk modulus of compressibility (depending on the surface properties of the spheres) and V_0 is the volume at $\alpha = 60°$ ($V_0 = \pi/6$).

The deformation caused by shearing stresses consists of the change of α, accompanied by the decrease of porosity. We know from the theory of liquids (FRENKEL, 1943) that the shearing stress is given by

$$\tau = -\frac{\partial F}{\partial \beta} \qquad (10)$$

(with $\beta = 90° - \alpha$),

where $F(\beta, T)$ is the free energy of the lattice referred to one particle. It is a function of variation of the height of the unit cell and of the temperature; since we are dealing with isothermic processes, it can be taken as the potential energy of one sphere with respect to the sphere immediately below. It is, then, given by $h - h_0$, where h is the height of the unit cell at α and h_0 is the minimum value at $\alpha = 60°$ $\left(h_0 = D\frac{\sqrt{2}}{3}\right)$.

Assuming again, as a first approximation, a linear relationship for $h = h(\alpha)$, (connecting $h = 1$ for $\alpha = 90°$ and $h = h_0$ for $\alpha = 60°$), we have $h = 1 - c_2\beta$, i.e.

$$\frac{\partial F}{\partial \beta} = \frac{\partial h}{\partial \beta} = c_2,$$

therefore

$$\tau = \text{const} = c_2.$$

The shearing resistance of the medium can be given as

$$\tan \phi = \frac{\tau}{\sigma}; \text{ i.e.,}$$

$$\tan \phi = \frac{c_2}{c_1(V - V_0)} = \frac{c}{V - V_0}, \tag{11}$$

in complete agreement with the formula of WINTERKORN which has been suggested by the analogy between liquids and grain assemblies (BATSCHINSKI, 1913).

Values, calculated on the basis of the concept of the solid and liquid state of macromeritic systems, i.e. using Eq. (12) have been compared with experimental data obtained by various dependable workers on the friction properties of granular materials (FAROUKI, 1963).

It must be emphasized that this formula represents the first approximation. Besides the substitution of the relations $V = f(p)$ and $h = = h(\beta)$ with straight lines — which actually, as it can be seen from the Fig. 5 and 6, may be considered as justified — it neglects an important factor.

This approximation is also involved in BATSCHINSKI's formula: namely, the activation energy for the diffusion of holes in the liquid and in the particle assembly has been disregarded. It should be borne in mind also that the application of the equation to higher pressures can hardly give exact values, since the dependence of the volume on the pressure and of the energy on the volume deviates from a linear law in this region.

Now, we try to arrive at a better approximation in the evaluation of the inner resistance of our macromeritic liquid. We consider the dependence of V on σ, instead of Eq. (9), according to an empirical equation which was proposed long ago by TAIT (FRENKEL, 1943).

It states, that there is a *strain-hardening* during the process of compression: for a given amount of compression we must apply greater all-around stresses if there is already a certain stress of this kind acting. This means that the bulk modulus of compressibility is not a constant, but a function of the all-around pressure itself. If we put

$$C = \frac{dp}{d\varepsilon} = \frac{p + b}{a\,b}, \tag{12}$$

we have for the volume change

$$V - V_0 = a \log\left(\frac{p+b}{b}\right). \tag{13}$$

It can be seen that the equation is of the same form as the equation for the compression of soils proposed by TERZAGHI (1925).

The term $(V - V_0)$ is given as a trigonometric function of β; it is justified, without affecting the accuracy, to replace it with a *parabola* of the second degree. The deviation from the exact value can be made smaller than 1%. Then

$$V - V_0 = C_1\beta^2 \tag{14}$$

and from Eq. (13)

$$p = b\left(e^{\frac{V-V_0}{a}} - 1\right) = b\left(e^{\frac{C_1\beta_1^2}{a}} - 1\right). \tag{15}$$

The formula for the height of the unit cell, giving a measure for the available potential energy [Eq. (6)] will be replaced again by a parabola of the second degree. Then

$$F = 1 - C_2\beta^2 \tag{16}$$

and

$$\tau = -\partial F/\partial\beta = 2C_2\beta. \tag{17}$$

The shearing resitance is given by

$$\tan\phi = \frac{\tau}{p} = \frac{2C_2\beta}{b\left(e^{\frac{C_1\beta^2}{a}} - 1\right)}. \tag{18}$$

According to Eq. (14) we have

$$\beta = \sqrt{\frac{V-V_0}{C_1}} = \sqrt{\frac{\varepsilon - \varepsilon_{\min}}{C_1}}$$

and

$$\tan\phi = C\frac{\sqrt{\varepsilon - \varepsilon_{\min}}}{e^{\frac{\varepsilon - \varepsilon_{\min}}{a}} - 1} \tag{19}$$

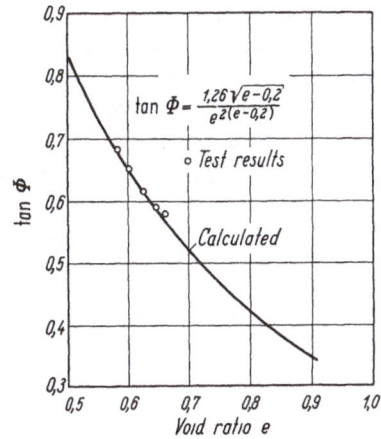

Fig. 12. Comparison of experimental and calculated data.

Fig. 12. gives an example: the evaluation of the measurements of D. W. TAYLOR carried out with Ottawa Standard Sand; we give also the curve according to Eq. (19) using the constants as given on the same figure. The deviations are very small.

We would like to mention that a *third approximation* is also available to evaluate the function $\tan\phi = f(\varepsilon)$. We consider the variation of the volume of the unit cell together with the variation of the *"imaginary"* coordination number, as established in section 2. As we have seen, there is a linear relationship between these terms. Now, we are justified to assume that the shearing stress, necessary to produce

movement, or, in other words, to overcome the friction on the surface
of the spheres, is directly porportional to $(N - 6)$. In the case of the
cubic arrangement of the spheres $(N = 6)$, the shearing resistance is
zero (melting point). Then, according to Eq. (9) we have

$$N - 6 = 6(2 + \sqrt{2})(1 - V)$$

and

$$\tau = C'(N - 6) = C_1[a - b(V - V_0)].$$

Assuming, as in the first approximation

$$\sigma = c_2(V - V_0),$$

we have

$$\tan \phi = \frac{c}{V - V_0} - B. \tag{20}$$

This equation differs from Winterkorn's equation in the additional
term $(-B)$ only. It may account for the "activation energy" neglected
in Batschinski's equation. It would be worthwhile to check also this
equation against experimental data; it is not impossible that cases
that turned out to yield negative values for ε_{min}, which is physically
difficult to visualize, would fit Eq. (20).

The volumetric relationships of particle packings will give place to
further very interesting statements if we consider the shear deformations
too. Thus we can get data on the critical density.

3. Compressibility of Packings

Les us investigate now the compressibility of dispersed systems,
of packing, by making use of analogies taken from the theory of liquids.

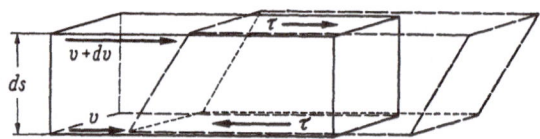

Fig. 13. Definition of viscosity.

If there is a translation in a liquid, then we have a shearing stress
acting ou the sides of an element of the liquid (Fig. 13):

$$\tau = \eta \frac{dv}{ds}, \tag{21}$$

where η is the viscosity of the liquid and τ is the shearing stress causing
a difference dv in the velocities on the upper and lower surface of the
element with the height ds.

Arrhenius and de Guzman, at the beginning of this century,
proposed a formula for the value of η that proved itself dependable

even in the light of modern research. It has the form

$$\eta = A \exp (E/kT), \qquad (22)$$

where

A   = constant,
E   = the activation energy necessary to induce viscous flow in
       the liquid,
T   = absolute temperature,
k   = gas constant,
kT   energy available for the activation of flow.

BATSCHINSKI (1913) derived an expression for the relation ship between the viscosity and the volume of the liquid. According to him:

$$\eta = \frac{B}{V - b}, \qquad (23)$$

where B and b are independent from the temperature and the pressure, respectively $b (= V_0)$ is the smallest possible volume of the liquid. Equating the formulae (22) and (23) we get

$$\boxed{V - V_0 = C e^{-E/kT}}, \qquad (24)$$

where $C = B/A = $ constant. This equation can be regarded as the *equation of state* of molecular *and* macromeritic liquids. We can derive the equation of compression from it.

The activation energy which is needed to start the viscous flow is given by

$$E = E_0 + p \varDelta V, \qquad (25)$$

where $\varDelta V$ is the smallest volume of a hole in the packing. After some shortcuts we get:

$$e = a + c \exp (-bp)$$

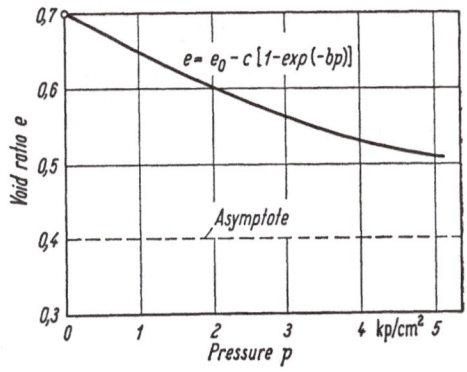

Fig. 14. Compression of soils due to a hydrostatic state of stresses.

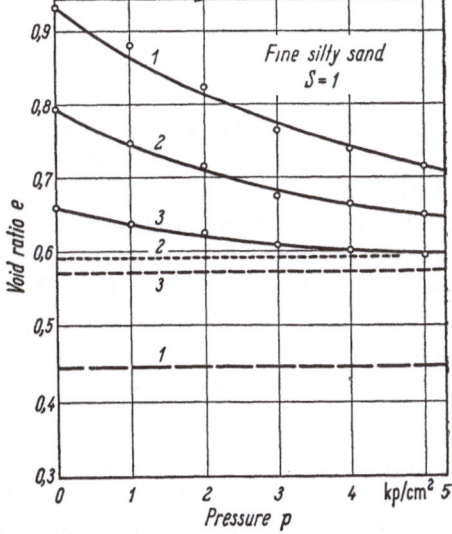

Fig. 15. Some numerical examples for the compression curve.

and with $p = 0$, $e = e_0 = a + c$,

$$\boxed{e = e_0 - C[1 - \exp(-bp)]} \ . \tag{26}$$

Fig. 14. shows the curve calculated by Eq. (2); Fig. 15 compares some test results with the calculated curves.

4. Lateral Pressures of Packings

The study of systematic packings can be extended to the investigation of lateral pressures exerted by them. This investigation furnishes some interesting results that may be of value for the better understanding of earth pressure phenomena (Kézdi, 1962).

We start with the problem of the cubic packing, that, according to Winterkorn's conception, represents the *"melting point"* of the grain assembly. If this is true, then the coefficient of lateral pressures, as

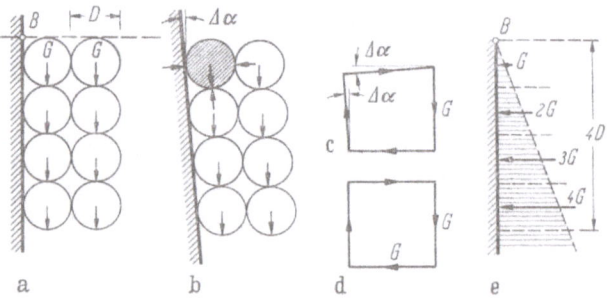

Fig. 16 a—e. Lateral pressure of cubic packings.

interpreted in the theory of earth pressure, has to be *unity* (cf. Fig. 1). This packing, however, gives at the first glance zero value (see Fig. 16); there are no horizontal forces between the spheres. The slightest tilting of the wall **AB** will produce, however, lateral forces; if there is no friction between the spheres, this horizontal force amounts, if $\Delta\alpha \to 0$, in the first row, to $H = W$. In the second row, we have $2W$, and so on, in the n-th row nW. The coefficient of earth pressure is given by

$$K_0 = \frac{2E_0}{h^2\gamma}, \quad \text{or} \quad K_0 = \frac{\sigma_x}{\sigma_z}. \tag{27}$$

This definition assumes a continuous distribution of the forces transmitted through the particles, that is, the number of contacts becomes infinity. To calculate, we must substitute the packing by a continuous mass. This can be accomplished in the following manner.

We have, for the width D of the back of the wall

$$D E_0 = W + 2W + \cdots + nW = \frac{1}{2} n(n+1) W,$$

$$W = \frac{\pi}{6} D^3 \gamma_s,$$

$$E_0 = \frac{1}{2} n(n+1) \frac{\pi}{6} D^2 \gamma_s. \tag{28}$$

This gives a triangular distribution for the horizontal stresses, with an intensity

$$\sigma_x = \frac{2 \Sigma E}{h} = (n+1) \frac{\pi}{6} D \gamma_s, \tag{29}$$

since $h = nD$.

Now, the vertical force exerted by a vertical row of spheres amounts to

$$N = nW = n \frac{\pi}{6} D^3 \gamma_s,$$

giving a uniform pressure

$$\sigma_z = \frac{N}{D^2} = nD \frac{\pi}{6} \gamma_s \tag{30}$$

and the coefficient of the earth pressure at rest:

$$K_0 = \frac{\sigma_x}{\sigma_z} = 1 + \frac{1}{n} \tag{31}$$

and, with $n \to \infty$, $K_0 = 1$.

Of course, it would have been simpler to say that this coefficient is given by

$$K_0 = \frac{H}{N} = \frac{nW}{nW} = 1.$$

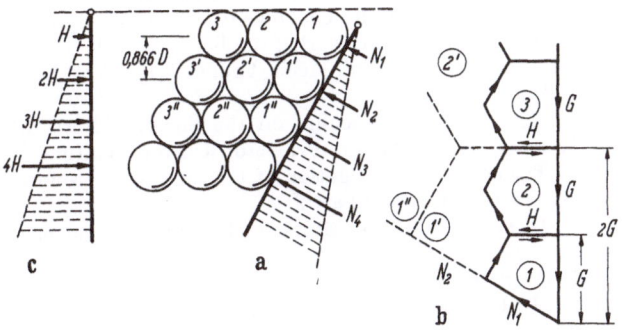

Fig. 17 a—c. Forces in orthorhombic system.

The more complicated way shown here will be useful for the treatment of other cases.

We consider the *orthorhombic* system. Its density is midway between that of the densest and the loosest system and, in fact, is the densest state in the case of a *plane* system. The determination of forces between the spheres and acting on a plane with the angle of $\alpha = 60°$, respectively, is shown in Fig. 17.

With the same procedure as before, we have for the resultant force on $\overline{AB}$:

$$DE_0 = \frac{1}{2} n(n+1) \frac{\pi}{6} D^3 \alpha_s \qquad (32)$$

and

$$E_0 = K_0 \frac{s^2 \gamma_s}{2}.$$

Therefore,

$$K_0 = \left(1 + \frac{1}{n}\right) \frac{\pi}{6} = 0.523 \; (n \to \infty)$$

and, for the vertical plane

$$DE_{0h} = \frac{1}{2} n(n+1) \frac{1}{0.866^2} \cdot H,$$

$$H = W \tan 30°,$$

$$K_{0h} = \left(1 + \frac{1}{n}\right) \frac{1}{0.866^2} \cdot \frac{\pi}{6} \cdot 0.577 = 0.404. \qquad (33)$$

These results are in very good agreement with the measured values for sand of middle density.

References

BATSCHINSKI (1913): Z. phys. Chem. **84**, 643.

FAROUKI, O. T.: Properties of Granular Systems. Department of Civil Engineering, Princeton University, March 1963.

FILEP, L.: Egyenlő gömbökből álló halmazok. Vízügyi Közlemények, Budapest, 1936.

FRENKEL, J.: Kinetic Theory of Liquids, New York: Dover Publications 1955.

KÉZDI, Á.: Erddrucktheorien, Berlin/Göttingen/Heidelberg: Springer 1962.

KÉZDI, Á.: Bodenmechanik, Berlin: Verlag für Bauwesen 1964.

PELTIER, J.: Proc. 4th Int. Conf. Soil Mech. Found. Engn. Vol. III, Discussions, London, 1958.

SJAASTAD, G. D.: The Effect of Vacuum on the Shearing Resistance of Ideal Granular Systems, Ph. D. Thesis, Princeton University, June 1963.

SLICHTER, G. S.: Theoretical Investigation of the motion of ground water. U.S. Geol. Survey, 19th ann. rept., Part 2.

SMITH, W. O., P. D. FOOTE and P. F. BUDANG: Packing of homogeneous spheres. Phys. Rev. **34**, 1271—1274 (1929).

TAYLOR, D. W.: Fundamentals of Soil Mechanics, New York: J. Wiley 1948.

TERZAGHI, K. V.: Erdbaumechanik auf bodenphysikalischer Grundlage, Leipzig u. Wien: Deuticke 1925.

WINTERKORN, H. F.: Macromeritic Liquids. ASTM Symposium on Dynamic Testing of Soils. 2. July 1953.

WINTERKORN, H. F.: Introduction to Engineering Soil Science. Princeton University, 1960. (Mimeographed.)

2.1 Problèmes de l'Équilibre Limite des Sols Non-Homogènes

Par

Z. Sobotka

1. Introduction

L'objet de la présente communication, c'est la théorie génerale de l'équilibre limite et des lignes de glissement des sols non-homogènes.

Partant de différentes conditions d'écoulement qui peuvent être exprimées par les courbes intrinsèques, par les critères énergétiques ou par les surfaces d'écoulement, l'auteur a dérivé les systèmes des équations de l'équilibre limite qui définissent le champ des contraintes et qui peuvent être résolues par la méthode de caractéristiques en différences finies.

Pour le champ des vitesses, l'auteur fait usage de l'équation de compressibilité et de la relation entre les directions principales du tenseur contraintes et du tenseur vitesses de déformation.

Dans les cas spéciaux, le champ des contraintes et le champ des vitesses peuvent être déterminés séparément.

2. Champ des Contraintes

Le champ des contraintes est défini par les équations d'équilibre et par la condition d'écoulement, qui, dans le cas de déformation plane, peut être exprimée sous la forme génerale suivante:

$$\phi(\sigma_x, \sigma_y, \tau_{xy}; x, y) = 0. \tag{2.1}$$

Après la différentiation de cette fonction par rapport à x et à y et après l'élimination des dérivées partielles de la contrainte de cisaillement au moyen des équations d'équilibre

$$\frac{\partial \tau_{yx}}{\partial y} = X - \frac{\partial \sigma_x}{\partial x}, \tag{2.2}$$

$$\frac{\partial \tau_{xy}}{\partial x} = Y - \frac{\partial \sigma_y}{\partial y}, \tag{2.3}$$

l'auteur a obtenu le système des équations de l'équilibre limite en termes des contraintes normales

$$\lambda \frac{\partial \sigma_x}{\partial x} + \eta \frac{\partial \sigma_y}{\partial x} - \chi \frac{\partial \sigma_y}{\partial y} + A = 0, \qquad (2.4)$$

$$\lambda \frac{\partial \sigma_x}{\partial y} + \eta \frac{\partial \sigma_y}{\partial y} - \chi \frac{\partial \sigma_x}{\partial x} + B = 0, \qquad (2.5)$$

où

$$\lambda = \frac{\partial \phi}{\partial \sigma_x}, \qquad \eta = \frac{\partial \phi}{\partial \sigma_y}, \qquad \chi = \frac{\partial \phi}{\partial \tau_{xy}}, \qquad (2.6)$$

$$A = \frac{\partial \phi}{\partial x} + Y \frac{\partial \phi}{\partial \tau_{xy}}, \qquad B = \frac{\partial \phi}{\partial y} + X \frac{\partial \phi}{\partial \tau_{xy}}. \qquad (2.7)$$

Le système des équations quasi-linéaires (2.4) et (2.5) a deux familles de lignes caractéristiques représentant les lignes de glissement, qui sont données par les équations suivantes

$$\omega_{1,2} \equiv \frac{dy}{dx} = \frac{1}{2\eta} \left(-\chi \pm \sqrt{\chi^2 - 4\lambda\eta} \right), \qquad (2.8)$$

$$\xi_{1,2}^2 \frac{d\sigma_x}{dx} - \eta^2 \frac{d\sigma_y}{dx} = A\eta + B\xi_{1,2}, \qquad (2.9)$$

où

$$\xi_{1,2} = \frac{1}{2} \left(\chi \pm \sqrt{\chi^2 - 4\lambda\eta} \right). \qquad (2.10)$$

La condition de réalité des lignes caractéristiques s'ensuit de l'équation (2.8)

$$\chi^2 \geqq 4\lambda\eta. \qquad (2.11)$$

3. Équations de l'Équilibre Limite aux Cas de Divers Critères d'Écoulement

Pour illustrer l'application des équations de l'équilibre limite, dérivées dans la section précédente, nous allons montrer les relations pour le champ des contraintes dans les divers types des matériaux. Les formes typiques des courbes intrinsèques sont représentées sur la fig. 1.

Dans le cas du matériau idéalement plastique, la courbe intrinsèque (fig. 1a) est définie par l'équation suivante:

$$\phi(\sigma_x, \sigma_y, \tau_{xy}; x, y) \equiv (\sigma_x - \sigma_y)^2 + 4\tau_{xy}^2 - 4\tau_T^2(x, y) = 0. \qquad (3.1)$$

Insérant cette relation dans les équations (2.4) et (2.5), nous avons

$$\frac{\partial \sigma_x}{\partial x} - \frac{\partial \sigma_y}{\partial x} - \frac{4\tau_{xy}}{\sigma_x - \sigma_y}\frac{\partial \sigma_y}{\partial y} + \frac{4\left(\tau_{xy}Y - \tau_T\dfrac{\partial \tau_T}{\partial x}\right)}{\sigma_x - \sigma_y} = 0, \qquad (3.2)$$

$$\frac{\partial \sigma_x}{\partial y} - \frac{\partial \sigma_y}{\partial y} - \frac{4\tau_{xy}}{\sigma_x - \sigma_y}\frac{\partial \sigma_x}{\partial x} + \frac{4\left(\tau_{xy}X - \tau_T\dfrac{\partial \tau_T}{\partial y}\right)}{\sigma_x - \sigma_y} = 0. \qquad (3.3)$$

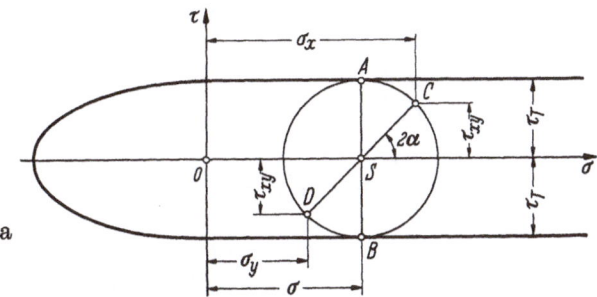

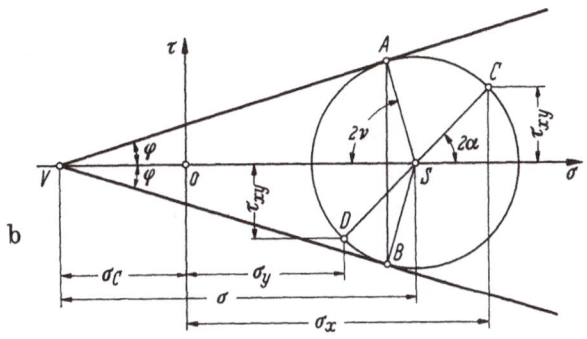

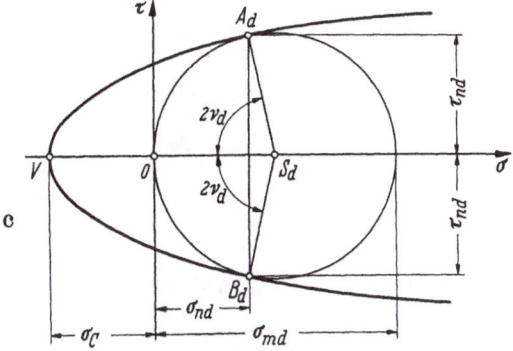

Fig. 1a–c. Courbes intrinsèques pour divers matériaux.

Les lignes caractéristiques sont définies par les équations qui résultent des relations (2.8) et (2.9):

$$\frac{dy}{dx} = \frac{2(\tau_{xy} \mp \tau_T)}{\sigma_x - \sigma_y},$$ (3.4)

$$4\left(\frac{\tau_{xy} \pm \tau_T}{\sigma_x - \sigma_y}\right)^2 \frac{d\sigma_x}{dx} - \frac{d\sigma_y}{dx} = 8\frac{\tau_{xy} \pm \tau_T}{(\sigma_x - \sigma_y)^2}\left(\tau_{xy} X - \tau_T \frac{\partial \tau_T}{\partial y}\right)$$

$$- \frac{4}{\sigma_x - \sigma_y}\left(\tau_{xy} Y - \tau_T \frac{\partial \tau_T}{\partial x}\right) = 0.$$ (3.5)

En tenant compte des relations géométriques sur le cercle de Mohr (fig. 1a), on a aussi:

$$\frac{dy}{dx} = \tan 2\alpha \pm \frac{1}{\cos 2\alpha},$$ (3.6)

d'où

$$\frac{dy}{dx} = \tan\left(\alpha \pm \frac{\pi}{4}\right),$$ (3.7)

α étant l'angle entre la direction principale et l'axe x.

La courbe intrinsèque pour les sols cohésifs — pulvérulents à l'angle constant du frottement interne est représentée sur la fig. 1b. Elle est définie par l'équation suivante

$$\phi(\sigma_x, \sigma_y, \tau_{xy}; x, y) \equiv (\sigma_x - \sigma_y)^2 + 4\tau_{xy}^2$$
$$- [\sigma_x + \sigma_y + 2\sigma_c(x, y)]^2 \sin^2 \varphi(x, y) = 0,$$ (3.8)

φ désignant l'angle du frottement interne,
σ_C la cohésion normale.

Portant l'expression (3.8) dans les équations (2.4) et (2.5), nous avons:

$$[\sigma_x - \sigma_y - (\sigma_x + \sigma_y + 2\sigma_c)\sin^2\varphi]\frac{\partial\sigma_x}{\partial x}$$

$$- [\sigma_x - \sigma_y + (\sigma_x + \sigma_y + 2\sigma_c)\sin^2\varphi]\frac{\partial\sigma_y}{\partial x} - 4\tau_{xy}\frac{\partial\sigma_y}{\partial y}$$

$$+ 4\tau_{xy} Y - 2(\sigma_x + \sigma_y + 2\sigma_c)\sin^2\varphi\frac{\partial\sigma_c}{\partial x}$$

$$- (\sigma_x + \sigma_y + 2\sigma_c)^2 \sin\varphi\cos\varphi\frac{\partial\varphi}{\partial x} = 0,$$ (3.9)

$$[\sigma_x - \sigma_y - (\sigma_x + \sigma_y + 2\sigma_c)\sin^2\varphi]\frac{\partial\sigma_x}{\partial y}$$

$$- [\sigma_x - \sigma_y + (\sigma_x + \sigma_y + 2\sigma_c)\sin^2\varphi]\frac{\partial\sigma_y}{\partial y} - 4\tau_{xy}\frac{\partial\sigma_x}{\partial x}$$

$$+ 4\tau_{xy} X - 2(\sigma_x + \sigma_y + 2\sigma_c)\sin^2\varphi\frac{\partial\sigma_c}{\partial y}$$

$$- (\sigma_x + \sigma_y + 2\sigma_c)^2 \sin\varphi\cos\varphi\frac{\partial\varphi}{\partial y} = 0.$$ (3.10)

D'après les équations (2.8) et (2.9), les lignes caractéristiques sont définies par

$$\frac{dy}{dx} = \frac{2\tau_{xy} \mp (\sigma_x + \sigma_y + 2\sigma_c)\sin\varphi\cos\varphi}{\sigma_x - \sigma_y + (\sigma_x + \sigma_y + 2\sigma_c)\sin^2\varphi}, \qquad (3.11)$$

$$\left[\frac{2\tau_{xy} \pm (\sigma_x + \sigma_y + 2\sigma_c)\sin\varphi\cos\varphi}{\sigma_x - \sigma_y + (\sigma_x + \sigma_y + 2\sigma_c)\sin^2\varphi}\right]^2 \frac{d\sigma_x}{dx} - \frac{d\sigma_y}{dx}$$

$$= \frac{2\tau_{xy} \pm (\sigma_x + \sigma_y + 2\sigma_c)\cos\varphi}{2\left[\sigma_x - \sigma_y + (\sigma_x + \sigma_y + 2\sigma_c)\sin^2\varphi\right]^2}\left[4\tau_{xy}X \right. \qquad (3.12)$$

$$\left. - 2(\sigma_x + \sigma_y + 2\sigma_c)\sin^2\varphi\,\frac{\partial\sigma_c}{\partial y} - (\sigma_x + \sigma_y + 2\sigma_c)^2 \sin\varphi\cos\varphi\,\frac{\partial\varphi}{\partial y}\right]$$

$$- \frac{4\tau_{xy}Y - 2(\sigma_x + \sigma_y + 2\sigma_c)\sin^2\varphi\,\dfrac{\partial\sigma_c}{\partial x} - (\sigma_x + \sigma_y + 2\sigma_c)^2 \sin\varphi\cos\varphi\,\dfrac{\partial\varphi}{\partial x}}{2\left[\sigma_x - \sigma_y + (\sigma_x + \sigma_y + 2\sigma_c)\sin^2\varphi\right]}.$$

En introduisant dans l'équation (3.11) l'égalité

$$\tau_0 = \frac{\tau_{xy}}{\sin 2\alpha} = \frac{\sigma_x - \sigma_y}{2\cos 2\alpha} = \frac{1}{2}(\sigma_x + \sigma_y + 2\sigma_c)\sin\varphi, \qquad (3.13)$$

résultant des relations géométriques représentées sur la fig. 1b, nous avons

$$\frac{dy}{dx} = \tan(\alpha \mp \nu), \qquad (3.14)$$

où

$$\nu = \frac{\pi}{4} - \frac{\varphi}{2}.$$

La courbe intrinsèque pour les matériaux cohésifs, représentée sur la fig. 1c par une parabole du deuxième degré est définie par l'équation suivante

$$\phi(\sigma_x, \sigma_y, \tau_{xy}; x, y) = (\sigma_x - \sigma_y)^2 + 4\tau_{xy}^2 \qquad (3.15)$$

$$- \frac{\sigma_{md}^2(x, y)}{\sigma_{md}(x, y) + 2\sigma_c(x, y)}\left[\sigma_x + \sigma_y + 2\sigma_c(x, y)\right] = 0,$$

où σ_{md} est la contrainte limite en compression simple et σ_c la cohésion idéale représentée aussi par la contrainte limite en traction biaxiale.

Portant l'expression (3.15) dans les équations (2.8) et (2.9), nous obtenons pour les lignes caractéristiques les relations suivantes

$$\frac{dy}{dx} = \frac{2\tau_{xy} \mp \sigma_{md}\sqrt{\dfrac{1}{\sigma_{md} + 2\sigma_c}(\sigma_x + \sigma_y + 2\sigma_c) - \dfrac{1}{4}\left(\dfrac{\sigma_{md}}{\sigma_{md} + 2\sigma_c}\right)^2}}{\sigma_x - \sigma_y + \dfrac{\sigma_{md}^2}{2(\sigma_{md} + 2\sigma_c)}}, \qquad (3.16)$$

$$\left[\frac{2\tau_{xy} \pm \sigma_{md}\sqrt{\dfrac{1}{\sigma_{md}+2\sigma_c}(\sigma_x+\sigma_y+2\sigma_c)-\dfrac{1}{4}\left(\dfrac{\sigma_{md}}{\sigma_{md}+2\sigma_c}\right)^2}}{\sigma_x-\sigma_y+\dfrac{\sigma_{md}^2}{2(\sigma_{md}+2\sigma_c)}}\right]^2 \frac{d\sigma_x}{dx}-\frac{d\sigma_y}{dy}$$

$$=\frac{2\tau_{xy}\pm\sigma_{md}\sqrt{\dfrac{1}{\sigma_{md}+2\sigma_c}(\sigma_x+\sigma_y+2\sigma_c)-\dfrac{1}{4}\left(\dfrac{\sigma_{md}}{\sigma_{md}+2\sigma_c}\right)^2}}{2\left[\sigma_x-\sigma_y+\dfrac{\sigma_{md}^2}{2(\sigma_{md}+2\sigma_c)}\right]^2} \tag{3.17}$$

$$\times\left[4\tau_{xy}X-(\sigma_x+\sigma_y+2\sigma_c)\frac{\partial}{\partial y}\left(\frac{\sigma_{md}^2}{\sigma_{md}+2\sigma_c}\right)\right]$$

$$-\frac{4\tau_{xy}Y-(\sigma_x+\sigma_y+2\sigma_c)\dfrac{\partial}{\partial x}\left(\dfrac{\sigma_{md}^2}{\sigma_{md}+2\sigma_c}\right)-\dfrac{2\sigma_{md}^2}{\sigma_{md}+2\sigma_c}\dfrac{\partial\sigma_c}{\partial x}}{2\left[\sigma_x-\sigma_y+\dfrac{\sigma_{md}^2}{2(\sigma_{md}+2\sigma_c)}\right]}.$$

Le critère général énérgétique pour un matériau non-homogène à l'anisotropie orthogonale qui fait intervenir la contrainte moyenne, est défini par

$$\phi(\sigma_x,\sigma_y,\tau_{xy};x,y)\equiv(\sigma_x-\sigma_y)^2+4K(\sigma_x,\sigma_y;x,y)\tau_{xy}^2-F(\sigma_x,\sigma_y;x,y)=0. \tag{3.18}$$

Puisque le coefficient K et la fonction F dépendent de la contrainte moyenne $\sigma_M=\frac{1}{2}(\sigma_x+\sigma_y)$, nous avons:

$$\frac{\partial K}{\partial\sigma_x}=\frac{\partial K}{\partial\sigma_y},\qquad\frac{\partial F}{\partial\sigma_x}=\frac{\partial F}{\partial\sigma_y}. \tag{3.19}$$

Introduisant équation (3.18) dans les relations (2.8) et (2.9), nous obtenons pour les lignes caractéristiques les équations suivantes:

$$\frac{dy}{dx}=\frac{2K\tau_{xy}\pm\sqrt{(\sigma_x-\sigma_y)^2+4K^2\tau_{xy}^2-\dfrac{1}{4}\left(4\tau_{xy}^2\dfrac{\partial K}{\partial\sigma_x}-\dfrac{\partial F}{\partial\sigma_x}\right)^2}}{\sigma_x-\sigma_y-\dfrac{1}{2}\left(4\tau_{xy}^2\dfrac{\partial K}{\partial\sigma_n}-\dfrac{\partial F}{\partial\sigma_x}\right)},\quad(3.20)$$

$$\left[\frac{2K\tau_{xy}\pm\sqrt{(\sigma_x-\sigma_y)^2+4K^2\tau_{xy}^2-\dfrac{1}{4}\left(4\tau_{xy}^2\dfrac{\partial K}{\partial\sigma_x}-\dfrac{\partial F}{\partial x}\right)^2}}{\sigma_x-\sigma_y-\dfrac{1}{2}\left(4\tau_{xy}^2\dfrac{\partial K}{\partial\sigma_x}-\dfrac{\partial F}{\partial\sigma_x}\right)}\right]^2\frac{d\sigma_x}{dx}-\frac{d\sigma_y}{dy}$$

$$=\frac{2K\tau_{xy}\pm\sqrt{(\sigma_x-\sigma_y)^2+4K^2\tau_{xy}^2-\dfrac{1}{4}\left(4\tau_{xy}^2\dfrac{\partial K}{\partial\sigma_x}-\dfrac{\partial F}{\partial\sigma_x}\right)^2}}{2\left[\sigma_x-\sigma_y-\dfrac{1}{2}\left(4\tau_{xy}^2\dfrac{\partial K}{\partial\sigma_x}-\dfrac{\partial F}{\partial\sigma_x}\right)\right]^2}\left(8\tau_{xy}KX+4\tau_{xy}^2\dfrac{\partial K}{\partial y}-\dfrac{\partial F}{\partial y}\right)$$

$$-\frac{8\tau_{xy}KY+4\tau_{xy}^2\dfrac{\partial K}{\partial x}-\dfrac{\partial F}{\partial x}}{2\left[\sigma_x-\sigma_y-\dfrac{1}{2}\left(4\tau_{xy}^2\dfrac{\partial K}{\partial\sigma_x}-\dfrac{\partial F}{\partial\sigma_x}\right)\right]}. \tag{3.21}$$

Les formules précédentes sont écrites en forme détaillée pour illustrer le procédé général. Pour les calculs pratiques, il vaut mieux faire usage des valeurs λ, η, χ, A et B, qui ont été définies par les équations (2.6) et (2.7), pour rendre les formules plus condensées.

4. Solution du Champ des Contraintes au Cas de la Déformation Plane

Le champ des contraintes peut être déterminé séparément si les conditions à la limite n'introduisent pas les vitesses et s'il n'est pas nécessaire de considérer les vitesses pour choisir entre plusieurs solutions possibles. Il faut que le problème soit statiquement déterminé.

L'auteur présente la solution générale du champ des contraintes par la méthode des lignes caractéristiques en différences finies.

En remplaçant les différentielles dans les équations (2.8) et (2.9) par les différences finies, nous obtenons les formules de récurrence pour les coordonnées des noeuds du réseau des lignes caractéristiques (fig. 2) et pour les composantes des contraintes dans ces noeuds:

$$x_k = \frac{\omega_{1i} x_i - \omega_{2j} x_j - y_i + y_j}{\omega_{1i} - \omega_{2j}}, \tag{4.1}$$

$$y_k = y_i + \omega_{1i}(x_k - x_i) = y_j + \omega_{2j}(x_k - x_j), \tag{4.2}$$

$$\sigma_{xk} = \frac{\eta_j^2[(A_i \eta_i + B_i \xi_{1i})(x_k - x_i) + \xi_{1i}^2 \sigma_{xi} - \eta_i^2 \sigma_{yi}]}{\xi_{1i}^2 \eta_j^2 - \xi_{2j}^2 \eta_i^2}$$
$$- \frac{\eta_i^2[(A_j \eta_j + B_j \xi_{2j})(x_k - x_j) + \xi_{2j}^2 \sigma_{xj} - \eta_j^2 \sigma_{yj}]}{\xi_{1j}^2 \eta_j^2 - \xi_{2j}^2 \eta_i^2}, \tag{4.3}$$

$$\sigma_{yk} = \sigma_{yi} + \frac{1}{\eta_i^2}[\xi_{1i}^2(\sigma_{xk} - \sigma_{xi}) - (A_i \eta_i + B_i \xi_{1i})(x_k - x_i)]$$
$$= \sigma_{yj} + \frac{1}{\eta_j^2}[\xi_{2j}^2(\sigma_{xk} - \sigma_{xj}) - (A_j \eta_j + B_j \xi_{2j})(x_k - x_j)]. \tag{4.4}$$

La contrainte de cisaillement s'ensuit de la condition d'écoulement

$$\phi(\sigma_{xk}, \sigma_{yk}, \tau_{xyk}; x_k, y_k) = 0$$

c'est à dire:

$$\tau_{xyk} = f(\sigma_{xk}, \sigma_{yk}; x_k, y_k). \tag{4.5}$$

Pour rendre le calcul plus précis, on peut faire usage des différences du deuxième ordre. En remplaçant les sections curvilinéaires des lignes caractéristiques par les arcs paraboliques (fig. 2) on obtient le système d'équations

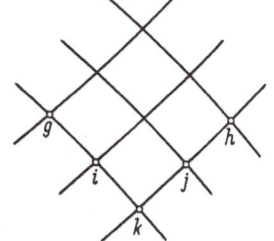

Fig. 2.   Réseau des lignes caractéristiques.

$$y_k - y_i = \omega_{1i}(x_k - x_i) + a_i(x_k - x_i)^2, \tag{4.6}$$

$$y_k - y_j = \omega_{2j}(x_k - x_j) + a_j(x_k - x_j)^2, \tag{4.7}$$

où

$$a_i = \frac{y_g - y_i - \omega_{1i}(x_g - x_i)}{(x_g - x_i)^2}, \tag{4.8}$$

$$a_j = \frac{y_h - y_j - \omega_{2j}(x_h - x_j)}{(x_h - x_j)^2}. \tag{4.9}$$

Pour les coordonnés des noeuds du réseau des lignes caratéristiques, nous avons les formules de réccurence

$$x_k = \frac{1}{2(a_j - a_i)} \Big[2a_j x_j - 2a_i x_i + \omega_{1i} - \omega_{2j}$$
$$\pm \sqrt{(2a_j x_j - 2a_i x_i + \omega_{1i} - \omega_{2j})^2 + 4(a_j - a_i)(y_i - y_j - }$$
$$\sqrt{- \omega_{2j} x_j - \omega_{1i} x_i + c_i x_i^2 - c_j x_j^2)} \Big], \tag{4.10}$$

$$y_k = y_i + \omega_{1i}(x_k - x_i) + a_i(x_k - x_i)^2$$
$$= y_j + \omega_{2j}(x_k - x_j) + a_j(x_k - x_j)^2. \tag{4.11}$$

Le formule (4.10) donne deux valeurs de x_k, dont il faut choisir la valeur ayant un sens physique.

Les formules de récurence pour les contraintes normales peuvent être obtenues d'une manière analogue:

$$\sigma_{xk} = \sigma_{xi} + r_{xi}(x_k - x_i) + s_{xi}(x_k - x_i)^2, \tag{4.12}$$

$$\sigma_{yk} = \sigma_{yi} + r_{yi}(x_k - x_i) + s_{yi}(x_k - x_i)^2, \tag{4.13}$$

où

$$r_{xi} = \frac{(A_i\eta_i + B_i\xi_{1i})\eta_j^2 - \left(A_j\eta_j + B_j\xi_{2j} + \xi_{2j}^2\frac{s_{xk}}{a_j} - \eta_j^2\frac{s_{yk}}{a_j}\right)\eta_i^2}{\xi_{1i}^2\eta_j^2 - \xi_{2j}^2\eta_i^2}, \tag{4.14}$$

$$r_{yi} = \frac{(A_i\eta_i + B_i\xi_{1i})\xi_{2j}^2 - \left(A_j\eta_j + B_j\xi_{2j} + \xi_{2j}^2\frac{s_{xk}}{a_j} - \eta_j^2\frac{s_{yk}}{a_j}\right)\xi_{1i}^2}{\xi_{1i}^2\eta_j^2 - \xi_{2j}^2\eta_i^2}, \tag{4.15}$$

$$s_{xk} = \sigma_{xj} - \sigma_{xi} + \left(\frac{x_k - x_j}{x_h - x_j}\right)^2(\sigma_{xh} - \sigma_{xj}) - \left(\frac{x_k - x_i}{x_g - x_i}\right)^2(\sigma_{xg} - \sigma_{xi}), \tag{4.16}$$

$$s_{yk} = \sigma_{yj} - \sigma_{yi} + \left(\frac{x_k - x_j}{x_h - x_j}\right)^2(\sigma_{yh} - \sigma_{yj}) - \left(\frac{x_k - x_i}{x_g - x_i}\right)^2(\sigma_{yg} - \sigma_{yi}), \tag{4.17}$$

$$s_{xi} = \frac{1}{(x_g - x_i)^2}[\sigma_{xg} - \sigma_{xi} - r_{xi}(x_g - x_i)], \tag{4.18}$$

$$s_{yi} = \frac{1}{(x_g - x_i)^2}[\sigma_{yg} - \sigma_{yi} - r_{yi}(x_g - x_i)]. \tag{4.19}$$

Les formules de la méthode du deuxième ordre donnent les résultats avec une précision beaucoup plus élevée que les formules dérivées des différences du premier ordre.

La solution des problèmes de l'équilibre limite en termes des contraintes normales présente quelques avantages en comparaison avec les autres procédés parce qu' elle rend possible, sans complications considérables, la détérmination du champ des contraintes pour les sols non-homogènes, anisotropes et à condition d'écoulement quelconque.

5. Champ des Vitesses

Pour la détermination du champ des viteses, nous avons, en premier lieu, l'équation de compressibilité, qui, dans le cas de déformation plane, a la forme suivante

$$\frac{\partial(\varrho v_x)}{\partial x} + \frac{\partial(\varrho v_y)}{\partial y} = 0, \tag{5.1}$$

v_x, v_y désignant les composantes de la vitesse de déplacement et ϱ la densité qui peut être fonction de la contrainte moyenne et des coordonnées. C'est pourquoi nous avons

$$\frac{\partial \varrho}{\partial \sigma_x} = \frac{\partial \varrho}{\partial \sigma_y}. \tag{5.2}$$

La condition de compressibilité peut être écrite sous la forme développée:

$$\frac{\partial v_x}{\partial x} + \frac{\partial v_y}{\partial y} \tag{5.3}$$

$$+ \frac{1}{\varrho}\left[v_x\left(\frac{\partial \varrho}{\partial x} + \frac{\partial \varrho}{\partial \sigma_x}\frac{\partial \sigma_x}{\partial x} + \frac{\partial \varrho}{\partial \sigma_y}\frac{\partial \sigma_y}{\partial x}\right) + v_y\left(\frac{\partial \varrho}{\partial y} + \frac{\partial \varrho}{\partial \sigma_x}\frac{\partial \sigma_x}{\partial y} + \frac{\partial \varrho}{\partial \sigma_y}\frac{\partial \sigma_y}{\partial y}\right)\right] = 0.$$

Dans le cas des sols non-homogènes incompressibles, la densité varie avec les coordonnées seulement. Cela s'exprime par

$$\frac{\partial v_x}{\partial x} + \frac{\partial v_y}{\partial y} + \frac{1}{\varrho}\left(v_x \frac{\partial \varrho}{\partial x} + v_y \frac{\partial \varrho}{\partial y}\right) = 0. \tag{5.4}$$

La plus simple, c'est la condition d'incompressibilité pour les matériaux homogènes dont la densité est constante:

$$\frac{\partial v_x}{\partial x} + \frac{\partial v_y}{\partial y} = 0. \tag{5.5}$$

L'autre équation fondamentale pour le champ des vitesses résulte de la relation entre les directions principales du tenseur contraintes et du tenseur vitesses de déformation.

L'auteur a découvert que, dans le cas de compressibilité et d'intervention de la contrainte moyenne sur la condition d'écoulement, les directions des contraintes principales et des vitesses principales de déformation ne coincident pas mais elles forment entre elles l'angle de déviation qui prend les valeurs entre le zéro et la moitié de l'angle du frottement interne.

Un effet analogue peut être observé aussi avec les matériaux anisotropes.

Les relations entre les vitesses de déformation et les vitesses de déplacement sont données par

$$\dot{\varepsilon}_x = \frac{\partial v_x}{\partial x}, \qquad \dot{\varepsilon}_y = \frac{\partial v_y}{\partial y}, \qquad \dot{\gamma}_{xy} = \frac{\partial v_x}{\partial y} + \frac{\partial v_y}{\partial x}. \qquad (5.6)$$

Les directions des contraintes principales σ_1 et σ_2 forment avec la direction positive de l'axe x les angles α et $\alpha + \frac{\pi}{2}$ et les directions des vitesses principales de déformation $\dot{\varepsilon}_1$ et $\dot{\varepsilon}_2$ font avec l'axe x les angles β et $\beta + \frac{\pi}{2}$ (fig. 3).

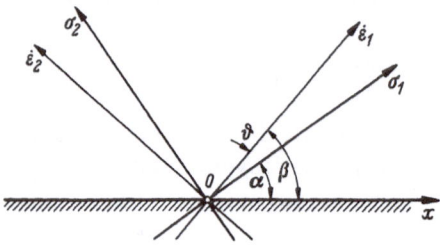

D'après la fig. 3, nous avons

$$\beta = \alpha + \vartheta, \qquad (5.7)$$

ϑ désignant l'angle de déviation.

Fig. 3. Directions des contraintes et des vitesses de déformation principales.

Les angles entre la direction positive de l'axe x et les directions des contraintes principales sont définis par

$$\tan 2\alpha = \frac{2\tau_{xy}}{\sigma_x - \sigma_y} = \frac{\dfrac{\partial v_x}{\partial y} + \dfrac{\partial v_y}{\partial x} - \left(\dfrac{\partial v_x}{\partial x} - \dfrac{\partial v_y}{\partial y}\right) \tan 2\vartheta}{\dfrac{\partial v_x}{\partial x} - \dfrac{\partial v_y}{\partial y} + \left(\dfrac{\partial v_x}{\partial y} + \dfrac{\partial v_y}{\partial x}\right) \tan 2\vartheta}. \qquad (5.8)$$

Pour les directions des vitesses principales de déformations, nous avons la relation suivante

$$\tan 2\beta = \frac{\dfrac{\partial v_x}{\partial y} + \dfrac{\partial v_y}{\partial x}}{\dfrac{\partial v_x}{\partial x} - \dfrac{\partial v_y}{\partial y}} = \frac{2\tau_{xy} + (\sigma_x - \sigma_y) \tan 2\vartheta}{\sigma_x - \sigma_y - 2\tau_{xy} \tan 2\vartheta}. \qquad (5.9)$$

Les relations (5.8) et (5.9) fournissent la seconde équation fondamentale pour le champ des vitesses

$$\frac{\partial v_x}{\partial x} \left[(\sigma_x - \sigma_y) \tan 2\vartheta + 2\tau_{xy}\right] - \frac{\partial v_x}{\partial y} (\sigma_x - \sigma_y - 2\tau_{xy} \tan 2\vartheta)$$

$$- \frac{\partial v_y}{\partial x} (\sigma_x - \sigma_y - 2\tau_{xy} \tan 2\vartheta) - \frac{\partial v_y}{\partial y} \left[(\sigma_x - \sigma_y) \tan 2\vartheta + 2\tau_{xy}\right] = 0. \qquad (5.10)$$

Pour $\vartheta = 0$, les équations (5.8), (5.9) ou (5.10) fournissent, comme un cas particulier, les relations de Lévy-Misés

$$\frac{\dfrac{\partial v_x}{\partial y} + \dfrac{\partial v_y}{\partial x}}{2\tau_{xy}} = \frac{\dfrac{\partial v_x}{\partial x} - \dfrac{\partial v_y}{\partial y}}{\sigma_x - \sigma_y}, \qquad (5.11)$$

ou

$$\frac{\dot{\gamma}_{xy}}{2\tau_{xy}} = \frac{\dot{\varepsilon}_x - \dot{\varepsilon}_y}{\sigma_x - \sigma_y}. \qquad (5.12)$$

Ces relations expriment la coïncidence des directions principales des contraintes et des vitesses de déformation.

Introduisant dans les relations (5.8) et (5.9) la valeur $\vartheta = 0$ et la condition d'incompressibilité (5.5), nous avons

$$\tan 2\alpha = \tan 2\beta = \frac{2\tau_{xy}}{\sigma_x - \sigma_y} = \frac{\dfrac{\partial v_x}{\partial y} + \dfrac{\partial v_y}{\partial x}}{2\dfrac{\partial v_x}{\partial x}}. \tag{5.13}$$

L'autre cas limite est défini par la valeur de l'angle de déviation égale à la moitié de l'angle du frottement interne: $\vartheta = \frac{\varphi}{2}$. Portant cette valeur dans les relation (5.8), (5.9) et (5.10), nous obtenons

$$\tan 2\alpha = \frac{2\tau_{xy}}{\sigma_x - \sigma_y} = \frac{\dfrac{\partial v_x}{\partial y} + \dfrac{\partial v_y}{\partial x} - \left(\dfrac{\partial v_x}{\partial x} - \dfrac{\partial v_y}{\partial y}\right)\tan\varphi}{\dfrac{\partial v_x}{\partial x} - \dfrac{\partial v_y}{\partial y} + \left(\dfrac{\partial v_x}{\partial y} + \dfrac{\partial v_y}{\partial x}\right)\tan\varphi}, \tag{5.14}$$

$$\tan 2\beta = \frac{\dfrac{\partial v_x}{\partial y} + \dfrac{\partial v_y}{\partial x}}{\dfrac{\partial v_x}{\partial x} - \dfrac{\partial v_y}{\partial y}} = \frac{2\tau_{xy} + (\sigma_x - \sigma_y)\tan\varphi}{\sigma_x - \sigma_y - 2\tau_{xy}\tan\varphi}, \tag{5.15}$$

$$\frac{\partial v_x}{\partial x}\left[(\sigma_x - \sigma_y)\tan\varphi + 2\tau_{xy}\right] - \frac{\partial v_x}{\partial y}(\sigma_x - \sigma_y - 2\tau_{xy}\tan\varphi)$$

$$- \frac{\partial v_y}{\partial x}(\sigma_x - \sigma_y - 2\tau_{xy}\tan\varphi) - \frac{\partial v_y}{\partial y}\left[(\sigma_x - \sigma_y)\tan\varphi + 2\tau_{xy}\right] = 0. \tag{5.16}$$

Les relations précédentes expriment la coïncidence de la direction de la vitesse maximum de déformation de cisaillement avec la direction des lignes actives de glissement.

6. Solution du Champ des Vitesses

Dans certains cas, le champ des vitesses peut être déterminé séparément du champ des contraintes.

Pour le calcul numérique, on peut faire usage des équations (5.3) et (5.10) qui peuvent être résolues en différences finies par la méthode des caractéristiques.

Le système des équations (5.3) et (5.10) a deux familles de lignes caractéristiques, qui sont données par les équations suivantes

$$\psi_{1,2} \equiv \frac{dy}{dx} = \frac{2\tau_{xy} + (\sigma_x - \sigma_y)\tan 2\vartheta \pm \sqrt{[(\sigma_x - \sigma_y)^2 + 4\tau_{xy}^2](1 + \tan^2 2\vartheta)}}{\sigma_x - \sigma_y - 2\tau_{xy}\tan 2\vartheta}, \tag{6.1}$$

$$\psi_{1,2}\frac{dv_x}{dx} - \frac{dv_y}{dx} = \mp C\frac{\sqrt{[(\sigma_x - \sigma_y)^2 + 4\tau_{xy}^2](1 + \tan^2 2\vartheta)}}{\sigma_x - \sigma_y - 2\tau_{xy}\tan 2\vartheta}, \tag{6.2}$$

où

$$C = \frac{1}{\varrho}\left[v_x\left(\frac{\partial\varrho}{\partial x} + \frac{\partial\varrho}{\partial\sigma_x}\frac{\partial\sigma_x}{\partial x} + \frac{\partial\varrho}{\partial\sigma_y}\frac{\partial\sigma_y}{\partial x}\right) + v_y\left(\frac{\partial\varrho}{\partial y} + \frac{\partial\varrho}{\partial\sigma_x}\frac{\partial\sigma_x}{\partial y} + \frac{\partial\varrho}{\partial\sigma_y}\frac{\partial\sigma_y}{\partial y}\right)\right].$$

En tenant compte de la relation (5.9), on peut écrire les équations (6.1) et (6.2) sous la forme

$$\frac{dy}{dx} = \tan\left(\beta \pm \frac{\pi}{4}\right), \tag{6.3}$$

$$\frac{dv_x}{dx} \tan\left(\beta \pm \frac{\pi}{4}\right) - \frac{dv_y}{dx} = \mp \frac{C}{\cos 2\beta}. \tag{6.4}$$

D'après l'équation (6.3), les lignes caractéristiques font un angle droit entre elles.

En remplaçant dans les équations (6.3) et (6.4) les différentielles par les différences finies, nous obtenons les formules de réccurence pour calculer les coordonnées des noeuds du réseau des caractéristiques du champ des vitesses et les composantes de vitesse de déplacement dans ces noeuds

$$x_k = \frac{x_i \tan\left(\beta_i + \frac{\pi}{4}\right) - x_j \tan\left(\beta_j - \frac{\pi}{4}\right) - y_i + y_j}{\tan\left(\beta_i + \frac{\pi}{4}\right) - \tan\left(\beta_j - \frac{\pi}{4}\right)}, \tag{6.5}$$

$$y_k = y_i + (x_k - x_i)\tan\left(\beta_i + \frac{\pi}{4}\right) = y_j + (x_k - x_j)\tan\left(\beta_j - \frac{\pi}{4}\right), \tag{6.6}$$

$$v_{xk} =$$
$$\frac{v_{xi}\tan\left(\beta_i + \frac{\pi}{4}\right) - v_{xj}\tan\left(\beta_j - \frac{\pi}{4}\right) - \frac{C_i}{\cos 2\beta_i}(x_k - x_i) - \frac{C_j}{\cos 2\beta_j}(x_k - x_j) - v_{yi} + v_{yj}}{\tan\left(\beta_i + \frac{\pi}{4}\right) - \tan\left(\beta_i + \frac{\pi}{4}\right)}$$

$$\tag{6.7}$$

$$v_{yk} = v_{yi} + \frac{C_i}{\cos 2\beta_i}(x_k - x_i) + (v_{xk} - v_{xi})\tan\left(\beta_i + \frac{\pi}{4}\right)$$
$$= v_{yj} - \frac{C_j}{\cos 2\beta_j}(x_k - x_j) + (v_{xk} - v_{xj})\tan\left(\beta_j - \frac{\pi}{4}\right). \tag{6.8}$$

7. Solution Complète des Problèmes de l'Équilibre Limite

Dans le cas général, il faut calculer les contraintes et les vitesses de déplacement simultanément.

Les équations (2.4), (2.5), (5.3) et (5.10) représentent un système à quatre inconnues σ_x, σ_y, v_x et v_y, qui peut s'écrire sous la forme générale :

$$a_{11}\frac{\partial\sigma_x}{\partial y} + a_{12}\frac{\partial\sigma_y}{\partial y} = A_1, \tag{7.1}$$

$$a_{21}\frac{\partial\sigma_x}{\partial y} + a_{22}\frac{\partial\sigma_y}{\partial y} = A_2, \tag{7.2}$$

$$a_{31}\frac{\partial\sigma_x}{\partial y} + a_{32}\frac{\partial\sigma_y}{\partial y} + a_{33}\frac{\partial v_x}{\partial y} + a_{34}\frac{\partial v_y}{\partial y} = A_3, \tag{7.3}$$

$$a_{43}\frac{\partial v_x}{\partial y} + a_{44}\frac{\partial v_y}{\partial y} = A_4, \tag{7.4}$$

résultant après l'élimination des dérivées partielles par rapport à x au moyen des relations differentielles

$$d\sigma_x = \frac{\partial \sigma_x}{\partial x}\,dx + \frac{\partial \sigma_x}{\partial y}\,dy\,,$$

$$d\sigma_y = \frac{\partial \sigma_y}{\partial x}\,dx + \frac{\partial \sigma_y}{\partial y}\,dy\,,$$

$$dv_x = \frac{\partial v_x}{\partial x}\,dx + \frac{\partial v_x}{\partial y}\,dy\,,$$

$$dv_y = \frac{\partial v_y}{\partial x}\,dx + \frac{\partial v_y}{\partial y}\,dy\,.$$

Les lignes caractéristiques sont déterminées par la nullité des trois déterminants suivants

$$\begin{vmatrix} a_{11} & a_{12} & 0 & 0 \\ a_{21} & a_{22} & 0 & 0 \\ a_{31} & a_{32} & a_{33} & a_{34} \\ 0 & 0 & a_{43} & a_{44} \end{vmatrix} = 0\,, \tag{7.5}$$

$$\begin{vmatrix} A_1 & a_{12} & 0 & 0 \\ A_2 & a_{22} & 0 & 0 \\ A_3 & a_{32} & a_{33} & a_{34} \\ A_4 & 0 & a_{43} & a_{44} \end{vmatrix} = 0\,, \tag{7.6}$$

$$\begin{vmatrix} a_{11} & a_{12} & A_1 & 0 \\ a_{21} & a_{22} & A_2 & 0 \\ a_{21} & a_{32} & A_3 & a_{34} \\ 0 & 0 & A_4 & a_{44} \end{vmatrix} = 0\,. \tag{7.7}$$

Le déterminant (7.5) peut être décomposé de la manière suivante

$$\begin{vmatrix} a_{11} & a_{12} \\ a_{21} & a_{22} \end{vmatrix} \cdot \begin{vmatrix} a_{33} & a_{34} \\ a_{43} & a_{44} \end{vmatrix} - \begin{vmatrix} a_{11} & 0 \\ a_{21} & 0 \end{vmatrix} \cdot \begin{vmatrix} a_{32} & a_{34} \\ 0 & a_{44} \end{vmatrix}$$

$$+ \begin{vmatrix} a_{11} & 0 \\ a_{21} & 0 \end{vmatrix} \cdot \begin{vmatrix} a_{32} & a_{33} \\ 0 & a_{43} \end{vmatrix} + \begin{vmatrix} a_{12} & 0 \\ a_{22} & 0 \end{vmatrix} \cdot \begin{vmatrix} a_{31} & a_{34} \\ 0 & a_{44} \end{vmatrix} \tag{7.8}$$

$$- \begin{vmatrix} a_{12} & 0 \\ a_{22} & 0 \end{vmatrix} \cdot \begin{vmatrix} a_{31} & a_{33} \\ 0 & a_{34} \end{vmatrix} + \begin{vmatrix} 0 & 0 \\ 0 & 0 \end{vmatrix} \cdot \begin{vmatrix} a_{31} & a_{32} \\ 0 & 0 \end{vmatrix} = 0$$

d'où

$$\begin{vmatrix} a_{11} & a_{12} \\ a_{21} & a_{22} \end{vmatrix} \cdot \begin{vmatrix} a_{33} & a_{34} \\ a_{43} & a_{44} \end{vmatrix} = 0\,. \tag{7.9}$$

Après l'introduction des expressions dérivées dans les sections 2 et 6, la relation précédente peut s'écrire sous la forme

$$\left[\eta \left(\frac{dy}{dx}\right)^2 + \chi \frac{dy}{dx} + \lambda \right] \cdot \left[\left(\frac{dy}{dx}\right)^2 \cos 2\beta - 2\frac{dy}{dx} \sin 2\beta - \cos 2\beta \right] = 0. \tag{7.10}$$

Il y a trois cas à distinguer:

1° Le premier terme du produit (7.10) est égal à zéro et le second terme est different de zéro. Nous avons les caractéristiques du champ des contraintes définies par la relation (2.8).

2° Le premier terme n'est pas nul et le second terme est égal à zéro et définit les caractéristiques du champ des vitesses données par l'équation (6.1) ou (6.3).

3° Les deux termes du produit (7.10) sont égaux à zéro. La relation (7.10) donne simultanément les caractéristiques du champ des contraintes et du champ des vitesses.

Le second déterminant (7.6) peut être décomposé en

$$\begin{vmatrix} A_1 & a_{12} \\ A_2 & a_{22} \end{vmatrix} \cdot \begin{vmatrix} a_{33} & a_{34} \\ a_{43} & a_{44} \end{vmatrix} - \begin{vmatrix} A_1 & 0 \\ A_2 & 0 \end{vmatrix} \cdot \begin{vmatrix} a_{32} & a_{34} \\ 0 & a_{44} \end{vmatrix}$$

$$+ \begin{vmatrix} A_1 & 0 \\ A_2 & 0 \end{vmatrix} \cdot \begin{vmatrix} a_{32} & a_{33} \\ 0 & a_{43} \end{vmatrix} + \begin{vmatrix} a_{12} & 0 \\ a_{22} & 0 \end{vmatrix} \cdot \begin{vmatrix} A_3 & a_{34} \\ A_4 & a_{44} \end{vmatrix} \tag{7.11}$$

$$- \begin{vmatrix} a_{12} & 0 \\ a_{22} & 0 \end{vmatrix} \cdot \begin{vmatrix} A_3 & a_{33} \\ A_4 & a_{43} \end{vmatrix} + \begin{vmatrix} 0 & 0 \\ 0 & 0 \end{vmatrix} \cdot \begin{vmatrix} A_3 & a_{32} \\ A_4 & 0 \end{vmatrix} = 0,$$

d'où

$$\begin{vmatrix} A_1 & a_{12} \\ A_2 & a_{22} \end{vmatrix} \cdot \begin{vmatrix} a_{33} & a_{34} \\ a_{34} & a_{44} \end{vmatrix} = 0. \tag{7.12}$$

Cette relation peut s'écrire sous la forme

$$\left(\xi_{1,2}^2 \frac{d\sigma_x}{dx} - \eta^2 \frac{d\sigma_y}{dx} - A\eta - B\xi_{1,2} \right)$$

$$\times \left[\left(\frac{dy}{dx}\right)^2 \cos 2\beta - 2\frac{dy}{dx} \cos 2\beta - \cos 2\beta \right] = 0. \tag{7.13}$$

Comme dans le cas précédent, il y aussi trois cas possibles qui sont exprimés par l'équation (7.13). Nous en pouvons obtenir les relations (2.9) pour les caractéristiques du champ des contraintes et (6.3) pour le caractéristiques du champ des vitesses.

La valeur zéro du troisième déterminant (7.7) fournit la relation

$$\begin{vmatrix} a_{11} & a_{12} \\ a_{21} & a_{22} \end{vmatrix} \cdot \begin{vmatrix} A_3 & a_{34} \\ A_4 & a_{44} \end{vmatrix} = 0, \tag{7.14}$$

d'où

$$\left[\eta \left(\frac{dy}{dx}\right)^2 + \chi \frac{dy}{dx} + \lambda \right] \left\{ \frac{dv_x}{dx} \frac{dy}{dx} \cos 2\beta - \frac{dv_y}{dx} \cos 2\beta \right.$$
$$\left. \pm \frac{1}{\varrho} \left[v_x \left(\frac{\partial \varrho}{\partial x} + \frac{d\sigma_x}{dx} \frac{\partial \varrho}{\partial \sigma_x} + \frac{d\sigma_y}{dx} \frac{\partial \varrho}{\partial \sigma_y} \right) + v_y \frac{\partial \varrho}{\partial y} \right] \right\} = 0. \qquad (7.15)$$

Le relation précendente donne l'équation (2.8) pour les caractéristiques du champ des contraintes et l'équation (6.4) pour les caractéristiques du champ des vitesses.

De tout ce qui a été prouvé, on peut formuler le théorème suivant:

Les caractéristiques du système complet déterminant simultanément le champ des contraintes et des vitesses sont exprimées par les mêmes équations que les caractéristiques du chaque champ considéré séparément avec l' exception des relations le long des caractéristiques du champ des vitesses, qui sont données par

$$\frac{dv_x}{dx} \tan\left(\beta \pm \frac{\pi}{4}\right) - \frac{dv_y}{dx} = \mp \frac{1}{\varrho \cos 2\beta} \left[v_x \left(\frac{\partial \varrho}{\partial x} + \frac{d\sigma_x}{dx} \frac{\partial \varrho}{\partial \sigma_x} + \frac{d\sigma_y}{dx} \frac{\partial \varrho}{\partial \sigma_y} \right) + v_y \frac{\partial \varrho}{\partial y} \right],$$

8. Conclusion

La solution du champ des contraintes en termes des contraintes normales et du champ des vitesses en termes des vitesses de déplacement dans la direction des axes orthogonales a certains avantages en comparaison avec les autres procédés parce qu'elle peut être employée sans difficulté pour les sols à la non-homogénéité et à l'anisotropie quelleconque ainsi que pour les plus complexes critères d'écoulement.

Dans certains cas, le champ des contraintes et le champ des vitesses peuvent être considérés séparément. Les vitesses sont ainsi calculées après la détermination du champ des contraintes.

L'auteur a prouvé que la solution simultané des deux champs fournit les mêmes caractéristiques que les solutions séparées du champ des contraintes et des vitesses.

Dans le cas général, les directions des contraintes principales et des vitesses principales de déformation ne coïncident pas mais elles font l'angle de déviation entre elles qui peut être déterminé par la voie expérimentale des lignes de glissement et les lignes d'écoulement.

La valeur de l'angle de déviation dépend de la compressibilité, de l'intervention de la contrainte moyenne sur la condition d'écoulement et de l'anisotropie du sol.

Bibliographie

[1] BIAREZ, J.: Mécanique des sols (en polonais), Wrocław/Warszawa/Krakow, 1962.
[2] MANDEL, J.: Équilibre par tranches planes des solides à la limite d'écoulement, Imprimerie Louis-Jean, Gap 1942.

[3] Mandel, J.: Problèmes de déformation plane (et de contrainte plane) pour les corps parfaitement plastiques. Séminaire de plasticité. École Polytechnique, 1961. Publications scientifiques et techniques du Ministère de l'Air, Paris 1962.

[4] Sobotka, Z.: The limiting equilibrium of non-homogeneous soils. Non-Homogeneity in Elasticity and Plasticity. Proc. of IUTAM Symposium, Warsaw, September 1958, London/New York/Paris/Los Angeles: Pergamon Press.

[5] Sobotka, Z.: On a new approach to the analysis of limit states in soils and in other continuous media. Bull. Acad. Polonaise Sci., Sér. Sci. Tech. **9**, No. 2 (1961).

[6] Sobotka, Z.: Some axially symmetrical and three-dimensional problems of the limit states of non-homogeneous continuous media. Proc. of the Xth Intern. Congress of Applied Mechanics, Stresa, Italy, 1960.

[7] Sobotka, Z.: The slip lines and slip surfaces in the theory of plasticity and soil mechanics. Appl. Mech. Rev. **14**, No. 10 (1961).

Discussion

Contribution de S. Irmay:

Problèmes de l'equilibre limite des sols non homogènes. Dr. Sobotka defines the equation of the yield or intrinsic curve in the two-dimensional case, in cohesive soils by:

$$(\sigma_x - \sigma_y)^2 + 4\tau_{xy}^2 = 2a(\sigma_x + \sigma_y) + b^2. \tag{1}$$

In *loose* soils this equations becomes:

$$(\sigma_x - \sigma_y)^2 + 4\tau_{xy}^2 = \sin^2 \varphi (\sigma_x + \sigma_y + c)^2 \tag{2}$$

a, b^2, c are positive constants.

The yield curve may advantageously be expressed by the stress invariants, in our case by:

$$\left. \begin{aligned} I_1 &= \sigma_x + \sigma_y = p_1 + p_2, \\ I_2 &= \sigma_x \sigma_y - \tau_{xy}^2 = p_1 p_2, \end{aligned} \right\} \tag{3}$$

p_1, p_2 are the principal stresses, positive in compression. Mohr's circle shows that:

$$\left. \begin{aligned} 2p_1 &= I_1 + (I_1^2 - 4I_2)^{1/2}, \\ 2p^2 &= I_1 - (I_1^2 - 4I_2)^{1/2}. \end{aligned} \right\} \tag{4}$$

This requires that:

$$I_1^2 \geq 4I_2. \tag{5}$$

If we plot the curve

$$4I_2 = I_1^2 \tag{6}$$

in the plane (I_1, I_2), we obtain a parabola with a vertical axis and vertex at the origin. Along it we have (6). Inequality (5) means that only the half-space below that parabola is physically meaningful.

On the other hand the yield curve (1) of *cohesive* soils becomes in the (I_1, I_2) plane:

$$4I_2 + (a^2 + b^2) = (I_1 - a)^2 \tag{7}$$

which is parabola (6) displaced to the right by a, and downwards by $(a^2 + b^2)/4$. Only the hashed zone of the figure is physically possible. The two parabolas intersect at the point P situated at:

$$\left.\begin{array}{l} I_1 = -b^2/2a < 0, \\ I_2 = b^4/16a^2 > 0. \end{array}\right\}\tag{8}$$

The points of $I_2 = 0$ correspond to $p_1 = 0$ (at R_1) and $p_2 = 0$ (at R_2).

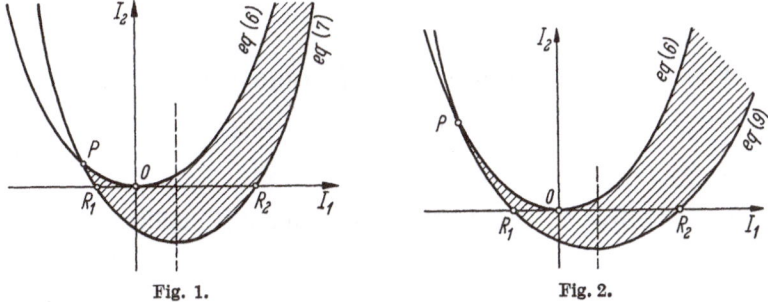

Fig. 1.   Fig. 2.

In *loose* soils the intrinsic curve (2) becomes in the (I_1, I_2) plane:

$$4I_2 + c^2 \tan^2 \varphi = \cos^2 \varphi (I_1 - c \tan \varphi)^2.\tag{9}$$

This is also a parabola, similar to (6), but displaced to the right by $c \tan \varphi$, downwards by $c^2 \tan^2 \varphi$, and the ordinates of which are decreased by the factor $\cos^2 \varphi < 1$. Only the zone between the two parabolas is physically possible. The parabolas intersect at a point P situated at:

$$I_1 = -c.\tag{10}$$

These remarks show the usefulness of the *invariants plane* (I_1, I_2) in the study of limiting states. There is no difficulty in extending these results to 3-dimensional space, using an *invariants space* (I_1, I_2, I_3).

Had we considered the invariants space (I_1, I_2'), where I_2' refers to the deviator stress tensor:

$$I_2' = p_1' p_2' = (p_1 - I_1/2)(p_2 - I_2/2) = I_2 - I_1^2/4.\tag{11}$$

In this system, the parabola (6) is replaced by $I_2' = 0$. Inequality (5) requires that:

$$I_2' \leq 0.\tag{12}$$

The equation of the yield curve (7) for *cohesive* soils becomes:

$$4I_2' = -(2aI_1 + b^2)\tag{13}$$

which is a straight line. This has also been shown by M. Sirieys in his contribution to the Symposium. The physical zone is the wedge to the right between the lines (13) and $I_2' = 0$.

For *loose* soils, the yield curve (9) becomes a parabola.

2.2 On the Consolidating Viscoelastic Layer under Quasi-Static Loading

By

Alfred M. Freudenthal and **William R. Spillers**

Abstract

The application of BIOT's theory of consolidating media to the problem of secondary consolidation of soils is discussed briefly. A solution is presented to the problem of the axisymmetric viscoelastic consolidating layer which is partially loaded on its surface. A numerical example is included.

List of Symbols

b	DARCY's coefficient	α	consolidation constant
f	porosity	η	shear viscosity
G	elastic shear modulus	λ, μ	LAMÉ constants
$H(t)$	HEAVISIDE step function	τ_{ij}	stress tensor (includes a
K	relaxed bulk modulus [$K = \frac{1}{3}(3\lambda + 2G)$]		fluid effect)
M	consolidation constant	τ	relaxation time
p	pore fluid pressure		($\tau = \eta/G$)
u_i	(average) displacement vector of the solid skeleton		

1. Introduction

The representation of the settlement of soils subject to quasi-static distributed vertical loads by assuming a porous elastic medium with pore water flow was first developed by TERZAGHI for the one-dimensional problem [1] and later generalized for the three-dimensional problem by BIOT in a series of comprehensive papers [2]. The resulting theory of "consolidation" predicts that under constant load the settlement asymptotically approaches a stable limit. This theory is not directly applicable to the settlement of clay as it was observed that the settlement of clay under quasi-static load proceeds at an almost constant rate which, for the usual conditions existing under structures, may reach the magnitude of 1/8 to 1/2 inch per year [3]. This type of settlement that is not predicted by the theory of consolidation is commonly referred to as settlement due to "secondary time effects" or as "secondary consolidation".

Tests under conditions of simple shear have shown that the long-time deformational response of clay can be represented by viscoelastic constitutive equations of the BURGERS type involving elasticity, creep recovery and quasiviscous flow. These results provide the basis for the representation of secondary consolidation as a viscoelastic phenomenon in shear superimposed on the primary consolidation by pore water flow. It appears sufficient, however, to use a MAXWELL response in shear rather than the more complicated BURGERS response until more test results become available, particularly since the short-time response predicted by the BURGERS type of constitutive equation is formally of a similar character as that arising in primary consolidation; it is therefore quite difficult to distinguish between the two phenomena during the initial range of settlement, while the long-time deformation in shear of the BURGERS type response is sufficiently closely approximated by the MAXWELL response. It appears therefore that the representation of the deformational response of soils by a linear viscoelastic consolidating medium provides the rational framework for the discussion of both primary and secondary consolidation.

There are relatively few solutions available to boundary value problems for a consolidating medium. For the elastic consolidating medium there are solutions by BIOT [2] (for a reduced form of the general equations), PARIA [4], FREUDENTHAL and SPILLERS [5], and SPILLERS [6]. For the viscoelastic consolidating medium there is the work of FREUDENTHAL and SPILLERS [5] and SPILLERS [6]. Solutions of the general equations have only been presented for the half-space, the spherical cavity, and the uniformly loaded layer (a one-dimensional problem) thus far. In many practical problems it is necessary to deal with a layer which is partially loaded in which the lateral flow of fluid is significant. In the following, a solution for such a problem is presented.

2. Axi-Symmetric Consolidating Viscoelastic Layer

Consider an axi-symmetric infinite layer $0 \leq z \leq L$, $0 \leq r < \infty$. For an elastic consolidating medium, the differential equations [2]

$$\mu \nabla^2 \boldsymbol{u} + (\lambda + \mu) \nabla \nabla \cdot \boldsymbol{u} - \alpha \nabla p = 0, \tag{1}$$

$$\nabla^2 p = b f^{-2} (\alpha \nabla \cdot \dot{\boldsymbol{u}} + M^{-1} p) \tag{2}$$

become

$$\mu (u_{r,rr} + r^{-1} u_{r,r} - r^{-2} u_r + u_{r,zz}) + (\lambda + \mu) \varepsilon_{kk,r} - \alpha p_{,r} = 0, \tag{3}$$

$$\mu (u_{z,rr} + r^{-1} u_{z,r} + u_{z,zz}) + (\lambda + \mu) \varepsilon_{kk,z} - \alpha p_{,z} = 0, \tag{4}$$

$$p_{,rr} + r^{-1} p_{,r} + p_{,zz} = b \alpha f^{-2} \dot{\varepsilon}_{kk} + b M^{-1} f^{-2} \dot{p}. \tag{5}$$

In transform (HANKEL, LAPLACE) space these equations take the form

$$\mu \bar{\bar{u}}_{r,zz} - m^2(\lambda + 2\mu)\,\bar{\bar{u}}_r - (\lambda + \mu)\,m\bar{\bar{u}}_{z,z} + \alpha m\bar{\bar{p}} = 0, \quad (6)$$

$$-\mu m^2 \bar{\bar{u}}_z + (\lambda + 2\mu)\,\bar{\bar{u}}_{z,zz} + (\lambda + \mu)\,m\bar{\bar{u}}_{r,z} - \alpha\bar{\bar{p}}_{,z} = 0, \quad (7)$$

$$\bar{\bar{p}}_{,zz} - (m^2 + b s M^{-1} f^{-2})\,\bar{\bar{p}} - b\alpha s f^{-2} m\bar{\bar{u}}_r - b\alpha s f^{-2}\bar{\bar{u}}_{z,z} = 0, \quad (8)$$

where

$$\bar{\bar{\tau}}_{rz}, \bar{\bar{u}}_r = \int_0^\infty \int_0^\infty \exp(-st)\,J_1(mr)\,r\,(\tau_{rz}, u_r)\,dr\,dt\,,$$

$$\bar{\bar{\varepsilon}}_{kk}, \bar{\bar{p}}, \bar{\bar{u}}_z, \bar{\bar{\tau}}_{zz} = \int_0^\infty \int_0^\infty \exp(-st)\,J_0(mr)\,r\,(\varepsilon_{kk}, p, u_z, \tau_{zz})\,dr\,dt$$

and where the medium is considered to be initially at rest. This form has previously been obtained by the authors [5]. The layer is subjected to the following boundary conditions

at $z = 0, L$ $p = 0$ or $\bar{\bar{p}} = 0$,

at $z = 0, L$ $u_r = 0$ or $\bar{\bar{u}}_r = 0$,

at $z = 0,$ $u_z = 0$ or $\bar{\bar{u}}_z = 0$,

at $z = L,$ $\tau_{zz} = -H(t)\,P(\pi a^2)^{-1}$ $r \leq a$,

                $= 0$ $r > a$

                          or $\bar{\bar{\tau}}_{zz} = -PJ_1(am)(\pi a s m)^{-1}$

Consider a solution in the form

$$\bar{\bar{u}}_r = \sum_n \bar{\bar{U}}_n \sin n\pi z L^{-1}, \quad\quad\quad\quad (9)$$

$$\bar{\bar{u}}_{z,z} = \sum_n \bar{\bar{V}}_n \sin n\pi z L^{-1} - PJ_1(am)[\pi a s m(\lambda + 2\mu)]^{-1}, \quad (10)$$

$$\bar{\bar{p}} = \sum_n \bar{\bar{P}}_n \sin n\pi z L^{-1} \quad\quad\quad\quad (11)$$

which satisfies the boundary conditions. Inserting these solutions in the differential equations (6)—(8) and using the sine series, $1 = \sum_{n\,\text{odd}} 4(n\pi)^{-1} \sin n\pi z L^{-1}$, it follows that the coefficients $\bar{\bar{J}}_n$, $\bar{\bar{V}}_n$, and $\bar{\bar{P}}_n$ are zero for even values of n and are determined by three simultaneous linear equations

$$
\begin{bmatrix}
\mu\dfrac{n^2\pi^2}{L^2}+m^2(\lambda+2\mu) & (\lambda+\mu)\,m & -\alpha m \\[2ex]
(\lambda+\mu)\,m & \mu m^2\dfrac{L^2}{n^2\pi^2}+(\lambda+2\mu) & -\alpha \\[2ex]
-m\alpha & -\alpha & \dfrac{-\dfrac{n^2\pi^2}{L^2}-\left(m^2+\dfrac{bs}{Mf^2}\right)}{bsf^{-2}}
\end{bmatrix}
\begin{bmatrix}
\overline{\overline{U}}_n \\[2ex] \overline{\overline{V}}_n \\[2ex] \overline{\overline{P}}_n
\end{bmatrix}
$$

$$
= \frac{-4PJ_1(am)}{n\pi^2 a s\,m(\lambda+2\mu)}
\begin{bmatrix}
m(\lambda+\mu) \\[2ex]
-\mu m^2\dfrac{L^2}{n^2\pi^2} \\[2ex]
\alpha
\end{bmatrix}
\tag{12}
$$

for odd values of n.

The determination of the above coefficients completes the formulation of the problem for an elastic consolidating medium. The formulation for a viscoelastic consolidating medium is obtained from Eq. (12) by replacing appropriate elastic constants by functions of the Laplace transform parameter s at a convenient time [7].

It is now possible to compute quantities of interest from Eqs. (9) to (11) by inverting the HANKEL-LAPLACE transforms which involve rather complex integrals. Attention here will be restricted to the computation of the surface displacement at the origin.

The surface displacement in transform space is

$$
\overline{\overline{u}}_z = +\sum_{n\,\mathrm{odd}} \overline{\overline{V}}_n \frac{L}{n\pi} - \frac{PLJ_1(am)}{\pi a s m(\lambda+2\mu)}
\tag{13}
$$

where, from Eq. (12)

$$
\overline{\overline{V}}_n = \frac{4}{n\pi}\,\frac{P}{\pi a s m}\,\frac{J_1(am)}{(\lambda+2\mu)}
\tag{14}
$$

$$
-\frac{4PJ_1(am)}{n\pi^2 a s m\mu}\,
\frac{\left[\mu\dfrac{n^2\pi^2}{L^2}+m^2(\lambda+2\mu)\right]\left[\dfrac{n^2\pi^2}{L^2}+m^2+\dfrac{bs}{Mf^2}\right]+\alpha^2 m^2 b s f^{-2}}
{\left(\dfrac{Lm^2}{n\pi}+\dfrac{n\pi}{L}\right)^2\left[\left(\dfrac{n^2\pi^2}{L^2}+m^2+\dfrac{bs}{Mf^2}\right)(\lambda+2\mu)+\alpha^2 b s f^{-2}\right]}.
$$

First the HANKEL transform will be inverted at the origin $(r=0)$, the integral

$$
\overline{u}_z(r) = \int_0^\infty J_0(mr)\,m\,\overline{\overline{u}}_z\,dm
$$

simplifies to

$$
\overline{u}_z(0) = \int_0^\infty m\,\overline{\overline{u}}_z\,dm.
$$

Then [8]

$$
\overline{u}_z(0) = +\sum_{n\,\mathrm{odd}} \overline{V}_n L(n\pi)^{-1} - PL[\pi a^2 s(\lambda+2\mu)]^{-1},
\tag{15}
$$

where

$$\overline{V}_n = 4P[n\pi^2 a^2 s(\lambda + 2\mu)]^{-1} - (\overline{\varphi}_1 + \overline{\varphi}_2 + \overline{\varphi}_3) \tag{16}$$

in which $\overline{\varphi}_1 + \overline{\varphi}_2 + \overline{\varphi}_3$ represents the last term of Eq. (14) which may be written, using partial fractions, as

$$\overline{\overline{\varphi}}_1 = \frac{AJ_1(am)}{m(m^2 + c_1^2)}, \qquad \overline{\overline{\varphi}}_2 = \frac{BJ_1(am)}{m(m^2 + c_1^2)^2}, \qquad \overline{\overline{\varphi}}_3 = \frac{CJ_1(am)}{m(m^2 + c_2^2)}, \tag{17}$$

$$c_1^2 = +n^2\pi^2 L^{-2}, \quad c_2^2 = +n^2\pi^2 L^{-2} + bs M^{-1}f^{-2} + \alpha^2 bs f^{-2}(\lambda + 2\mu)^{-1},$$

$$A = \frac{4nP}{as\mu L^2}\left[1 + \frac{n^2\pi^2}{L^2}\frac{\alpha^2 bsf^{-2}\mu}{(\lambda + 2\mu)^2\left(\dfrac{bs}{Mf^2} + \dfrac{\alpha^2 bsf^{-2}}{\lambda + 2\mu}\right)}\right],$$

$$B = \frac{-4nP}{as\mu L^2}\frac{n^2\pi^2}{L^2}\frac{(\lambda + \mu)\dfrac{bs}{Mf^2} + bsf^{-2}\alpha}{\dfrac{bs}{Mf^2} + \dfrac{\alpha^2 bsf^{-2}}{\lambda + 2\mu}},$$

$$C = -\frac{4nP}{as\mu L^2}\frac{n^2\pi^2}{L^2}\frac{\alpha^2 bsf^{-2}\mu}{(\lambda + 2\mu)^2\left(\dfrac{bs}{Mf^2} + \dfrac{\alpha^2 bsf^{-2}}{\lambda + 2\mu}\right)}.$$

From [8]

$$\overline{\varphi}_1 = \frac{A}{c_1^2}\left[\frac{1}{a} - c_1 k_1(c_1 a)\right],$$

$$\overline{\varphi}_2 = \frac{B}{c_1^2}\left[\frac{1}{c_1^2 a} - \frac{a k_2(c_1 a)}{2}\right], \tag{18}$$

$$\overline{\varphi}_3 = \frac{C}{c_2^2}\left[\frac{1}{a} - c_2 k_1(c_2 a)\right],$$

where $k_1(x)$, $k_2(x)$ are modified BESSEL functions of the third kind.

Viscoelasticity is now introduced, using the viscoelastic analogy, by replacing

$$\mu \quad \text{by} \quad Gs/(s + \tau^{-1})$$

$$\text{and} \quad \lambda \quad \text{by} \quad [k\tau^{-1} + (k - 2/3G)s]/(s + \tau^{-1}).$$

Since c_1 is not a function of s, all terms of $\overline{u}_z(o)$ except $\overline{\varphi}_3$ contain s as a rational function and are consequently relatively simple to invert. Let

$$\overline{\varphi}_3 = \overline{\Psi} + \overline{\theta},$$

$$\overline{\Psi} = C c_2^{-2} a^{-1}, \quad \overline{\theta} = -C c_2^{-1} k_1(c_2 a).$$

Since $\overline{\Psi}$ contains the transform parameter as a rational function it will not be given further consideration. $\overline{\theta}$ will be inverted using the theorem of LAPLACE transformations which states that [9]

$$\mathcal{L}^{-1}\left\{\frac{F[h(s)]}{g(s)}\right\} = \int_0^\infty \varphi(t, \eta) f(\eta) \, d\eta, \tag{19}$$

where $F(s) = \mathcal{L}\{f(t)\}$ and $\varphi(t,\eta) = \mathcal{L}^{-1}\{\exp[-\eta h(s)]/g(s)\}$. It is known that $[10]$ $\mathcal{L}^{-1}\{\sqrt{\beta}\,s^{-1}\,k_1(\sqrt{\beta s})\} = \exp[-\beta/(4t)]$. This may be applied to $\bar{\theta}$ if $\beta = a^2$, $f(t) = \exp[-a^2/(4t)]$, $g(s) = -C^{-1}a$, and $h(s) = c_2^2$. Since c_2^2 can be written in the form $c_2^2 = a_1 + a_2 s + \dfrac{a_3}{s + a_4}$,

$$\varphi(t, \eta) = -\mathcal{L}^{-1}\left\{Ca^{-1}\exp\left[-\eta\left(a_1 + a_2 s + \frac{a_3}{s + a_4}\right)\right]\right\},$$

$$= -a^{-1}e^{-\eta a_1} H(t - \eta a_2)\, f_1(t - \eta a_2), \tag{20}$$

$$f_1(t) = \mathcal{L}^{-1}\{C \exp[-\eta a_3/(s + a_4)]\},$$

$$= \mathcal{L}^{-1}\left\{C \sum_{n=0} [-\eta a_3/(s + a_4)]^n/n!\right\}, \tag{21}$$

$$= \sum_{n=0}^{\infty} \varphi_n(t),$$

in which the computation of φ_n involves only the inversion of a rational function of the transform parameter s. From Eq. (19)

$$\theta = -\int_0^{\infty} a^{-1} e^{-\eta a_1} H(t - \eta a_2) \sum_{n=0}^{\infty} \varphi_n(t - \eta a_2)\, e^{\frac{-a^2}{4\eta}}\, d\eta,$$

$$= -a^{-1}\int_0^{t/a_2} e^{-\eta a_1 - \frac{a^2}{4\eta}} \sum_{n=0}^{\infty} \varphi_n(t - \eta a_2)\, d\eta. \tag{22}$$

The expansion and term-by-term inversion in Eq. (21) can be justified $[11]$. Eq. (22) completes the description of the computation of $\bar{u}_z(o)$.

3. Numerical Example

Fig. 1 shows the results of numerical calculations performed for specific values of the parameters. The non-dimensional surface displacement at the origin, $\delta = u_z(o)a^{-1}$, is plotted on the non-dimensional time, $T = t\{L^2bf^{-2}[M^{-1} + \alpha^2(k + 4/3\,G)^{-1}]\}^{-1}$ for $\tau = 1000$, $G/k = 1$, $a = 1$, $b = Mf^2$, $\alpha^2 b = f^2$, $4P\pi^{-2}a^{-2} = G$ and $L = \pi$. These particular values of the parameters were chosen to give a case in which the pore water flow and the shear flow are separated similar to some of the examples shown in $[5]$. The calculations are rather lengthy and no attempt was made to determine specifically the dependence of the surface displacement on the parameters involved.

It should be emphasized that a meaningful application of the theory of the consolidating medium to problems in soil mechanics must await the experimental determination of the necessary material properties since there is presently very little data available.

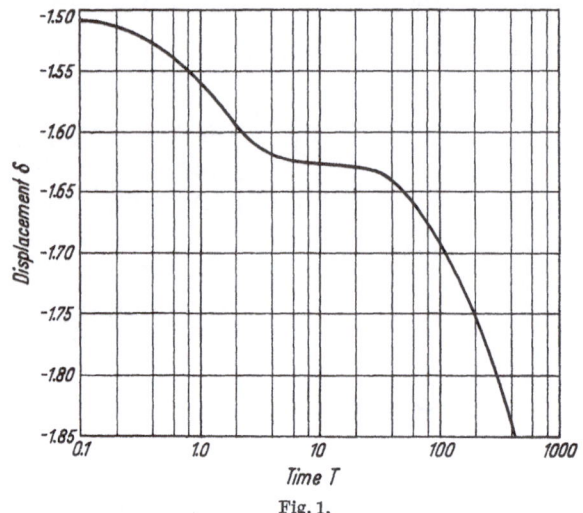

Fig. 1.

References

[1] TERZAGHI, K.: Erdbaumechanik auf bodenphysikalischer Grundlage, Leipzig, 1925.

[2] BIOT, M. A.: Bruxelles Society Scientifique, Annales, B55, 1935. — BIOT, M. A., and D. G. WILLIS: J. Appl. Mech., Dec. 1957, pp. 594—601. (References are given in this paper to BIOT's other papers.)

[3] TERZAGHI, K., and R. B. PECK: Soil Mechanics in Engineering Practice, New York: J. Wiley 1948, pp. 76, 241.

[4] PARIA, G.: J. Math. Phys. **36**, 4 (1958).

[5] FREUDENTHAL, A. M., and W. R. SPILLERS: J. Appl. Phys. **33**, 9, pp. 2661—2668 (1962).

[6] SPILLERS, W. R.: Inter. J. Mech. Sci. **4** (1962).

[7] FREUDENTHAL, A. M., and H. GEIRINGER: The Mathematical Theory of the Inelastic Continuum, Handbuch der Physik, Vol. 6, Berlin/Göttingen/Heidelberg: Springer 1958, p. 272.

[8] ERDELYI, A. (editor): Tables of Integral Transform, Vol. II, New York: McGraw Hill 1954.

[9] MCLACHLAN, N. W.: Modern Operational Calculus, New York: Cambridge Univ. Press 1948, p. 43.

[10] ERDELYI, A. (editor): loc. cit., I. Tables of Integral Transform, Vol. I, New York: McGraw Hill 1954.

[11] FREUDENTHAL, A. M., and W. R. SPILLERS: Some Quasi-Static Boundary Value Problems of Consolidating Elastic and Viscoelastic Media, Office of Ordnance Research, Tech. Report 1, Columbia University, February 1961.

2.3 Calcul des Solutions Approchées dans les Problèmes d'Équilibre Limite Plan et de Révolution des Milieux Pesants Obéissant à la Loi de Coulomb

Julien Kravtchenko et Robert Sibille

I. Principe de la Méthode de Calcul dans le Cas Plan

La répartition des contraintes d'un état d'équilibre plan pour un milieu cohérent, homogène, isotrope et pesant, obéissant à la loi de Coulomb est décrite par le système quasi linéaire de 2 équations aux dérivées partielles, du type hyperbolique: (Ox étant orienté suivant la verticale descendante).

$$
\left.
\begin{aligned}
(1 + \sin \varrho \cos 2\varphi) \frac{\partial \sigma}{\partial x} + \sin \varrho \sin 2\varphi \frac{\partial \sigma}{\partial y} - 2\sigma \sin \varrho \sin 2\varphi \frac{\partial \varphi}{\partial x} & \\
+ 2\sigma \sin \varrho \cos 2\varphi \frac{\partial \varphi}{\partial y} = \gamma, & \\
\sin \varrho \sin 2\varphi \frac{\partial \sigma}{\partial x} + (1 - \sin \varrho \cos 2\varphi) \frac{\partial \sigma}{\partial y} + 2\sigma \sin \varrho \cos 2\varphi \frac{\partial \varphi}{\partial x} & \\
- 2\sigma \sin \varrho \sin 2\varphi \frac{\partial \varphi}{\partial y} = 0, &
\end{aligned}
\right\}
\tag{1}
$$

que l'on peut, avec Sokolovski, écrire sous la forme:

$$
\begin{aligned}
\frac{\partial \xi}{\partial x} + \tan(\varphi + \mu) \frac{\partial \xi}{\partial y} &= \frac{-\gamma \sin(\varphi - \mu)}{2\sigma \sin \varrho \cos(\varphi + \mu)}, \\
\frac{\partial \eta}{\partial x} + \tan(\varphi - \mu) \frac{\partial \eta}{\partial y} &= \frac{\gamma \sin(\varphi + \mu)}{2\sigma \sin \varrho \cos(\varphi - \mu)}
\end{aligned}
\tag{2}
$$

avec:

$$
\begin{aligned}
\xi &= \frac{\cot \varrho}{2} \log \frac{\sigma}{\sigma_0} + \varphi, \\
\eta &= \frac{\cot \varrho}{2} \log \frac{\sigma}{\sigma_0} - \varphi,
\end{aligned}
\tag{2'}
$$

en désignant par:

ϱ = angle de frottement interne du matériau,
γ = poids spécifique du matériau par unité d'aire,

$$
\mu = \frac{\pi}{4} - \frac{\varrho}{2},
$$

φ = angle d'inclinaison de la contrainte principale majeure sur Ox,

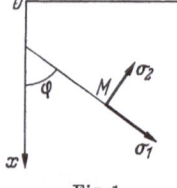

σ = contrainte moyenne fictive:

$$\sigma = \frac{\sigma_1 + \sigma_2 + 2H}{2}.$$

σ_1, σ_2 = contraintes principales, cf. fig. 1,

$H = c \cot \varrho$, c = cohésion,

σ_0 = contrainte de référence.

Fig. 1.

Sauf cas exceptionnels, les problèmes aux limites qui se posent en pratique, relativement aux systèmes (1) ou (2), ne peuvent être résolus explicitement. Mais on sait en former des solutions approchées par des méthodes numériques, basées principalement sur l'emploi des multiplicités bicaractéristiques. Or ce procédé exige la mise en oeuvre de moyens de calculs importants. Il y a donc intérêt à chercher des méthodes approchées, plus aisément maniables. L'objet du présent travail est de décrire une méthode permettant d'obtenir une solution approchée par un procédé assez rapide. En voici le principe: nous avons observé que dans de nombreux cas usuels les données frontières sont analytiques et d'ailleurs très simples. Dans ce cas, les multiplicités bicaractéristiques sont, dans le voisinage des éléments réguliers des frontières, constituées par des courbes analytiques régulières, différant peu, dans ce voisinage, des multiplicités surosculatrices. Nous nous sommes alors proposés de tester l'approximation obtenue pour les solutions de (1) ou (2) en utilisant les développements limités de Cauchy-Kovalewska. Pratiquement, nous limiterons ces développements aux termes de degré 2 ou 3. Il est évident que la solution approchée, ainsi obtenue, n'est valable que localement. Si la précision atteinte au voisinage de la frontière est excellente, elle diminue lorsqu'on s'éloigne de cette frontière. Et, faute de connaître exactement le rayon de convergence des développements limités utilisés, il est difficile de chiffrer la borne supérieure de l'erreur commise en un point donné du massif lorsqu'on néglige les termes de rang supérieur à 3. D'ailleurs, il est difficile, en général, de déterminer le rayon de convergence des développements limités de Cauchy-Kovalewska: il s'en suit que, dans un cas concret, nous ne pouvons affirmer à priori, que notre développement limité s'obtient à partir d'une série convergente. C'est là les points faibles de notre procédé. Notons, toutefois, que dans de nombreux problèmes, dont les solutions numériques sont connues, notre méthode a donné d'excellents résultats, valables dans toute la gamme des profondeurs utiles au technicien. Signalons également qu'il est possible d'accroître la précision obtenue au moyen de la méthode du prolongement analytique.

L'étude au voisinage des singularités est beaucoup moins banale. Nous avons obtenu des résultats concernant ces singularités qui nous semblent nouveaux. Ici encore, on arrive, au point de vue pratique, à des conclusions optimistes.

Nous exposons ici, tout d'abord, l'application de notre méthode à quelques problèmes classiques particuliers, (problème de CAUCHY régulier du type I, singularité de PRANDTL, problème de CAUCHY du type III); puis nous indiquons l'extension de notre procédé aux problèmes d'équilibre limite de révolution; nous donnons enfin les résultats numériques obtenus dans le cas des fondations et des murs de soutènement. On précise à la fin du mémoire l'économie de calcul que l'emploi de notre méthode — dans la mesure où elle est applicable — permet de réaliser.

Cette rédaction s'inspire largement des travaux antérieurs qu'on trouvera cités dans la bibliographie.

II. Applications à des Problèmes Particuliers

Dans tout ce qui suit nous opérerons en variables adimensionnelles.

1. Problème de Cauchy de Type I

Le massif est limité par une frontière OA (que nous supposerons être un segment de droite, pour simplifier (fig. 2). Le long de OA le massif est soumis à des charges données à priori ($\sigma(O, y)$ et $\varphi(O,y)$ sont,

dans les cas pratiques, des fonctions analytiques simples de y, en général, des constantes ou des fonctions linéaires de y). Les traces sur le plan xOy des multiplicités bicaractéristiques de (1) ou (2) vérifient l'équation différentielle:

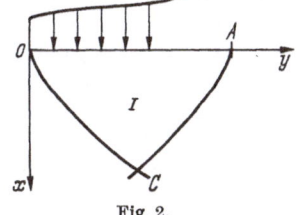

Fig. 2.

$$\frac{dy}{dx} = \tan(\varphi \pm \mu), \qquad (3)$$

équation que nous pouvons dériver par rapport à x autant de fois qu'il est nécessaire. Par exemple, on a:

$$\frac{d^2y}{dx^2} = [1 + \tan^2(\varphi \pm \mu)]\left(\frac{\partial\varphi}{\partial x} + \frac{\partial\varphi}{\partial y} \cdot \frac{dy}{dx}\right). \qquad (4)$$

$\sigma(O,y)$ et $\varphi(O,y)$ (ou ξ et η) étant connus ainsi que leurs dérivées en y sur OA (que nous supposerons n'être pas la trace d'une multiplicité bicaractéristique), nous pouvons calculer en un point quelconque de OA, et en particulier en O, toutes les dérivées en x et y de σ et φ (ou ξ et η): il suffit de résoudre le système (1) ou (2) en $\left(\frac{\partial\sigma}{\partial x}\right)_0$ et $\left(\frac{\partial\varphi}{\partial x}\right)_0$; puis de résoudre le système obtenu en dérivant par rapport à x et y les 2 équations de (1) ou (2) en $\left(\dfrac{\partial^2\sigma}{\partial x_2}\right)_0$ $\left(\dfrac{\partial^2\sigma}{\partial x\,\partial y}\right)_0$ $\left(\dfrac{\partial^2\varphi}{\partial x^2}\right)_0$ $\left(\dfrac{\partial^2\varphi}{\partial x\,\partial y}\right)_0$. De même, (3) et (4) permettent le calcul de $\left(\dfrac{dy}{dx}\right)_0$ $\left(\dfrac{d^2y}{dx^2}\right)_0 \ldots$

C'est le calcul classique de Cauchy-Kovalewska, valable dans le voisinage d'un point régulier de la frontière. La solution est alors déterminée dans le domaine I limité par OA et les 2 caractéristiques OC et AC de familles différentes. Nous donnons sous forme de développements limités aux termes en x^3 inclusivement l'équation d'une caractéristique, OC par exemples:

$$y = \left(\frac{dy}{dx}\right)_0 x + \left(\frac{d^2y}{dx^2}\right)_0 \frac{x^2}{2} + \left(\frac{d^3y}{dx^3}\right)_0 \frac{x^3}{6}. \tag{5}$$

Les expressions approchées de σ et φ dans ce domaine I s'écrivent: (en se limitant à l'approximation du deuxième degré en x et y, ce qui suffit en général).

$$\begin{aligned}
\sigma &= \sigma_0 + \left(\frac{\partial\sigma}{\partial x}\right)_0 x + \left(\frac{\partial\sigma}{\partial y}\right)_0 y + \left(\frac{\partial^2\sigma}{\partial x^2}\right)_0 \frac{x^2}{2} + \left(\frac{\partial^2\sigma}{\partial y^2}\right)_0 \frac{y^2}{2} + \left(\frac{\partial^2\sigma}{\partial x\,\partial y}\right)_0 xy, \\
\varphi &= \varphi_0 + \left(\frac{\partial\varphi}{\partial x}\right)_0 x + \left(\frac{\partial\varphi}{\partial y}\right)_0 y + \left(\frac{\partial^2\varphi}{\partial x^2}\right)_0 \frac{x^2}{2} + \left(\frac{\partial^2\varphi}{\partial y^2}\right)_0 \frac{y^2}{2} + \left(\frac{\partial^2\varphi}{\partial x\,\partial y}\right)_0 xy.
\end{aligned} \tag{6}$$

On consultera pour plus de détails [1] et [2].

2. Singularité de Prandtl

Le point O n'est plus un point régulier: par hypothèse toutes les bicaractéristiques supposées analytiques et régulières, d'une même famille passent par ce point. Rappelons qu'il s'agit de résoudre le problème suivant. On se donne:

a) Une bicaractéristique OC du système (1) ou (2), de la 1ère famille par exemple (fig. 3). Le long de OC, σ et φ (ou ξ et η) sont connus et constituent des données; on pose $\theta_1 = (\widehat{Ox,\,OC'})$, OC' tangente en O à OC

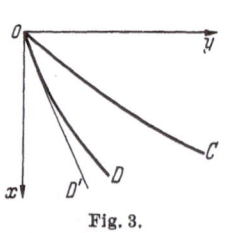

Fig. 3.

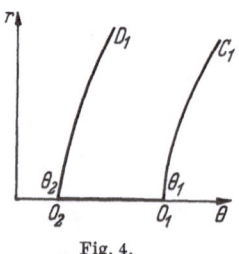

Fig. 4.

b) Une direction OD' issue de O et non tangente à OC en O. $\theta_2 = (Ox, OD')$. Il s'agit de construire la bicaractéristique OD, de la même famille que OC, tangente en O à OD' et de définir ensuite dans le domaine limité par OC et OD une solution de (1) (ou (2)) telle que toutes les bicaractéristiques de la 1ère famille (supposées régulières à priori) passent par O. En introduisant les coordonnées polaires:

$$x = r \cos \theta, \qquad y = r \sin \theta,$$

le système (2) s'écrit:

$$\left.\begin{aligned}
\cos (\psi + \mu) \frac{\partial \xi}{\partial r} + \frac{1}{r} \sin (\psi + \mu) \frac{\partial \xi}{\partial \theta} &= - \frac{\sin (\varphi - \mu)}{2 \sigma \sin \varrho}, \\
\cos (\psi - \mu) \frac{\partial \eta}{\partial r} + \frac{1}{r} \sin (\psi - \mu) \frac{\partial \eta}{\partial \theta} &= \frac{\sin (\varphi + \mu)}{2 \sigma \sin \varrho},
\end{aligned}\quad\right\} \tag{7}$$

avec: $\qquad \psi = \varphi - \theta, \qquad 2\psi + 2\theta = \xi - \eta.$

Plaçons nous dans le plan r, θ (fig. 4). La bicaractéristique OC est représentée par $O_1 C_1$; le point singulier O correspond au segment $O_1 O_2$ défini par les valeurs θ_1 et θ_2. Nous cherchons une solution $\xi(r, \theta)$, $\eta(r, \theta)$ du système (7) dans la région $C_1 O_1 O_2 D_1$ limitée par les bicaractéristiques de la 1ère famille, par exemple, $O_1 C_1$ et $O_2 D_1$ (cette dernière étant inconnue à priori) et le segment $O_1 O_2$ représentant la bicaractéristique dégénérée de la 2ème famille $r = 0$. Le long de $O_1 C_1 C_2$, ξ et η sont des données. Sur $r = 0$, $\theta_2 \le \theta \le \theta_1$ on a, comme le montre un calcul élémentaire:

$$\begin{aligned}
\eta(0, \theta) &= C^{\text{ste}} = \frac{\cot \varrho}{2} \log \sigma_0^* + \mu = \bar{\xi}_0(\theta), \\
\xi(0, \theta) &= \frac{\cot \varrho}{2} \log \sigma_0^* + 2\theta - \mu = \bar{\eta}_0(\theta),
\end{aligned} \tag{8}$$

ou σ_0^* est une constante, calculée au point O_1 et définie par

$$\sigma_0^* = \sigma_{01} \exp (- 2 \tan \varrho \cdot \mu).$$

Du fait de la présence du terme $\frac{1}{r}$, le système (7) n'est pas régulier pour $r = 0$; mais nous allons montrer qu'il existe une solution et une seule du problème, analytique et régulière dans le voisinage du segment $O_1 O_2$.

En effet, si une telle solution existe, nous pouvons l'écrire sous la forme:

$$\begin{aligned}
\xi(r, \theta) &= \bar{\xi}_0(\theta) + r \xi_1(r, \theta), \\
\eta(r, \theta) &= \bar{\eta}_0(\theta) + r \eta_1(r, \theta).
\end{aligned} \tag{9}$$

ξ_1 et η_1 sont solutions analytiques et régulières du système:

$$\cos (\psi + \mu) \frac{\partial \xi_1}{\partial r} + \frac{1}{r} \sin (\psi + \mu) \frac{\partial \xi_1}{\partial \theta} = \frac{1}{r} \left[- \frac{\sin (\varphi - \mu)}{2 \sigma \sin \varrho} - \frac{2}{r} \sin (\psi + \mu) \right.$$
$$\left. - \xi_1 \cos (\psi + \mu) \right], \tag{10}$$

$$\cos (\psi - \mu) \frac{\partial \eta_1}{\partial r} + \frac{1}{r} \sin (\psi - \mu) \frac{\partial \eta_1}{\partial \theta} = \frac{1}{r} \left[\frac{\sin (\varphi + \mu)}{2 \sigma \sin \varrho} - \eta_1 \cos (\psi + \mu) \right].$$

La solution régulière cherchée n'existe que si $\xi_1(0,\theta) = \bar{\xi}_1(\theta)$ et $\eta_1(0,\theta) = \bar{\eta}_1(\theta)$ satisfont à :

$$\bar{\eta}_1 - 2\bar{\xi}_1 = \frac{\sin(\theta - 2\mu)}{2\bar{\sigma}\sin\varrho} \quad \text{avec} \quad \bar{\sigma} = \sigma_0^* e^{2\tan\varrho\theta},$$

$$\frac{d\bar{\eta}_1}{d\theta} = \eta_1 \tan\varrho - \frac{\sin\theta}{2\bar{\sigma}\sin\varrho\cos\varrho}. \tag{11}$$

On montre alors facilement que $O_1 C_1$ est une bicaractéristique de la 1ère famille de (10), le long de laquelle les valeurs de ξ_1 et η_1 sont données par (9), et que $O_1 O_2$ est une bicaractéristique de la 2ème famille de (10).

Moyennant (11), les $2^{\text{èmes}}$ membres de (10) sont alors des séries entières et convergentes en τ, ξ_1, η_1.

Ce qui précède montre que ξ_1 et η_1 sont solutions d'un problème régulier de CAUCHY de type II et on sait qu'un tel problème admet une solution analytique et régulière et une seule dans le voisinage de $O_1 O_2$.

Cette conclusion ouvre la voie au calcul d'une solution approchée du système (10) sous la forme :

$$\xi_1(r,\theta) = \sum_{n=1}^{\infty} r^{n-1} \bar{\xi}_n(\theta),$$

$$\eta_1(r,\theta) = \sum_{n=1}^{\infty} r^{n-1} \bar{\eta}_n(\theta). \tag{12}$$

Nous avons explicité les fonctions $\bar{\xi}_n(\theta)$ et $\bar{\eta}_n(\theta)$, pour $n = 1, 2$, à la fois dans le cas où $O_1 C_1$ est une bicaractéristique de la 1ère famille et dans le cas où $O_1 C_1$ est une bicaractéristique de la 2ème famille.

Nous donnons ci-après les formules pour $n = 1$ dans le 1er cas seulement :

$$\bar{\eta}_1(\theta) = \frac{e^{-2\tan\varrho\theta}(3\tan\varrho\sin\theta + \cos\theta)}{2\sigma_0^*\sin\varrho\cos\varrho(9\tan^2\varrho + 1)} + C_1 e^{\tan\varrho\theta},$$

$$\bar{\xi}_1(\theta) = \frac{\bar{\eta}_1}{2} - \frac{\sin(\theta - 2\mu)}{4\sigma_0^* e^{2\tan\varrho\theta}\sin\varrho}, \tag{13}$$

A partir de $n = 3$ les calculs nécessaires pour expliciter les inconnues $\xi_n(\theta)'$ et $\eta_n(\theta)$ deviennent trop laborieux pour être pratiquement utilisables. Mais les raisonnements offrent l'intérêt de justifier les conclusions en toute rigueur, d'etablir directement l'analyticité des $\xi_n(\theta)$ et $\eta_n(\theta)$ en θ pour n quelconque et d'appliquer la méthode des majorantes d'une manière relativement simple.

La constante C_1 se calcule en O_1 en écrivant :

$$\bar{\xi}_1(\theta_1) = \left(\frac{\partial\xi}{\partial r}\right) \quad \text{connu en } O_1 \text{ de la zone II} \tag{14}$$

On notera que les formules de [*18*] relatives aux valeurs explicites de ξ_n et η_n sont inexactes.

Dans les cas pratiques, intéressant l'art de l'ingénieur, les développements limités à $n = 2$ fournissent déjà une bonne approximation des solutions connues. Mais là encore nous ignorons le rayon de convergence de (12) et il n'est pas possible de donner une borne théorique assez précise de l'erreur commise en négligeant les termes r^n pour $n > 2$. Pour les détails, cf. [*9*], [*10*] et [*11*]. On notera seulement ici qu'à notre connaissance, la théorie qui précède est nouvelle et elle permet d'étudier en toute rigueur le voisinage d'une singularité de PRANDTL. Les mêmes remarques valent encore pour l'étude du cas de révolution ci-après.

3. Extension au Cas de la Symétrie Axiale

Nous avons étendu cette étude relative à la singularité de PRANDTL au cas de révolution cf. [*11*]. L'espace étant rapporté au système d'axes Orz, Oz, axe de révolution (fig. 5), étant orienté suivant la verticale descendante, les équations de l'équilibre limite s'écrivent pour un milieu pesant:

$$\frac{\partial \xi}{\partial r} + \tan(\varphi + \mu)\frac{\partial \xi}{\partial \theta} = -\frac{1}{2r}\frac{\sin(\varphi + \mu) - x\sin(\varphi - \mu)}{\cos(\varphi + \mu)}$$
$$+ \frac{\gamma\cos(\varphi - \mu)}{2\sigma\sin\varrho\cos(\varphi + \mu)},$$
$$\frac{\partial \eta}{\partial r} + \tan(\varphi - \mu)\frac{\partial \eta}{\partial \theta} = \frac{1}{2r}\frac{\sin(\varphi - \mu) - x\sin(\varphi + \mu)}{\cos(\varphi - \mu)}$$
$$- \frac{\gamma\cos(\varphi + \mu)}{2\sigma\sin\varrho\cos(\varphi - \mu)},$$

$$\tag{15}$$

$\chi = \pm 1$ selon le type d'équilibre le long de OS. $OS = a$ est le rayon du poinçon.

Après passage aux coordonnées polaires d'origine S origine de la singularité de PRANDTL

$$r = s\cos\theta,$$
$$z = s\sin\theta,$$

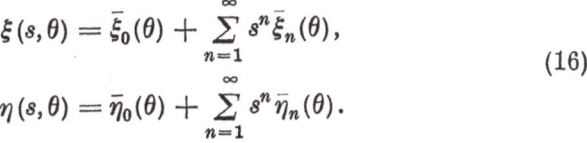

Fig. 5.

la recherche de la solution analytique de (15) dans la zone CSD revient à expliciter les fonctions $\bar{\xi}_n$ et $\bar{\eta}_n$ permettant de développer ξ et η en série de la forme:

$$\xi(s,\theta) = \bar{\xi}_0(\theta) + \sum_{n=1}^{\infty} s^n\bar{\xi}_n(\theta),$$
$$\eta(s,\theta) = \bar{\eta}_0(\theta) + \sum_{n=1}^{\infty} s^n\bar{\eta}_n(\theta).$$

$$\tag{16}$$

Nous donnons ci-dessous les formules pour $n = 1$ dans le cas de la poussée sur SA et butée sur SO ($\varkappa = +1$) (cf. pour les détails [11])

$$\bar{\eta}_1 = \frac{\sin(\theta - \mu) - \tan\varrho\cos(\theta - \mu)}{2a^1\cos\mu(\tan^2\varrho + 1)} - \frac{e^{-2\tan\theta}(\sin\theta - 3\tan\varrho\cos\theta)}{2\sigma_0^*\sin\varrho\cos\varrho(9\tan^2\varrho + 1)} + C_1 e^{\tan\varrho\theta}$$

$$C_1 \quad \text{calculé par} \quad (\bar{\eta}_1)_{s,\text{II}} = \left(\frac{\partial\eta}{\partial r}\right)_{s,\text{II}},$$

$$\bar{\xi}_1 = \frac{\bar{\eta}_1}{2} - \frac{1}{2}\left[\frac{\sin\mu\cos(\theta - \mu)}{a^1} - \frac{\cos(\theta - 2\mu)}{2\sigma_0^* e^{2\tan\varrho\theta}\sin\varrho}\right], \tag{17}$$

Nous avons explicité les formules pour $n = 2$ (non données ici). A partir de $n = 3$ les calculs sont trop laborieux pour permettre une utilisation pratique, mais la remarque faite précédemment conserve sa valeur.

4. Problème de Cauchy du Type III

Il s'agit de trouver une solution de (1) dans un domaine limité d'une part par une frontière OB sur laquelle des conditions sont

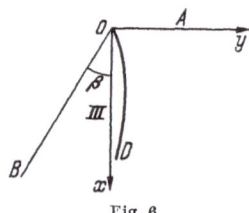

Fig. 6.

imposées, d'autre part par la bicaractéristique OD précédente (fig. 6). Dans cette zone — où la solution est supposée régulière — nous reprenons x et y comme variables, les fonctions inconnues pouvant être soit ξ, η, soit σ et φ.

φ est connu par les conditions aux limites le long de l'écran; dans les applications $\varphi = C^{\text{ste}}$ et OB est une droite $(\widehat{Ox, OB}) = \beta$. Nous déterminons σ et φ (ou ξ et η) au moyen des développements limités, les dérivées utiles de σ et φ en O_2 étant calculées en utilisant:

1° l'équation de la caractéristique OD et les valeurs des fonctions σ et φ (ou ξ et η) le long de celle-ci, ainsi que les valeurs des dérivées au point O_2, zone BOD,

2° les conditions limites sur OB; $\varphi = C^{\text{ste}}$, $\dfrac{\partial\varphi}{\partial x} + \dfrac{\partial\varphi}{\partial y} \cdot \dfrac{dy}{dx}\tan\beta = 0$,

3° les équations (1) ou (2) de l'équilibre indéfini et les équations obtenues par dérivation par rapport à x et à y.

On consultera pour des exemples d'application [2] et [3].

5. Problème avec Ligne de Discontinuité des Contraintes

Dans les problèmes que nous venons d'examiner la répartition des contraintes était continue dans tout le massif, point singulier O exclu. Cette solution n'est pas toujours valable, elle exige que β reste inférieur à un angle limite β_0. Si $\beta > \beta_0$ il faut imaginer un schéma à 2 zones séparées par une ligne de discontinuité des contraintes OR, de forme inconnue à priori (Fig. 7). Dans chacune de ces 2 zones la solution est, par hypothèse, analytique et régulière, origine comprise. Le problème revient ici à:

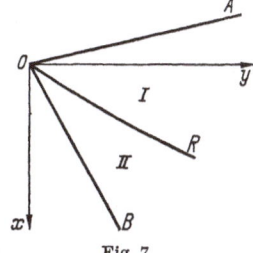

a) déterminer l'équation approchée de OR sous forme d'un développement limité;

b) déterminer les coefficients des développements limités des fonctions inconnues σ et φ (ou ξ et η) autour de O.

Le long de la ligne de discontinuité, les composantes σ_n et τ_{nt} sont égales de part d'autre. Par contre, σ_t subit une discontinuité.

Fig. 7.

On écrit l'équation approchée de la ligne de discontinuité; la solution dans la zone II est alors la même que dans la zone III du schéma précédent à 3 zones. Dans la zone I la solution est ou bien explicite, ou bien peut être déterminée par la méthode d'approximation exposée ci-dessus. On se reportera pour plus de détail à [3] et [4].

III. Exemples d'Application

1. Problème des Fondations

Le massif, en état d'équilibre limite, est limité par 2 plans de traces OA et OB sur le plan vertical (fig. 8).

Le matériau cohérent et pesant obéit à la loi de COULOMB. Le long de OA s'exerce une charge normale, variant linéairement. L'équilibre le long de OA est un équilibre de butée (OC est alors une bicaractéristique de la 2e famille).

Données numériques:

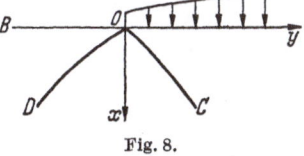

$$\varrho = 20°, \quad \varphi(0,y) = 0 \qquad \text{pour } y > 0,$$
$$\varphi(0,y) = -\frac{\pi}{2} \quad \text{pour } y < 0$$

sur OA: $\sigma_x = 40 + 4y$.

Fig. 8.

On trouvera dans les tableaux ci-après quelques résultats numériques obtenus par notre méthode comparés à ceux donnés par SOKOLOVSKI qui a construit point par point le réseau des caractéristiques.

Zone I. *Détermination de OC et des Inconnues sur OC*

x	Méthode décrite			Résultats de SOKOLOVSKI		
	y	ξ	η	y	ξ	η
0,14	0,097	4,758	4,784	0,10	4,76	4,78
0,71	0,466	4,777	4,897	0,47	4,78	4,90
1,42	0,879	4,802	5,028	0,88	4,80	5,03

Zone II. *Détermination de OD et des Inconnues sur OD*

x	Méthode décrite			Résultats de Sokolovski		
	y	ξ	η	y	ξ	η
0,18	− 0,254	1,658	4,773	− 0,25	1,65	4,77
0,86	− 1,174	1,783	4,840	− 1,18	1,78	4,85
1,66	− 2,189	1,924	4,916	− 2,20	1,92	4,94

Zone III. *Contrainte sur OB*

x	y	Méthode de décrite		Résultats de Sokolovski	
		ξ	η	ξ	η
0	− 0,50	1,596	4,737	1,59	4,74
0	− 2,45	1,541	4,682	1,53	4,67
0	− 4,73	1,511	4,653	1,46	4,60

2. Mur de Soutènement avec Surface Inclinée

La surface du sol est inclinée de l'angle $(\widehat{Ox, Ou}) = (Oy, Ov) = \varepsilon$; on utilise les axes Ouv, déduits par rotation de ε de Oxy (fig. 9)

Le massif est homogène, isotrope de poids spécifique γ, de cohésion c; sur la surface inclinée agit la charge normale et uniforme p. Le calcul est fait dans le cas des 3 zones (cf. fig. 3) et en variables adimensionnelles:

$$\sigma' = \frac{\sigma}{k}, \quad u' = \frac{\gamma}{k}u, \quad v' = \frac{\gamma}{k}v, \quad \sigma_0 = k, \quad k = p + H,$$

$$\varrho = 20°, \quad \varepsilon = 6°40, \quad \beta = -40°. \quad \text{Butée sur } OB.$$

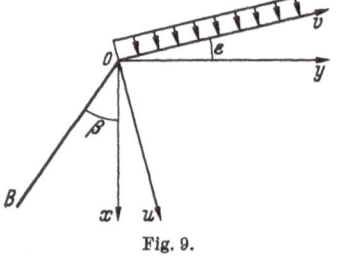

Fig. 9.

On trouvera dans les tableaux ci-après les résultats obtenus par notre méthode comparés à ceux, donnés par une méthode numérique, de Hajal. Remarquons que dans la zone I la solution explicite est connue et a permis de calculer le rayon de convergence des développements limités dans cette zone. On trouve ici $u \leq 0,41$; nous nous sommes bornés aux résultats pour $u < 0,4$ mais le prolongement analytique a été fait dans tout le domaine et a donné une bonne précision.

Zone I. *Caractéristique OC — Valeurs de ξ et η sur OC*

u	Méthode décrite			Résultats théoriques		
	v	ξ	η	v	ξ	η
0	0	2,2619	− 1,1123	0	2,2621	− 1,1112
0,1288	0,1865	2,4400	− 0,9643	0,1867	2,4403	− 0,9594
0,2955	0,4344	2,6432	− 0,7836	0,4331	2,6405	− 0,7848

Zone II. *Caractéristique OD — Valeurs de ξ et η sur OD*

u	Méthode décrite			Résultats HAJAL		
	v	ξ	η	v	ξ	η
0	0	2,2619	1,2237	0	2,2621	1,2239
0,1577	— 0,0243	2,5374	1,2862	— 0,0250	2,5193	1,2811
0,3819	— 0,0355	2,909	1,3911	— 0,0380	2,8008	1,3552

Zone III. *Valeurs de ξ et η le long de OB*

u	Méthode décrite		Résultats HAJAL	
	ξ	η	ξ	η
0	2,2619	1,2237	2,2621	1,2239
0,1080	2,5198	1,4681	2,5179	1,4797
0,4947	3,1552	1,9403	3,1216	2,0833

On voit donc que la méthode décrite fournit toutes les données utiles à l'ingénieur avec toute la précision souhaitable. Il convient d'observer que notre procédé fournit directement la répartition des contraintes le long de OB, seule donnée utile à l'ingénieur, sans qu'il soit nécessaire de déterminer les équations de OC et de OD et la répartition des contraintes le long de ces courbes. Nous n'avons calculé ces éléments intermédiaires qu'à titre de test de contrôle. Au contraire, les méthodes numériques habituelles exigent le tracé des réseaux de bicaractéristiques dans les trois zones AOC, COD, DOB. On voit l'économie de calcul que permet de réaliser l'emploi de notre méthode.

3. Répartition des Contraintes sous un Poinçon Cylindrique

Signalons le problème suivant dont l'étude est actuellement en cours. Nous nous proposons de chercher la répartition des contraintes sous un poinçon cylindrique de diamètre $2a$, le sol à l'extérieur étant chargé uniformément et normalement (fig. 10). Par notre méthode nous avons déterminé exactement la tangente en A et la parabole osculatrice à la courbe $\sigma = \sigma(r)$. Ces résultats ont permis de choisir correctement le pas d'intégration lors d'un calcul numérique par la méthode des différences finies. Il est à noter, en effet, que l'étude rigoureuse locale, a priori, de la singularité de PRANDTL peut être utile au calculateur. Dans le cas traité, il se trouve que la courbure du graphique Γ de $\sigma = \sigma(r)$ en A est très petite. Le choix d'un pas trop grand peut alors entrainer des erreurs considérables pour le résultat; par exemple, le calcul peut donner un faux signe à la courbure

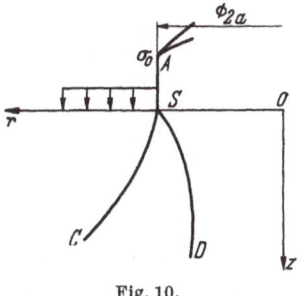

Fig. 10.

de Γ en A et l'allure de Γ obtenue à partir d'un pas trop grand peut être entièrement fausse. C'est ce qui s'est effectivement produit au cours de plusieurs calculs publiés dans la littérature. Dans le cas présent donc, pour appliquer notre méthode, il faut pousser les calculs au moins jusqu'àu troisième ordre d'approximation, ce qui exige des calculs assez laborieux mais qui, en revanche, ont un caractère universel et qui s'appliquent à tous les cas.

Conclusion

En conclusion, il nous semble que notre procédé présente encore un intérêt en dépit du développement des méthodes de calcul à la machine. Son emploi n'exige qu'une programmation très simple. Dans de nombreux problèmes, il offre l'avantage de pouvoir être mis en oeuvre avec des moyens de calcul rudimentaires et être dès lors, utilisé avec succès par ceux qui ne disposent pas de calculatrices électroniques. Dans d'autres cas, il donne localement une solution très voisine de la solution exacte et permet de ce fait de choisir une méthode correcte de calcul numérique à la machine.

Bibliographie

[1] KRAVTCHENKO, J., et R. SIBILLE: Comptes-Rendus Académie des Sciences **255**, p. 79.
[2] SIBILLE, R.: Comptes-Rendus **255**, 365.
[3] DEMBICKI, E.:, et R. SIBILLE: Comptes-Rendus **256**, 593.
[4] DEMBICKI, E.: Comptes-Rendus **256**, 2537.
[5] DEMBICKI, E.: Comptes-Rendus **256**, 2778.
[6] DEMBICKI, E.: Comptes-Rendus **256**, 3579.
[7] NEGRE, R.: Comptes-Rendus **256**, 4358.
[8] NEGRE, R.: Archivum Hydrotechnici X (1963).
[9] KRAVTCHENKO, J., et R. SIBILLE: Comptes-Rendus **257**, 3297.
[10] SIBILLE, R.: Comptes-Rendus **258**, 75.
[11] SIBILLE, R.: Comptes-Rendus **258**, 2017.
[12] SOKOLOVSKI, V. V.: Statics of Soil Media, London: Butterworths Scientific Publications, 1960, p. 40.
[13] HAJAL, M.: Etude générale de la butée d'un écran plan contre un massif cohérent par la théorie des caractéristiques. Thèse de 3e cycle, Grenoble 1961.
[14] HEURTAUX, J.: Calcul des contraintes en poussée exercées par un massif pulvérulent sur un écran plan. Recueil de l'Ingénieur, édition de l'Académie des Sciences de l'URSS 61, Tome I, Fac. 1, pp. 135, 144.
[15] BEREZANCEW, B. G.: Problème de l'équilibre limite d'un milieu pulvérulent en symétrie axiale (en russe), Moscou: Edition d'état de la littérature technique et théorique 1952.
[16] COX, A. D., G. EASON et H. G. HOPKINS: Axially symetric plastic deformation in soils. Math. Phys. Sci. **254**, No. 1036, 1, 45 (1961).
[17] DEMBICKI, E.: Proceedings of the seminar on Soil Mechanics and foundation engineering, Łódź. juin 1964, p. 47—60.

[*18*] Dembicki, E., J. Kravtchenko et R. Sibille: Journal de Mécanique **3**, No. 3, Septembre 1964.

[*19*] Dembicki, E.: Thése d' Ingénieur-Docteur, Grenoble 1962.

Discussion

Question posée par E. de Beer: On peut être particulièrement heureux que M. le Professeur Kravtchenko ait établi une solution explicite approchée pour le problème de l'équilibre limite plan des milieux pesants, ce qui simplifie fortement les calculs par rapport aux méthodes numériques de Sokolovsky ou graphiques de Josselin de Jong.

M. Kravtchenko a notamment cité comme exemples les problèmes de poussée et de butée sur les murs de soutènement et les problèmes de fondations. Toutefois les exemples donnés concernent le cas de l'existence d'un seul point singulier.

Comment peut-on appliquer la nouvelle méthode explicite pour résoudre le problème d'un mur de hauteur limitée, où il a une perturbation de l'état de contrainte au pied D.

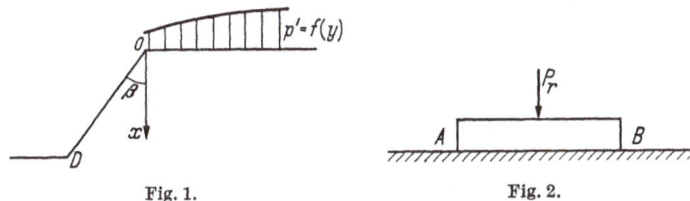

Fig. 1.   Fig. 2.

Comment peut-on appliquer la nouvelle méthode pour déterminer la charge limite centrée ou excentrée sur une fondation rigide rugueuse AB établie à la surface ou à une certaine profondeur dans un milieu pesant. Ce sont les deux problèmes pratiques pour l'ingénieur.

Réponse de J. Kravtchenko: Je réponds que la méthode décrite ci-dessus peut être appliquée au cas de deux singularités là où il existe un schema résolutif. Par exemple, dans le cas symétrique, on peut en concevoir plusieurs. Commençons par un schéma où tout le massif est à l'état d'équilibre limite. Alors on sait appliquer la méthode. Mais son exactitude n'a pas été testée par le calcul.

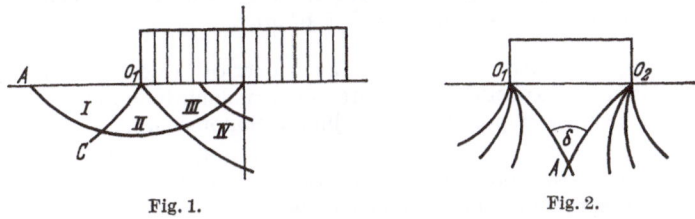

Fig. 1.   Fig. 2.

Pour le schéma de M. Josselin de Jong il suffit de déterminer les caractéristiques O_1A et O_2A, symétriques, qui se coupent suivant un angle donné à priori. Cela est aisé. La question est de savoir seulement si la courbe O_1A approche suffisamment la courbe exacte.

Question posée par J. Verdeyen: Le conférencier a attiré l'attention sur les divergences obtenues dans certains cas par la méthode simplifiée qu'il propose. Nous croyons utile de faire remarquer que des divergences analogues se sont produites lors de l'application de la méthode originale de Sokolovsky à certains cas particuliers. Le changement de variables $\chi = \frac{1}{2} \cot \varrho \ln \frac{\sigma}{\sigma_0}$ fait intervenir la constante σ_0 que l'on peut choisir arbitrairement. Suivant le choix de cette constante, le rapport $\frac{\sigma}{\sigma_0}$ peut être très faible ou très grand. Il en est résulté des divergences surtout pour les valeurs élevées du rapport. Les calculs qui ont toujours donné des résultats positifs correspondent à des valeurs de $\frac{\sigma}{\sigma_0}$ comprises entre 1 et 2. Tout ce qui précède n'est bien entendu valable que dans le stade actuel de nos calculs.

Réponse de J. Kravtchenko:

1. La méthode proposée par M. Sibille et moi-même est *localement* très supérieure à la méthode des caractéristiques. Je précise sur le graphique ci-après: la bicaractéristique M_N est remplacée par la ligne polygonale $MPQN$ qui est d'autant plus voisine de la courbe exacte que le pas MP est plus petit.

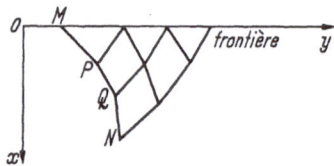

Notre méthode, au contraire, donne la tangente, la courbure, plus généralement, les dérivées de tous les ordres de $y = y(x)$, ($y = y(x)$ étant l'équation de la bicaractéristique) au point M. Donc, dans le voisinage de M, la méthode décrite donne des résultats très supérieurs. Mais au delà d'un voisinage qu'on ne sait d'ailleurs délimiter, la courbe réelle peut s'écarter beaucoup de la courbe approchée.

2. En ce qui concerne l'«ennui» de calcul, je l'attribue au choix défectueux du pas d'intégration. A ce sujet, MM. Sibille et Nègre peuvent donner beaucoup d'indications utiles. Signalons que nous avons eu des difficultés analogues avec le problème du poinçon de révolution. Dans ce cas, le procédé de Sokolovsky converge mal; si on diminue le pas, la courbe approchée change sensiblement — ce que la méthode décrite permet de prévoir à priori.

Question posée par Bent Hansen: The numerical calculation method presented by Messrs. Kravtchenko and Sibille is undoubtly an improvement when compared with the step-by-step procedure of Sokolovsky.

However, in practice it is usually only the stress resultants that are required such as total earth pressures or bearing capacities, not the detailed stress distributions. In this case there exists two even simpler methods based on the use of Kötter's equations along the zone boundaries and the conditions of equilibrium for the entire zone elements. The zone boundaries may be approximated with circle arcs or continuous chains of circle arcs. The geometrical conditions and the stress conditions along the zone boundaries must be satisfied strictly, but the equations of projection and the moment equation for the zone elements are more or less optional. If they are not satisfied some assumption must be made about a linear stress distribution.

The first method is indicated by BRINCH HANSEN. It uses only the extreme zone boundary, not the internal boundaries between zone elements. Then at least one of the equations of projection must be used to give good approximations. In the second method the secondary slip lines are also considered and then very good approximations can be obtained by using only the geometrical and the stress conditions. If necessary the major zone elements may be subdivided (giving then also a stress distribution) but one need hardly ever do that.

Réponse de R. SIBILLE: Nous remercions M. BENT HANSEN de ses remarques relatives à la méthode de calcul de M. BRINCH HANSEN. Toutefois nous remarquons qu'il s'agit d'écrire l'équilibre d'une zone limitée par des lignes de glissement choisies arbitrairement comme arcs de cercles. Il nous apparaît plus logique d'utiliser les schémas habituels (à 3 zones avec singularité de PRANDTL, à 2 zones avec ligne de discontinuité de contraintes) et de rechercher dans tout le domaine une solution analytique par la méthode décrite.

Notons d'ailleurs que la détermination des contraintes inconnues le long d'un écran ou sous une fondation, sous forme d'un polynome, permet par une intégration simple, le calcul de la force portante utile à l'Ingénieur.

2.4 Rhéologie des Problèmes de Déformation Plane des Milieux Continus

Par

Z. Sobotka

1. La Notion des Contraintes et des Déformations Équivalentes

L'auteur a introduit la notion des contraintes et des déformations équivalentes afin de transformer les complexes équations rhéologiques en plus simples relations entre les composantes des deux variables équivalentes.

En faisant quelques restrictions concernant le comportement rhéologique, on peut exprimer la relation déformations-contraintes pour un matériel anisotrope comme il suit:

$$\varphi\left[\varepsilon_{ij}, \frac{\partial \varepsilon_{ij}}{\partial t}, \frac{\partial^2 \varepsilon_{ij}}{\partial t^2}, \dots \frac{\partial^p \varepsilon_{ij}}{\partial t^p}, \int_0^t K_1(t, \tau_1)\, \varepsilon_{ij}(\tau_1)\, d\tau_1, \right.$$

$$\left. \int_0^t \int_0^{\tau_1} K_2(t, \tau_2)\, \varepsilon_{ij}(\tau_2)\, d\tau_1\, d\tau_2, \dots, t \right]$$

$$= f_{ij}\left\{ A_{klmn}, \psi\left[\sigma_{mn}, \frac{\partial \sigma_{mn}}{\partial t}, \frac{\partial^2 \sigma_{mn}}{\partial t^2}, \dots \frac{\partial^r \sigma_{mn}}{\partial t^r}, \right.\right.$$

$$\left.\left. \int_0^t L_1(t, \tau_1)\, \sigma_{mn}(\tau_1)\, d\tau_1, \int_0^t \int_0^{\tau_1} L_2(t, \tau_2)\, \sigma_{mn}(\tau_2)\, d\tau_1 d\tau_2, \dots t \right], \right\}$$

(1.1)

$\varepsilon_{ij}, \frac{\partial \varepsilon_{ij}}{\partial t}, \frac{\partial^2 \varepsilon_{ij}}{\partial t^2}, \dots$ désignant les composantes du tenseur des déformations et de leurs dérivées par rapport au temps,

$$\int_0^t K_1(t, \tau_1)\, \varepsilon_{ij}(\tau_1)\, d\tau_1, \int_0^t \int_0^{\tau_1} K_2(t, \tau_2)\, \varepsilon_{ij}(\tau_2)\, d\tau_1\, d\tau_2, \dots$$

les fonctions intégrales des composantes du tenseur des déformations et du temps,

$\sigma_{mn}, \frac{\partial \sigma_{mn}}{\partial t}, \frac{\partial^2 \sigma_{mn}}{\partial t^2}, \dots$ les composantes du tenseur des contraintes et de leurs dérivées,

$$\int_0^{t_1} L_1(t, \tau_1)\, \sigma_{mn}(\tau_1)\, d\tau_1, \int_0^t \int_0^{\tau_1} L_2(t, \tau_2)\, \sigma_{mn}(\tau_2)\, d\tau_1\, d\tau_2, \dots$$

les fonctions intégrales des composantes du tenseur des contraintes et du temps,

A_{klmn} le tenseur de l'anisotropie, qui est du quatrième ordre,

t le temps.

Portant les déformations et les contraintes équivalentes, données par:

$$\bar{\varepsilon}_{ij} = \varphi\left[\varepsilon_{ij}, \frac{\partial \varepsilon_{ij}}{\partial t}, \frac{\partial^2 \varepsilon_{ij}}{\partial t^2}, \ldots \frac{\partial^p \varepsilon_{ij}}{\partial t^p}, \int\limits_0^t K_1(t,\tau_1)\,\varepsilon_{ij}(\tau_1)\,d\tau_1, \right.$$
$$\left. \int\limits_0^t \int\limits_0^{\tau_1} K_2(t,\tau_2)\,\varepsilon_{ij}(\tau_2)\,d\tau_1\,d\tau_2, \ldots t\right], \tag{1.2}$$

$$\bar{\sigma}_{mn} = \psi\left[\sigma_{mn}, \frac{\partial \sigma_{mn}}{\partial t}, \frac{\partial^2 \sigma_{mn}}{\partial t^2}, \ldots \frac{\partial^r \sigma_{mn}}{\partial t^r}, \int\limits_0^t L_1(t,\tau_1)\,\sigma_{mn}(\tau_1)\,d\tau_1, \right.$$
$$\left. \int\limits_0^t \int\limits_0^{\tau_1} L_2(t,\tau_2)\,\sigma_{mn}(\tau_2)\,d\tau_1\,d\tau_2, \ldots t\right], \tag{1.3}$$

dans l'équation (1.1), nous obtenous:

$$\bar{\varepsilon}_{ij} = f_{ij}(A_{klmn}, \bar{\sigma}_{mn}). \tag{1.4}$$

En introduisant la notion des déformations et des contraintes équivalentes, l'auteur a fait les réserves suivantes:

1° Le tenseur de l'anisotropie ne varie pas avec le temps mais il peut dépendre des coordonnées.

2° Les fonctions φ et ψ ne dépendent pas des coordonnées mais elles peuvent varier avec le temps.

3° Les fonctions φ et ψ sont linéaires par rapport aux composantes de déformation et de contrainte, de leurs dérivées et de leurs intégrales mais elles peuvent être non-linéaires par rapport au temps.

4° Le chargement externe change avec le temps par sa quantité mais sa forme reste constante. Il peut être représenté par le produit de deux fonctions dont l'une dépend des coordonnées et l'autre du temps.

5° Les conditions aux limites ne changent pas avec le temps.

La notion des déformations et des contraintes équivalentes rend possible l'intégration séparée par rapport aux coordonnées de celle par rapport au temps, ce qui conduit aux simplifications considérables dans la solution des problèmes concernant les déformations rhéologiques.

2. Divers Types Particuliers des Relations pour les Déformations et les Contraintes Équivalentes

Dans certaines conditions, l'équation (1.4) peut être développée au moyen des règles de l'algèbre tensorielle en formule suivante

$$\bar{\varepsilon}_{ij} = \phi_0 B_{ijkl}\delta_{kl} + \phi_1 B_{ijkl}\bar{\sigma}_{kl} + \phi_2 B_{i\alpha kl} B_{\alpha jmn}\bar{\sigma}_{kl}\bar{\sigma}_{mn}, \tag{2.1}$$

δ_{kl} étant le delta de Kronecker, B_{ijkl} le tenseur de l'anisotropie, qui est du quatrième ordre, et ϕ_0, ϕ_1, ϕ_2 les fonctions d'invariants, qui sont définies par les systèmes des trois équations suivantes

$$\bar{\varepsilon}_{ij}\delta_{ij} = \phi_0 B_{ijkl}\delta_{ij}\delta_{kl} + \phi_1 B_{ijkl}\delta_{ij}\bar{\sigma}_{kl} + \phi_2 B_{i\alpha kl}B_{\alpha jmn}\delta_{ij}\bar{\sigma}_{kl}\bar{\sigma}_{mn}, \quad (2.2)$$

$$\begin{aligned}
\bar{\varepsilon}_{ij}\bar{\varepsilon}_{ij} = {}& \phi_0^2 B_{ijkl}B_{ijmn}\delta_{kl}\delta_{mn} + \phi_1^2 B_{ijkl}B_{ijmn}\bar{\sigma}_{kl}\bar{\sigma}_{mn} \\
& + \phi_2^2 B_{i\alpha kl}B_{\alpha jmn}B_{i\beta pq}B_{\beta jrs}\bar{\sigma}_{kl}\bar{\sigma}_{mn}\bar{\sigma}_{pq}\bar{\sigma}_{rs} \\
& + 2\phi_0\phi_1 B_{ijkl}B_{ijmn}\delta_{kl}\bar{\sigma}_{mn} + 2\phi_0\phi_2 B_{ijkl}B_{i\alpha mn}B_{\alpha jpq}\delta_{kl}\bar{\sigma}_{mn}\bar{\sigma}_{pq} \\
& + 2\phi_1\phi_2 B_{ijkl}B_{i\alpha mn}B_{\alpha jpq}\bar{\sigma}_{kl}\bar{\sigma}_{mn}\bar{\sigma}_{pq},
\end{aligned} \quad (2.3)$$

$$\begin{aligned}
\bar{\varepsilon}_{ij}\bar{\varepsilon}_{i\alpha}\bar{\varepsilon}_{\alpha j} = {}& \phi_0^3 B_{ijkl}B_{i\alpha mn}B_{\alpha jpq}\delta_{kl}\delta_{mn}\delta_{pq} + \phi_1^3 B_{ijkl}B_{i\alpha mn}B_{\alpha jpq}\bar{\sigma}_{kl}\bar{\sigma}_{mn}\bar{\sigma}_{pq} \\
& + \phi_2^3 B_{i\alpha kl}B_{\alpha jmn}B_{i\beta pq}B_{\beta jrs}B_{\gamma \delta tu}B_{\delta jab}\bar{\sigma}_{kl}\bar{\sigma}_{mn}\bar{\sigma}_{pq}\bar{\sigma}_{rs}\bar{\sigma}_{tu}\bar{\sigma}_{ab} \\
& + 3\phi_0^2\phi_1 B_{ijkl}B_{i\alpha mn}B_{\alpha jpq}\delta_{kl}\delta_{mn}\bar{\sigma}_{pq} \\
& + 3\phi_0\phi_1^2 B_{ijkl}B_{i\alpha mn}B_{\alpha jpq}\delta_{kl}\bar{\sigma}_{mn}\bar{\sigma}_{pq} \\
& + 3\phi_0^2\phi_2 B_{ijkl}B_{i\alpha mn}B_{\alpha\beta pq}B_{\beta jrs}\delta_{kl}\delta_{mn}\bar{\sigma}_{pq}\bar{\sigma}_{rs} \\
& + 3\phi_0\phi_2^2 B_{ijkl}B_{i\alpha mn}B_{\alpha\beta pq}B_{\beta jrs}B_{\gamma jtu}\delta_{kl}\bar{\sigma}_{mn}\bar{\sigma}_{pq}\bar{\sigma}_{rs}\bar{\sigma}_{tu} \\
& + 3\phi_1^2\phi_2 B_{ijkl}B_{i\alpha mn}B_{\alpha\beta pq}B_{\beta jrs}\bar{\sigma}_{kl}\bar{\sigma}_{mn}\bar{\sigma}_{pq}\bar{\sigma}_{rs} \\
& + 3\phi_1\phi_2^2 B_{ijkl}B_{i\alpha mn}B_{\alpha\beta pq}B_{\beta jrs}B_{\gamma jtu}\bar{\sigma}_{kl}\bar{\sigma}_{mn}\bar{\sigma}_{pq}\bar{\sigma}_{rs}\bar{\sigma}_{tu} \\
& + 6\phi_0\phi_1\phi_2 B_{ijkl}B_{i\alpha mn}B_{\alpha\beta pq}B_{\beta jrs}\delta_{kl}\bar{\sigma}_{mn}\bar{\sigma}_{pq}\bar{\sigma}_{rs}.
\end{aligned} \quad (2.4)$$

L'auteur a dérivé les autres relations entre les déformations et les contraintes équivalentes de la fonction énergétique des contraintes équivalentes, qui, pour la déformation plane d'un milieu à l'anisotropie orthogonale, est donnée par l'expression suivante:

$$\begin{aligned}
\Psi_0 = {}& \frac{1}{2}\big(c_{11}\bar{\sigma}_x^2 + c_{22}\bar{\sigma}_y^2 + c_{33}\bar{\sigma}_z^2 \\
& + 2c_{12}\bar{\sigma}_x\bar{\sigma}_y + 2c_{13}\bar{\sigma}_x\bar{\sigma}_z + 2c_{23}\bar{\sigma}_y\bar{\sigma}_z + c_{44}\bar{\tau}_{xy}^2\big) \\
& + \sum_{n=2}^{\infty}\sum_{k=0}^{n+1}\frac{2}{n+1+|n+1-2k|}\big(c_{(n+1-k)(1)}\bar{\sigma}_x^{n+1-k}\bar{\sigma}_y^k \\
& + c_{(n+1-k)(1)k(3)}\bar{\sigma}_x^{n+1-k}\bar{\sigma}_z^k + c_{(n+1-k)(2)k(3)}\bar{\sigma}_y^{n+1-k}\bar{\sigma}_z^k\big) \\
& + \sum_{n=2}^{\infty}\sum_{k=0}^{n+1}\sum_{j=0}^{n+1-k}\frac{2c_{(n+1-k-j)(1)(2)k(2)j(3)}}{n+1+|n+1-2k-2j|}\bar{\sigma}_x^{n+1-k-j}\bar{\sigma}_y^k\bar{\sigma}_z^j \\
& + \sum_{n=2}^{\infty}c_{(n+1)(4)}\bar{\tau}_{xy}^{n+1},
\end{aligned} \quad (2.5)$$

$c_{(n+1-k)(1)k(2)}$ désignant le coefficient avec l'index 1 répété $n + 1 - k$ fois et avec l'index 2 qui est répété k-fois.

Après la différentiation de la fonction (2.5) par rapport aux composantes $\bar{\sigma}_x$, $\bar{\sigma}_y$, $\bar{\sigma}_z$ et $\bar{\tau}_{xy}$, on obtient les expressions pour les déformations équivalentes

$$
\begin{aligned}
\bar{\varepsilon}_x = {}& c_{11}\bar{\sigma}_x + c_{12}\bar{\sigma}_y + c_{13}\bar{\sigma}_z \\
& + \sum_{n=2}^{\infty}\sum_{k=0}^{n+1} \frac{2(n+1-k)}{n+1+|n+1-2k|}\left(c_{(n+1-k)(1)k(2)}\,\bar{\sigma}_x^{n-k}\bar{\sigma}_y^k\right. \\
& \left. + c_{(n+1-k)(1)k(3)}\,\bar{\sigma}_x^{n-k}\bar{\sigma}_z^k\right) \\
& + \sum_{n=2}^{\infty}\sum_{k=0}^{n+1}\sum_{j=0}^{n+1-k} \frac{2(n+1-k-j)}{n+1+|n+1-2k-2j|} \\
& \times c_{(n+1-k-j)(1)k(2)j(3)}\bar{\sigma}_x^{n-k-j}\bar{\sigma}_y^k\bar{\sigma}_z^j,
\end{aligned}
\tag{2.6}
$$

$$
\begin{aligned}
\bar{\varepsilon}_y = {}& c_{12}\bar{\sigma}_x + c_{22}\bar{\sigma}_y + c_{23}\bar{\sigma}_z \\
& + \sum_{n=2}^{\infty}\sum_{k=0}^{n+1} \frac{2}{n+1+|n+1-2k|}\left[k\,c_{(n+1-k)(1)k(2)}\,\bar{\sigma}_x^{n+1-k}\bar{\sigma}_y^{k-1}\right. \\
& \left. + (n+1-k)\,c_{(n+1-k)(2)k(3)}\,\bar{\sigma}_y^{n-k}\bar{\sigma}_z^{k-1}\right] \\
& + \sum_{n=2}^{\infty}\sum_{k=0}^{n+1}\sum_{j=0}^{n+1-k} \frac{2k}{n+1+|n+1-2k-2j|} \\
& \times c_{(n+1-k)(1)k(2)j(3)}\bar{\sigma}_x^{n+1-k-j}\bar{\sigma}_y^{k-1}\bar{\sigma}_z^j,
\end{aligned}
\tag{2.7}
$$

$$
\begin{aligned}
\bar{\varepsilon}_z = {}& c_{13}\bar{\sigma}_x + c_{23}\bar{\sigma}_y + c_{33}\bar{\sigma}_z \\
& + \sum_{n=2}^{\infty}\sum_{k=0}^{n+1} \frac{2k}{n+1+|n+1-2k|}\left(c_{(n+1-k)(1)k(3)}\,\bar{\sigma}_x^{n+1-k}\bar{\sigma}_z^{k-1}\right. \\
& \left. + c_{(n+1-k)(2)k(3)}\,\bar{\sigma}_y^{n+1-k}\bar{\sigma}_z^{k-1}\right) \\
& + \sum_{n=2}^{\infty}\sum_{k=0}^{n+1}\sum_{j=0}^{n+1-k} \frac{2j}{n+1+|n+1-2k-2j|} \\
& \times c_{(n+1-k-j)(1)k(2)j(3)}\bar{\sigma}_x^{n+1-k}\bar{\sigma}_y^k\bar{\sigma}_z^{j-1},
\end{aligned}
\tag{2.8}
$$

$$
\bar{\gamma}_{xy} = c_{44}\bar{\tau}_{xy} + \sum_{n=2}^{\infty} c_{(n+1)(4)}\bar{\tau}_{xy}^n.
\tag{2.9}
$$

Dans le cas de déformation plane, nous avons $\bar{\varepsilon}_z = 0$ et c'est pourquoi la contrainte équivalente $\bar{\sigma}_z$ peut être exprimée de l'équation non-linéaire (2.8) au moyen des composantes $\bar{\sigma}_x$ et $\bar{\sigma}_y$. L'élimination de $\bar{\sigma}_z$ de l'équation (2.8) peut présenter, dans le cas général, de grandes difficultés. C'est pourquoi, nous allons borner les considérations aux cas avec un nombre fini des membres.

La solution se simplifie beaucoup si la déformation dans la direction de l'axe z peut être prise, même approximativement, pour linéaire. Dans ce cas, nous avons

$$\bar{\varepsilon}_z \equiv c_{13}\bar{\sigma}_x + c_{23}\bar{\sigma}_y + c_{33}\bar{\sigma}_z, \qquad (2.10)$$

d'où

$$\bar{\sigma}_z = -\frac{1}{c_{33}}(c_{13}\bar{\sigma}_x + c_{23}\bar{\sigma}_y). \qquad (2.11)$$

Après l'élimination de la contrainte équivalente $\bar{\sigma}_z$, la fonction énérgétique des contraintes équivalentes peut s'écrire de la manière suivante

$$\Psi_0 = \frac{1}{2}(C_{11}\bar{\sigma}_x^2 + 2C_{12}\bar{\sigma}_x\bar{\sigma}_y + C_{22}\bar{\sigma}_y^2 + c_{44}\bar{\tau}_{xy}^2)$$

$$+ \sum_{n=2}^{m}\sum_{k=0}^{n+1}\frac{2}{n+1+|n+1-2k|}C_{(n+1-k)(1)k(2)}\bar{\sigma}_x^{n+1-k}\bar{\sigma}_y^k$$

$$+ \sum_{n=2}^{m}c_{(n+1)(4)}\bar{\tau}_{xy}^{n+1}, \qquad (2.12)$$

où $C_{(n+1-k)(1)k(2)}$ sont les nouveaux modules de déformation. Pour la déformation linéaire, nous avons

$$C_{11} = c_{11} - \frac{c_{13}^2}{c_{33}}, \quad C_{12} = c_{12} - \frac{c_{13}c_{23}}{c_{33}}, \quad C_{22} = c_{22} - \frac{c_{23}^2}{c_{33}}. \quad (2.13)$$

Pour simplification, la fonction (2.12) est prise avec le nombre fini des membres.

Après la differentiation de la fonction (2.12) par rapport aux composantes $\bar{\sigma}_x$, $\bar{\sigma}_y$ et $\bar{\tau}_{xy}$, on obtient pour les déformations équivalentes les formules suivantes:

$$\bar{\varepsilon}_x = C_{11}\bar{\sigma}_x + C_{12}\bar{\sigma}_y$$
$$+ \sum_{n=2}^{m}\sum_{k=0}^{n+1}\frac{2(n+1-k)}{n+1+|n+1-2k|}C_{(n+1-k)(1)k(2)}\bar{\sigma}_x^{n-k}\bar{\sigma}_y^k, \qquad (2.14)$$

$$\bar{\varepsilon}_y = C_{12}\bar{\sigma}_x + C_{22}\bar{\sigma}_y$$
$$+ \sum_{n=2}^{m}\sum_{k=0}^{n+1}\frac{2k}{n+1+|n+1-2k|}C_{(n+1-k)(1)k(2)}\bar{\sigma}^{n+1-k}\bar{\sigma}_y^k, \qquad (2.15)$$

$$\bar{\gamma}_{xy} = c_{44}\bar{\tau}_{xy} + \sum_{n=2}^{m}\sum_{k=0}^{n+1}c_{(n+1)(4)}\bar{\tau}_{xy}^n. \qquad (2.16)$$

Les relations entre le valeurs équivalentes et les valeurs réelles peuvent être exprimées par les polynômes:

$$\bar{\varepsilon}_{ij} = a_0 \varepsilon_{ij} + a_1 \frac{\partial \varepsilon_{ij}}{\partial t} + a_2 \frac{\partial^2 \varepsilon_{ij}}{\partial t^2} + \cdots + a_p \frac{\partial^p \sigma_{ij}}{\partial t^p} \tag{2.17}$$

$$+ a_{-1} \int_0^t K_1(t, \tau_1) \varepsilon_{ij}(\tau_1)\, d\tau_1 + a_{-2} \int_0^t \int_0^{\tau_1} K_2(t, \tau_2) \varepsilon_{ij}(\tau_2)\, d\tau_1 d\tau_2 + \cdots,$$

$$\bar{\sigma}_{kl} = b_0 \sigma_{kl} + b_1 \frac{\partial \sigma_{kl}}{\partial t} + b_2 \frac{\partial^2 \sigma_{kl}}{\partial t^2} + \cdots + \frac{\partial^r \sigma_{kl}}{\partial t^r} \tag{2.18}$$

$$+ b_{-1} \int_0^t L_1(t, \tau_1) \sigma_{kl}(\tau_1)\, d\tau_1 + \int_0^t \int_0^{\tau_1} L_2(t, \tau_2) \sigma_{kl}(\tau_2)\, d\tau_1 d\tau_2 + \cdots,$$

où les coefficients a_k et b_1 sont constantes ou fonctions du temps.

3. Fonction des Contraintes Équivalentes

Insérant les relations (2.14), (2.15) et (2.16) dans l'équation de compatibilité des déformations équivalentes

$$\frac{\partial^2 \bar{\varepsilon}_x}{\partial y^2} + \frac{\partial^2 \bar{\varepsilon}_y}{\partial x^2} = \frac{\partial^2 \bar{\gamma}_{xy}}{\partial x\, \partial y}, \tag{3.1}$$

on obtient

$$C_{11} \frac{\partial^2 \bar{\sigma}_x}{\partial y^2} + C_{12} \frac{\partial^2 \bar{\sigma}_x}{\partial x^2} + C_{12} \frac{\partial^2 \bar{\sigma}_y}{\partial y^2} + C_{22} \frac{\partial^2 \bar{\sigma}_y}{\partial x^2} - c_{44} \frac{\partial^2 \bar{\tau}_{xy}}{\partial x\, \partial y}$$

$$+ \sum_{n=2}^{m} \sum_{k=0}^{n+1} \frac{2\, C_{(n+1-k)(1)k(2)}}{n+1+|n+1-2k|} \left\{ k \left[(n+1-k)\, \bar{\sigma}_x^{n-k} \bar{\sigma}_y^{k-1} \frac{\partial^2 \bar{\sigma}_x}{\partial x^2} \right. \right.$$

$$+ (n+1-k)(n-k)\, \bar{\sigma}_x^{n-1-k} \bar{\sigma}_y^{k}\, {}^1 \left(\frac{\partial \sigma_x}{\partial x} \right)^2$$

$$+ 2(n+1-k)(k-1)\, \bar{\sigma}_x^{n-k} \bar{\sigma}_y^{k-2} \frac{\partial \bar{\sigma}_x}{\partial x} \frac{\partial \bar{\sigma}_y}{\partial x} \tag{3.2}$$

$$+ (k-1)(k-2)\, \bar{\sigma}_x^{n+1-k} \bar{\sigma}_y^{k-3} \left(\frac{\partial \sigma_y}{\partial x} \right)^2 + (k-1)\, \bar{\sigma}_x^{n+1-k} \bar{\sigma}_y^{k-2} \frac{\partial^2 \sigma_y}{\partial x^2} \right]$$

$$+ (n+1-k) \left[(n-k)\, \bar{\sigma}_x^{n-1-k} \bar{\sigma}_y^{k} \frac{\partial^2 \bar{\sigma}_x}{\partial y^2} \right.$$

$$+ (n-k)(n-1-k)\, \bar{\sigma}_x^{n-2-k} \bar{\sigma}_y^{k} \left(\frac{\partial \bar{\sigma}_x}{\partial y} \right)^2 + 2k(n-k)\, \bar{\sigma}_x^{n-1-k} \bar{\sigma}_y^{k-1} \frac{\partial \bar{\sigma}_x}{\partial y} \frac{\partial \bar{\sigma}_y}{\partial y}$$

$$\left. \left. + k(k-1)\, \bar{\sigma}_x^{n-k} \bar{\sigma}_y^{k-2} \left(\frac{\partial \bar{\sigma}_y}{\partial y} \right)^2 + k \bar{\sigma}_x^{n-k} \bar{\sigma}_y^{k-1} \frac{\partial^2 \bar{\sigma}_y}{\partial y^2} \right] \right\}$$

$$- \sum_{n=2}^{m} c_{(n+1)(4)} \left[n \bar{\tau}_{xy}^{n-1} \frac{\partial^2 \bar{\tau}_{xy}}{\partial x\, \partial y} + n(n-1)\, \bar{\tau}_{xy}^{n-2} \frac{\partial \bar{\tau}_{xy}}{\partial x} \frac{\partial \bar{\tau}_{xy}}{\partial y} \right] = 0.$$

Après avoir introduit la fonction ϕ des contraintes équivalentes suivant les relations

$$\bar{\sigma}_x = \frac{\partial^2 \phi}{\partial y^2}, \quad \bar{\sigma}_y = \frac{\partial^2 \phi}{\partial x^2}, \quad \bar{\tau}_{xy} = -\frac{\partial^2 \phi}{\partial x\, \partial y} + Xy + Yx, \tag{3.3}$$

où X et Y sont les constantes forces de masse,

satisfaisant aux équations d'équilibre, dans l'équation (3.2), l'auteur a obtenu l'équation non-linéaire aux dérivées partielles:

$$
C_{22}\frac{\partial^4 \phi}{\partial x^4} + (2\,C_{12} + c_{44})\frac{\partial^4 \phi}{\partial x^2\,\partial y^2} + C_{11}\frac{\partial^4 \phi}{\partial y^4}
$$

$$
+ \sum_{n=2}^{m}\sum_{k=0}^{n+1} \frac{2\,C_{(n+1-k)(1)k(2)}}{n+1+|n+1-2k|}\left\{ k\left[(k-1)\left(\frac{\partial^2 \phi}{\partial x^2}\right)^{k-2}\left(\frac{\partial^2 \phi}{\partial y^2}\right)^{n+1-k}\frac{\partial^4 \phi}{\partial x^4} \right.\right.
$$

$$
+ (k-1)(k-2)\left(\frac{\partial^2 \phi}{\partial x^2}\right)^{k-3}\left(\frac{\partial^2 \phi}{\partial y^2}\right)^{n+1-k}\left(\frac{\partial^3 \phi}{\partial x^3}\right)^2
$$

$$
+ 2(n+1-k)(k-1)\left(\frac{\partial^2 \phi}{\partial x^2}\right)^{k-2}\left(\frac{\partial^2 \phi}{\partial y^2}\right)^{n-k}\frac{\partial^3 \phi}{\partial x^3}\frac{\partial^3 \phi}{\partial x\,\partial y^2}
$$

$$
+ (n+1-k)(n-k)\left(\frac{\partial^2 \phi}{\partial x^2}\right)^{k-1}\left(\frac{\partial^2 \phi}{\partial y^2}\right)^{n-1-k}\left(\frac{\partial^3 \phi}{\partial x\,\partial y^2}\right)^2
$$

$$
\left. + (n+1-k)\left(\frac{\partial^2 \phi}{\partial x^2}\right)^{k-1}\left(\frac{\partial^2 \phi}{\partial y^2}\right)^{n-k}\frac{\partial^4 \phi}{\partial x^2\,\partial y^2}\right]
$$

$$
+ (n+1-k)\left[k\left(\frac{\partial^2 \phi}{\partial x^2}\right)^{k-1}\left(\frac{\partial^2 \phi}{\partial y^2}\right)^{n-k}\frac{\partial^4 \phi}{\partial x^2\,\partial y^2}\right. \tag{3.4}
$$

$$
+ k(k-1)\left(\frac{\partial^2 \phi}{\partial x^2}\right)^{k-2}\left(\frac{\partial^2 \phi}{\partial y^2}\right)^{n-k}\left(\frac{\partial^3 \phi}{\partial x^2\,\partial y}\right)^2
$$

$$
+ 2k(n-k)\left(\frac{\partial^2 \phi}{\partial x^2}\right)^{k-1}\left(\frac{\partial^2 \phi}{\partial y^2}\right)^{n-1-k}\frac{\partial^3 \phi}{\partial x^2\,\partial y}\frac{\partial^3 \phi}{\partial y^3}
$$

$$
+ (n-k)(n-1-k)\left(\frac{\partial^2 \phi}{\partial x^2}\right)^{k}\left(\frac{\partial^2 \phi}{\partial y^2}\right)^{n-2-k}\left(\frac{\partial^3 \phi}{\partial y^2}\right)^2
$$

$$
\left.\left. + (n-k)\left(\frac{\partial^2 \phi}{\partial x^2}\right)^{k}\left(\frac{\partial^2 \phi}{\partial y^2}\right)^{n-1-k}\frac{\partial^4 \phi}{\partial y^4}\right]\right\}
$$

$$
+ \sum_{n=2}^{m} c_{(n+1)(4)}\left[n\left(-\frac{\partial^2 \phi}{\partial x\,\partial y} + X\,y + Y\,x\right)^{n-1}\frac{\partial^4 \phi}{\partial x^2\,\partial y^2}\right.
$$

$$
\left. - n(n-1)\left(-\frac{\partial^2 \phi}{\partial x\,\partial y} + X\,y + Y\,x\right)^{n-2}\frac{\partial^3 \phi}{\partial x^2\,\partial y}\frac{\partial^3 \phi}{\partial x\,\partial y^2}\right] = 0.
$$

La relation précédente est l'équation de Lagrange-Euler du fonctionnel suivant

$$
\Psi = \iint\limits_{F} \left\{ \frac{1}{2}\left[C_{22}\left(\frac{\partial^2 \phi}{\partial x^2}\right)^2 + 2\,C_{12}\frac{\partial^2 \phi}{\partial x^2}\frac{\partial^2 \phi}{\partial y^2} + C_{11}\left(\frac{\partial^2 \phi}{\partial y^2}\right)^2 \right.\right.
$$

$$
\left. + c_{44}\left(\frac{\partial^2 \phi}{\partial x\,\partial y}\right)^2 - 2c_{44}(X\,y + Y\,x)\frac{\partial^2 \phi}{\partial x\,\partial y}\right] \tag{3.5}
$$

$$
+ \sum_{n=2}^{m}\sum_{k=0}^{n+1} \frac{2\,C_{(n+1-k)(1)k(2)}}{n+1+|n+1-2k|}\left(\frac{\partial^2 \phi}{\partial x^2}\right)^{k}\left(\frac{\partial^2 \phi}{\partial y^2}\right)^{n+1-k}
$$

$$
\left. + \sum_{n=2}^{m} c_{(n+1)(4)}\left(-\frac{\partial^2 \phi}{\partial x\,\partial y} + X\,y + Y\,x\right) + c_{44}(X\,y + Y\,x)^2 \right\}dx\,dy.
$$

Pour la déformation linéaire, l'équation (3.4) s'écrit, après l'introduction des relations (2.13), sous la forme:

$$\left(c_{22} - \frac{c_{23}^2}{c_{33}}\right)\frac{\partial^4 \phi}{\partial x^4} + 2\left(c_{12} - \frac{c_{13}c_{23}}{c_{33}} + \frac{1}{2}\,c_{44}\right)\frac{\partial^4 \phi}{\partial x^2\,\partial y^2} + \left(c_{11} - \frac{c_{13}^2}{c_{33}}\right)\frac{\partial^4 \phi}{\partial y^4} = 0.$$

(3.6)

Le fonctionnel de cette équation est

$$\Psi = \frac{1}{2}\iint\limits_F \left[\left(c_{22} - \frac{c_{23}^2}{c_{33}}\right)\left(\frac{\partial^2 \phi}{\partial x^2}\right)^2 + 2\left(c_{12} - \frac{c_{13}c_{23}}{c_{33}}\right)\frac{\partial^2 \phi}{\partial x^2}\frac{\partial^2 \phi}{\partial y^2}\right.$$

$$+ \left(c_{11} - \frac{c_{13}^2}{c_{33}}\right)\left(\frac{\partial^2 \phi}{\partial y^2}\right)^2 + c_{44}\left(\frac{\partial^2 \phi}{\partial x\,\partial y}\right)^2$$

(3.7)

$$\left. - 2c_{44}(Xy + Yx)\frac{\partial^2 \phi}{\partial x\,\partial y} + c_{44}(Xy + Yx)^2\right]dx\,dy.$$

Il y a encore une autre fonction des contraintes équivalentes qui satisfait aux équations d'équilibre

$$\frac{\partial \bar{\sigma}_x}{\partial x} + \frac{\partial \bar{\tau}_{xy}}{\partial y} = X,$$

$$\frac{\partial \bar{\tau}_{xy}}{\partial x} + \frac{\partial \bar{\sigma}_x}{\partial y} = Y$$

(3.8)

et qui est définie par les relations

$$\bar{\sigma}_x = \frac{\partial^2 \phi}{\partial y^2} + Xx, \qquad \bar{\sigma}_y = \frac{\partial^2 \phi}{\partial x^2} + Yy, \qquad \bar{\tau}_{xy} = -\frac{\partial^2 \phi}{\partial x\,\partial y}.$$

(3.9)

L'équation fondamentale est dans ce cas plus complexe.

Si les forces de masse varient avec les avec les coordonnées et ont le potentiel donné par:

$$X = \frac{\partial V}{\partial x}, \qquad Y = \frac{\partial V}{\partial y},$$

(3.10)

la fonction des contraintes équivalentes est définie par les relations

$$\bar{\sigma}_x = \frac{\partial^2 \phi}{\partial y^2} + V, \qquad \bar{\sigma}_y = \frac{\partial^2 \phi}{\partial x^2} + V, \qquad \bar{\tau} = -\frac{\partial^2 \phi}{\partial x\,\partial y}.$$

(3.11)

Dans ce cas, nous avons pour la déformation linéaire la suivante équation fondamentale

$$\left(c_{22} - \frac{c_{23}^2}{c_{33}}\right)\frac{\partial^4 \phi}{\partial x^4} + 2\left(c_{12} - \frac{c_{13}c_{23}}{c_{33}} + \frac{1}{2}\,c_{44}\right)\frac{\partial^4 \phi}{\partial x^2\,\partial y^2} + \left(c_{11} - \frac{c_{13}^2}{c_{33}}\right)\frac{\partial^4 \phi}{\partial y^4}$$

(3.12)

$$= -\left[\left(c_{12} + c_{22} - \frac{c_{13}c_{23} + c_{23}^2}{c_{33}}\right)\frac{\partial^2 V}{\partial x^2} + \left(c_{11} + c_{12} - \frac{c_{13}^2 + c_{13}c_{23}}{c_{33}}\right)\frac{\partial^2 V}{\partial y^2}\right]$$

avec le fonctionnel

$$\Psi = \frac{1}{2}\iint\limits_F \left[\left(c_{22} - \frac{c_{23}^2}{c_{33}}\right)\left(\frac{\partial^2 \phi}{\partial x^2} + V\right)^2 + 2\left(c_{12} - \frac{c_{13}c_{23}}{c_{33}}\right)\left(\frac{\partial^2 \phi}{\partial x^2} + V\right)\left(\frac{\partial^2 \phi}{\partial y^2} + V\right)\right.$$

$$\left. + \left(c_{11} - \frac{c_{13}^2}{c_{33}}\right)\left(\frac{\partial^2 \phi}{\partial y^2} + V\right) + c_{44}\left(\frac{\partial^2 \phi}{\partial x\,\partial y}\right)^2\right]dx\,dy.$$

(3.13)

Pour la déformation non-linéaire, le fonctionnel de l'équation fondamentale s'écrit sous la forme:

$$
\Psi = \iint_F \left\{ \frac{1}{2} \left[C_{22} \left(\frac{\partial^2 \phi}{\partial x^2} + V \right)^2 + 2 C_{12} \left(\frac{\partial^2 \phi}{\partial x^2} + V \right) \left(\frac{\partial^2 \phi}{\partial y^2} + V \right) \right. \right.
$$
$$
\left. + C_{11} \left(\frac{\partial^2 \phi}{\partial y^2} + V \right)^2 + c_{44} \left(\frac{\partial^2 \phi}{\partial x\, \partial y} \right)^2 \right]
$$
$$
+ \sum_{n=2}^{m} \sum_{k=0}^{n+1} \frac{2 C_{(n+1-k)(1)k(2)}}{n+1+|n+1-2k|} \left(\frac{\partial^2 \phi}{\partial x^2} + V \right)^k \left(\frac{\partial^2 \phi}{\partial y^2} + V \right)^{n+1-k}
$$
$$
\left. + \sum_{n=2}^{m} c_{(n+1)(4)} \left(- \frac{\partial^2 \phi}{\partial x\, \partial y} \right)^{n+1} \right\} dx\, dy . \tag{3.14}
$$

4. Solution de l'Équation Linéaire

Considérons, en qualité d'exemple, l'état des contraintes dans le mi-plan avec l'anisotropie orthogonale avec les axes x et y (fig. 1), qui

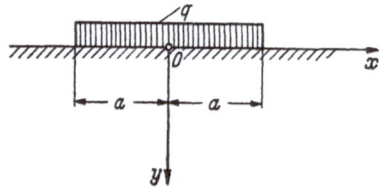

est uniformément chargé dans la section de longeur de $2a$.

Le poids par l'unité de volume est constant.

Pour exprimer la charge uniforme dépendant du temps dans la section limitée, on peut faire usage de l'intégrale de Fourier

Fig. 1. Le mi-plan avec la charge uniforme dans la section de longeur de $2a$.

$$
q(x)\, \omega(t) = \frac{2}{\pi} \int_0^\infty \cos \alpha x\, d\alpha \int_0^a q\, \omega(t) \cos \alpha \lambda\, d\lambda
$$
$$
= \frac{2 q \omega(t)}{\pi} \int_0^\infty \frac{\sin \alpha a}{\alpha} \cos \alpha x\, d\alpha , \tag{4.1}
$$

où $\omega(t)$ est la fonction continue du temps.

Portant dans l'équation (3.6), qui peut s'écrire sous la forme abrégée

$$
C_{22} \frac{\partial^4 \phi}{\partial x^4} + (2 C_{12} + c_{44}) \frac{\partial^4 \phi}{\partial x^2\, \partial y^2} + C_{11} \frac{\partial^4 \phi}{\partial y^4} = 0, \tag{4.2}
$$

la fonction des contraintes équivalentes, qui a la forme générale

$$
\phi = \int_0^\infty \frac{1}{\alpha^2} f(y) \cos \alpha x\, d\alpha , \tag{4.3}
$$

nous obtenons pour la solution générale la suivante équation caractéristique

$$
C_{11} \varkappa^4 - (2 C_{12} + c_{44})\, \alpha^2 \varkappa^2 + C_{22} \alpha^4 = 0, \tag{4.4}
$$

d'où

$$\varkappa_{1,2,3,4} = \pm \sqrt{\frac{2C_{12} + c_{44}}{2C_{11}} \pm \sqrt{\left(\frac{2C_{12} + c_{44}}{2C_{11}}\right)^2 - \frac{C_{22}}{C_{11}}}} = \pm \alpha k_{1,2}. \quad (4.5)$$

La fonction des contraintes équivalentes prend alors la forme générale:

$$\phi = \int_0^\infty \frac{1}{\alpha^2} (A e^{k_1 \alpha y} + B e^{-k_1 \alpha y} + C e^{k_2 \alpha y} + D e^{-k_2 \alpha y}) \cos \alpha x \, d\alpha. \quad (4.6)$$

Nous allons retenir les termes avec les puissances négatives seulement parce que les composantes des contraintes ne peuvent pas tendre vers l'infinité. Nous avons:

$$\phi = \int_0^\infty \frac{1}{\alpha^2} (B e^{-k_1 \alpha y} + D e^{-k_2 \alpha y}) \cos \alpha x \, d\alpha. \quad (4.7)$$

Les composantes des contraintes équivalentes sont données par:

$$\bar{\sigma}_x = \int_0^\infty (B k_1^2 e^{-k_1 \alpha y} + D k_2^2 e^{-k_2 \alpha y}) \cos \alpha x \, d\alpha, \quad (4.8)$$

$$\bar{\sigma}_y = -\int_0^\infty (B e^{-k_1 \alpha y} + D e^{-k_2 \alpha y}) \cos \alpha x \, d\alpha, \quad (4.9)$$

$$\bar{\tau}_{xy} = -\frac{\partial^2 \phi}{\partial x \, \partial y} = -\int_0^\infty (B k_1 e^{-k_1 \alpha y} + D k_2 e^{-k_2 \alpha y}) \sin \alpha x \, d\alpha. \quad (4.10)$$

Pour la limite du mi-plan $y = 0$, nous avons:

$$\bar{\sigma}_y = -\int_0^\infty (B + D) \cos \alpha x \, d\alpha, \qquad \bar{\tau}_{xy} = -\int_0^\infty (B k_1 + D k_2) \sin \alpha x \, d\alpha. \quad (4.11)$$

Pour $y = 0$, il faut satisfaire aux conditions aux limites:

$$\bar{\sigma}_y = q \, \bar{\omega}(t), \qquad \bar{\tau}_{xy} = 0, \quad (4.12)$$

où

$$\bar{\omega}(t) = b_0 \, \omega(t) + b_1 \frac{\partial \omega(t)}{\partial t} + b_2 \frac{\partial^2 \omega(t)}{\partial t^2} + \cdots + b_r \frac{\partial^r \omega(t)}{\partial t^r}$$

$$+ b_1 \int_0^t L_1(t_1 \tau_1) \, \omega(\tau_1) \, d\tau_1 + \cdots.$$

Ces conditions peuvent s'écrire sous la forme:

$$-\int_0^\infty (B + D) \cos \alpha x \, d\alpha = \frac{2q \bar{\omega}(t)}{\pi} \int_0^\infty \frac{\sin \alpha a}{\alpha} \cos \alpha x \, d\alpha, \quad (4.13)$$

$$-\int_0^\infty (B k_1 + D k_2) \sin \alpha x \, d\alpha = 0. \quad (4.14)$$

On y satisfait pour chaque x par:

$$B + D = \frac{-2q\,\bar{\omega}(t)}{\pi} \cdot \frac{\sin \alpha a}{\alpha}, \qquad Bk_1 + Dk_2 = 0, \qquad (4.15)$$

d'où

$$B = \frac{2k_2 q\,\bar{\omega}(t)}{\pi(k_1 - k_2)} \cdot \frac{\sin \alpha a}{\alpha}, \qquad (4.16)$$

$$D = -\frac{2k_1 q\,\bar{\omega}(t)}{\pi(k_1 - k_2)} \cdot \frac{\sin \alpha a}{\alpha}. \qquad (4.17)$$

En portant les valeurs précédentes dans les expressions (4.8), (4.9) et (4.10), on obtient pour les contraintes équivalentes les formules suivantes

$$\bar{\sigma}_x = \frac{2q\,\bar{\omega}(t)}{\pi(k_1 - k_2)} \int_0^\infty k_1 k_2 (k_1 e^{-k_1 \alpha y} - k_2 e^{-k_2 \alpha y}) \frac{\sin \alpha a}{\alpha} \cos \alpha x \, d\alpha, \quad (4.18)$$

$$\bar{\sigma}_y = -\frac{2q\,\bar{\omega}(t)}{\pi(k_1 - k_2)} \int_0^\infty (k_2 e^{-k_1 \alpha y} - k_1 e^{-k_2 \alpha y}) \frac{\sin \alpha a}{\alpha} \cos \alpha x \, d\alpha, \quad (4.19)$$

$$\bar{\tau}_{xy} = -\frac{2q\,\bar{\omega}(t)}{\pi(k_1 - k_2)} \int_0^\infty k_1 k_2 (e^{-k_1 \alpha y} - e^{-k_2 \alpha y}) \frac{\sin \alpha a}{\alpha} \sin \alpha x \, d\alpha. \quad (4.20)$$

Ces équations peuvent être exprimées sous la forme finie:

$$\bar{\sigma}_x = \frac{k_1 k_2 q\,\bar{\omega}(t)}{\pi(k_1 - k_2)} \left(k_1 \arctan \frac{2k_1 a y}{k_1^2 y^2 + x^2 - a^2} - k_2 \arctan \frac{2k_2 a y}{k_2^2 y^2 + x^2 - a^2} \right), \quad (4.21)$$

$$\bar{\sigma}_y = -\frac{q\,\bar{\omega}(t)}{\pi(k_1 - k_2)} \left(k_2 \arctan \frac{2k_1 a y}{k_1^2 y^2 + x^2 - a^2} - k_1 \arctan \frac{2k_2 a y}{k_2^2 y^2 + x^2 - a^2} \right), \quad (4.22)$$

$$\bar{\tau}_{xy} = -\frac{k_1 k_2 q\,\bar{\omega}(t)}{\pi(k_1 - k_2)} \ln \frac{k_1^2 y^2 + (a+x)^2}{k_2^2 y^2 + (a+x)^2} \cdot \frac{k_2^2 y^2 + (a-x)^2}{k_1^2 y^2 + (a-x)^2}. \quad (4.23)$$

5. Relations pour les Contraintes et pour les Déformations Actuelles

Après avoir déterminé les contraintes équivalentes, les valeurs actuelles s'ensuivent de l'équation intégrodifférentielle (2.18).

Dans un cas plus simple, si la contrainte équivalente est donnée par l'équation différentielle aux coefficients constants

$$\bar{\sigma}_{ij} = a_0 \sigma_{ij} + a_1 \frac{\partial \sigma_{ij}}{\partial t} + a_2 \frac{\partial^2 \sigma_{ij}}{\partial t^2} + \cdots + a_p \frac{\partial^p \bar{\sigma}_{ij}}{\partial t^p}, \qquad (5.1)$$

la contrainte actuelle s'exprime par

$$\sigma_{ij} = C_1 e^{\lambda_1 t} + C_2 e^{\lambda_2 t} + \cdots + e^{\lambda_r t}(C_r + C_{r+1} t$$
$$+ \cdots + C_{r+s-1} t^{s-1}) + e^{at}(C_u \cos \beta t + C_{u+1} \sin \beta t) + \sigma_{ij0}, \qquad (5.2)$$

où $\lambda_1, \lambda_2, \ldots \lambda_{r-1}$, sont les simples, λ_r multiples et $\lambda_u = \alpha + \beta i$, $\lambda_{u+1} = \alpha - \beta i$ les complexes racines de l'équation caractéristique

$$a_p \lambda^p + a_{p-1} \lambda^{p-1} + \cdots + a_2 \lambda^2 + a_1 \lambda + a_0 = 0 \qquad (5.3)$$

et σ_{ij0} est l'intégrale particulière qui dépend de la fonction des contraintes équivalentes.

Les constants C_i peuvent etre déterminés des conditions initiales.

Dans le cas de la relaxation, la contrainte équivalente ne dépend pas du temps et nous avons

$$\sigma_{ij0} = \frac{\bar{\sigma}_{ij}}{a_0}. \qquad (5.4)$$

Si la contrainte actuelle est donnée par l'équation linéaire du premier ordre

$$a_1 \frac{\partial \sigma_{ij}}{\partial t} + a_0 \sigma_{ij} = \bar{\sigma}_{ij}(t), \qquad \bar{\sigma}_{ij} = p(x,y)\,\bar{\omega}(t), \qquad (5.5)$$

elle peut être déterminée par la formule asymptotique

$$\sigma_{ij} = \frac{p(x,y)}{a_0} \sum_{k=0}^{\infty} (-1)^k \left(\frac{a_1}{a_0}\right)^k \left\{ \frac{\partial^k \bar{\omega}(t)}{\partial t^k} - \frac{a_0}{a_1}(t - t_0) \left[\frac{\partial^k \bar{\omega}(t)}{\partial t^k}\right]_0 \right\}$$
$$+ \bar{\omega}(0)\, e^{-\frac{a_0}{a_1}(t-t_0)}, \qquad (5.6)$$

qui a été dérivée par le procedé per partes répété. Dans ce cas les coefficients a_k peuvent être fonctions du temps.

Pour les contraintes équivalentes définies à la fin de la section précédente les valeurs actuelles peuvent être experimées sous la forme suivante

$$\sigma_x = \frac{k_1 k_2 q}{\pi(k_1 - k_2)\, a_0} \left(k_1 \arctan \frac{2 k_1 a y}{k_1^2 y^2 + x^2 - a^2} - k_2 \arctan \frac{2 k_2 a y}{k_2^2 y^2 + x^2 - a^2} \right)$$
$$\times \left(\sum_{k=0}^{\infty} (-1)^k \left(\frac{a_1}{a_0}\right)^k \left\{ \frac{\partial^k \bar{\omega}(t)}{\partial t^k} - e^{-\frac{a_0}{a_1}(t-t_0)} \left[\frac{\partial^k \bar{\omega}(t)}{\partial t^k}\right]_0 \right\} \right. \qquad (5.7)$$
$$\left. + \bar{\omega}(0)\, e^{-\frac{a_0}{a_1}(t-t_0)} \right),$$

$$\sigma_y = -\frac{q}{\pi(k_1 - k_2)\, a_0} \left(k_2 \arctan \frac{2 k_1 a y}{k_1^2 a^2 + x^2 - a^2} - k_1 \arctan \frac{2 k_2 a y}{k_2^2 y^2 + x^2 - a^2} \right)$$
$$\times \left(\sum_{k=0}^{\infty} (-1)^k \left(\frac{a_1}{a_0}\right)^k \left\{ \frac{\partial^k \bar{\omega}(t)}{\partial t^k} - e^{-\frac{a_0}{a_1}(t-t_0)} \left[\frac{\partial^k \bar{\omega}(t)}{\partial t^k}\right]_0 \right\} \right. \qquad (5.8)$$
$$\left. + \bar{\omega}(0)\, e^{-\frac{a_0}{a_1}(t-t_0)} \right),$$

$$\tau_{xy} = -\frac{k_1 k_2 q}{\pi(k_1 - k_2)\, a_0} \ln \frac{k_1^2 y^2 + (a+x)^2}{k_2^2 y^2 + (a+x)^2} \cdot \frac{k_2^2 y^2 + (a+x)^2}{k_1^2 y^2 + (a+x)^2}$$

$$\times \left(\sum_{k=0}^{\infty} (-1)^k \left(\frac{a_1}{a_0}\right)^k \left\{ \frac{\partial^k \overline{\omega}(t)}{\partial t^k} - e^{-\frac{a_0}{a_1}(t-t_0)} \left[\frac{\partial^k \overline{\omega}(t)}{\partial t^k}\right]_0 \right\} \right. \tag{5.9}$$

$$\left. + \overline{\omega}(0)\, e^{-\frac{a_0}{a_1}(t-t_0)} \right).$$

Les déformations équivalentes peuvent être déterminées au moyen des équations (2.14), (2.15) et (2.16).

Les déformations actuelles peuvent être déterminées par une manière analogue que les contraintes actuelles.

Bibliographie

[1] Biezeno, C. B., u. R. Grammel: Technische Dynamik, Bd. I, Berlin/ Göttingen/Heidelberg: Springer 1953.

[2] Freudenthal, A. M., and H. Geiringer: The Mathematical Theories of the Inelastic Continuum. Handbuch der Physik, herausgegeben von S. Flügge, Bd. VI, Elastizität und Plastizität, Berlin/Göttingen/Heidelberg: Springer 1958.

[3] Mandel, J.: Application du calcul opérationnel á l'étude des corps viscoélastiques. Cahiers du Groupe Français d'Etudes de Rhéologie, Tome III, Paris 1958.

[4] Reiner, M.: Elasticity Beyond the Elastic Limit. Amer. J. Math. **70**, 443–446 (1948).

[5] Reiner, M.: Deformation, Strain and Flow, London: Pewis 1960.

[6] Rivlin, R. S.: Large Elastic Deformations. Rheology. Theory and Applications, edited by F. R. Eirich, Vol. I, New York: Academie Press 1956.

[7] Sobotka, Z.: Théorie de Plasticité et des États Limites des Constructions, Vol. I (en tchéque), Prague: Czechoslovak Academy of Sciences 1954.

[8] Sobotka, Z.: Some Problems of Non-Linear Rheology. The International Symposium of the IUTAM on Second-Order Effects in Elasticity Plasticity and Fluid Dynamics, Haifa 1962.

[9] Sobotka, Z.: On Rheology of Non-Linear Shell Problems IASS Symposium on Non-Elastical Shell Problems, Warsaw 1963.

2.5 Soil Consolidation under the Influence of an External Load Normal to the Boundary of Half-Space[1]

By

N. N. Verigin

1. Let the instantaneous external load $q(x,y)$ be applied for the moment $t = 0$ to a part of half-space boundary, the area of which is ω (Fig. 1a). The half space consists of soil all the pores of which are

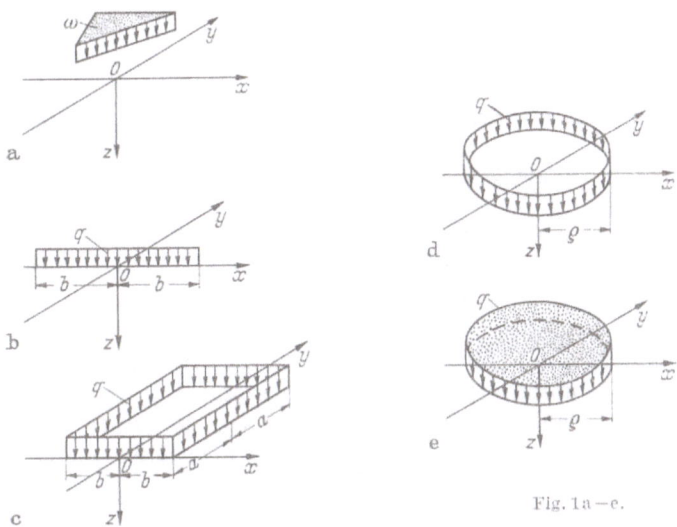

Fig. 1a—e.

filled with water. The area ω is considered to be permeable (foundation with drainage).

Referring now to TERZAGHI, GERSEVANOV, FLORIN and some others it will be apparent that for water and mineral grains of soil both assumed to be incompressible, the hydrodynamical water pressure $P(x,y,z,t)$ developed in the soil as a result of load q applied thereto will satisfy the

---

[1] This report in a more complete form will be published as part of the proceedings of the Soil Mechanics International Congress, held in Canada in 1965.

FOURIER's equation.

$$C \triangle P = \frac{\partial P}{\partial t}, \tag{1}$$

where: c = factor of consolidation,
          t = time,
         x, y, z = coordinates,
         $\triangle$ = LAPLACE' operator

Because of water incompressibility at the moment $t = 0$ some original pressure distribution $P_0(x, y, z)$ instantaneously appears, causing water percolation from the area ω to the unloaded boundary of the half-space. Function P_0 satisfies the LAPLACE equation or Eq. (1) at $\partial P / \partial t = 0$.

Now to solve the problem of soil consolidation in halfspace, Eq. (1) has to be integrated at:

$$P(x, y, z, 0) = P_0(x, y, z), \quad P(x, y, 0, t) = 0,$$
$$P(r, t) = P(\infty, t) = 0, \quad \left(r = \sqrt{x^2 + y^2 + z^2} \right), \tag{2}$$

where: P_0 is determined from Eq. (1) at $\partial P / \partial t = 0$ and having assumed the following conditions:

$$P_0(x, y, 0) = q, \quad \text{(inside area } \omega\text{)};$$

$$P_0(x, y, 0) = 0, \quad \text{(outside the area } \omega\text{)}.$$

We shall have:

$$P_0(r) = P(\infty) = 0. \tag{3}$$

Solution of Eq. (1) under the conditions (2) can be found by superpositioning of pointed stationary and nonstationary dipoles disposed on the area ω. With the load q-const being uniformly loaded the general solution of the problem will be expressed by the following integral:

$$p = \frac{qz}{2\pi} \iint\limits^{\omega} \frac{M[\lambda(m, n)]}{r^3(m, n)} \, dm \, dn, \tag{4}$$

$$M(\lambda) = \operatorname{erf} \lambda - \frac{2}{\sqrt{\pi}} \lambda \exp\left(-\lambda^2\right), \quad \operatorname{erf} \lambda = \frac{2}{\sqrt{\pi}} \int\limits_0^\lambda e^{-\zeta^2} \, d\zeta, \tag{5}$$

$$\lambda = \frac{r}{2\sqrt{ct}}, \quad r = \sqrt{(x-m)^2 + (y-n)^2 + z^2}. \tag{6}$$

Application of integral (4) is spread all over the area of ω.

2. Uniform load qt/m along a straight line of finite length $2b$ (strip foundation).

From Eq. (4) we obtain (Fig. 1, b)

$$p = \frac{qz}{2\pi} \int\limits_{-b}^{+b} \frac{M\left[\lambda(m)\right]}{r^3(m)}\,dm = \frac{qz}{2\pi R^2}\,N\left(v, u_1, u_2\right), \tag{7}$$

$$N = A\left(v, u_1\right) - A\left(v, u_2\right), \tag{8}$$

$$A\left(v, u\right) = u\,\frac{\operatorname{erf}\left(\sqrt{v^2 + u^2}\right)}{\sqrt{v^2 + u^2}} - \operatorname{erf}\left(u\right)\exp\left(-v^2\right), \tag{9}$$

$$v = \frac{R}{2\,\sqrt{ct}}, \qquad u_{1,2} = \frac{x \pm b}{2\,\sqrt{ct}}, \qquad R = \sqrt{y^2 + z^2}. \tag{10}$$

Below the centre of the straight line $(x = y = 0)$ the pressure will be:

$$p = \frac{q}{\pi z}\left[\frac{b}{\sqrt{b^2 + z^2}}\operatorname{erf}\frac{\sqrt{b^2 + z^2}}{2\,\sqrt{ct}} - \operatorname{erf}\left(\frac{b}{2\,\sqrt{ct}}\right)\exp\left(-\frac{z^2}{4ct}\right)\right]. \tag{7a}$$

In case of a straight line of infinite lenght $(b = \infty)$ the pressure will be:

$$p = \frac{qz}{\pi R^2}\left[1 - \exp\left(-\frac{R^2}{4ct}\right)\right]. \tag{7b}$$

The particular case has found a good agreement with the FLAMAN's problem. V. J. KOROTKIN was the first to obtain the expression for P.

3. Uniform load qt/m along the perimeter of a rectangle having sides $2a$ and $2b$ (strip foundation of building).

Here (Fig. 1c) the pressure P is calculated by adding the four solutions (7).

Below the centre of the rectangular side $2b$ it will be:

$$p = \frac{qz}{2\pi}\,P, \qquad P = \frac{N_{0,0}}{z^2} + \frac{N_{0,2a}}{4a^2 + z^2} + 2\,\frac{N_{a,b}}{b^2 + z^2}, \tag{11}$$

where $N_{0,0}$; $N_{0,2a}$; $N_{a,b}$ — values of function N from (8) to (10) respectively at $x = y = 0$; $x = 0, y = 2a$; and $x = a, y = b$ respectively. By this for $N_{a,b}$ in the expression $u_{1,2}$ from 10 "b" is replaced by "a".

4. Uniform distribution of load qt/m along the periphery of the circle with radius ϱ (Fig. 1d).

From Eq. (4)

$$p = \frac{q\varrho z}{2\pi} \int\limits_{0}^{2\pi} \frac{M\left[\lambda(\varphi)\right]}{r^3(\varphi)}\,d\varphi, \tag{12}$$

where M and λ are expressed as in (5) and (6) whereas r is found:

$$r = \sqrt{\varrho^2 + z^2 + \varrho_b^2 + 2\varrho\varrho_b \cos\left(\varphi - \varphi_b\right)}, \qquad \varrho_b = \sqrt{x^2 + y^2},$$

$$\varphi_b = \operatorname{arc\,tan}\frac{y}{x}. \tag{13}$$

For the points below the centre of the periphery of the circle $(\varrho_b = 0)$ we have

$$p = \frac{q\varrho z}{r^3} M(\lambda), \qquad \lambda = \frac{r}{2\sqrt{ct}}, \qquad r = \sqrt{\varrho^2 + z^2}. \qquad (12\,\mathrm{a})$$

By this M is determined from (5).

5. Uniform distribution of load qt/m^2 along (through) the area of the circle with radius ϱ (column footing).

From Eq. (4) we obtain (Fig. 1e)

$$p = \frac{qz}{2\pi} \int\limits_0^\varrho \int\limits_0^{2\pi} \frac{M[\lambda(\varphi, \varrho)]}{r^3(\varphi, \varrho)} \varrho \, d\varrho \, d\varphi, \qquad (14)$$

where M, λ are expressed as in (5), (6) whereas "r" is determined as in (13).

For the points below the centre f circle $(\varrho_b = 0)$ the pressure will be

$$p = qS(\lambda, w), \qquad S(\lambda, w) = \mathrm{erf}\,(\lambda w) - w\,\mathrm{erf}\,(\lambda), \qquad w = \frac{z}{r}, \qquad (14\,\mathrm{a})$$

wherein "λ" and "r" are determined from (12a).

6. By applying the load "q" in accordance with any low $q = f(t)$, the curve $q = f(t)$ can be substituted for a broken line, consisting of vertical and horizontal straight lines (stepped line). With such an instantaneous-stepped n — multiple applying of load, we shall have for all the considered cases:

$$p = \sum_{k=0}^{k=n} q_k F(t - t_k), \qquad (15)$$

where $q_k = q_0, q_1 \dots q_n$ denote the loads, applied instantaneously to the area "ω" at the time moments $t_k = 0, t_1 \dots t_n$, F-function $\frac{p}{q}$ from (8), from (11), from (5) or (14a), wherein "$t - t_k$" is accepted for all the cases instead of "t".

2.6 Consideration of Heterogenity and Non-Linear Character in Analysis of Bed Creep

By

J. K. Zaretsky and N. A. Tsytovitch

To-day's methods of bed computation are based on the assumption of bed homogeneity and are true of the resilient stage of the work alone. However, the immediate observations and experiments show that natural soil beds are in the overwhelming majority heterogenous, their deformation changing in depth uniformly or in stages. The simpliest example of the latter is the scheme of the layer of compressible soil having limited thickness, that lies on the in-compressible bed. This computation scheme already gives results that are closer to those observed. However, in fact, a far more complicated heterogeneity of soils in depth is found. This is caused by their different density in depth under the effect of their own weight, dispersion of packing pressures under local load, heterogeneity of temperature field in erecting, for example, heated structures on permafrost soils, etc.

Apart from that, in the wide range of pressures in general case the natural beds are to be regarded as non-linearly deformed, in creep stage, for instance.

This has brought forth the need to consider the non-linear creep of the bed's material in its heterogeneity in the computation of foundations. In our efforts to satisfy this need we in the first place were to face the problem of intensely-deformed state of bed under the effect of a concentrated force and a certain distribution load. Then we could pass over to the problem of contact pressures that arise at the foot of foundations.

The carried-out research is presented according to the following plan:

1. Setting the task.

2. Radial distribution of intensity under the effect of a concentrated force.

3. Deformation in time of the bed boundary under the effect of the distribution load.

4. The computation of hard foundations on a non-linearly deformed and heterogenous bed (a contact task).

Part 1 gives grounds to the principal assumptions taken to solve the tasks set.

Part 2 makes it clear in what cases and within which laws of deformation radial distribution of bed pressures under flat deformation becomes possible. For the problems of non-linear creep of bed that suffers the effect of a concentrated force and distribution load, and some other problems that are close to the first allow a comparatively simple and closed decision in case of radial distribution of intensity.

The intensely-deformed state of bed the heterogeneity of which (as has been in detail investigated by J. K. ZARETSKY) is characterized by the parameter of the deformation law changing with depth according to the stage law. Besides the possibility of radial intensity distribution in case the heterogenous bed material possesses the properties of volumetric compressibility, the changing of its form being influenced by the average normal pressure, is investigated.

On the basis of the results obtained in Part 2. Part 3 defines boundary under the effect of the distribution load. The time factor that influences the development of bed deformations was introduced with the help of BOLZMAN-VOLTERRA theory of hereditory creep.

The results of this part of work have been used in practice. The nature of bed settling under the effect of a uniform distribution load is found out. The paper presents an investigation of the influence that is being exerted by the non-linear nature of the deformation law and the heterogeneity of bed upon the quantity of its settling and the bowl of flexure.

In computing hard foundations (see Part 4) on a non-linearly deforming and heterogenous bed the nature of contact pressures distribution becomes evident. If so, it presents no difficulty to compute the foundation itself using the usual method of construction mechanics.

The defining of reaction pressures under the foot of a hard foundation gave rise to a contact problem solved by J. K. ZARETSKY. He has applied a new method to solving the Fredholm singular integral equation of the first type with the nucleus $K(x,y) = |x - y|^{-\varkappa}$; $\varkappa < 1$.

This method allows to reduce the equation with finite limits to that of infinite limits and a nucleus depending on the arguments difference. It makes it possible to use effectively the double transformation of LAPLACE. The suggested method is simple enough and gives a closed solving.

The solving of the flat contact task has produced a highly important practical result. That is, the consideration of the bed real properties (its heterogeneity and the non-linear character of the deformation law) leads to a more uniform distribution of reaction pressures, and therefore to diminution of the curving moment that has an effect upon the foundation. In other words, it makes the construction of the foundation easier.

2.7 Sur Certains Problèmes Rhéologiques de la Mécanique des Sols

Par

A. S. Stroganov

1. Propriétés Mécaniques et Équations Rhéologiques des Sols

Les recherches expérimentales de l'Auteur (cf. [1], [2], [3]) conduisent à représenter les propriétés mécaniques des sols par les relations suivantes:

$$\sigma_i = \frac{\overline{G} \tan \psi}{\tan \psi + \overline{G}\varepsilon_i} (H + \sigma)\, \varepsilon_i + \mu \xi_i, \tag{1.1}$$

$$\sigma_0 + \sigma = \alpha (\varepsilon_0 + \varepsilon_k)^\beta + \chi \xi_i, \tag{1.2}$$

$$\varepsilon_s = -\nu \varepsilon_i, \tag{1.3}$$

où: σ_i est l'intensité des contraintes tangentielles; $\overline{G}$ est le module initial de plasticité; $\tan \psi$ est le coefficient de frottement suivant la coupe octaédrique; H est la pression hydrostatique équivalente à la cohésion; σ la contrainte moyenne normale; ε_i est l'intensité des déformations de cisaillement; μ est le coefficient de viscosité plastique; ξ_i est l'intensité des vitesses de déformation de cisaillement; σ_0 est une pression hydrostatique constante, traduisant la cohésion du matériau; ε_k est la déformation volumique de compressibilité; ε_0 est la déformation volumique correspondant à σ_0; χ est le coefficient de viscosité volumique; ξ est la vitesse de déformation volumique; ε_s est la déformation volumique de dilatance; ν est le coefficient de dilatance.

La fonction:

$$\overline{G}(\varepsilon_i) = \frac{\overline{G} \tan \psi}{\tan \psi + \overline{G}\varepsilon_i} \tag{1.4}$$

définit le module réduit de plasticité pour $\xi_i = 0$ (c'est-à-dire pour les valeurs nulles des vitesses de déformation de cisaillement); elle est déterminée par une approximation homographique de la courbe expérimentale (cf. fig. 1), qui donne la corrélation entre l'intensité réduite

des contraintes tangentielles $\dfrac{\sigma_\iota}{H + \sigma}$ et les déformations de cisaillement[1].
La fonction:

$$K\left(\varepsilon_k\right) = \frac{\sigma}{\varepsilon_k} = \frac{\alpha\left(\varepsilon_0 + \varepsilon_k\right)^\beta - \sigma_0}{\varepsilon_k} \tag{1.5}$$

définit le module de compressibilité volumique pour $\xi = 0$; elle est déterminée par l'approximation en loi de puissance de la courbe expérimentale qui donne la corrélation entre la contrainte moyenne normale et la compressibilité volumique (cf. fig. 2).

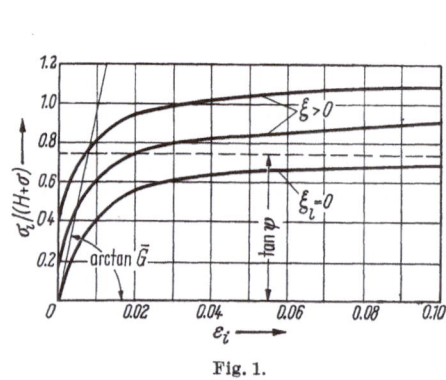

Fig. 1.

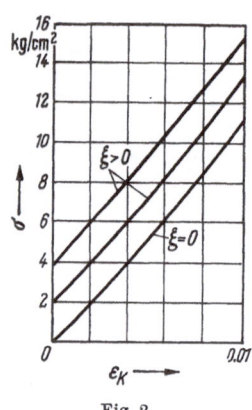

Fig. 2.

Le coefficient de dilatance ν est une constante rhéologique indépendante de la contrainte moyenne normale (cf. [1]), déterminée expérimentalement, par exemple au moyen des essais de cisaillement pur avec mesure simultanée des déformations volumiques.

La relation entre déviateurs:

$$D_\sigma = \frac{2\overline{G}\tan\psi}{\tan\psi + \overline{G}\varepsilon_\iota}\left(H + \sigma\right) D_\varepsilon + 2\mu D_\xi, \tag{1.6}$$

obtenue à partir de (1.1) et de la théorie des déformations de Hencky, est l'équation fondamentale d'état, non linéaire, des milieux élastovisqueux, généralisant le matériau de Kelvin-Voigt. Eu égard à la valeur de la déformation volumique totale:

$$\varepsilon = \varepsilon_k - \nu\varepsilon_\iota \tag{1.7}$$

l'équation d'état mécanique écrite, par exemple, en axes cartésiens devient, en vertu de (1.6) et après élimination de la contrainte moyenne

---

[1] A. I. Botkine (cf. [4]) est le premier à avoir obtenu en 1939 cette dépendance sous forme d'une courbe invariante.

normale grâce à (1.2):

$$\left.\begin{array}{l}
\sigma_x = \left[1 + \dfrac{2\overline{G}\tan\psi}{\tan\psi + \overline{G}\varepsilon_i}\left(\varepsilon_x - \dfrac{1}{3}\,\varepsilon\right)\right] \\[2mm]
\qquad \times \{H + \alpha[(\varepsilon_0 + \varepsilon) + \nu\varepsilon_i]^\beta - \sigma_0 + \chi\xi\} \\[2mm]
\qquad + 2\mu\left(\xi_x - \dfrac{1}{3}\,\xi\right) - H
\end{array}\right\} \quad (1.8)$$

et les équations analogues pour σ_y, σ_z et

$$\tau_{xy} = \frac{\overline{G}\tan\psi}{\tan\psi + \overline{G}\varepsilon_i}\{H + \alpha[(\varepsilon_0 + \varepsilon) + \nu\varepsilon_i]^\beta - \sigma_0 + \chi\xi\}\gamma_{xy} + \mu\eta_{xy}$$

et les équations analogues pour τ_{yz} et τ_{zx}.

Dans le cas du matériau incompressible ($\varepsilon_k = 0$), les équations (1.8) se simplifient.

Pour l'état visco-plastique rigide ($\overline{G} = \infty$), la relation (1.1) prend la forme:

$$\sigma_i = (H + \sigma)\tan\psi + \mu\xi_i \qquad (1.9)$$

ce qui généralise la relation de HENCKY — laquelle constitue déjà une extension du schéma du matériau visco-plastique de BINGHAM.
La relation entre les déviateurs s'écrit dans ce cas (cf. [5]):

$$D_\sigma = 2\left[\frac{(H + \sigma)\tan\psi}{\xi_i} + \mu\right]D_\xi. \qquad (1.10)$$

Eu égard à l'hypothèse d'incompressibilité du milieu ($\xi_k = 0$) les équations d'état s'écrivent en axes cartésiens:

$$\left.\begin{array}{l}
\sigma_x - \sigma = 2\left[\dfrac{(H + \sigma)\tan\psi}{\xi_i} + \mu\right]\left(\xi_x - \dfrac{1}{3}\,\nu\xi_i\right) \\[2mm]
\text{et les équations analogues pour } \sigma_y \text{ et } \sigma_z, \\[2mm]
\qquad \tau_{xy} = \left[\dfrac{(H + \sigma)\tan\psi}{\xi_i} + \mu\right]\eta_{xy}
\end{array}\right\} \quad (1.11)$$

et les équations analogues pour τ_{yz} et τ_{zx}.

Dans le cas particulier du milieu non visqueux, ($\mu = 0$), ces équations généralisent les formules de DRUCKER et PRAGER (cf. [6]) que l'on peut aussi obtenir en introduisant la notion de potentiel plastique (cf. [7]) sous la forme nouvelle:

$$\theta \equiv \sigma_i - (H + \sigma)\,\nu = 0 \qquad (1.12)$$

et qui se réduit à l'expression classique dans le cas où $\nu = \tan\psi$.

Les lois de comportement ainsi obtenues doivent encore être précisées au moyen des recherches expérimentales nouvelles. Mais d'ores et déjà elles peuvent servir de point de départ à l'étude d'une série de problèmes de Mécanique des Sols et, en particulier, dans le cas de certains problèmes non linéaires relatifs aux milieux visco-élastiques qui font l'objet de la présente communication.

2. Déformation Unidimensionnelle du Sol

Considérons une déformation unidimensionnelle du sol, considéré comme matériau visco-élastique non linéaire[1]. Moyennant les conditions suivantes: compression suivant une direction (compression mono-axiale); absence des dilatations latérales, nous aurons un état caractérisé par la tension et les déformations homogènes, défini par:

$$\left.\begin{aligned}
&\varepsilon_2 = \varepsilon_3 = 0, \\
&\varepsilon = \varepsilon_1 + \varepsilon_2 + \varepsilon_3 \doteq \varepsilon_1, \\
&\varepsilon_i = \frac{2}{\sqrt{3}}\,\varepsilon, \\
&\xi_2 = \xi_3 = 0, \\
&\xi = \xi_1 + \xi_2 + \xi_3 = \xi_1.
\end{aligned}\right\} \tag{2.1}$$

En substituant (2.1) dans les équations d'état (1.8) rapportées aux axes principaux, on aura:

$$\sigma_1 = \left(1 + \frac{4}{3}\,\frac{\overline{G}\tan\psi}{\tan\psi + \frac{2}{\sqrt{3}}\overline{G}\varepsilon}\,\varepsilon\right)\left\{H + \alpha\left[\varepsilon_0 + \left(1 + \frac{2}{\sqrt{3}}\nu\right)\varepsilon\right]^\beta - \sigma_0\right\}$$

$$+ \left[\chi\left(1 + \frac{4}{3}\,\frac{\overline{G}\tan\psi}{\tan\psi + \frac{2}{\sqrt{3}}\overline{G}\varepsilon}\,\varepsilon\right) + \frac{4}{3}\,\mu\right]\xi - H, \tag{2.2}$$

$$\sigma_2 = \sigma_3 = \left(1 - \frac{2}{3}\,\frac{\overline{G}\tan\psi}{\tan\psi + \frac{2}{\sqrt{3}}\overline{G}\varepsilon}\,\varepsilon\right)\left\{H + \alpha\left[\varepsilon_0 + \left(1 + \frac{2}{\sqrt{3}}\nu\right)\varepsilon\right]^\beta - \sigma_0\right\}$$

$$+ \left[\chi\left(1 - \frac{2}{3}\,\frac{\overline{G}\tan\psi}{\tan\psi + \frac{2}{\sqrt{3}}\overline{G}\varepsilon}\,\varepsilon\right) - \frac{2}{3}\,\mu\right]\xi - H. \tag{2.3}$$

Pour l'état stabilisé du sol, les relations (2.2) et (2.3) définissent ce qu'on appelle la courbe de compression monoaxiale et la courbe de pression latérale.

---

[1] Ce problème est traité dans [8] en négligeant l'influence de la viscosité.

La relation entre les pressions verticale (σ_1) et latérale (σ_2 et σ_3), admise en Mécanique des Sols, peut, pour l'état stabilisé du matériau ($\xi = 0$) être obtenue à partir des formules (2.2) et (2.3) en utilisant la déformation axiale ε comme paramètre.

Il est intéressant d'obtenir la courbe de fluage du matériau soumis à une déformation homogène pour une valeur donnée de la charge axiale. Posant $\sigma_1 = p =$ cte dans (2.2), nous obtenons, eu égard à $\xi = \dfrac{d\varepsilon}{dt}$, une équation différentielle, dont l'intégrale s'écrit:

$$t = \int_0^\varepsilon \frac{\chi\left(1 + \dfrac{4}{3}\,\dfrac{\overline{G}\tan\psi}{\tan\psi + \dfrac{2}{\sqrt{3}}\overline{G}\varepsilon}\,\varepsilon\right) + \dfrac{4}{3}\mu}{p - \left(1 + \dfrac{4}{3}\,\dfrac{\overline{G}\tan\psi}{\tan\psi + \dfrac{2}{\sqrt{3}}\overline{G}\varepsilon}\,\varepsilon\right)\left\{H + \alpha\left[\varepsilon_0 + \left(1 + \dfrac{2}{\sqrt{3}}\nu\right)\varepsilon\right]^\beta - \sigma_0\right\} - H}\,d\varepsilon.$$

$$(2.4)$$

L'intégration numérique de (2.4) donne la courbe de fluage; pour un sol caractérisé par les constantes ci-après: $\overline{G} = 50$; $\tan\psi = 0{,}364$; $H = 3{,}0\ \text{kg/cm}^2$; $\alpha = 3500\ \text{kg/cm}^2$; $\beta = 1{,}3$; $\nu = 0{,}2$; $\sigma_0 = 3{,}0\ \text{kg/cm}^2$; $\varepsilon_0 = 0{,}00437$; $\mu = 500\ \text{kg. heure/cm}^2$; $\varkappa = 1000\ \text{kg. heure/cm}^2$, la courbe correspondante est représentée sur la fig. 3. L'asymptote du graphique du fluage (tracée en pointillé sur la fig. 3) s'obtient en annulant le dénominateur de l'expression sous le signe d'intégration. En utilisant la courbe de fluage et son asymptote, il est aisé de définir la pression latérale (2.3) en fonction du temps (fig. 4), ainsi que la valeur asymptotique de la pression latérale (indiquée en pointillé sur la fig. 4).

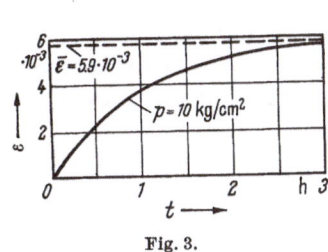

Fig. 3.

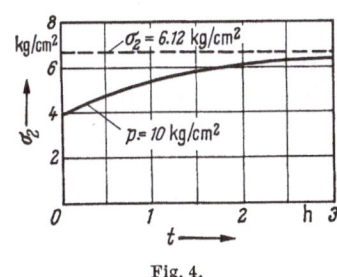

Fig. 4.

Notons que le coefficient de pression latérale pour l'état stabilisé ($\xi = 0$) du sol, déterminé à partir de (2.2) et (2.3) sous forme du quotient $\zeta = \dfrac{\sigma_2}{\sigma_1} = \dfrac{\sigma_3}{\sigma_1}$, est fonction de la déformation axiale avec les valeurs limites:

$$\zeta|_{\varepsilon \to 0} = \frac{1 - \dfrac{2}{3}\,\overline{G}H\,\dfrac{\varepsilon_0}{\sigma_0\left(1 + \dfrac{2}{\sqrt{3}}\,\nu\right)}}{1 + \dfrac{4}{3}\,\overline{G}H\,\dfrac{\varepsilon_0}{\sigma_0\left(1 + \dfrac{2}{\sqrt{3}}\,\nu\right)}}, \tag{2.5}$$

$$\zeta|_{\varepsilon \to \infty} = \frac{1 - \dfrac{1}{\sqrt{3}}\,\tan\psi}{1 + \dfrac{2}{\sqrt{3}}\,\tan\psi}. \tag{2.6}$$

Pour le sol pulvérulent ($H = 0$), le coefficient de pression latérale se réduit à la valeur 1 [cf. (2.5)]; l'état correspondant des tensions est alors hydrostatique. A l'état plastique, [cf. (2.6)], les coefficients de pression latérale pour les sols pulvérulents et cohérents ne dépendent que du coefficient de frottement.

3. Déformation à Symétrie Axiale d'un Cylindre Creux de Révolution

Considérons une déformation axisymétrique d'un cylindre circulaire creux, dont le matériau est visco-élastique non linéaire, et se déforme à volume constant.

La condition d'invariance du volume permet alors d'expliciter les déplacements radiaux et leurs vitesses[1],

$$U_r(t) = U_b(t)\,\frac{b}{r}\,\operatorname{sign} U, \tag{3.1}$$

$$V_r(t) = V_b(t)\,\frac{b}{r}\,\operatorname{sign} V \tag{3.2}$$

lorsqu'on se donne, par exemple, le déplacement U_b et sa vitesse V_b sur le contour extérieur $r = b$ du cylindre, les fonctions $\operatorname{sign} U = \operatorname{sign} V$ étant définies par les égalités: $\operatorname{sign} U = +1$ dans le cas ($U > 0$) de la compression du cylindre et $\operatorname{sign} U = -1$ dans le cas ($U < 0$) de la dilatation.

Utilisant les équations d'état mécaniques (1.8), écrites en coordonnées polaires et eu égard aux expressions (3.1) et (3.2) des composantes de la déformation et de sa vitesse, nous aurons, quand le cylindre travaille à la compression ($\operatorname{sign} U = \operatorname{sign} V = +1$)

$$\left.\begin{aligned}
\sigma_r &= \left(1 - \frac{2\overline{G}\tan\psi}{\tan\psi + 2\overline{G}\,U_b\,\dfrac{b}{r^2}}\,U_b\,\frac{b}{r^2}\right)(H + \sigma) - 2\mu V_b\,\frac{b}{r^2} - H, \\[4mm]
\sigma_\varphi &= \left(1 + \frac{2\overline{G}\tan\psi}{\tan\psi + 2\overline{G}\,U_b\,\dfrac{b}{r^2}}\,U_b\,\frac{b}{r^2}\right)(H + \sigma) + 2\mu V_b\,\frac{b}{r^2} - H. \\[4mm]
\sigma_z &= \sigma
\end{aligned}\right\} \tag{3.3}$$

[1] Dans la suite nous omettrons d'écrire l'argument (t).

En portant (3.3) dans l'équation de l'équilibre,

$$\frac{\partial \sigma_r}{\partial r} + \frac{\sigma_r - \sigma_\psi}{r} = 0 \qquad (3.4)$$

on obtient l'équation différentielle :

$$\frac{\partial \sigma}{H + \sigma} - \frac{8(\overline{G}\,U_b\,b)^2}{r\left(r^2 + \dfrac{2\overline{G}\,U_b\,b}{\tan \psi}\right)\left[r^2 + 2\overline{G}\,U_b\,b\,\dfrac{1 - \tan \psi}{\tan \psi}\right]\tan \psi}\,\partial r = 0. \qquad (3.5)$$

En intégrant cette relation pour une valeur donnée de la pression radiale $\sigma_r = p_a$ le long de la surface interne du cylindre $r = a$, on trouve :

$$H + \sigma = \left[(H + p_a) + 2\mu\,V_b\,\frac{b}{a^2}\right]\left(\tan \psi + 2\overline{G}\,U_b\,\frac{b}{r^2}\right)$$

$$\times \frac{\left[\tan \psi + 2\overline{G}\,U_b\,\dfrac{b}{a^2}\,(1 - \tan \psi)\right]^{\frac{\tan\psi}{1-\tan\psi}}}{\left[\tan \psi + 2\overline{G}\,U_b\,\dfrac{b}{r^2}\,(1 - \tan \psi)\right]^{\frac{1}{1-\tan\psi}}}. \qquad (3.6)$$

En portant (3.6) dans les équations d'état (3.3), on trouve :

$$\sigma_r = \left[(H + p_a) + 2\mu\,V_b\,\frac{b}{a^2}\right]\left[\frac{\tan \psi + 2\overline{G}\,U_b\,\dfrac{b}{a^2}\,(1 - \tan \psi)}{\tan \psi + 2\overline{G}\,U_b\,\dfrac{b}{r^2}\,(1 - \tan \psi)}\right]^{\frac{\tan\psi}{1-\tan\psi}} \qquad (3.7)$$

$$- 2\mu\,V_b\,\frac{b}{r^2} - H.$$

$$\sigma_\varphi = \left[(H + p_a) + 2\mu\,V_b\,\frac{b}{a^2}\right]\left[\tan \psi + 2\overline{G}\,U_b\,\frac{b}{r^2}\,(1 + \tan \psi)\right]$$

$$\times \frac{\left[\tan \psi + 2\overline{G}\,U_b\,\dfrac{b}{a^2}\,(1 - \tan \psi)\right]^{\frac{\tan\psi}{1-\tan\psi}}}{\left[\tan \psi + 2\overline{G}\,U_b\,\dfrac{b}{r^2}\,(1 - \tan \psi)\right]^{\frac{1}{1-\tan\psi}}} + 2\mu\,V_b\,\frac{b}{r^2} - H. \qquad (3.8)$$

$$\sigma_z = \sigma. \qquad (3.9)$$

Ainsi, les formules (3.7), (3.8) et (3.9) permettent de définir l'état des contraintes du cylindre creux lorsque l'échantillon cylindrique est soumis sur sa face interne à des pressions normales données, alors que les déplacements et leurs vitesses sont connus sur la surface externe.

Si l'on pose maintenant, dans (3.7), (3.8) et (3.9) $\overline{G} = \infty$ on retrouve la solution du problème de l'écoulement visco-plastique du cylindre creux déjà obtenue dans [5], en suivant une autre voie, sous la forme:

$$\sigma_r = (H + p_a)\left(\frac{r}{a}\right)^{\frac{2\tan\psi}{1-\tan\psi}} + 2\mu\, V_b \frac{b}{a^2}\left[\left(\frac{r}{a}\right)^{\frac{2\tan\psi}{1-\tan\psi}} - \left(\frac{a}{r}\right)^2\right] - H, \quad (3.10)$$

$$\sigma_\varphi = (H + p_a)\frac{1+\tan\psi}{1-\tan\psi}\left(\frac{r}{a}\right)^{\frac{2\tan\psi}{1-\tan\psi}}$$
$$+\, 2\mu\, V_b \frac{b}{a^2}\left[\frac{1+\tan\psi}{1-\tan\psi}\left(\frac{r}{a}\right)^{\frac{2\tan\psi}{1-\tan\psi}} + \left(\frac{a}{r}\right)^2\right] - H, \quad (3.11)$$

$$\sigma_z = \sigma = (H + p_a)\frac{1}{1-\tan\psi}\left(\frac{r}{a}\right)^{\frac{2\tan\psi}{1-\tan\psi}}$$
$$+\, 2\mu\, V_b \frac{b}{a^2}\frac{1}{1-\tan\psi}\left(\frac{r}{a}\right)^{\frac{2\tan\psi}{1-\tan\psi}} - H \quad (3.12)$$

d'ou on peut déduire, compte tenu de (3.10) — ($r = b$) — la force portante du cylindre (lorsque la vitesse des déplacements de la face externe est donnée à priori) ainsi que la valeur limite de cette force:

$$\bar{p}_b = (H + p_a)\left(\frac{b}{a}\right)^{\frac{2\tan\psi}{1-\tan\psi}} - H \quad (3.13)$$

correspondant au cas $V_b = 0$ (absence d'un écoulement visco-plastique). Il est intéressant d'utiliser la solution ainsi obtenue en vue du calcul du déplacement de la face externe de l'échantillon en fonction du temps, connaissant les pressions radiales sur les faces externes et internes ($r = a$, $r = b$). A cet effet, posons: $H + \sigma_r|_{r=b} = H + p_b > H + p_a$ et introduisons dans (3.7) ces conditions en tenant compte de la relation $V_b = \frac{dU_b}{dt}$. On obtient une équation différentielle dont l'intégration conduit à la formule:

$$t = 2\mu\frac{b}{a^2}\int\limits_0^{U_b}\frac{\left[\dfrac{\tan\psi + 2\bar{G}\,U_b\frac{b}{a^2}(1-\tan\psi)}{\tan\psi + 2\bar{G}\,U_b\frac{1}{b}(1-\tan\psi)}\right]^{\frac{\tan\psi}{1-\tan\psi}} - \left(\dfrac{a}{b}\right)^2}{(H+p_b) - (H+p_a)\left[\dfrac{\tan\psi + 2\bar{G}\,U_b\frac{b}{a^2}(1-\tan\psi)}{\tan\psi + 2\bar{G}\,U_b\frac{1}{b}(1-\tan\psi)}\right]^{\frac{\tan\psi}{1-\tan\psi}}}\,dU_b. \quad (3.14)$$

L'intégration a un sens sur l'intervalle $0 \leq \bar{U}_b \leq U_b$, la valeur limite $\bar{U}_b$ correspondant à la vitesse du déplacement nulle $\left(\frac{dU_b}{dt} = 0\right)$ et qui se calcule en égalant à zéro le dénominateur de l'expression sous le signe d'intégration de (3.14).

En intégrant numériquement (3.14), on trouve la relation $U_b = U_b(t)$; on obtient aussi la courbe de fluage de la surface extérieure du cylindre en fonction du temps, dont l'asymptote est $U_b = \overline{U}_b$.

Pour obtenir la courbe de fluage d'une surface cylindrique quelconque à l'intérieur de l'épaisseur du cylindre ($U_a \leq U_r \leq U_b$), on calculera U_b à partir de (3.1) et on portera le résultat dans (3.14), où on remplacera la limite d'intégration U_b par U_r.

Le cas particulier ou $b = \infty$, la pression p_b étant constante, offre un intérêt spécial. Dans ce cas (3.14) devient, en y faisant $U_b = U_a \dfrac{a}{b}$ et en effectuant le passage à la limite $b = \infty$:

$$
t = 2 \frac{\mu}{a} \int\limits_{0}^{U_a} \frac{dU_a}{(H + p_b)\left[\dfrac{\tan\psi}{\tan\psi + 2\overline{G} U_a \dfrac{1}{a}(1 - \tan\psi)}\right]^{\frac{\tan\psi}{1-\tan\psi}} - (H + p_a)} .
$$

(3.15)

On obtient ainsi la courbe de fluage de la surface intérieure du cylindre. Les composantes des tensions sont définies au moyen de (3.7), (3.8) et (3.9) en y remplaçant U_b et $V_b = \dfrac{dU_b}{dt}$ par leurs valeurs à l'instant t déterminées à partir de la courbe du fluage.

A titre d'exemple, considérons le cas d'un cylindre dont la face externe a un rayon infini; on a ainsi le schéma de l'état du sol autour d'une galerie de mine. Les données numériques sont les suivantes:

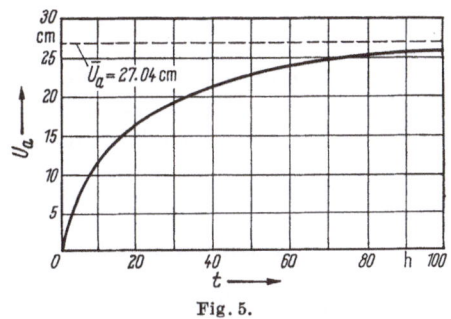

Fig. 5.

1° dimensions du cylindre: $2a = 8{,}0$ m; $2b = \infty$,

2° pression radiale à la face interne: $p_a = 0$,

3° pression radiale à la face externe $p_b = 10$ kg/cm²,

4° déplacement de la surface interne du cylindre à l'instant initial: $U_a|_{t=0} = 0$,

5° paramètres mécaniques du sol: $\overline{G} = 50$; $\tan\psi = 0{,}364$; $H = 3{,}0$ kg/cm²; $\mu = 500$ kg · heure/cm².

La courbe de fluage de la face interne du cylindre, obtenue par calcul numérique, est représentée sur la figure 5; l'asymptote, tracée en pointillé correspond à l'amortissement complet des déplacements en fonction du temps.

La répartition des contraintes, définies par les formules (3.7), (3.8) et (3.9), est donneé par les graphiques de la figure 6 correspondant à l'instant initial $t = 0$ et pour $t = \infty$ (lorsque les déformations sont complètement amorties).

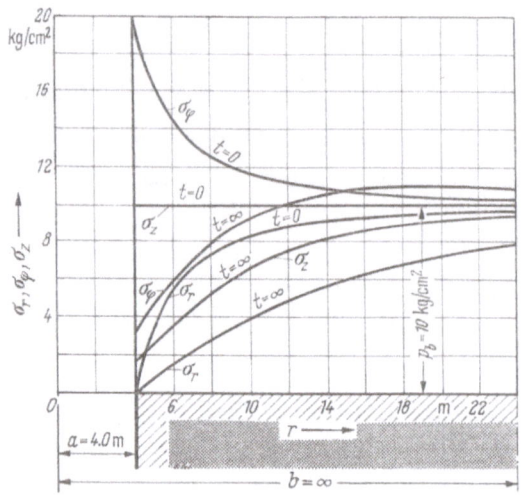

Fig. 6.

Les solutions ainsi obtenues ont été utilisées pour le calcul des galeries de mines, des batardeaux résistants à la pression des glaces etc.; à cet effet, on a combiné de façon appropriée les résultats précédents (cf. [9]).

4. Deformation à Symétrie Axiale d'une Sphère Creuse

Considérons la déformation d'une sphère creuse à symétrie axiale, dont le matériau, élastique non linéaire ($\mu = \chi = 0$), est linéairement compressible ($K(\varepsilon_k) = K = $ Cte) et à dilatance nulle: $\nu = 0$.

Les équations d'état (1.8), s'écrivent, en coordonnées sphériques r, $\varphi (a \leq r \leq b)$ (la troisième variable restant sans influence en vertu des hypothèses faites), dans les conditions que l'on vient de définir:

$$\sigma_r = \frac{4}{3} \frac{\overline{G} \tan \psi}{\tan \psi + \dfrac{2}{\sqrt{3}} \overline{G}(\varepsilon_r - \varepsilon_\varphi)} (H + K\varepsilon)(\varepsilon_r - \varepsilon_\varphi) + K\varepsilon,$$

$$\sigma_\varphi = -\frac{2}{3} \frac{\overline{G} \tan \psi}{\tan \psi + \dfrac{2}{\sqrt{3}} \overline{G}(\varepsilon_r - \varepsilon_\varphi)} (H + K\varepsilon)(\varepsilon_r - \varepsilon_\varphi) + K\varepsilon. \tag{4.1}$$

Introduisons (cf. [10]) la nouvelle fonction inconnue au moyen de la relation:

$$\varepsilon_r - \varepsilon_\varphi = \frac{dU_r}{dr} - \frac{U_r}{r} = \Phi(r).$$ (4.2)

En intégrant (4.2), on trouve:

$$U_r = r\left(C_1 + \int \frac{\Phi}{r}\,dr\right),$$ (4.3)

$$\varepsilon = \Phi + 3\left(C_1 + \int \frac{\Phi}{r}\,dr\right),$$ (4.4)

$$\varepsilon_i = \frac{2}{\sqrt{3}}\,|\Phi|.$$ (4.5)

Moyennant ces relations (4.1) prend la forme:

$$\left.\begin{aligned}
\sigma_r &= \frac{4}{3}\,\frac{\overline{G}\tan\psi}{\tan\psi + \dfrac{2}{\sqrt{3}}\overline{G}\Phi}\left\{H + K\left[\Phi + 3\left(C_1 + \int \frac{\Phi}{r}\,dr\right)\right]\right\}\Phi \\
&\qquad + K\left[\Phi + 3\left(C_1 + \int \frac{\Phi}{r}\,dr\right)\right], \\
\sigma_\varphi &= -\frac{2}{3}\,\frac{\overline{G}\tan\psi}{\tan\psi + \dfrac{2}{\sqrt{3}}\overline{G}\Phi}\left\{H + K\left[\Phi + 3\left(C_1 + \int \frac{\Phi}{r}\,dr\right)\right]\right\}\Phi \\
&\qquad + K\left[\Phi + 3\left(C_1 + \int \frac{\Phi}{r}\,dr\right)\right].
\end{aligned}\right\}$$ (4.6)

Posons:

$$F(r) = \frac{4}{3}\,\frac{\overline{G}\tan\psi}{\tan\psi + \dfrac{2}{\sqrt{3}}\overline{G}\Phi}\left\{H + K\left[\Phi + 3\left(C_1 + \int \frac{\Phi}{r}\,dr\right)\right]\right\}\Phi + K\Phi$$ (4.7)

ou $F(r)$ est une nouvelle fonction inconnue. Portant (4.7) dans (4.6), il vient:

$$\sigma_r = F(r) + 3K\left(C_1 + \int \frac{\Phi}{r}\,dr\right),$$ (4.8)

$$\sigma_\varphi = \sigma_r - \frac{3}{2}\,[F(r) - K\Phi].$$ (4.9)

En substituant (4.8) et (4.9) dans l'équation d'équilibre:

$$\frac{d\sigma_r}{dr} + 2\,\frac{\sigma_r - \sigma_\varphi}{r} = 0$$ (4.10)

on obtient l'équation différentielle:

$$\frac{dF(r)}{dr} + 3\,\frac{F(r)}{r} = 0$$ (4.11)

dont l'intégration générale s'écrit:

$$F(r) = \frac{C_2}{r^3}.$$ (4.12)

Posons $F_a = F(a)$, $F_b = F(b)$; d'aprés (4.12), il vient:

$$F(r) = F_a \frac{a^3}{r^3}, \qquad F(r) = F_b \frac{b^3}{r^3}. \qquad (4.13)$$

La formule (4.8), eu égard à (4.3), peut être écrite:

$$\sigma_r = F(r) + 3K \frac{U_r}{r}. \qquad (4.14)$$

Par exemple, les valeurs de la pression à la surface intérieure de la sphère creuse $r = a$, et du déplacement radial $U_r = -U_a$ (rupture) étant supposées données à priori, on a, en vertu de (4.14) ($\sigma_r = p_a$ étant une donnée):

$$F_a = p_a + 3K \frac{U_a}{a}. \qquad (4.15)$$

D'un autre coté, (4.7) donne pour $r = a$, eu égard à (4.3)

$$F_a = \frac{4}{3} \frac{\overline{G} \tan \psi}{\tan \psi + \frac{2}{\sqrt{3}} \overline{G} \Phi} \left[H + \left(\Phi_a + 3 \frac{U_a}{a} \right) K \right] \Phi_a + K \Phi_a \qquad (4.16)$$

d'où l'on déduit la valeur frontière:

$$\Phi_a = \frac{1}{2} \frac{3 \left(\frac{2}{\sqrt{3}} \overline{G} F_a - K \tan \psi \right) - 4 \overline{G} \tan \psi \left(H - 3K \frac{U_a}{a} \right)}{\overline{G} K (2\sqrt{3} + 4 \tan \psi)} + \qquad (4.17)$$

$$\sqrt{\frac{1}{4} \left[\frac{3 \left(\frac{2}{\sqrt{3}} \overline{G} F_a - K \tan \psi \right) - 4 \overline{G} \tan \psi \left(H - 3K \frac{U_a}{a} \right)}{\overline{G} K (2\sqrt{3} + 4 \tan \psi)} \right]^2 + \frac{3 F_a \tan \psi}{\overline{G} K (2\sqrt{3} + 4 \tan \psi)}}$$

ou le signe $+$ devant le radical a été adopté parce qu'on a supposé positif le premier membre de (4.2).

Pour définir $\Phi(r)$, utilisons (4.2), (4.3) et (4.7); (4.7) donne, eu égard à (4.3) et (4.13):

$$U_r = \frac{r}{3K} \left[\frac{F_a \frac{a^3}{r^3} - K\Phi}{\frac{4}{3} \frac{\overline{G} \tan \psi}{\tan \psi + \frac{2}{\sqrt{3}} \overline{G} \Phi} \Phi} - K\Phi - H \right]. \qquad (4.18)$$

En combinant (4.2) et (4.18), on trouve l'équation différentielle non linéaire:

$$\frac{d\Phi}{dr} = - \frac{3 \left[\frac{F_a}{K} \frac{a^3}{r^3} \left(\tan \psi + \frac{2}{\sqrt{3}} \overline{G} \Phi \right) + \frac{4}{3} \overline{G} \tan \psi \Phi^2 \right]}{r \left[\frac{F_a}{K} \frac{a^3}{r^3} \tan \psi + \frac{4}{3} \overline{G} \left(\frac{\sqrt{3}}{2} + \tan \psi \right) \Phi^2 \right]} \Phi \qquad (4.19)$$

dont l'intégration numérique permet de définir la fonction inconnue dans tout le domaine $a \le r \le \le b$ et, en particulier, de fixer Φ_b à la surface extérieure de la sphère creuse.

En définitive, (4.18) donne U_b, alors que (4.14) conduit à:

$$p_b = F_a \frac{a^3}{b^3} + 3K \frac{U_b}{b}. \qquad (4.20)$$

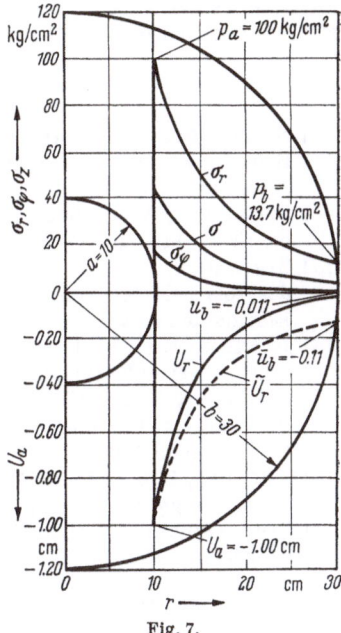

Fig. 7.

Les valeurs frontières obtenues de U_b et p_b à la surface extérieure de la sphère creuse ne coïncident pas, en général, avec les valeurs données à priori; en effet, nous sommes partis uniquement des conditions aux limites pour $r = a$. Lorsque U_a est donné, il faut donc utiliser la méthode de tir; calculant la solution pour une suite de valeurs de p_a, on choisira celle des valeurs de ce paramètre qui conduit à la valeur de p_b fixée à priori.

Une fois obtenue une telle solution, les déplacements sont définis par (4.18), ainsi que les composantes des contraintes:

$$\sigma_r = F_a \frac{a^3}{r^3} + 3K \frac{U_r}{r}, \left.\begin{array}{l} \\ \\ \end{array}\right\} \qquad (4.21)$$
$$\sigma_\varphi = \sigma_r - \frac{3}{2}\left(F_a \frac{a^3}{r^3} - K\Phi\right),$$

$$\sigma = K\left(\Phi + 3\frac{U_r}{r}\right). \qquad (4.22)$$

Pour illustrer ce qui précède, on donne (cf. 1a fig. 7) les résultats des calculs numériques, relatifs au cas d'une sphère creuse, les paramètres mécaniques du milieu étant les suivants:

$$\bar{G} = 100; \quad \tan\psi = 0,9; \quad H = 10\,\text{kg/cm}^2; \quad K = 1000\,\text{kg/cm}^2.$$

Un problème analogue a été résolu pour un matériau incompressible doué de visco-élasticité non linéaire. La méthode utilisée est analogue à celle décrite au § 3 ci-dessus. Voici les conclusions. Lorsqu'on néglige la vicsosité, la condition d'incompressibilité n'a pas d'influence sensible sur la répartition des contraintes; seules les déformations $\tilde{U}_r$ changent — d'ailleurs dans le sens de l'accroissement.

Pour finir, nous attirons l'attention sur l'intérêt qu'il y a à étendre la classe des problèmes de mécanique des sols traités avec la prise en compte des effets visco-élastiques non linéaires et à perfectionner les méthodes numériques de résolution de ces problèmes au moyen des calculatrices électroniques. On peut espérer obtenir dans cette voie des résultats pratiques importants.

Bibliographie

[1] STROGANOV, A. S.: Méthodes de prévision des tassements finis des fondations des constructions. Travaux de l'Institut d'Energétique de Moscou. Fasc. XIX 1956 (en russe).

[2] STROGANOV, A. S.: Plane plastic deformation of soil. Brussels Conference on Earth Pressure Problems, Proceedings V. I., Brussels 1958.

[3] STROGANOV, A. S., et J. B. LABZOV: Fluage et écoulement visco-plastique des sols gelés en compression triaxiale. Journal de l'Ingénieur, fasc. 3. Edition de l'Académie des Sciences de l'URSS, 1964 (sous presse, en russe).

[4] BOTKINE, A. I.: Recherches sur l'état des contraintes dans les milieux cohérents et pulvérulents. Bulletin de VNIIG, t. XXIV, 1939 (en russe).

[5] STROGANOV, A. S.: Visco-plastic flow of soils. Proceedings of the Fifth International Conference on Soil Mechanics and Foundation Engineering. Paris: Dunod 1961.

[6] DRUCKER, D. C., and W. PRAGER: Soil Mechanics and Plastic Analysis on Limit Design. Quart. Appl. Math. 10, 2 (1952).

[7] STROGANOV, A. S.: Analyse de la déformation plane des sols. Journal de l'Ingénieur, Éditions de l'Ac. des Sciences de l'URSS, 1964, (sous presse, en russe).

[8] STROGANOV, A. S.: Lateral Earth Pressure at Linear Deformation. Brussels Conference on Earth Pressure problems. Proceedings, Vol. 3, Brussels, 1958.

[9] STROGANOV, A. S.: Méthodes de calcul simultané des galeries de mines et des pressions dans les roches et les matériaux obéissant à des lois visco-élastiques non linéaires. Recueil: Pression des roches et renforcements des batardeaux verticaux, Moscou: Gosgortechizdat 1963.

[10] KATCHANOV, L. M.: Mécanique des milieux plastiques, Moscou: Editions techniques de l'état 1948.

Rapport Général
Relatif à la 1ᵉ et à la 2ᵉ Sous-Section

Rapporteur: D. Radenkovic

Introduction

En raison de la maladie du Professeur SOKOLOVSKY au moment du Symposium, le travail des deux premières sous-sections se trouve présenté dans un seul rapport.

Dans ce rapport l'ensemble des communications est classé en trois groupes principaux:

1° études générales;
2° comportement du milieu pulvérulent;
3° fluage des argiles.

Les deux derniers groupes comportent chacun trois divisions:

a) études microrhéologiques; b) équations de comportement; c) étude des équations (solutions des problèmes aux limites).

1. Études Générales

Sous cette rubrique sont classés les travaux où l'attention n'est pas centrée sur l'idéalisation d'un matériau particulier, mais plutôt sur la discussion des différentes définitions possibles du comportement mécanique.

La communication la plus générale dans ce sens est celle d'ANGLES D'AURIAC qui avance une ébauche de classification des lois de comportement rhéologique. L'auteur énonce des principes censés régir les comportements possibles et propose une terminologie adaptée aux différentes possibilités. En fait il s'agit ici d'une tentative d'axiomatisation de la rhéologie; son lien avec la mécanique des sols semble résider dans la suggestion implicite que les idéalisations classiques du comportement des sols sont trop simplistes pour rendre compte de la complexité réelle des phénomènes.

L'idée fondamentale de l'étude de GREEN et RIVLIN est que pour des matériaux de structure complexe, comme le sont par exemple des sols, la distribution des vitesses et des forces ne peut pas être représentée d'une manière adéquate par les champs vectoriels classiques. Donc il faut introduire, pour rendre compte de la déformation d'un tel milieu, des champs tensoriels d'ordre $\alpha + 1$ (multipolar displacement fields); les forces correspondantes, dites de $(\alpha + 1)$ ème espèce sont définies par l'expression du travail. Les auteurs établissent les équations fondamentales pour un tel milieu. Le matériau élastique dont l'énergie intérieure dépend des dérivées des déplacements d'ordre supérieur à un en serait un cas particulier.

OLSZAK et PERZYNA s'occupent du comportement dynamique des sols qu'ils assimilent à un milieu élastique/visco-plastique. Le sol est considéré comme élastique jusqu'à la limite d'écoulement; au-delà de cette limite la composante anélastique de la vitesse de déformation est proportionnelle à une fonction de la différence entre l'état de contraintes actuel et l'état qui correspond à la limite d'écoulement statique. Cette fonction reste à déterminer par des expériences. Le problème de la propagation des ondes dans un tel matériau est étudié par la méthode des caractéristiques.

Le travail de DRUCKER garde encore un caractère très général, mais le problème du comportement des sols y est abordé de bien plus près. Plusieurs idéalisations élémentaires du sol considéré comme un matériau plastique écrouissable sont discutées. L'influence du temps (comportement visqueux) est prise en considération dans certaines formes de la loi de l'écoulement, mais les états asymptotiques correspondants sont supposés dépendre uniquement du parcours de charge et non de la vitesse de déformation. Par là l'étude des sols se rattache à la théorie de la plasticité dans laquelle l'auteur attribue le rôle essentiel à la notion de stabilité du matériau. Les thèses théoriques sont confrontées avec les intéressantes expériences de HENKEL et de ROSCOE et autres.

L'étude de MANDEL a pour objet de discuter les conditions de stabilité mécanique d'un élément du sol et notamment le postulat de DRUCKER. L'auteur montre que le postulat de DRUCKER est une condition suffisante de stabilité, mais qu'il n'est pas une condition nécessaire lorsqu'il y a des frottements intérieurs du type de COULOMB. Une condition nécessaire est proposée, qui découle du fait qu'un matériau stable doit pouvoir propager une perturbation sous forme d'ondes réelles. Toutefois, la stabilité individuelle d'un élément n'est pas une condition nécessaire pour la stabilité d'un massif. Ceci autorise l'existence de caractéristiques réelles dans un domaine limité.

2. Comportement du Milieu Pulvérulent

Les études présentées concernent en général le sable, mais les travaux sur le comportement des roches (cf. sous-section 3) concernant les problèmes de l'équilibre limite se rattachent à ce groupe.

a) **Études Microrhéologiques.** KÉZDI dans son travail (présenté par WINTERKORN) cherche à déterminer le coefficient du frottement interne à partir de l'étude des empilages réguliers, partant d'une idée de WINTERKORN, selon laquelle les assemblages des grains peuvent être traités comme des liquides *macroméritiques*. Trois formules donnant le coefficient de frottement en fonction de la densité sont proposées, correspondant aux différentes possibilités d'approximation utilisée.

L'étude de HAYTHORNTHWAITE concerne un milieu bidimensionnel (par ex. empilage des rouleaux). Une relation entre les contraintes principales valable pour les empilages réguliers est établie sans faire l'hypothèse de coïncidence des directions principales des contraintes et des déformations. En partant de cette relation et en définissant un matériau statiquement isotrope comme un amoncellement de petits paquets réguliers, l'auteur cherche à construire une théorie rationnelle du milieu pulvérulent.

Dans sa communication MURAYAMA présente de nombreux résultats d'expériences et en recherche une interprétation probabiliste.

Enfin le travail de LITWINISZYN se situe en marge des essais précédents, le problème n'étant pas traité du point de vue de la mécanique proprement dit. L'auteur assimile le milieu pulvérulent à une collection de particules dont le mouvement correspond à un processus stochastique de MARKOV, indépendant (sauf d'une manière indirecte) des forces. Il est intéressant de constater la parenté des résultats du calcul et des observations portant sur le champ des vitesses d'un écoulement.

b) Équations de Comportement. Le problème difficile des équations de comportement d'un milieu pulvérulent (relations complètes tridimensionnelles de la théorie macroscopique) n'a pas été abordé directement dans les communications présentées au Symposium. La question n'en était pas moins présente dans la majorité des exposés. Les idées fondamentales ont été discutées par DRUCKER et MANDEL (cf. 1).

A l'heure actuelle il est, peut-on dire, presque unanimement admis que l'extension directe de la théorie du potentiel plastique aux milieux à frottement interne n'est pas satisfaisante. Il y a une tendance nette à admettre l'invariance du volume et à abandonner l'hypothèse de la coïncidence des directions principales (JOSSELIN DE JONG, HAYTHORNTHWAITE, SOBOTKA). Pourtant de nouvelles et meilleures solutions ne semblent pas encore mûres. A cet égard le problème de trouver les théorèmes limites correspondants (s'ils existent) et (même avant) le problème de l'unicité des solutions sont inquiétants.

L'étude de JOSSELIN DE JONG reflète parfaitement l'état actuel des idées sur ce sujet.

c) Étude des Équations. Des solutions nouvelles des problèmes aux limites n'ont pas été présentées, quoique certaines études expérimentales (cf. sous-section 4) suggèrent des idées intéressantes à ce sujet.

Par contre des améliorations des méthodes effectives de calcul ont été proposées.

SOBOTKA introduit pour un matériau non-homogène et anisotrope à la place des paramètres naturels classiques, de nouvelles variables exprimées dans les coordonées cartésiennes.

L'auteur s'occupe à la fois du champ des contraintes et du champ des vitesses. Les deux systèmes correspondants de caractéristiques sont distincts, l'hypothèse du potentiel plastique n'étant pas admise (cf. 2 b).

L'étude de KRAVTCHENKO et SIBILLE consiste à rechercher des solutions approchées en utilisant les développements limités de CAUCHY-KOVALEVSKA. C'est un procédé qui n'a pas été utilisé auparavant dans les calculs de l'équilibre limite et son application pourrait être intéressante à double titre: d'une part la possibilité de donner des solutions approchées analytiques, au moins pour certains cas, doit faciliter le travail de l'ingénieur qui utilise la théorie; d'autre part ce procédé permettra de préciser le comportement des solutions au voisinage des points singuliers.

3. Fluage des Argiles

a) Études Microrhéologiques. Les idéalisations présentées de la structure des argiles s'inspirent du modèle de la maison de cartes introduit par TAN (dans son travail avec GEUZE 1954). Les particules solides sont reliées dans les contacts (ponctuels, linéaires et de surface) par des forces de natures différentes dont l'interaction provoque une certaine rigidité. Les liaisons assurent un comportement quasi-élastique pour de faibles déformations; la consolidation secondaire est caractérisée par la rupture des liaisons actuelles et la formation de liaisons nouvelles, partiellement équivalentes.

Geuze étudie un tel modèle de la structure en se plaçant dans le cas plan où les liaisons sont des charnières du type bout-contre-face. La discussion détaillée des forces de contact, de la résistance des joints et du rôle du liquide interstitiel conduit à des conclusions sur le comportement (non-linéaire) de ce modèle, qui sont confirmées qualitativement par des expériences de l'auteur.

Murayama et Shibata s'occupent du fluage des argiles en partant d'un modèle analogue de la structure et en s'appuyant sur des considérations statistiques. Ceci les conduit à représenter l'argile comme un matériau de Bingham généralisé (élasticité, viscosité non-Newtonnienne et frottement intérieur liés en parallèle). Les résultats théoriques sont comparés avec de nombreuses expériences.

Le travail de Irmay et Zaslavky est à citer sous cette rubrique, car l'eau interstitielle et son écoulement sont un élément essentiel de toute étude physique concernant le comportement des argiles.

b) Équations de Comportement. Il est difficile de répartir les communications qui concernent la consolidation des argiles entre les sous-sections 1, 2 ou 3 car les auteurs s'occupent à la fois du problème de la définition du matériau, du calcul effectif et de la vérification expérimentale. La présente rubrique englobe donc un certain nombre de travaux où l'équivalence de ces trois directions de recherche est la plus marquée. Le trait commun de ces travaux est la recherche du compromis entre la non-linéarité essentielle des phénomènes et le désir de trouver des définitions du comportement suffisamment simples, pour permettre les calculs dans l'application de l'ingénieur.

L'étude de Tan concerne surtout l'influence du premier invariant de déformation sur les opérateurs rhéologiques, toutes les caractéristiques du matériau dépendant à chaque instant en premier lieu de la densité actuelle. L'auteur à conçu plusieurs types d'appareils pour vérifier ses conclusions théoriques dans des expériences aussi pures que possible.

Schiffman, Ladd et Chen étudient la possibilité de représenter les phénomènes de la consolidation secondaire par des modèles rhéologiques simples (à un nombre de paramètres limité). Les solutions analytiques pour plusieurs modèles sont développées et comparées à la fois entre elles et avec les expériences.

Krizek et Kondner (travail présenté par Schiffman) cherchent à exprimer le spectre de relaxation sous forme de la fonction d'erreur de Gauss. Les données expérimentales obtenues dans les expériences dynamiques (vibrations) et statiques (fluage) couvrent dans le temps un intervalle de neuf décades.

Le travail de Šuklje se rapporte à l'application directe des résultats obtenus avec des essais oedométriques ou triaxiaux dans le calcul de la consolidation des couches cylindriques, en se basant sur la méthode semi-graphique des isotaches développée antérieurement par l'auteur.

c) Étude des Équations. Dans sa communication sur la rhéologie des problèmes de la déformation plane, Sobotka introduit la notion de contraintes et de déformations équivalentes en essayant de généraliser les applications des opérateurs linéaires à certains problèmes de la viscoélasticité non-linéaire.

Freudenthal et Spillers (étude présentée par Perzyna) discutent l'application de la théorie de consolidation de Biot du problème de la consolidation secondaire des argiles. Ils présentent une solution du problème en symétrie axiale d'une couche viscoélastique chargée partiellement sur la surface. Un exemple numérique est donné.

Conclusion

La revue rapide des communications, qui précède, montre que le travail du Symposium rend fidèlement compte de l'état actuel des recherches concernant les applications de la rhéologie dans la mécanique des sols: tous les domaines sont représentés et les questions fondamentales ont été soulevées.

On assiste dans l'évolution actuelle de cette discipline à un progrès constant de nos connaissances, pourtant un tournant spectaculaire ne semble pas s'annoncer.

En ce moment, à part les études microrhéologiques qui peuvent beaucoup apporter à la compréhension des phénomènes, une appréciation exacte du rôle de la friction de COULOMB dans la déformation des milieux pulvérulents et un approfondissement de la théorie de viscoélasticité marqueraient le progrès décisif que nous souhaitons tous.

3.1 Determination of the Rheological Parameters and the Hardening Coefficients of Clays

By

Tan Tjong-kie

Summary

A micro-rheological analysis of the process of consolidation, hardening and flow is presented, which is based on the concept, that clay particles form a rigid card-house. Basing on an analysis of the various time effects the author suggests a method for the determination of the rheological parameters and hardening coefficients as a function of the volume dilatation. Shear test results show that the superposition principle is valid for the clay investigated, when the porosity is kept constant. The stress-strain relationships for confined compression, determined after the waterpressure has elapsed, however are non linear, whereas the permeability decreases exponentially with the volume deformation. A mathematical analysis of the oedometer tests is given and the theoretical formulas presented can describe the test results satisfactorily.

In order to study the possibility of building engineering constructions on soft clay-layers, which are known to cause many troubles, a systematic research has been started by our Institute to investigate their typical complex rheological properties. These type of soils has the following characteristics: high porosity, water content, and compressibility; low permeability, elastic modulus, viscosity and strength. One of the most fundamental features which distinguishes clay from other rheological materials, is that this substance hardens, i.e. it increases considerably in rigidity and strength with the decrease in porosity, whereas its permeability decreases simultaneously. Especially in soft clays this property is of primary importance. In spite of its great economical significance, the hardening of clays, however, has been studied only little. Theoretical studies have been started independently by FLORIN (1953) and TAN (1954), who consider the influence of hardening on one dimensional consolidation, but as yet no reliable method for the measurements of the hardening quantities is known. A non linear empirical formula for the prediction of settlements from oedometer tests has been suggested by KOPPEJAN (1948).

Micro Structural Analysis

For a better insight into the rheological properties of clays, it is desirable to make an analysis of the deformation mechanism of its internal structure. A study has been presented for the special case that deviations from the initial statistical state is small (TAN 1957a) and

here follows its brief summary: The clay network is regarded as a three dimensional cardhouse, which is built up of clay platelets of various shapes and sizes, mutually interconnected in point contacts (corner to flat surface), line contacts (edge to flat surface) and surface contacts (flat surface to flat surface) (Fig. 1). With the help of an electron microscope ROSENQVIST has investigated in a very ingenious way that such a structure indeed exists. So this cardhouse concept gives us a realistic starting point for our micro rheological analysis.

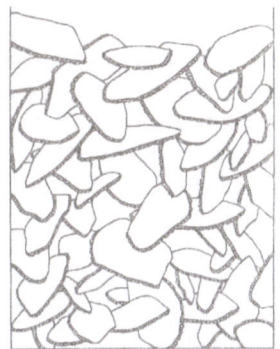

Fig. 1. Three dimensional cardhouse structure for clay after TAN.

The clay used for our experiments belongs to the 2:1 lattice type, which is regarded as having a dualistic character. The flat surfaces of the particles bear under the circumstances of our experiments a negative charge, the edges and corners on the contrary a positive one. The double layers are developed in the water phase starting from these fixed layers with opposite signs. So mutually repulsing COULOMB forces exist between the flat surfaces of the micelles as a result of double layer interpenetration of equal sign. The particles are bonded in the contact hinges by attractive COULOMB and LONDON- VAN DER WAALS forces and also by bonding forces of cations and hydrogen bridges. Due to the interaction of these forces the clay structure has a certain rigidity. Every deformation requires a flow of the pore water, which exerts a retarding action. During small shearing deformations this migration of water particles will be such that their total volume remains constant. Flow is ascribed to the sliding and disruption of bonds whereas simultaneously equivalent bonds are formed in other places. This phenomenon of the jumping of bonds occurs also in one dimensional consolidation; as this is equivalent to a decrease in rigidity of the soil skeleton with the time, the deformations will be larger than for a non flowing soil structure (secondary time effects), whereas the transfer of stress from the fluid to the solid phase requires much more time resulting into a maintaining of the water-pressure over a longer period. As long as the statistical state of the clay remains nearly constant it may be described by a set of equations containing operator forms with constant coefficients. However soft soils should be compressed considerably for being suitable as a foundation soil and then the structural changes no longer can be neglected. With the compression, the soil will gain in rigidity and strength as the number of contacts increases and especially the formation of more line and surface contacts will give a considerable contribution to the hardening. Further, the repulsive forces will increase as the planar surfaces are approaching nearer. Simultaneously the tortuous channels formed by the pores will be smaller, where the narrowing effect will cause a decrease in permeability. So the action of hardening will be measured in an increase of the shear and compression moduli and also of the yield values, viscosity and strength, but in a decrease of the permeability. Whereas the gain in rigidity causes a quicker transfer of stress from the fluid to the solid phase, thus in a quicker elapse of the waterpressure, the decrease in permeability just shows the opposite effect. It is assumed usually that both contradicting processes are compensating one and another completely, but as both processes are originating from quite different mechanisms, it can hardly be expected to occur. Even in our oedometer tests on clays, which can not be considered as soft clays of high porosity, a definite change of the coefficient of consolidation with the volume dilatation has been measured. So the complicated process of hardening can quite well be understood on the basis of above cardhouse structure. Another interesting phenomenon can be predicted when a permeability test under constant porosity will be performed. It can be understood that a part of the particles will reorient themselves gradually in relation to the streamlines. As the particles will change their mutual orientations the physico chemical and mechanical balance will be disturbed and this also will cause a time effect. Preliminary experiments show indeed that the discharge under constant gradient and porosity increases with the time attaining a certain maximum.

From this analysis it follows that the rheological parameters should be studied as a function of the first strain invariant; of course they also will be influenced by the second and third strain invariants in some complicated manner. In this paper only the change with the first invariant will be considered. In the case of a homogeneously isotropic soil for stresses below the upper yield limit f_3, three independent parameters viz. the operator of hydrostatic compression $\theta(\varepsilon)$ and the shear operator $\psi(\varepsilon)$, or the operator for uniaxial compression $E(\varepsilon)$ in combination with the operator for lateral expansion $\nu(\varepsilon)$, and further the permeability parameter $K(\varepsilon)$, which all are functions of the volume deformation, are sufficient for its characterisation.

Systematic Measurement of the Rheological Parameters and Hardening Coefficients

So long as only negligible deviations from the initial statistical state are considered, a complex interaction of the following time effects can be expected:

1. the shearing creep and flow under deviatoric stresses;

2. the volume creep under hydrostatic stresses and also during large shearing deformations;

3. the retarding time effect of the consolidation process.

For considerable structural changes, in addition, account should be taken of the degree of transformation of state, which may be of two quite different characters:

4. the time effect of hardening, which is dependent mainly on the decrease in porosity, thus the volume dilatation ε;

5. the time effect of structural disintegration, as soon as the stresses exceed the upper yield limit f_3.

Engineering practice requires a thorough study of soils under the most complicated three dimensional stress fields; however when one performs a test under arbitrary loading and drainage conditions, a completely obscure interaction of all time effects will take place and the test results will be of confusing complexity. Hence in order to meet the engineering demands, not only a systematic testing method should be found for the unique measurements of the time factors, but also their combined effect and interaction should be understood satisfactorily. For this purpose it is desirable to start with keeping the state of stress and its variation with the time as simply as possible; as far as possible the time effects should better be studied separately; when a series of samples is required, then the samples should be taken as replicable as possible and remoulding effects should be reduced. In a more advanced stage of research, also more complicated stress fields and drainage conditions can be studied.

For these investigations the author has devised some laboratory and field plastometers, which are now in regular use in our Institute (see p. 260).

In the apparatuses type 1, 4 and 8 the deformation can be studied carefully after unloading; the types 2, 3 ,4, 6 and 7 are suitable for the measurement of the hardening coefficients. This paper reports only on test results performed with the apparatuses type 2, 3, 5, 6 and 7. It has been proved usefull to increase the loading stepwisely with the time; in this manner the validity of the superposition theorem can be investigated, further the rheological parameters can be studied as a function of the volume dilatation. For the first stage of this research, replicable

17*

Type of plastometer	Stressfield or principle	Drainage
1. Torsion plastometer (1953)[1]	Pure shear	undrained
2. Oedometer with measurement of permeability TYPE I (1957)	Laterally confined compression	drained
3. Oedometer with measurement of lateral stress TYPE II (1957)[2]	Laterally confined compression	drained
4. Compression plastometer with pore pressure gauge (1957, 1964)[2]	Axially symmetric compression	drained; undrained
5. Dynamic triaxial apparatus (1958)[3]	Axially symmetric compression	drained; undrained
6. Oedometer with measurement of pore pressure TYPE III (1963)	Laterally confined compression	drained
7. Simple shear plastometer (1963)	Homogeneously simple shear	drained
8. Field vane plastometer (1964)	Vane principle	undrained

samples have been moulded carefully from a reddish clay at various constant watercontents ranging from 44 to 47%. The clay has the following physico chemical properties; total base exchange capacity 28.4 m.eq/100 gr; specific surface 211 m²/gr both determined for the clay fraction less than 2μ; $W_L = 52\%$; $W_P = 26\%$; $I_p = 26\%$; porosity $n = 0.55$, degree of saturation $\delta = 99-100\%$; clay fraction $< 2\mu = 21\%$; fraction $< 5\mu = 36\%$. The samples have been stored for at least two weeks in order to reduce remoulding effects. This clay still belongs to the category of normal clays and can not yet be considered as soft clay.

Here follows a brief description of the tests and their results:

In contradistinction with the direct shear apparatus, where the sample is sheared under a complicated, unknown stress distribution varying with the deformation, the sample in the simple shear plastometer is subjected to homogeneously simple shearing deformations. Drainage is applied at its upper and lower surfaces. The sample is first consolidated under a constant vertical load σ_v and when the largest part of the settlement has taken place, the shearing stress τ is applied according to the step function. In this manner, the shear modulus ψ describing the first time effect can be determined and moreover the validity of the superposition theorem can also be investigated. In the tests the results of which are shown in Fig. 2, $\sigma_v = 0.5$ kg/cm², and every step of τ is 0.025 kg/cm². The curves in Fig. 2b have been derived from the shear time curve (Fig. 2a) by means of the superposition

[1] Proc. 2nd Int. Conf. Rheol. 1953, p. 249.

[2] Proc. 4th Int. Conf. Soil Mech. 1957, Vol. III, p. 87, 140; Scientia Sinica 1958, p. 86, 1060.

[3] 1st Scient. Session Techn. Univers. Wroclaw 1958, p. 325; Scientia Sinica 1959, p. 1169.

theorem; its application here is quite correct as the stress strain relationship up to $f_3 = 0.16\ \text{kg/cm}^2$ is linear (Fig. 2c). The flow curve relating the rate of shear to the shearing stress, is shown in Fig. 2d; the viscosity is calculated $\eta = 1.11 \times 10^{13}$ poises. The shear modulus as a function of the time is shown in Fig. 2e.

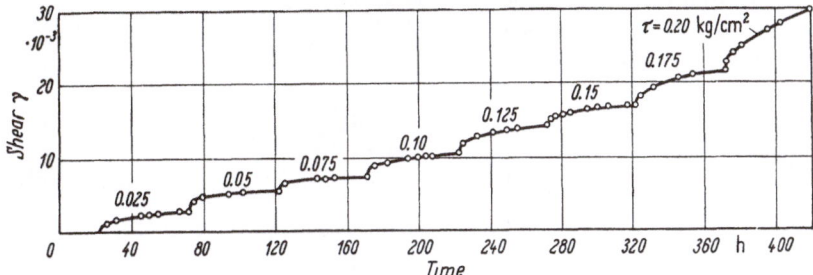

Fig. 2a. Shear-time curve for stepwise increase of shear stress τ.

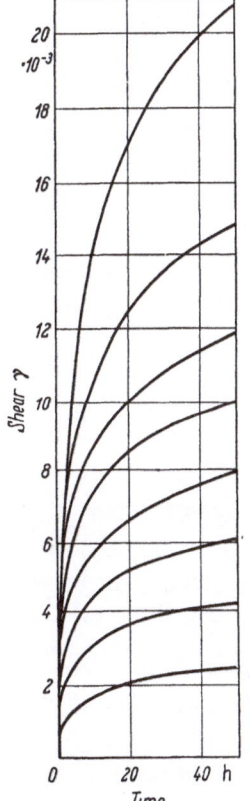

Fig. 2b. Shear time curves derived from Fig. 2a by means of superposition theorem.

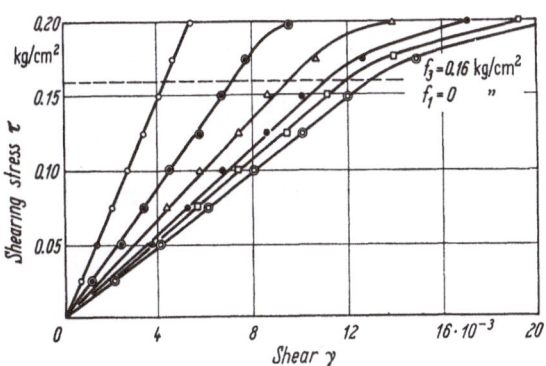

Fig. 2c. Stress strain curves for various time intervals.

The tests can be continued for higher vertical loads, increased according to the step-function; during every step, the above testing program can be repeated. In this way the change of the shearing modulus with the volume dilatation can be studied.

The linearity of the stress strain relationship has also been found in dynamic tests on clays with the help of the dynamic triaxial apparatus[1]. The stress strain relationships are plotted on the screen of an oscilloscope in the

[1] 1st Scient. Session Techn. Univers. Wroclaw 1958, p. 325; Scientia Sinica 1959, p. 1169.

form of LISSAJOUS figures
(Fig. 3). In the linear range
the figures are by approximation ellipses, but as soon
as the clay becomes non linear at higher stresses the
figures become complicated
(Fig. 3j), and Fig. 3k shows

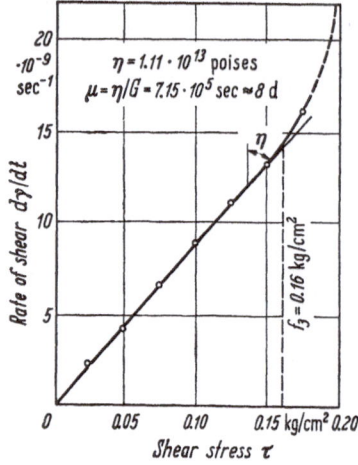

Fig. 2d. Flow curve.

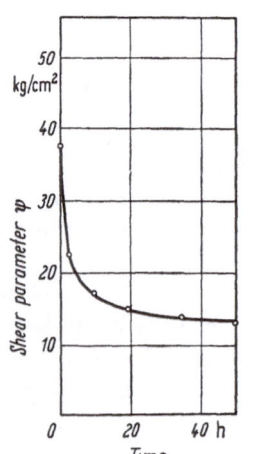

Fig. 2e. Shear parameter ψ as function
of time t.

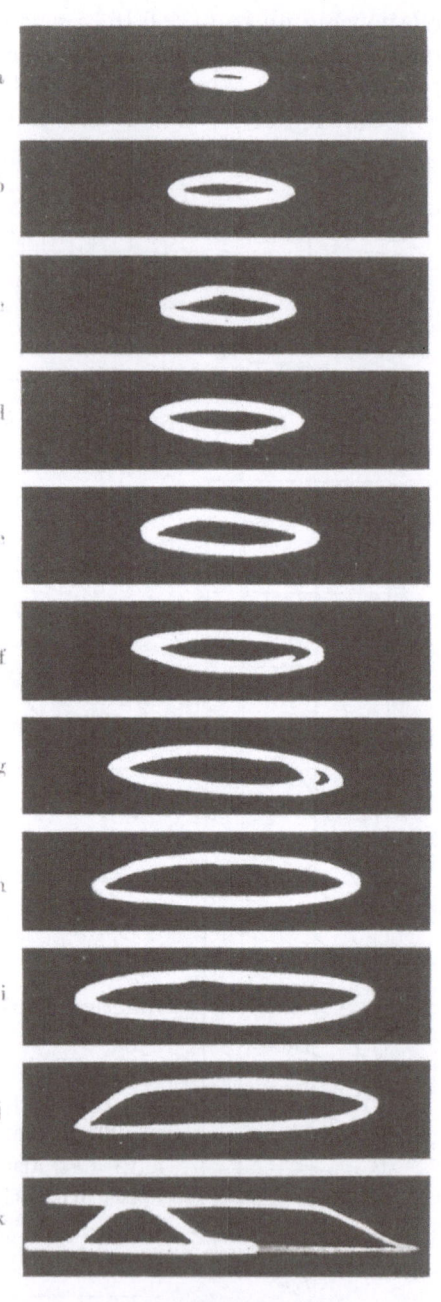

Fig. 3a—k. Typical LISSAJOUS figures for the dynamic stress strain relationships for clays at
frequencies up to 50 Hertz.

the relationship at the moment of failure whereby the sinusoidal stress applied just can be seen clearly.

In oedometer tests, where the sample is subjected to laterally confined compression (drained), an interaction of the first 4 time effects will be measured. The influence of the fourth time effect of hardening is negligible for small volume deformations; in this case the modulus of confined compression θ_c varies with the time governed by the first and second time effects. The resulting time function of θ_c is not directly measurable, as in oedometer tests unavoidably its complex interrelation with the third time function of consolidation is measured. By studying the settlement as a function of the time under constant loads, however, it is possible to determine θ_c at the beginning and the end of every loading interval. As it will be shown later the variation of $1/\theta_c(t)$ with the time can be obtained approximately by plotting the difference between the measured settlement and the theoretical settlement after TERZAGHI as a function of the time. This testing procedure can be repeated for successive steps of the stepwise loading and in this way the modulus of compression θ_c, the coefficient of consolidation C_v, and the permeability coefficient K can be determined as a function of the volume deformation ε. Every step should best be taken over such a long interval, that the waterpressure has elapsed practically and the secondary time effect can be measured satisfactorily.

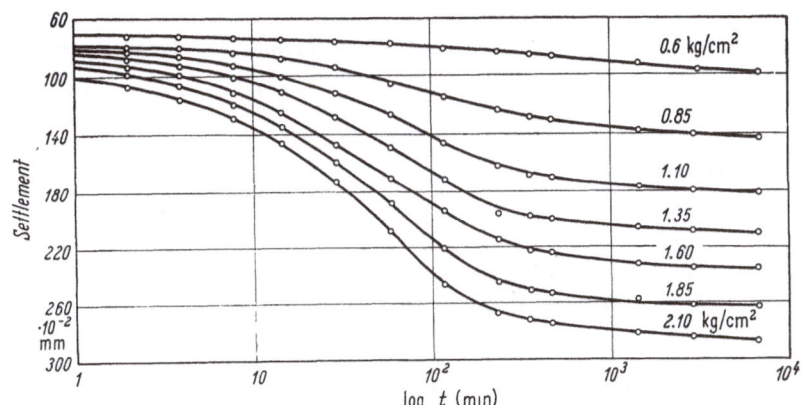

Fig 4ㄱ Settlement-time curves.

Fig. 4 shows the results of tests on a series of 7 identical samples, which have been tested in the oedometers type I. These samples have been preloaded first with $\sigma_v = 0.15 \text{ kg/cm}^2$ and then with $\sigma_v = 0.35 \text{ kg/}$ cm^2; hereafter every sample has been subjected once to a constant loading of 0.60, 0.85, . . ., 1.85, 2.10 kg/cm² for 120 hours. The permeability K and the compression modulus θ_c have been measured at the end of the preloading and loading intervals. The settlement time curves are shown in Fig. 4a; Fig. 4b shows the permeability measured, plotted against the volume deformation, and Fig. 4c shows the experimental relationship between the vertical stress σ_v and the deformation

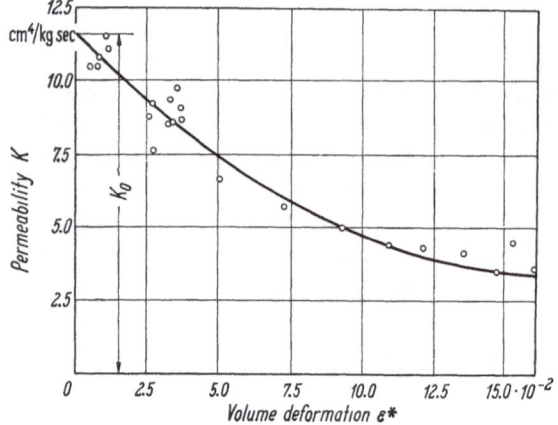

Fig. 4b.   Permeability $K(\varepsilon^*)$/Volume deformation ε^* after 120 hours.

Fig. 4c.   Effective vertical stress σ'/volume deformation ε^* after 120 hours.

ε^*, measured after 120 hours. As the waterpressure has elapsed practically at this moment this curve shows the relationship of the effective stress σ'_v with ε^*. These curves can be described satisfactorily by the equations:

$$K(\varepsilon^*) = K_0 \exp(-\varkappa_K \varepsilon^*) \qquad (a)$$

$$\sigma'_v(\varepsilon^*) = \theta_c(1 + \varkappa_c \varepsilon^*)\varepsilon^* \qquad (b)$$

where: ε^* = deformation after 120 hours, $K_0 = 1.15 \times 10^{-5}$ cm^4 kg^{-1} sec^{-1} (corresponding to 1.15×10^{-8} cm sec^{-1}, when the waterpressure is expressed in cm water), $\varkappa_K = 8.8$; $\theta_c = 8.4$ kg/cm^2 measured after 120 hours and $\varkappa_c = 3.5$. Formula (b) can be regarded as an integration for very large values of the time of the following non linear integral equation:

$$\varepsilon(t) + \int_0^t \phi(t-t')\,\varepsilon(t')^2\,dt' = \frac{\sigma(t)}{\theta_{ci}} + \frac{1}{\theta_{ci}}\int_0^t \varphi(t-t')\,\sigma(t')\,dt'$$

where $\phi(t')$ and $\varphi(t')$ are continuously differentiable decreasing functions of t', which become zero for $t' \to \infty$; θ_{ci} = instantaneous value of θ_c.

Clay samples have been tested also in oedometer type III, which is provided by a capacitance gauge, connected to an electronical measuring instrument for the measurement of static and dynamic waterpressures. According to Miss WEN (1964), who has designed and constructed this instrument, a degree of stability of ± 0.003 kg/cm^2 can be obtained, when the change in temperature will not exceed ± 0.1 °C. Due to the short response time of the oedometer the first correct readings of the water pressure can be taken already after 5 seconds.

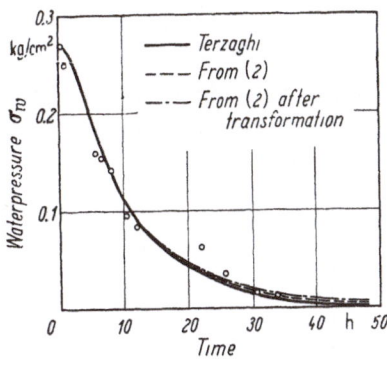

Fig. 5 a.   Waterpressure-time curve. Step I.

Fig. 5 b.   Waterpressure-time curve. Step II.

After being preloaded with 0.15 kg/cm², the sample whose test results are shown in Fig. 5 has been subjected to a stepwise loading increase of 0.30 kg/cm², which during every step is maintained constant for

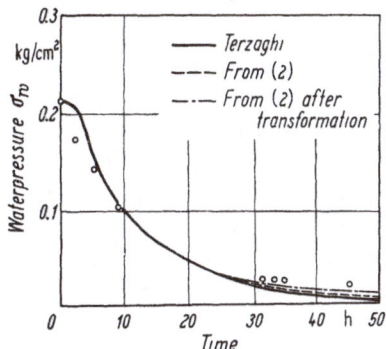

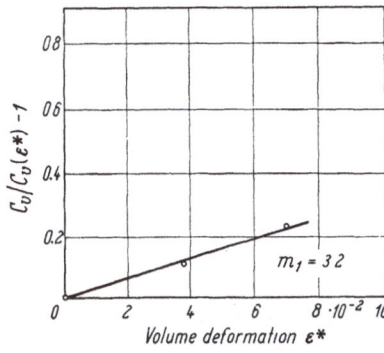

Fig. 5 c.   Waterpressure-time curve. Step III.

Fig. 5 d $C_v/C_v(\varepsilon^*) - 1$/volume deformation ε^* after 60 hours.

70 hours. The waterpressures measured during the first, second and third step are shown in Fig. 5a, b, c respectively. Practically the water-pressure has elapsed after 70 hours. The coefficient of consolidation C_v and the coefficient of volume deformation m_v have been calculated at the beginning of every new step, applying TAYLOR's square root fitting method; from the nature of the process of consolidation, hardening and flow, C_v and m_v should be functions of the coordinates and the time. Hence with the help of this method only the initial values of these coefficients can be determined; they give approximate

values only, as hardening and flow always occur, which factors are not considered in TAYLOR's method. The plot $C_v/C_v(\varepsilon^*)-1$ versus the volume deformation after 60 hours gives a linear relationship (Fig. 5 d):

$$\frac{C_v}{C_v(\varepsilon^*)} - 1 = m_1\varepsilon^* \quad \text{or} \quad C_v(\varepsilon^*) = \frac{C_v}{1+m_1\varepsilon^*}, \tag{c}$$

where C_v = initial value of the coefficient of consolidation and $C_v(\varepsilon^*)$ its value in the deformed state; $C_v = 4.5\times10^{-5}$ cm^2 sec^{-1} and $m_1 = 3.2$.

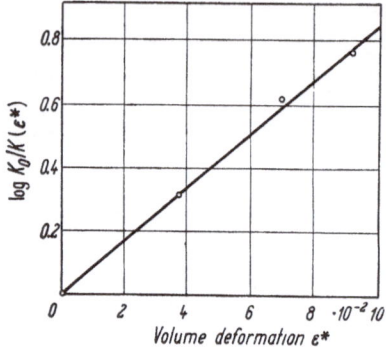

Fig. 5 e. log $K_0/K(\varepsilon^*)$/volume deformation ε^* after 60 hours.

Fig. 5 f. Effective stress σ_v'/volume deformation ε^* after 60 hours.

The experimental relationship of $K(\varepsilon^*)$ with ε^* follows formula (a) with $K_0 = 8.3\times10^{-6}$ cm^4 kg^{-1} sec^{-1} (= 8.3×10^{-9} cm sec^{-1}) and $\varkappa_K = 8.4$. The curve effective stress σ_v' versus ε^* (Fig. 5 f) follows formula (b) with $\theta_c = 7$ kg cm^{-2}; $\varkappa_c = 2.3$. The coefficient of volume deformation has been computed $m_v = 13\times10^{-2}$ cm^2 kg^{-1}. For a marine clay from the coastal areas, the following values have been determined: $C_v = 2.08\times10^{-3}$ cm^2 sec^{-1}; $m_1 = 17$; $m_v = 4.6\times10^{-2}$ cm^2 kg^{-1}; $K_0 = 1.67\times10^{-4}$ cm^4 kg^{-1} sec^{-1}; $\varkappa_K = 25.4$; $\theta_c = 10.5$ kg cm^2; $\varkappa_c = 3.5$, whereby the deformation is taken after 17 hours when the waterpressure has elapsed practically. The theoretical curves after TERZAGHI and TAN (1957b) fit the experimental dots quite well in the early and middle period of consolidation, except in Fig. 5 c (third step), where the experimental values remain below the theoretical ones. This phenomenon is often ascribed to wall friction. This idea is quite correct, but another important factor should not be overlooked: under relatively large effective stresses, the lower yield value f_1 can readily be formed and this internal friction may contribute considerably to the reduction of the waterpressure. The formation of the lower yield value has been observed earlier in triaxial and cell tests after long duration consolidation under relatively high hydrostatic pressures (GEUZE-TAN, 1950). At the end period, the waterpressure measured is always higher than predicted from both theories, but corresponds better with the curve obtained after transformation, by replacing C_v by $C_v/(1 + m_1 \varepsilon^*)$. Fig. 6 shows a clear deviation from the TERZAGHI curve, but formula

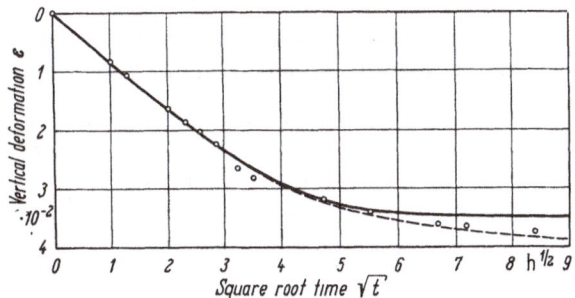

Fig. 6. Settlement-time curve. —— TERZAGHI; – – – from (1a–b); $\mu = 8$ days; $g^2 = 0.6$.

(1a—b) gives a more satisfactory agreement, when it is taken: $1/\theta_c(t) = \{1 - \exp(-g^2 t/\mu)\}/\theta$; the values for $\mu = 8$ days and $g^2 = 0.60$ have been determined with the help of the simple shear plastometer and the oedometer type II respectively.

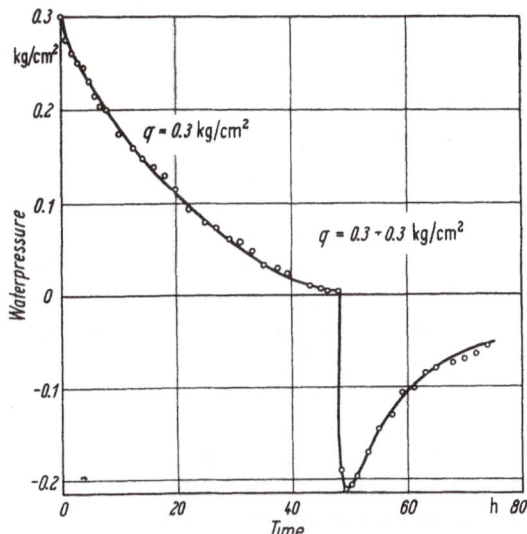

Fig. 7. Positive waterpressure during loading and negative waterpressure after unloading.

The waterpressure during hydrostatic loading and after unloading has been measured in the compression plastometer of the triaxial type; one of the results is shown in Fig. 7. It may be mentioned that the swelling of the sample is resisted to a certain degree by some friction and the weight of the plunger. Yet the magnitude of the instantaneously occuring negative pore pressure is a clear indication that the swelling modulus of clay is much higher than it is generally believed in Soil Mechanics.

Mathematical Analysis of One Dimensional Problems of Consolidation, Hardening and Secondary Time Effects

As it is described above, the value of the oedometer as a plastometer for the determination of the rheological parameters of clays can be increased, when provisions will be taken for the measurements of the lateral stress and the waterpressure as a function of the time. For the analysis of the test results the following formulas have been derived (TAN 1964):

$$w = w_p + w_s \tag{1}$$

$$w_p = \frac{hqg^2}{\theta} \left\{ 1 - \frac{8}{\pi^2} \sum_0^\infty \frac{1}{(2n+1)^2} e^{-\frac{C_v t}{h^2} \left(\frac{2n+1}{2}\pi\right)^2 \{1+g^2/\lambda\}} \right\}; \tag{1a}$$

$$w_s = \frac{hq(1-g^2)}{\theta_c(t)}; \tag{1b}$$

$$\sigma_w = \frac{4q}{\pi} \sum_0^\infty \frac{(-1)^n}{2n+1} e^{-\frac{C_v t}{h^2}\left(\frac{2n+1}{2}\pi\right)^2 \{1+g^2/\lambda\}} \cos\left(\frac{h-z}{h}\frac{2n+1}{2}\pi\right), \tag{2}$$

when

$$\lambda = -\frac{C_v \mu}{h^2}\left(\frac{2n+1}{2}\pi\right)^2 \tag{2a}$$

and

$$g^2 = \frac{1+\nu_e}{3(1-\nu_e)}; \qquad \frac{1}{3} \leq g^2 \leq 1. \tag{3}$$

In above formulas: w = settlement; w_p = primary settlement; w_s = secondary settlement; h = thickness clay layer; q = loading intensity; $m_v^{-1} = \theta/g^2$ = initial value of the modulus of confined compression θ_c; θ = modulus of hydrostatic compression; C_v = initial value of the coefficient of consolidation; $\mu = \eta/G$ = time of relaxation, $1/\theta_c(t)$ = creep time function; ν = Poisson's ratio of elastic part of deformation; t = time.

In the derivation, a linear stress strain relationship is assumed, the function $1/\theta_c(t)$ is choosen such that $1/\theta_c(t) = 0$ for $t = 0$ and constant for $t = \infty$; for the three parameter models, for instance, $1/\theta_c(t) = \{1 - \exp(-g^2 t/\mu)\}/\theta$. In the oedometer tests g^2/λ is small usually, and when it is neglected, then (1a) and (2) are reduced into the corresponding formulas after TERZAGHI. Hence the settlement w_s and thus the creep function $1/\theta_c(t)$ completely separated from the time effect of consolidation, can be obtained by subtracting the TERZAGHI settlement from the experimental settlement-time curve. For small values of the time the following formulas, which can be used for the computation of the initial values of the coefficients C_v, m_v, ν_e and the modulus θ, are convenient:

$$\sigma_w = q \left\{ 1 - e^{-(1-g^2)t/\mu} \operatorname{erfc} \frac{z}{2\sqrt{C_v t}} \right\} \tag{4}$$

$$w = 2qm_v \sqrt{\frac{C_v t}{\pi}} \left\{ 1 + \frac{1}{3} \left(\frac{1}{g^2} - 1 \right) \frac{g^2 t}{\mu} + \cdots \right\} \tag{5}$$

$$\sigma_h = hq - \frac{2(1 - 2\nu_e)}{1 - \nu_e} \sqrt{\frac{C_v t}{\pi}} q \tag{6}$$

when σ_h = lateral pressure over the height h.

All above formulas have been obtained for a non-hardening soil skeleton, but still hold approximately for the early stage of consolidation and hardening. The problem of hardening has been studied for some special cases. When it is assumed that the MAXWELL viscosity increases with the volume deformation according to

$$\eta(\varepsilon) = \eta_0(1 + \varkappa_\eta \varepsilon), \tag{7}$$

where η_0 = initial value of the viscosity and $\varkappa_\eta$ = hardening coefficient for viscous flow, whereas $K(\varepsilon)$ varies such that $K(\varepsilon)\eta(\varepsilon) = $ constant, and θ also remains constant, then identical formulas have been obtained as above, except that t should be replaced by T. Thereby the following transformation of the time should be applied:

$$t = \int_0^T (1 + \varkappa_\eta \bar\varepsilon)\, dT, \tag{8}$$

when $\bar\varepsilon$ denotes the average deformation:

$$\bar\varepsilon = \frac{w}{h}.$$

This transformation is reasonable for small values of the time. For large values of the time, the formulas (1), (2) and (3) still can be used as an approximation, whereby for the rheological parameters and hardening coefficients should be taken their ultimate values at the end of the loading period. The replacements of C_v by $C_v/(1 + m_1\varepsilon^*)$ is an affine transformation of the time axis; replacing θ by $\theta(\varepsilon^*)$ means an affine transformation of the settlement axis. Settlements and waterpressures for intermediate values of the time can be obtained by means of graphical interpolation.

All the rheological parameters and hardening coefficients required in above equations can be determined with the help of the oedometers and the simple shear plastometer according to the methods described above; the constants ν_e and g^2 can be computed from (3), (5) and (6), when the lateral-pressure-time and the settlement-time curves have been measured with the help of oedometer type II.

For clay layers of a few meters thickness and higher as they are met in engineering practice, all the above formulas still hold, except (1) and (2), which should be replaced by:

$$w = \frac{hq}{\theta} \left\{ 1 - \frac{8}{\pi^2} \sum_0^\infty \frac{1}{(2n+1)^2} e^{\frac{\omega g^2 t}{\mu}} \right\} P(\omega, \lambda); \qquad (9)$$

$$\sigma_w = \frac{4q}{\pi} \sum_0^\infty \frac{(-1)^n}{2n+1} e^{\frac{\omega g^2 t}{\mu}} \cos\left(\frac{h-z}{h} \frac{2n+1}{2} \pi \right) P(\omega, \lambda); \qquad (10)$$

when

$$P(\omega, \lambda) = \frac{(\omega+1)\lambda}{g^2 \omega^2 + \lambda}; \qquad \omega = \frac{-(1-\lambda) + \sqrt{(1-\lambda)^2 + 4g^2\lambda}}{2g^2}$$

and λ is given in (2a), which formulas are valid for values of the time in the order of magnitude of at least a month and longer.

References

FLORIN, V. A. (1953): Proc. Ac. Science USSR, Techn. Scient. Div. No. 9.

GEUZE, E. C. W. A., and TAN TJONG-KIE (1950): Géotechnique **2**, 141.

KOPPEJAN, A. W. (1948): Proc. 2nd ICOSOMEF III, 32.

ROSENQVIST, I. TH. (1959, 1963): Publications Norwegian Geotechnical Institute.

TAN TJONG-KIE (1954): Investigations on the rheological properties of clays, Ch. 6 and 9, Diss. Delft Techn. University (in Dutch with English summary); (1957a) Structure Mechanics of clays, (1957b) Secondary time effects and consolidation of clays, both Publ. Ac. Sin. Inst. for Civil Eng., also published in Scientia Sinica 1958, pp. 86, 1060; (1961) Proc. 5th ICOSOMEF I, 367; III, 141 (disc); (1964) Theoretical and experimental study of consolidation, hardening and secondary time effects, Forthcoming.

WEN HSUAN-MAY (1964): A new oedometer with electronical measurement of the waterpressure, (in Chin.) forthcoming.

Discussion

Question posée par S. IRMAY: The experimental formula of the DARCY coefficient $\log Ko/K$, can be computed, if we replace its KOZENY-CARMAN dependence on porosity n, i.e. $n^3 (1 - n)^2$ by IRMAY's formula $(n - n_0)^3/(1 - n)^2$, where n_0 is the ineffective (stagnand, irreducible) porosity occupied by stagnant water. Assuming $n_0 = 0.65n$, which is not unusual for clay, the authors experimental dependence on volume deformation [Eq. (a), Fig. 4b, Fig. 5e] is obtained.

Réponse de TAN TJONG-KIE: I have considered also the KÁRMÁN-KOZENY equation in plotting my experimental data, but it has been derived for granular materials. Its validity as far as I know has been found for some sands but not for clays. That is the reason why I did not give further consideration to this law.

Question posée par L. ŠUKLJE: When discussing at the European Conference 1963 in Wiesbaden the papers by STROGANOV and ANAGNOSTI, I made some objections to Professor TAN's interpretation of the conclusions of his theory of consolidation. As Professor TAN was not present in Wiesbaden, I should like to give him the opportunity of an answer on this occasion.

The principal resulting formulas of Dr. TAN's theory are reproduced in his paper presented to our Symposium. In the limit case $t \to \infty$ the Eq. (1) yields the final settlement

$$\varrho_{\infty \text{TAN}} = \frac{h\,q}{\theta}. \tag{A}$$

By limiting the coefficient of viscosity $\eta \to \infty$ the Eq. (1) reduces to TERZAGHI's classical solution with the and settlement

$$\varrho_{\infty \text{TERZAGHI}} = \frac{h\,q}{\theta}\,g^2. \tag{B}$$

Now in Professor TAN's interpretation the quotient

$$\frac{\varrho_{\infty \text{TAN}}}{\varrho_{\infty \text{TERZAGHI}}} = \frac{1}{g^2} = \frac{3\,(1-\nu)}{1+\nu} \genfrac{}{}{0pt}{}{\leq 3}{\geq 1} \tag{C}$$

yields the POISSON ratio ν whereby $\varrho_{\infty \text{TAN}}$ is related to the "final settlement" and $\varrho_{\infty \text{TERZAGHI}}$ to the "end value of the primary consolidation".

My objection was that the solution $\eta \to \infty$ with the end value (B) cannot be applied to viscous soils either to their full or to their primary consolidation. Consequently, the ratio (C) cannot be used to get the POISSON ratio of the soil.

I think that we must be very careful in defining and determining the so-called end values of the primary consolidation and the final settlement. In my opinion there is no essential difference in the character of the primary and of the secondary consolidation of saturated soils. The only difference is in the intensity of the filtration flow on the one and of the "plastic resistance" on the other hand.

Réponse de TAN TJONG-KIE:

1. I have made the most simple assumptions possible: HOOKEan behaviour under hydrostatic and MAXWELLian response under deviatoric stresses. In general the modulus of confined compression can be written: $\theta_c = \theta + 4\psi/3$. In my case $\theta =$ constant and $\psi =$ operatorform; if we assume unrestricted flow under deviatoric stresses then ultimately $\theta_c \to \theta$ and $\psi \to 0$. So the ratio of elastic (TERZAGHI) settlement and the visco-elastic settlement can never contain viscosity terms as in the ultimate state the rate of deformation $= 0$: hence it can only contain elasticity terms, which is $1/g^2 = \dfrac{3(1-\nu_e)}{1+\nu_e}$. For a refinement of the theory account should be taken of the following factors: (a) hardening, (b) the possibility that the deviatoric behaviour in confined compression may be different from the state of unrestricted shearing deformations, (c) change of isotropy into orthotropy. In a new theory I also take account of (a) under special cases. A further refinement will be of no practical meaning, as the rheological parameters can not be measured for the present. I agree that my formula is simple, but it can predict anyhow the upper bound of the settlements. As far as I have measured the prediction is quite reasonable for ν_e varying from $0{,}3 \to 0{,}42$, thus for rather stiff clays. For soft clays the ratio $1/g^2$ has been measured $2{,}7$ after a few days and my theoretical limit gives $3{,}0$ ($\nu_e = 0$).

2. Concerning the limits 1. $\eta \to \infty$ and then $t \to \infty$ or 2. $t \to \infty$ and then $\eta \to \infty$ they can never give the same result as you can see from the 3 parameter model. The case 1. means that we consider an elastic (TERZAGHI) material and then takes the settlement for $t \to \infty$. The case 2. means that we first assume a 3-parameter model, consider the settlement for $t \to \infty$, but then assume the

material to be elastic ($\eta \to \infty$). This latter assumption is contradictory to our original assumption of a 3 parameter model. Hence case 2. is physically meaningless. Professor SUKLJE has obviously overlooked this point and hence he incorrectly believes that as both limits are not equal the solution $\eta \to \infty$ (which only means that we consider a non-viscous soil skeleton) can not be applied for viscous soils with his consequent conclusion about the POISSON's ratio. This coefficient for soils of course is not a constant but an operatorform[1] containing as well elastic as viscous terms and where ν_s only represents the POISSON's ratio for the *elastic* part of the deformation. In the ultimate state the viscosity terms disappear and only the elastic terms are retained. Professor SUKLJE's interesting idea that "the only difference in the primary and secondary consolidation of satured soils is in the intensity of the filtration flow on the one and viscous resistance on the other hand" is just the basis of my theory of consolidation which basis as *I* have pointed out earlier is no longer satisfactory for hardening soils and hence this theory needs the necessary improvements for the non linear hardening effects

---

[1] See: References TAN (1954).

3.2 The Secondary Consolidation of Clay

By

Robert L. Schiffman, Charles C. Ladd and Albert T.-F. Chen

Synopsis

A study is made of various types of viscoelastic effective stress-strain relations, and their influence on the time-settlement relations in secondary consolidation. This study examines both deviatoric and volumetric components of the
effective stress-strain relations. It is shown that the effective stress-strain
relationship for a clay layer, is derivable from several combinations of volumetric and deviatoric behavior.

A five-parameter relationship is discussed and generalized to a continuous
one by functionals. The influence of various rheological parameters are discussed.

Experimental evidence is introduced for several normally consolidated and
over consolidated clay samples. The rate of secondary consolidation is approximately the same for both isotropically consolidated triaxial tests and for oedometer tests. The rate of secondary consolidation is lower for the over-consolidated
samples.

An explanation of behavior is presented in terms of the physio-chemistry
of the clay-water system. This study considers the particulate nature of clays
and two types of effective stress transmission; these being contact stresses and
double layer stresses.

Introduction

The theory of primary consolidation, as originally developed
(TERZAGHI, 1923; TERZAGHI and FRÖHLICH, 1936) concerned itself
with the one-dimensional vertical compression of loaded clay layers.

Observations concerning the process of secondary consolidation
(secondary compression) were first reported by GRAY (1936) and
BUISMAN (1936). This process, as reported, took note of the observations that many soils, the organic soils in particular, exhibited a
one-dimensional (oedometer) compression behavior which, in the later
stages of compression, became proportional to the logarithm of time.
An extensive analysis of field observations (BJERRUM, 1963, 1964) has
shown that there are two characteristic displacement-time effects in
secondary consolidation, and that these are distinguished by the load-
time relationships. In one case, where the load is constant with time
the deformation is directly proportional to the logarithm of time. In

the second case, where the load is oscillatory, such as tanks, or wind loads on buildings, the displacement is directly proportional to time; i.e., the structure settles at a constant rate.

Analyses leading to predictions of secondary consolidation effects have followed two paths of investigation. Engineering studies have universally followed the empirical approach of Buisman (1936), in which primary and secondary consolidation displacements are considered as separate and serial. In these analyses the secondary settlement is considered to be proportional to the logarithm of time.

Rational theories of secondary consolidation have treated the process of both primary and secondary consolidation as occurring in parallel. The over-all process consists of a fluid flow mechanism coupled with a visco-elastic effective stress-strain relationship. This conception, first proposed by Taylor (1942), assumed a Kelvin (spring and dash-pot in parallel) model for the effective stress-strain relationship of a clay layer. An improved model coupled an additional elastic element (spring) in series with a Kelvin unit (Ishii, 1951; Tan, 1957; Gibson and Lo, 1961; Florin, 1961) enabling a phasing of primary and secondary effects. A general linear visco-elastic theory, in terms of differential or integral operators have been developed (Schiffman, 1963) for discrete model representations of the effective stress-strain relations. Continuous distributions of effective stress-strain properties have been introduced by Mandel (1957), It is interesting to note that, with some simplification, the logarithmic time relationship can be approximated by using the first term of a logarithmic series of a power function in time.

Three-dimensional theories of secondary consolidation have been formulated, (Biot, 1956), upon the assumption that the effective stress-strain relationships are visco-elastic. Some theories have assumed incompressibility of pore fluid (Tan, 1961), and specific types of effective stress-strain relationships, (Tan, 1954, 1957), such as elasticity of volume change and Maxwell (spring and dash-pot in series) deviatoric behavior. The general theories of both primary and secondary consolidation (Biot, 1941, 1956), permit volume change in both the fluid and the soil skeleton. While this consideration of fluid compressibility is of considerable importance in the compression of rocks and porous concrete, its influence in the compression of saturated clays is negli- gible.

Experimental evidence as to the deviatoric behavior of clays Bingham, 1916; Taylor, 1942; Tan, 1954; Geuze and Tan, 1954) has indicated that a Maxwell model is a reasonable first approximation to the deviatoric effective stress-strain relationship.

Information on the volumetric effective stress-strain relationship is not nearly as conclusive as that for the deviator. A few studies

(TERZAGHI, 1953; TAYLOR, 1942; FANG, 1956; LO, 1961) indicate that the volumetric behavior exhibits somewhat stronger elasticity components than the deviatoric behavior. The experimental portions of this paper study this question. Based upon physical-reasoning alone, the form of the volumetric strain-time relationship under constant stress, must approach a finite limit asymptotically with time (SCHIFFMAN, 1959). A parallel arrangement of linear elastic and viscous elements is then required. An analysis of the triaxial compression, using a KELVIN volumetric relationship coupled with a MAXWELL deviatoric relationship (ANAGNOSTI, 1963), has shown good agreement with experiment.

The macroscopic analyses, described above, are directed at the numerical predication of the behavior of compressible soils in secondary consolidation. A microscopic (physio-chemical) study of secondary consolidation is directed towards an understanding of the fundamental mechanisms which govern the behavior of clays in secondary consolidation. TAYLOR (1942) considered that the adsorbed water layers (which he referred to as the "surface layer phase"), were of different order of viscosity than the pore water and that this "surface layer phase" was responsible for the viscous components in the effective stress-strain relationship. LAMBE (1953) suggested that secondary consolidation was due, in part, to the changes in interparticle forces brought about by the extrusion and/or diffusion of the adsorbed water films. The secondary time effects come about by the time required to change the adsorbed water films, and thus alter the interparticle forces. Shifting of particles and aggregate breakdown are also contributory factors.

TAN (1954, 1959) has likened the "edge-to-face" interparticle forces as mechanical linkages. The secondary consolidation effect is then considered as a continuous process of breaking and reforming of bonds putting the soil skeleton into a denser, more stable configuration.

In this paper, the secondary consolidation of compressible layers is studied first as an analytical theory in which the effective stress-strain relationships are constructed as linear visco-elastic entities. Experimental evidences as to the effects of volume change and deviatoric states of stress, and the state of preconsolidation of the clay are presented and related to the theory. Finally this paper advances a hypothesis of the mechanism of secondary consolidation effects, based upon physio-chemical evidence.

Formulation of Theory

The theory of secondary consolidation as applied to clay and organic soils is based upon the following physical assumptions.

18*

1. The soil is completely saturated with water.

2. The soil particles and the pore water is incompressible.

3. The flow of pore water follows Darcy's law; the coefficient of permeability being a constant.

4. The strains of the soil skeleton are controlled exclusively by the effective stresses, and the relationship between effective stresses and strains is time dependent.

5. The strains, velocities, and stress increments are small and the theory is quasi-static.

Within the framework of the Terzaghi-Rendulic theory of consolidation of clay layers, the process of compression of clay layers is governed by,

$$\frac{k}{\gamma_w} \frac{\partial^2 \sigma'}{\partial z^2} = \frac{\partial \varepsilon}{\partial t},$$ (1)

where (k) is the coefficient of permeability, (γ_w) is the unit weight of water, (σ') is the effective stress, and (ε) is the strain (dilatation) of the compressible layer.

The relationship between the effective stress (σ') and the vertical strain (ε) can be constructed in two ways. By means of the Boltzman superposition principle a functional relationship (Volterra, 1929; Mandel, 1957) can be established as follows:

$$\varepsilon(z, t) = \int\limits_0^t \frac{\partial \sigma'(z, \tau)}{\partial \tau} m_v(t - \tau) \, d\tau.$$ (2)

This equation when substituting into Eq. (1) will result in the following integro-differential equation governing the secondary consolidation of a clay layer,

$$\frac{k}{\gamma_w} \frac{\partial^2 \sigma^1(z, t)}{\partial z^2} = m_v(0) \frac{\partial \sigma'(z, t)}{\partial t} + \int\limits_0^t \frac{\partial \sigma'(z, \tau)}{\partial \tau} \frac{\partial m_v(t - \tau)}{\partial t} \, d\tau.$$ (3)

An alternate manner of representing secondary consolidation is by means of linear operators (Lee, 1955). The effective stress-strain relationship can be written in terms of operator equations (Schiffman, 1963) as follows.

$$3 R P(\sigma') = (P S + 2 Q R)(\varepsilon),$$ (4)

where P, R, Q and S are differential operators; P applying to the deviatoric stress, Q applying to the deviatoric strain, R applying to the volumetric stress, and S applying to the dilatation.

The governing differential equation in this mode of representation is then,

$$\frac{k}{\gamma_w} (P S + 2 Q R) \left(\frac{\partial^2 \varepsilon}{\partial z^2} \right) = 3 R P \left(\frac{\partial \varepsilon}{\partial t} \right).$$ (5)

Eq. (3) or (5) together with the appropriate boundary and initial conditions, constitutes a complete mathematical statement of the problem of the secondary consolidation of a homogeneous clay layer.

The coupling relationship of Eq. (4) can be applied to a variety of situations. As shown in Fig. 1 an elastic volume change will couple

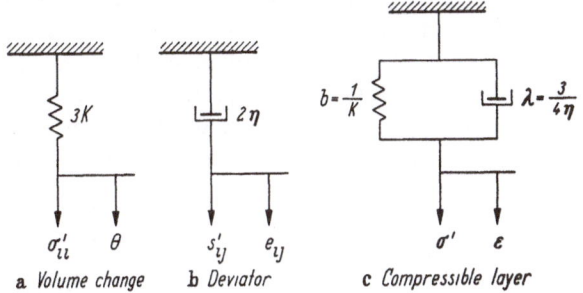

Fig. 1 a—c. Two-parameter effective stress-strain system.

with a viscous deviatoric component to form a KELVIN effective stress-strain relationship for the compressible layer. This coupling is, within the definition of the constants, the same as TAYLOR's Theory B (1942).

An elastic volume change is coupled to a MAXWELL type of deviatoric behavior, and shown in Fig. 2. The resulting effective stress-strain

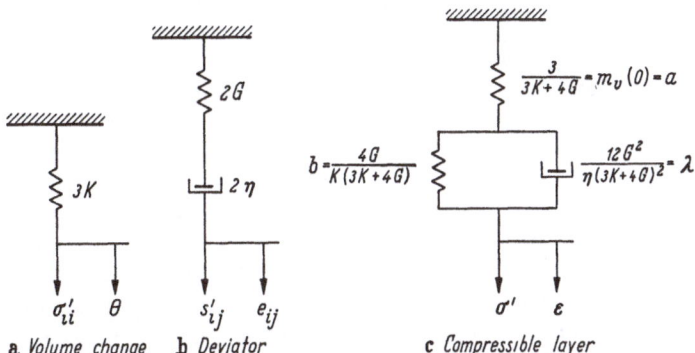

Fig. 2 a—c. Three-parameter effective stress-strain system.

relationship for the layer is a rheological model composed of an elastic element in series with a KELVIN model. This is the model relationship studied by ISHII (1951), TAN (1957, 1958), GIBSON and LO (1961) and FLORIN (1961).

The analysis of visco-elastic effective stress-strain relationships (SCHIFFMAN, 1963) has traditionally considered that the volumetric components to be elastic. Certainly one reason for this consideration, has been the traditional arguments derived from studies of metals and plastics, that the volume change is, at best, small, and therefore can be ignored. As indicated previously, this argument has some basis in experimental observations and deductions. On the other hand, there is an experimental basis to consider that the volumetric deformations of a soft clay or organic soil has a rheological component (Gibson and Henkel, 1954; Thompson, 1962; Roscoe, Schofield and Thurai-rajah, 1963). These studies indicate that there is a component of volumetric compression which is supplemental to the dissipation of excess pore pressure. In addition to experimental evidence, when one considers the complex particulate nature of clays, and organic soils, it would be overly fortuitous if the volume change behavior were independent of time.

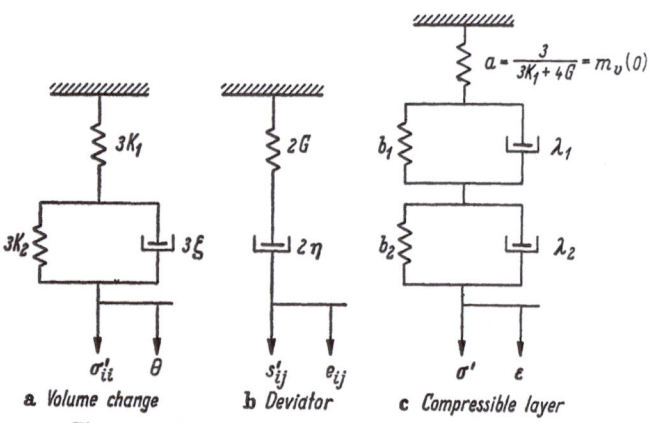

a Volume change   **b** Deviator   **c** Compressible layer

Fig. 3 a—c. Five-parameter effective stress-strain system.

The choice of a first order effect model is dictated by physical considerations. In the first place there should be an isolated elastic response. Secondly, the viscous response must be of such a nature as to approach a stable limit with time. Thus the choice of a first order model is as shown in Fig. 3a. This model consists of an elastic element coupled in series with a Kelvin model. The deviatoric relationship will, in the first order study, be assumed to behave as a Maxwell material as shown in Fig. 3b. As shown in Fig. 3c, the effective stress-strain relationship for the layer consists of two Kelvin models and an elastic element in series. The model parameters are a, b_1 and b_2 representing compressibilities, and λ_1 and λ_2 representing fluidities (reciprocal of the viscosity).

A more general relationship is established if the deviatoric model includes a retardation element as shown in Fig. 4b. The volumetric model is unchanged, with an elastic element serially connected to a retarded elastic element as shown in Fig. 4a. As shown in Fig. 4c the rheological model representing this visco-elastic law is a serial connection between an elastic element and three retardation elements, forming a seven parameter model system.

Depending upon how one wished to represent the rheology of the soil skeleton, a model system could be built providing as many parameters as necessary. It is evident, however, that in the pattern as indi-

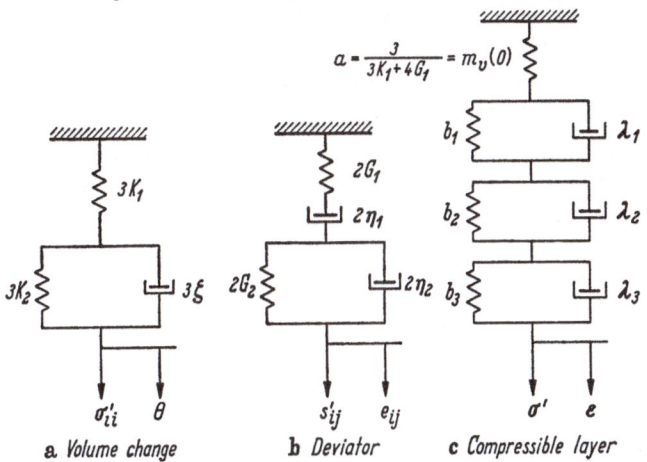

a. Volume change   b. Deviator   c. Compressible layer

Fig. 4 a—c. Seven-parameter effective stress-strain system.

cated in this analysis, the effective stress-strain relationship for a compressible layer consists of an elastic element defining the primary consolidation, coupled with a number of KELVIN models, representing the retardation of the soil skeleton. Thus a general effective stress-strain relationship is,

$$\varepsilon(z, t) = a\sigma'(z, t) + \sum_{i=1}^{N} \lambda_i \int_0^t \sigma'(z, \tau)\, e^{-(\lambda_i/b_i)(t-\tau)}\, d\tau, \tag{6}$$

where a is the instantaneous elastic compressibility, b_i the retarded elastic compressibility and λ_i the fluidity of the soil skeleton.

The compressibility, $m_v(t)$ is,

$$m_v(t) = a + \sum_{i=1}^{N} b_i(1 - e^{-(\lambda_i/b_i)t}), \tag{7}$$

where the parameter a exists only if the model has an isolated elastic element. For models with an even number of parameters the compressibility a is zero. From Eq. (7) it is seen that a continuous distribution of models would have the form,

$$m_v(t) = a + \int_0^{\infty} J(\xi)(1 - e^{-t/\xi})\, d\xi, \tag{8}$$

where $J(\xi)$ is the retardation spectrum.

It can be proved (Gurtin and Sternberg, 1962) that a given set of volumetric and deviatoric equations couple to a single differential equation for the compressible layer. The converse, however, is not true. That is, several physically different combinations of volume change and deviator rheology will result in the same differential equation for the compressible layer. True the constants will have different meanings, when extended back to the parent model; however, the parent models themselves can be physical representations of different volumetric and/ or deviatoric behavior. In Table 1, possible couplings are presented for the parameter systems discussed.

As will be discussed later, this lack of both necessity and sufficiency in the coupling relationships, presents a certain amount of ambiguity in the interpretation of test data. As a result testing procedures become somewhat more complex in that more detailed experimentation becomes necessary to define the phenomena.

Table 1. *Effective Stress-Strain Relations*

Volume change	Deviator	Compressible layer
E	V	K
		(2-Parameter System)
K	K	
E	M	E—K
		(3-Parameter System)
E—K	E	
E	E—K—K	
E—K	E—K	E—K—K
		(5-Parameter System)
E—K	M	
K—K	E	
E—K—K	E	
E	E—K—K—K	
E—K	E—K—K	E—K—K—K
		(7-Parameter System)
K—K	E—K	
E—K—K	E—K	
E—K—K	M	
K—K—K	E	
E—K—K—K	E	

E = Elastic element (Spring)
V = Viscous element (Dash-pot)
M = Relaxation element (Maxwell Model; E —V)
K = Retardation element (Kelvin Model; E ‖ V)
— = Series coupling
‖ = Parallel coupling

Five Parameter Secondary Consolidation System

The relationships previously discussed form the framework within which specific boundary value problems of secondary consolidation can be solved. The boundary conditions are those of a compressible layer, of thickness h loaded at the surface by a uniform load, q_0. The layer is free draining at the upper surface and impervious to flow at the lower rock surface. Following TERZAGHI's effective stress concept, the effective stresses are zero at the instant of loading.

The five parameter system shown in Fig. 3, has for its solution,

$$\frac{\sigma'(z, T_p)}{q_0} = 1 + 4\pi \sum_{n=1,3,5}^{\infty} n I_1 \sin\left(\frac{n\pi z}{2h}\right), \tag{9a}$$

$$U(T_p) = 1 + 8M \sum_{n=1,3,5}^{\infty} I_2, \tag{9b}$$

$$
I_1 = \frac{(x_1 - 4A_1)(x_1 - 4A_2)}{x_1(x_1 - x_2)(x_3 - x_1)} e^{-x_1 T_p} + \frac{(x_2 - 4A_1)(x_2 - 4A_2)}{x_2(x_2 - x_3)(x_1 - x_2)} e^{-x_2 T_p}
$$
$$
+ \frac{(x_3 - 4A_1)(x_3 - 4A_2)}{x_3(x_3 - x_1)(x_2 - x_3)} e^{-x_3 T_p}, \quad (x_1 \neq x_2 \neq x_3), \tag{9c}
$$

$$
I_1 = -\frac{(x_1 - 4A_1)(x_1 - 4A_2)}{x_1(x_1 - x_2)^2} + \frac{1}{x_2(x_2 - x_1)}\left\{(x_2 - 4A_1)(x_2 - 4A_2) T_p\right.
$$
$$
\left. + \frac{x_2(x_1 + 4A_1 + 4A_2)}{x_2 - x_1} + \frac{16 A_1 A_2(2x_2 - x_1)}{x_2(x_2 - x_1)}\right\} e^{-x_2 T_p}, \tag{9d}
$$

$$(x_1 \neq x_2, x_2 = x_3),$$

$$
I_2 = \frac{x_1^2 - a_0 x_1 + a_1}{x_1(x_1 - x_2)(x_3 - x_1)} e^{-x_1 T_p} + \frac{x_2^2 - a_0 x_2 + a_1}{x_2(x_2 - x_3)(x_1 - x_2)} e^{-x_2 T_p}
$$
$$
+ \frac{x_3^2 - a_0 x_3 + a_1}{x_3(x_3 - x_1)(x_2 - x_3)} e^{-x_3 T_p}, \quad (x_1 \neq x_2 \neq x_3), \tag{9e}
$$

$$
I_2 = -\frac{x_1^2 - a_0 x_1 + a_1}{x_1(x_1 - x_2)^2} e^{-x_1 T_p} + \frac{1}{x_2(x_2 - x_1)}\left\{(x_2^2 - a_0 x_2 + a_1) T_p\right.
$$
$$
\left. + \frac{x_2^2(x_1 - a_0) + a_1(2x_2 - x_1)}{x_2(x_2 - x_1)}\right\} e^{-x_2 T_p}, \quad (x_1 \neq x_2, x_2 = x_3), \tag{9f}
$$

$$A_1 = \frac{h^2 \lambda_1}{b_1} \frac{a\gamma_w}{k} = \frac{h^2 \lambda_1}{b_1 c_v(0)}, \tag{9g}$$

$$A_2 = \frac{h^2 \lambda_2}{b_2} \frac{a\gamma_w}{k} = \frac{h^2 \lambda_2}{b_2 c_v(0)}, \tag{9h}$$

$$a_0 = \frac{4(A_1 + A_2)}{M} - 4(A_1 m_2 - A_2 m_1), \tag{9i}$$

$$a = \frac{16 A_1 A_2}{M}, \tag{9j}$$

$$M = \frac{a}{a + b_1 + b_2} = \frac{1}{1 + m_1 + m_2}, \tag{9k}$$

$$m_1 = \frac{b_1}{a} \equiv \frac{b_1}{m_v(0)}, \tag{9l}$$

$$m_2 = \frac{b_2}{a} \equiv \frac{b_2}{m_v(0)}, \tag{9m}$$

$$T_p = \frac{c_v(0)}{4h^2} t, \tag{9n}$$

where x_1, x_2, and x_3 are the roots of the cubic equation,

$$p^3 + (a_0 + n^2\pi^2) p^2 + [a_1 + 4n^2\pi^2(A_1 + A_2)] p + 16n^2\pi^2 A_1 A_2 = 0, \tag{10a}$$

or

$$(p + x_1)(p + x_2)(p + x_3) = 0. \tag{10b}$$

The effects of specific model entities are studied by examining the effects of the following parameters,

$$A = \frac{\lambda_1 h^2}{b_1 c_v(0)}, \tag{11a}$$

$$L = \frac{\lambda_1}{\lambda_2}, \tag{11b}$$

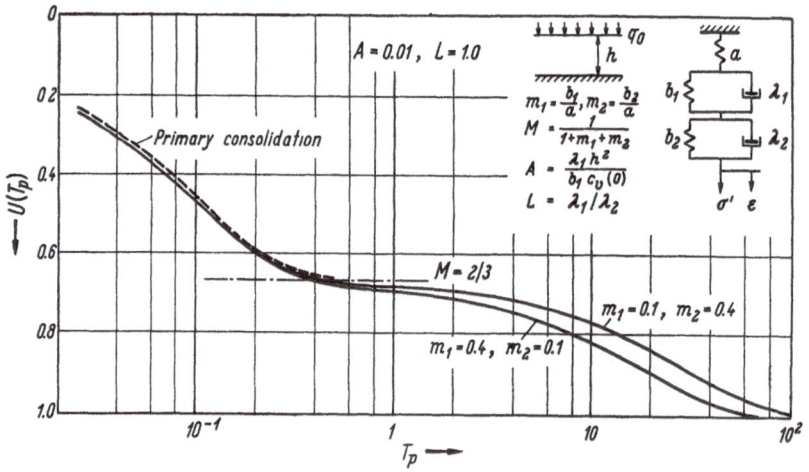

Fig. 5. Effect of soil skeleton compressibility ratios on time-consolidation relations — five-parameter system.

along with m_1 and m_2. It is noted that this affects a slight change in notation. Specification of these four parameters will completely specify the secondary consolidation of the compressible layer.

Fig. 5 presents a parametric study in which the effects of m_1 and m_2 are studied for fixed values of M, L, and A. It is seen that the variations affected by unit compressibilities are slight.

Fig. 6 shows the effect of variations of M for equal m_1 and m_2 and fixed values of A and L. However, the specific value of A and L tend to present a constructed (by intersection of tangents) value of M dif-

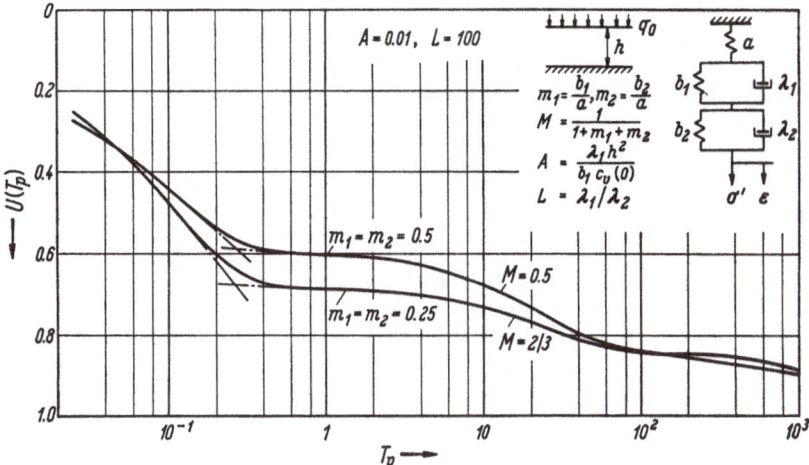

Fig. 6. Effect of compressibility factor on time-consolidation relations — five-parameter system.

ferent from the true value. This effect is more pronounced the lower the value of M. That is, the influence of viscosity and viscosity ratios on the time-consolidation relationship is more profound than the influence of compressibility and compressibility ratios.

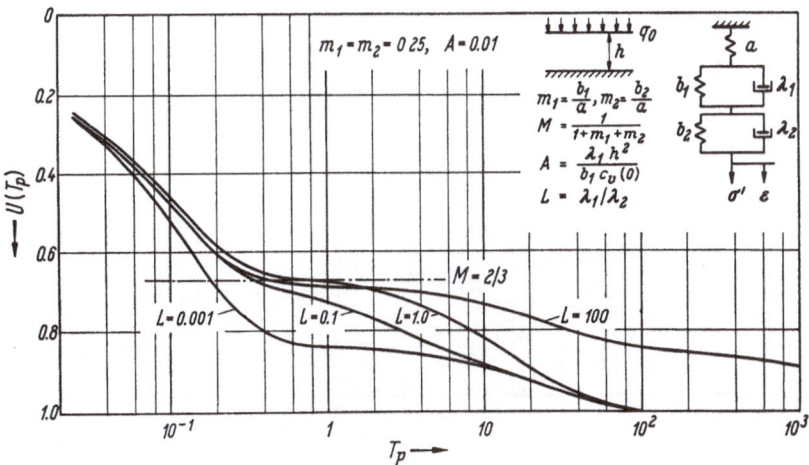

Fig. 7. Effect of soil skeleton viscosity ratios on time-consolidation relations — five-parameter system.

Fig. 7, 8 and 9 presents the effect of changes in A and L for fixed values of m_1 and m_2. In a more general sense, these figures present the overall effect of the use of multi-parameter rheological models in analysizing secondary consolidation. Furthermore these figures show the effect of differences in the time-consolidation relationships as influenced by the effective stress-strain relationships.

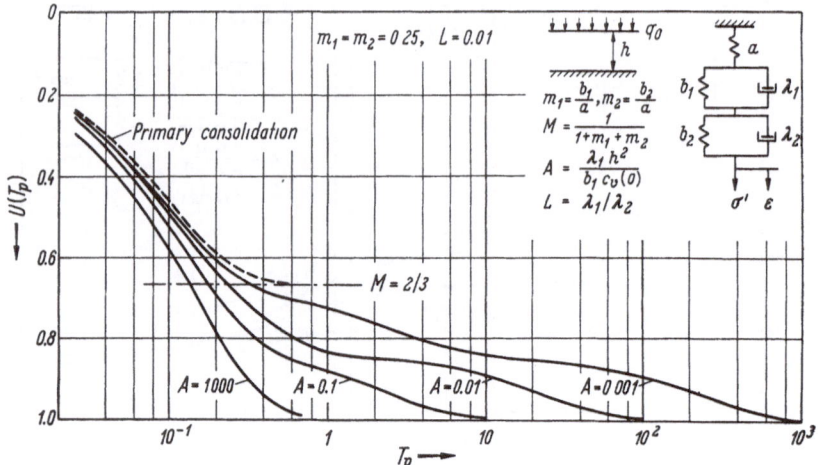

Fig. 8. Effect of viscosity factor on time-consolidation relations — five-parameter system.

A small value of A and a large value of L is indicative of a distribution of structural viscosities starting with λ_1 and decreasing. This presents a secondary consolidation phase which has two humps. More

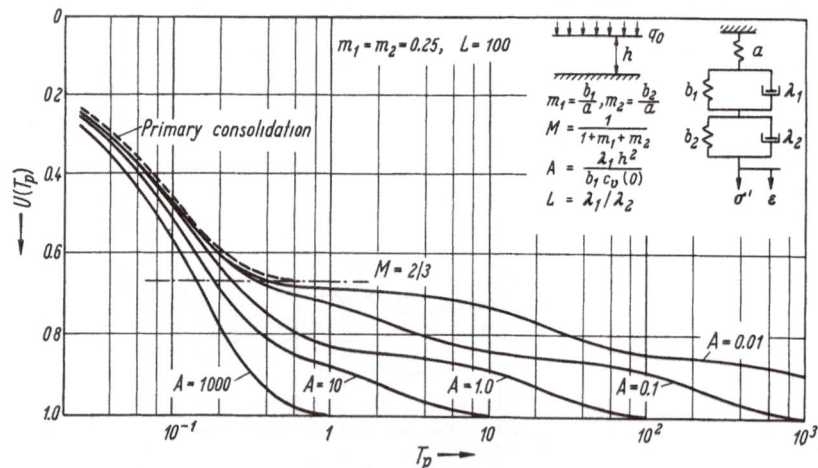

Fig. 9. Effect of viscosity factor on time-consolidation relations — five-parameter system.

parameters in the rheological system would provide more humps in the secondary phase of consolidation. An infinite number of parameters (continuous distribution) would provide a secondary curve which is linear in the logarithm of time. This is in accord with numerous test observations of static load behavior.

The number of rheological units and the relative magnitudes of the parameters depends upon the nature of the clay. As seen by Figs. 7, 8 and 9, the relationship between parameters can provide a series of distinctively shaped time-consolidation relationships. LEONARDS and GIRAULT (1961) have proposed various types of time-compression curves for a single clay; the types of behavior being dependent on the load increment ratio. The results presented here suggest that the shape of the time-compression curve is also profoundly affected by the compliance of the soil skeleton, and the relationship between the compliance and the coefficient of consolidation, $c_v(0)$.

An analysis of multi-parameter rheological models in which the loading is oscillatory, results in a time-consolidation behavior which is substantially linear, in the secondary phases. This type of behavior is confirmed by an analysis of field behavior (BJERRUM, 1964).

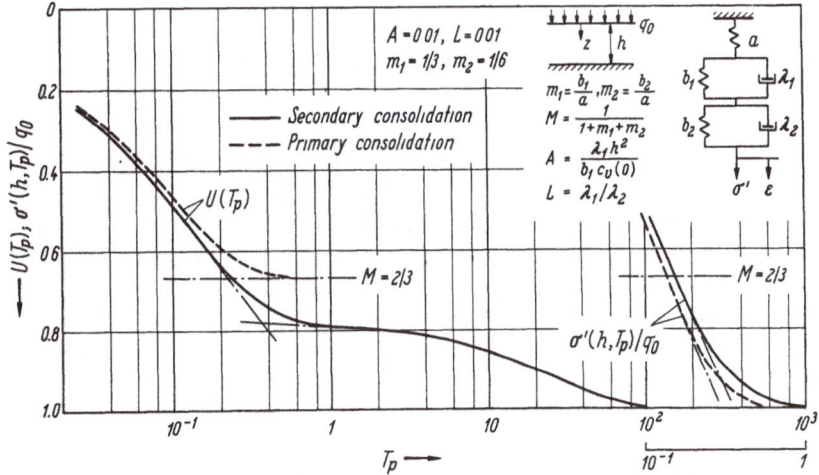

Fig. 10. Time-consolidation and effective stress-time relations — five-parameter system.

Fig. 10 presents a typical time-consolidation curve in which the viscosity factor A is of the same order or larger than the viscosity ratio L. This is the opposite extreme as that discussed previously. In this case the consolidation behavior appears to be similar to that predicted by a threeparameter system. However, the apparent distinction between the primary and secondary phases (as determined by intersection of tangents) is substantially different from the actual value of M. This is a typical case of the intermediate areas where part of the structural viscosity behaves like primary consolidation; resulting in an apparent value of M greater than the actual value. Fig. 10

also shows the effective stresses as generated at the bottom of the compressible layer. At the point of intersection of primary and secondary phases ($U = M$), there is still 17% of the excess pore pressure to be dissipated. It appears that for cases such as this, neither time-compression, nor pore pressure behavior can correctly distinguish between primary and secondary phases. It is true, that rheological model fitting starting at the end of excess pore pressure dissipation is practically feasible. This fit, however, is purely empirical and has little relationship to the rheology of the effective stress-strain relations.

Experimental Program

Experimental data are presented on the rate of secondary consolidation R_s for two saturated remolded clays. This rate is defined as the change in void ratio Δe per logarithmic time cycle, in a logarithmically linear relationship between Δe and time, and is presumed to exist after the excess pore pressure has been effectively dissipated. The variables to be studied are; (a) the magnitude of the consolidation pressure, (b) the degree of over-consolidation, and (c) the influence of the type of stress system imposed on the sample. Two types of tests are studied, the first being the oedometer test, which is theoretically equivalent to the compressible layer (one-dimensional consolidation). The second type of test is the triaxial test under an isotropic applied stress, which theoretically is equivalent to volumetric behavior. The boundary conditions of one-dimensional strain in the oedometer test require that the horizontal and vertical stresses be of different magnitudes. The ratio of horizontal to vertical stress is the coefficient of earth pressure at rest K_0, and is generally equal to (0.6 ± 0.2) for normally consolidated clays. It increases with unloading and will become greater than unity at over-consolidation ratios (maximum past consolidation stress (σ'_{cm}) divided by existing consolidation stress (σ'_c) exceeding about (3.5 ± 1.0) (Bishop, 1958; Skempton, 1961; Henkel and Sowa, 1963).

The notation σ'_c is used to designate the consolidation stress in the experimental studies, where as σ' is the consolidation stress in the theoretical studies.

Two clays were used in this study. The Boston Blue Clay (BBC) from Cambridge, Massachusetts is a silty clay of moderate plasticity and compressibility. The Vicksburg Buckshot clay (VBC) from Vicksburg, Mississippi is a silty clay of high plasticity and compressibility. Classification and mineralogical data on the two clays investigated are presented in Table 2.

Samples were prepared by placing a de-aired clay slurry (water content of two to four times the liquid limit) in a 9.5 inch diameter oedometer and consolidating the soil under a vertical stress of one to

Table 2. *Classification and Mineralogical Data*

Boston Blue Clay

Liquid limit	$= 33\%$
Plastic limit	$= 18\%$
Plasticity index	$= 15\%$
Activity	$= (\text{P. I.})/\% - 2\,\mu) = 0.5$
Composition:	Quartz, illite and chlorite clay minerals
Unified soil classification:	CL
Virgin compression index, $C_c = \Delta e/\Delta \log \sigma'_c = 0.25$	

Vicksburg Buckshot Clay

Liquid limit	$= 64\%$
Plastic limit	$= 25\%$
Plasticity index	$= 39\%$
Activity	$= (\text{P, I.})/(\% - 2\,\mu) = 0.7$
Composition:	40% Quartz and feldspar,
	50% Illite and montmorillonite clay minerals,
	1% Organic matter, 2% Ironoxides
Unified soil classification:	CH
Virgin compression index, $C_c = \Delta e/\Delta \log \sigma'_c = 0.55$	

one and one-half kilogram per square centimeter. The large sample was then extruded from the oedometer and cut into smaller specimens for subsequent triaxial and oedometer testing.

Some of the batches of BBC were mixed with 16 g/l of NaCl; the other batches used fresh water. The compression-time properties of the salt and fresh water batches appeared to be the same and are therefore treated as the same soil.

Brass fixed ring oedometers with diameters of 2.75 in. and heights of about 0.75 in. were loaded via platform scales (LAMBE, 1951) and the vertical deformations measured with extensiometers having an accuracy of 0.0001 in. The vertical stresses on the samples were doubled thus yielding a "pressure increment ratio" of unity. During those increments of interest, the stress was usually left on for a period of 6 to 10 days, primary consolidation being substantially completed within several hours.

One of the tests was run in a constant temperature room (temperature $= 19 \pm 1\,°C$). The others were not and consequently compression-time curves from some of the increments were uninterpretable because of large temperature fluctuations.

Standard English (CLOCKHOUSE and WYKEHAM-FARRANCE) triaxial cells with sample dimensions of 1.41 in. diameter by 3.15 in. long were employed. Vertical filters strips placed around the specimens led to a bottom drainage line. The specimens were encased with two pyrophylactics with a layer of silicone grease between them. These were fastened to the top cap and bottom pedestal with several rubber O-rings. The cell water was deaired and the cell pressure applied with self-compensating mercury columns (Bishop and Henkel, 1962).

Volume changes during secondary consolidation were read from 1 cc capacity burettes (interpolation of 0.001 cc possible) which replaced 5 to 10 cc capacity burettes used during primary consolidation. Companion dummy burettes were used to record the amount of daily evaporation of water from the burettes. The rate of evaporation averaged about 0.002 cc per day.

Length changes were recorded for the three triaxial tests on BBC. A small load was applied to the loading piston via a dead load hanger and an extensiometer used to measure piston travel. A comparison of the values of length changes and burette readings allowed the detection of membrane leakage if it became excessive.

All triaxial tests were run in controlled temperature rooms where the maximum daily temperature variation was generally less than 2—3 °C. During the increments of interest, volume change readings were taken for 6 to 10 or more days. As with the oedometer tests, all data are for a "pressure increment ratio" of unity.

Test Results

Rates of secondary consolidation R_s are summarized in Table 3 for triaxial and oedometer tests on normally consolidated (N. C.) and over-consolidated (O. C.) samples of Boston Blue Clay. Representative curves are plotted in Figs. 11 and 12, from oedometer and tri-

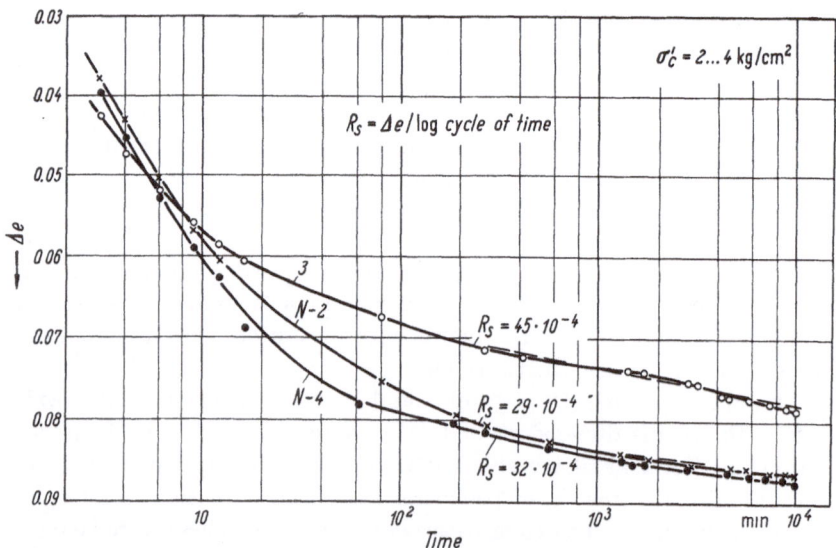

Fig. 11. Time-compression curves — oedometer tests — normally consolidated Boston Blue Clay.

axial tests respectively, on N. C. samples for a consolidation stress σ'_c going from 2 to 4 kg/cm². These curves show that while the use of a straight line approximation may sometimes be an over-simplication, it

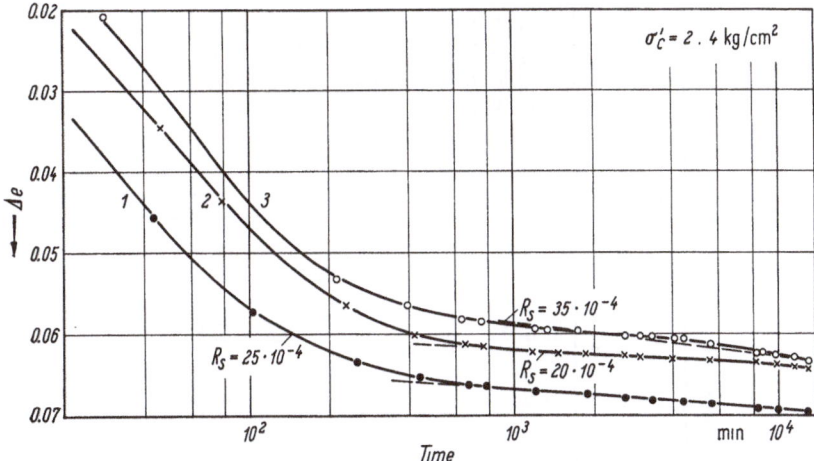

Fig. 12. Time-compression curves — triaxial tests — normally consolidated Boston Blue Clay.

is often useful in studying comparative rates of secondary consolidation. This is the purpose here.

Compression-time curves from oedometer and triaxial tests are compared in Fig. 13 for the increment of 4 to 8 kg/cm² on a N. C. clay. The curve from the triaxial test indicates the sensitivity of such data

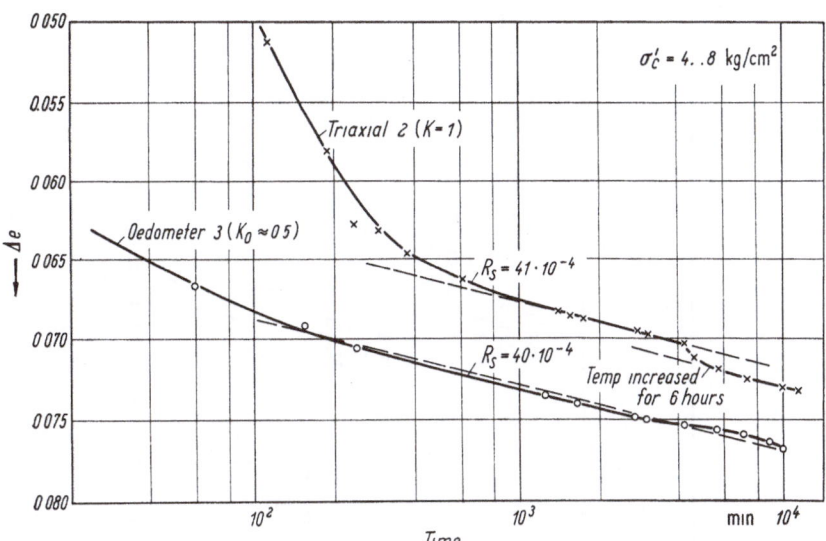

Fig. 13. Comparison of triaxial and oedometer tests — normally consolidated Boston Blue Clay.

19 Kravtchenko/Sirieys, Soil Mechanics

Table 3. *Rates of Secondary Consolidation Boston Blue Clay*

Type of test	Test No.	Pore fluid	Batch consolidation pressure (kg/cm²)	Initial void ratio	$R_s = \Delta e/\log$ cycle time, units of $10^{-4}\ \Delta e$							Remarks
					Normally consolidated			$\sigma'_{cm} = 8^2$		$\sigma'_{cm} = 16^2$		
					2–4[1]	4–8[1]	8–16[1]	2–4[1]	4–8[1]	1–2[1]	2–4[1]	
Triaxial (K = 1)	1	Salt water (16 g/l NaCl)	1.5	0.89	25 ± 4	39	—	11	27	—	—	Tests run in controlled temperature room
	2		1.5	0.93	20 ± 4	41	—	10	26	—	—	
	3		1.5	0.88	35 ± 12	43	—	11	28	—	—	
	Average				27	41	—	11	27	—	—	
	3	Salt water (16 g/l NaCl)	1.5	0.77	45	40	43	15	18	4	~5	Solids height, 2H₀ = 0.346 in.
Oedometer (K = K₀)	N-2	Fresh water	1.0	0.79	29	29	40	16	24	3	5	In constant temp. room. T = 19 ± 1°C. 2H₀ = 0.340 in.
	N-4		1.0	0.98	32	39 ± 11	41	11	21	8	~5	2H₀ = 0.338 in.
	12		1.0	1.16	25	36	50	16	23	6	~5	2H₀ = 0.373 in.
	Average				33	36	43	14	21	5	5	

[1] Pressure increment in kg/cm².
[2] Max. past pressure in kg/cm².

to temperature changes. The room housing the test was heated some
5 to 10 °C for a 6 hour period by arc lights. Fig. 14 compares compression-time curves from oedometer and triaxial tests on over-consolidated
samples.

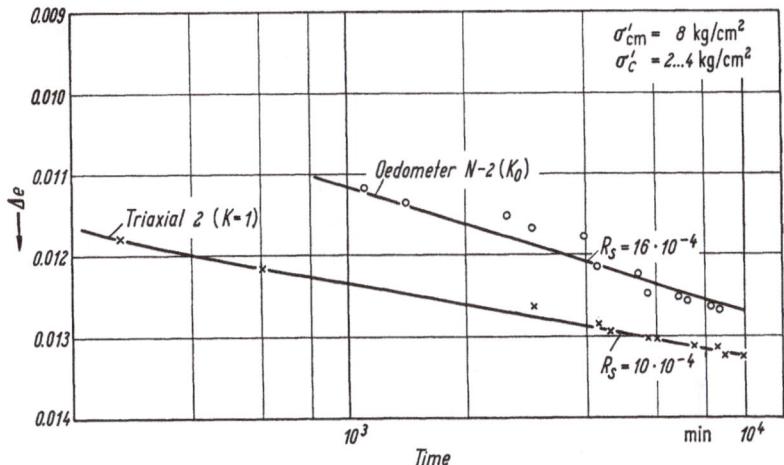

Fig. 14. Comparison of triaxial and oedometer tests — overconsolidated Boston Blue Clay.

The data in Table 3 and Fig. 11 through 14 indicate the following:

1. The rate of secondary consolidation R_s is approximately constant for N. C. samples, independent of consolidation stress. This has
also been noted by others (LEONARDS and GIRAULT, 1961; LO, 1961;
NEWLAND and ALLELEY, 1960).

2. The rate of secondary consolidation R_s decreases as the clay
becomes more highly precompressed; i.e., as the over-consolidation
ratio increases. This is widely known.

3. The rates of secondary consolidation R_s derived from the oedometer and triaxial tests are approximately equal. As was discussed
earlier the effective stress-strain model for the compressible layer
(oedometer) could be derived from a variety of volumetric and
deviatoric models. Within a given time range, (as for these particular
tests), it is quite conceivable that the same retardation times (modulus/
viscosity) are operative for volume change and one-dimensional
systems, resulting in the same rates of compression. Another possible
explanation may lie in an inherant anisotropy in the clay. In this case
an isotropic stress system will produce shear strains along with volume
change. The resulting strain rates could well be the same as in the
oedometer test. Added to the inherant anisotropy is the anisotropy
generated as the sample compresses. This strain anisotropy is a reflection of the inter-particle forces, which are discussed later.

19*

Oedometer test data for VBC are presented in Figs. 15 and 16 for
N. C. and O. C. samples respectively. Figs. 17 and 18 present similar
data from a triaxial test. The result are summarized in Table 4.

Data from the oedometer tests show that:

1. The rate of secondary consolidation R_s for N. C. samples decreased as the consolidation stress increased from 2 to 8 kg/cm², and then

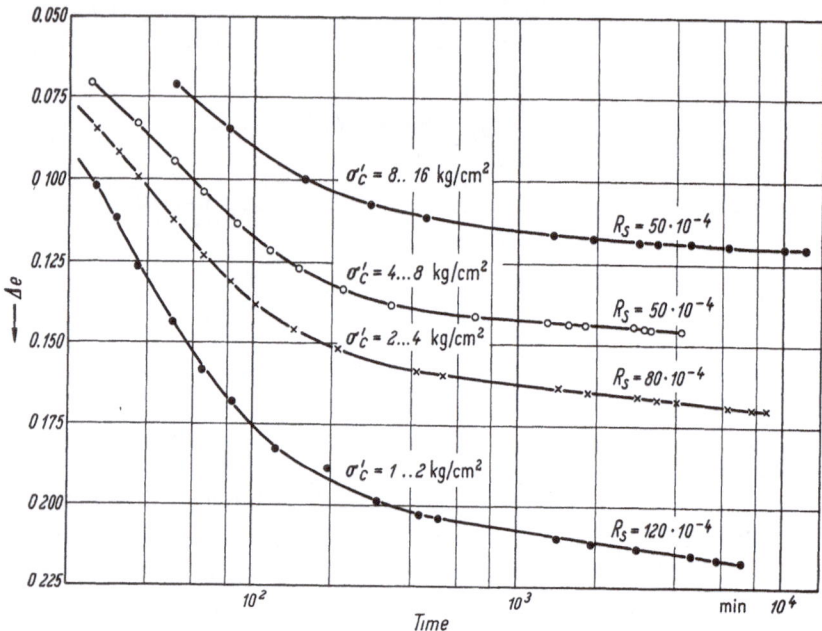

Fig. 15. Time-compression curves — oedometer tests — normally consolidated Vicksburg Buckshot
Clay.

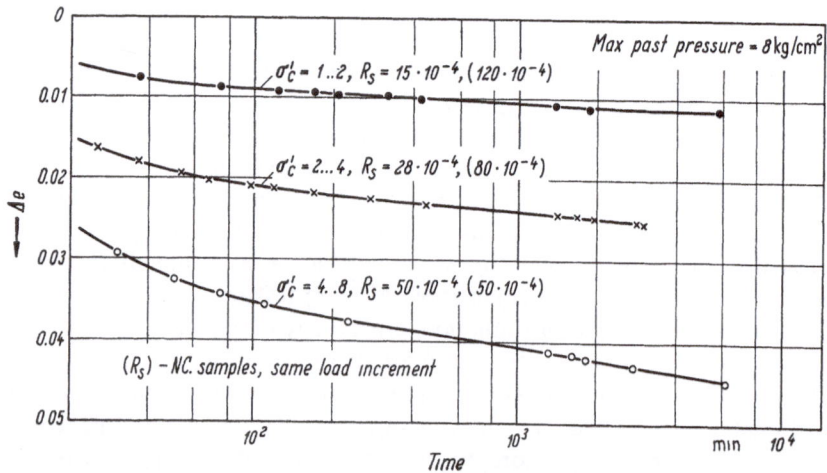

Fig. 16. Time-compression curves — oedometer tests — overconsolidated Vicksburg Buckshot Clay.

was unchanged at a consolidation stress of $16 \ \mathrm{kg/cm^2}$. The amount of primary consolidation also decreased with increasing stress, i. e., the void ratio — log pressure curve was concave upward. These and data

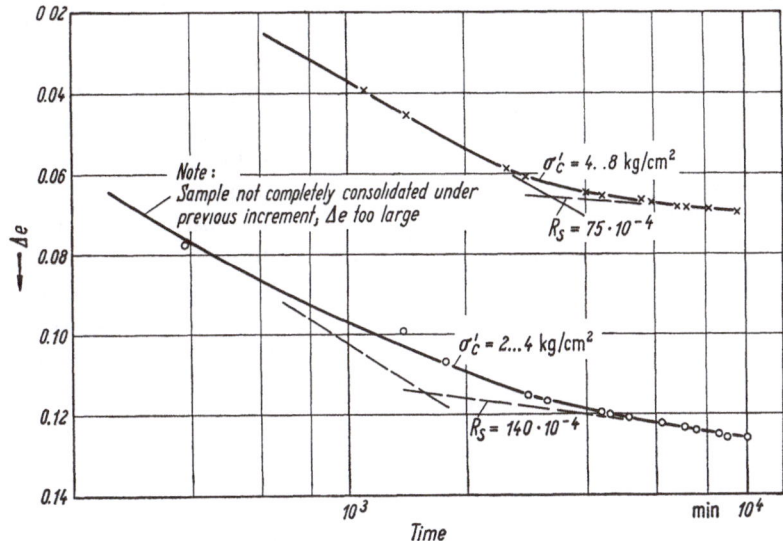

Fig. 17. Time-compression curves — triaxial tests — normally consolidated Vicksburg Buckshot Clay.

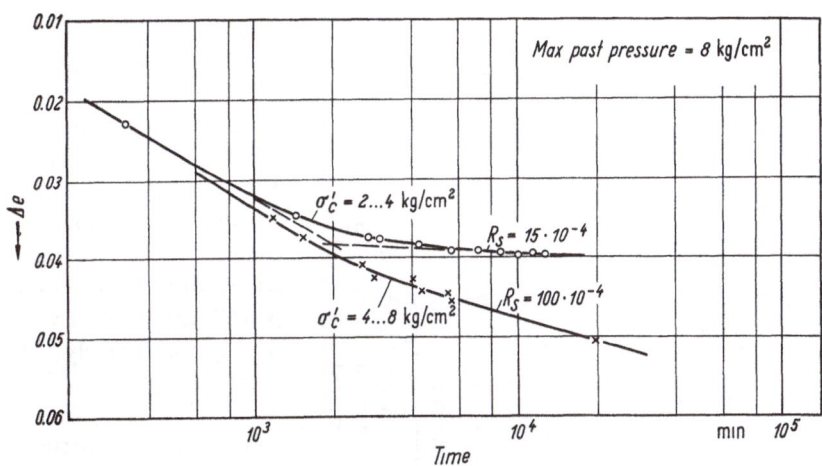

Fig. 18. Time-compression curves — triaxial tests — overconsolidated Vicksburg Buckshot Clay.

on other clays show that R_s generally increases with increasing compressibility of the soil, as would be expected.

2. The rate of secondary consolidation R_s decreases as the overconsolidation ratio increases, similar to the Boston Blue Clay.

Table 4. Rates of Secondary Consolidation Vicksburg Buckshot Clay

Type of test	Test No.	Batch consolidation pressure (kg/cm²)	Initial void ratio	$R_s = \Delta e/\log$ cycle time, units of $10^{-4}\Delta e$							Remarks
				Normally consolidated				$\sigma'_{cm} = 8^2$			
				$1-2^1$	$2-4^1$	$4-8^1$	$8-16^1$	$1-2^1$	$2-4^1$	$4-8^1$	
Triaxial ($K = 1$)	1	1.0	1.40	—	$140^3 \pm 10$	$75^3 \pm 25$	—	—	15 ± 5	100 ± 5	Tests run in controlled temperature room
Oedometer ($K = K_0$)	NFB	1.0	1.39	120	80	50	50	15	28	50	$2H_0 = 0.252$ in.
	CCL	1.0	1.25	140^4	—	75^5	—	—	—	—	$2H_0 = 0.450$ in.

[1] Pressure increment in kg/cm².
[2] Max. past pressure in kg/cm².
[3] Δe vs log t plot still curving at end of test so that value of R_s quoted may be too high.
[4] For pressure increment of 1.0—1.55 kg/cm².
[5] For pressure increment of 3.1—6.2 kg/cm².

The triaxial data on O. C. samples (Fig. 18) appear to show fairly reliable values of R_s but the data on the N. C. samples (Fig. 17) probably do not extend for sufficient time to obtain a linear e-log time relationship, if one exists. The last data points were obtained at an elapsed time equal to only 3 to 6 times that required for primary consolidation. The oedometer data in Fig. 15 shows that the plots are still curving after such a time interval (i. e., at 3 to 6 times the time required for primary consolidation). Consequently the value of R_s obtained form the triaxial tests on N. C. samples of VBC may be too large.

The limited data on VBC indicate that the rate of secondary consolidation from triaxial tests (with isotropic stresses) and from oedometer tests (with large shear stresses) is approximately the same.

Role of Interparticle Forces

Although there are scatter in the test data and the absolute magnitudes of the rates of secondary consolidation are liable to some doubt, the data do show that the values of R_s from triaxial tests with isotropic stresses are of the same relative magnitude as those from oedometer tests where there are large shear stresses; at least with normally consolidated and heavily over-consolidated samples. The following offers an explanation in terms of interparticle forces and the nature of clay fabric for this behavior.

Let us first look at the physio-chemical nature of effective stress as applied to normally consolidated clays. The normal effective stress is transmitted by two distinctly different components; namely via "contacts" wherein stress is carried by adsorbed water films and mineral to mineral contacts, and via double-layers wherein stress is carried by a net doublelayer repulsion. Calculations (LADD, 1961), though at best approximate, show that the effective stress in most normally consolidated clays (well dispersed sodium montmorillonite being an outstanding exception) is carried almost wholly via contacts and that the stresses at these contacts approach hundreds to thousands of atmospheres. Because clay is a particulate system composed of plately shaped particles, contacts between edges of one particle and the face of another will predominate (ROSENQVIST, 1961). Such a contact must of necessity carry shear stresses which will be of the same order of magnitude as the above quoted contact stresses. It is emphasized that these extremely large shear stresses at contacts between particles will exist even though the clay mass as a whole is under an isotropic stress system.

During the process of straining (both volumetric and deviatoric), the particles move relative to one another seeking positions of more stable equilibrium as particle contacts are slipping, bonds being broken and being remade. The same types of movements are believed to occur during secondary consolidation, but on a reduced scale. In other words, secondary consolidation is due to the movement of particle contacts to positions of more stable equilibrium, this movement being caused by the high shear stresses at the contacts which carry most of the effective stress acting on the clay. The movements continue until each of the contacts has established a "shear strength" exceeding its "shear stress".

The principle difference between volumetric strains and shear strains is that in the latter the particles begin to orient themselves into a preferred direction due to the externally applied shear stresses. The amount of preferred orientation is probably very small until failure is approached with the accompanying large shear strains. In an oedometer test on a normally consolidated clay where K_0 equals (0.6 ± 0.2) and the corresponding stress difference is only $1/2$ to $2/3$ of that corresponding to failure (for the same value of major effective principal stress), it is believed that the amount of preferential orientation is relatively small, and hence the nature of relative movements between particle contacts is very similar to that which occurs during isotropic compression. Hence the rate of secondary consolidation under isotropic and K_0 stress systems would not be appreciably different. However, as failure is approached and the particles assume a preferred orientation, the rate of secondary consolidation (creep may be a better term) increases since the particles can not easily find positions of stable equilibrium.

In heavily over-consolidated clays, it is hypothesized that most of the effective stress is transmitted via double-layer stresses and that the contact stresses are relatively small or even negative (i. e., particles stick together at their areas of contact). Since the contact stresses are low, the shear stresses at contacts would also be low and the tendency for relative movements between contacts greatly reduced. This picture of the difference in the physio-chemical nature of effective stress in over-consolidated as contrasted to normally consolidated clay ties in with the observed fact that the rate of secondary consolidation is much lower in over-consolidated clay samples than in normally consolidated samples of the same clay.

Acknowledgements

The theoretical study was supported, in part, by the National Science Foundation and the Land Locomotion Laboratory.

The computations were performed at the Rensselaer Computer Laboratory.

The authors are grateful to Dr. Robert E. Gibson of Imperial College for this many suggestions and to Professor George H. Handelman of Rensselaer Polytechnic Institute for his aid.

The experimental work was conducted in the Soil Mechanics Division of the Department of Civil Engineering, Massachusetts Institute of Technology, headed by Professor T. William Lambe. Mr. William Preston, Research Assistant, performed the tests.

The experimental study was supported, in part, by the Waterways Experiment Station, U. S. Army Corps of Engineers under the Research in Earth Physics Project.

References

Anagnosti, P. (1963): Stresses, Deformations and Pore Pressure in Triaxial Test Obtained by a Suitable Rheological Model. Proceedings of the European Conference on Soil Mechanics and Foundation Engineering, Wiesbaden.
Bingham, E. C. (1916): An Investigation of the Laws of Plastic Flow. Bulletin U.S. Bureau of Standards 18, 309—353.

BIOT, M. A. (1941): General Theory of Three-Dimensional Consolidation. J. Appl. Phys. **12**, 155—164.

BIOT, M. A. (1956): Theory of Deformation of a Porous Visco-elastic Anisotropic Solid. J. Appl. Phys. **27**, 459—467.

BISHOP, A. W. (1958): Test Requirements for Measuring the Coefficient of Earth Pressure at Rest. Proceedings, Brussels Conference on Earth Pressure Problems, Vol. 1, pp. 2—14.

BISHOP, A. W., and D. J. HENKEL (1962): The Measurement of Soil Properties in the Triaxial Test, 2nd Ed. Arnold, London: Arnold.

BJERRUM, L. (1963): Opening Address. Proceedings of the European Conference on Soil Mechanics and Foundation Engineering, Wiesbaden.

BJERRUM, L. (1964): Relasjon Mellom Malte Og Beregnede Setninger Av Byggverk Pa Leire Og Sand (Relation Between Measured and Estimated Settlements of Structures on Clay and Sand). Norwegian Geotechnical Institute, Oslo, Norway.

BUISMAN, A. S. K. (1936): Results of Long Duration Settlement Tests. Proceedings 1st International Conference on Soil Mechanics and Foundation Engineering, Vol. 1, pp. 103—106.

FANG, H. S. (1956): Three-Dimensional Consolidation Characteristics of Clays. Master of Science Thesis, School of Civil Engineering, Purdue University, Lafayette, Indiana.

FLORIN, V. A. (1961): Soil Mechanics and Foundation Engineering, Vol. 2 (In Russian), Moscow: Gosstroyizdat.

GEUZE, E. C. W. A., and T. K. TAN (1954): The Mechanical Behavior of Clays. Proceedings, Second International Congress on Rheology, New York: Academic Press, pp. 247—259.

GIBSON, R. E., and D. J. HENKEL (1954): Influence of Duration of Tests at Constant Rate of Strain on Measured 'Drained' Strength. Géotechnique **4**, 6—15.

GIBSON, R. E., and K. Y. LO (1961): A Theory of Consolidation for Soils Exhibiting Secondary Compression. Norwegian Geotechnical Institute Publication No. 41, (also Acta Polytechnica Scandinavica, 296/191, Ci 10).

GRAY, H. (1936): Progress Report on Research on the Consolidation of Fine-Grained Soils. Proceedings, First International Conference on Soil Mechanics and Foundation Engineering, Vol. 2, pp. 138—141.

GURTIN, M. E., and E. STERNBERG (1962): On the Linear Theory of Viscoelasticity. Archive for Rational Mechanics and Analysis **11**, No. 4, 291—356.

HENKEL, D. J., and V. A. SOWA (1963): The Influence of Stress History on the Stress Paths Followed in Undrained Triaxial Tests. Proceedings, ASTM-NRC Symposium on Laboratory Shear Testing of Soils.

ISHII, Y. (1951): General Discussion. ASTM Symposium on Consolidation Testing of Soils, STP No. 126, pp. 103—109.

LADD, C. C. (1961): Physical-Chemical Analysis of the Shear Strength of Salurated Clays. D. Sc. Thesis, Department of Civil Engineering, Massachusetts Institute of Technology.

LAMBE, T. W. (1951): Soil Testing for Engineers, New York: J. Wiley.

LAMBE, T. W. (1953): The Structure of Inorganic Soil. Proceedings of the ASCE **79**, Separate No. 315, October.

LEE, E. H. (1955): Stress Analysis in Viscoelastic Bodies, Quart. Appl. Math. **13**, 183—190.

LEONARDS, G. A., and P. GIRAULT (1961): A Study of the One-Dimensional

Consolidation Test. Proceedings, Fifth International Conference on Soil Mechanics and Foundation Engineering, Vol. 1, pp. 213—218.

Lo, K. Y. (1961): Secondary Compression of Clays. ASCE, Journal of the Soil Mechanics and Foundations Division **87**, No. SM 4, pp. 61—82.

Mandel, J. (1957): Consolidation de Couches d'Argiles. Proceedings, Fourth International Conference on Soil Mechanics and Foundation Engineering, Vol. 1, pp. 360—367.

Newland, P. L., and B. H. Allely (1960): A Study of the Consolidation Characteristics of a Clay. Géotechnique **10**, 62—74.

Roscoe, F. E., A. N. Schofield and A. Thurairajah (1963): Yielding of Clays in States Wetter than Critical. Géotechnique **13**, 211—240.

Rosenqvist, I. Th. (1961): Physico-Chemical Properties of Soils: Soil-Water Systems. Trans. ASCE **126**, Part 1, pp. 745—765.·

Schiffman, R. L. (1959): The Use of Visco-Elastic Stress-Strain Laws in Soil Testing. ASTM, STP. 254, Papers on Soils, 1959 Meetings, pp. 131—155.

Schiffman, R. L. (1963): The Visco-Elastic Compression of Soil-Water Systems. Proceedings, Fourth International Congress on Rheology.

Skempton, A. W. (1961): Horizontal Stresses in an Over-Consolidated Eocene Clay. Proceedings, Fifth International Conference on Soil Mechanics and Foundation Engineering, Vol. 1, pp. 351—357.

Tan, T. K. (1954): Investigations on the Rheological Properties of Clays. (In Dutch.) Ph. D. Thesis, Delft Technical University, The Netherlands.

Tan, T. K. (1957): Three-Dimensional Theory on the Consolidation and Flow of the Clay-Layers. Scientia Sinica **6**, No. 1, 203—215.

Tan, T. K. (1958): Secondary Time Effects and Consolidation of Clays. Scientia Sinica **7**, No. 11, 1060—1075.

Tan, T. K. (1959): Structure Mechanics of Clays. Scientia Sinica **8**, No. 1, pp. 83—97.

Tan, T. K. (1961): Consolidation and Secondary Time Effect of Homogeneous, Anisotropic, Saturated Clay Strata. Proceedings, Fifth International Conference on Soil Mechanics and Foundation Engineering, pp. 367—373.

Taylor, D. W. (1942): Research on Consolidation of Clays. Massachusetts Institute of Technology, Department of Civil Engineering, Serial 82.

Terzaghi, K. (1923): Die Berechnung der Durchlässigkeitsziffer des Tones aus dem Verlauf der Hydrodynamischen Spannungserscheinungen. Akademie der Wissenschaften in Wien. Sitzungsberichte. Mathematisch-naturwissenschaftliche Klasse. Part 11a, Vol. 132, 3/4, pp. 125—138.

Terzaghi, K. (1953): Discussion, Proceedings, Third International Conference on Soil Mechanics and Foundation Engineering, Vol. 3, pp. 158—159.

Terzaghi, K., and O. K. Fröhlich (1936): Theorie der Setzung von Tonschichten, Leipzig: Deuticke.

Thompson, W. J. (1962): Some Deformation Characteristics of Cambridge Gault Clay. Ph. D. Thesis, Cambridge University.

Volterra, V. (1929): Theory of Functionals and of Integral and Integro-Differential Equations, New York: Dover Publications.

Discussion

Question posée par S. Irmay: 1. The basic assumptions could be improved if Darcy's law were expressed in terms of relative flow velocities with respect to the consolidating soil solids. This was done by Gersevanov in the early thirties and by M. Biot in 1956. Florin has applied this assumption to the consolidation of unsaturated soils in 1948 with remarkable results. This might also explain the unusual behaviour of clays at high consolidation rates.

2. It might be interesting to plot $1/\Delta e$ versus t, to see whether a straight line is obtained.

Question posée par J. M. PIERRARD: 1. Le modèle rhéologique utilisé pour la consolidation secondaire est-il le même pour deux états de contrainte différents (Triaxial isotrope et oedomètre). Est-ce une hypothèse a priori, ou bien était-elle vérifiée expérimentalement?

2. Avez-vous pu déterminer, et si oui comment, les coefficients des modèles?

Question posée par L. ŠUKLJE: L'intérêt que le Professeur SCHIFFMAN et ses collègues ont consacré au traitement des modèles rhéologiques linéaires est bien justifié par le grand avantage des systèmes BOLTZMANNiens de pouvoir appliquer les solutions élastiques en utilisant la transformation de CARSON. La question se pose, cependant, quelles sont les possibilités d'exprimer les propriétés des sols par des modèles linéaires et quelles en sont les limitations. A mon avis, les possibilités en sont très restreintes.

Deux rapports qui ont été présentés au Congrès Européen de la Mécanique des Sols à Wiesbaden en 1963, traitent les relations rhéologiques des sols d'une manière comparable à celle du Professeur SCHIFFMAN et de ses collègues: le rapport de M. STROGANOV et celui de M. ANAGNOSTI. Dans les deux, les relations entre les composantes sphériques des tenseurs de contraintes et de déformations correspondent au modèle rhéologique du corps de KELVIN-VOIGT. Dans la solution de STROGANOV le sol est traité comme un médium visco-élastique non linéaire. La non-linéarité se rapporte aux modules élastiques de compression et de cisaillement tandis que les coefficients de la viscosité sphérique et deviatorique sont supposés d'être constants; les relations contraintes — déformations dévia-

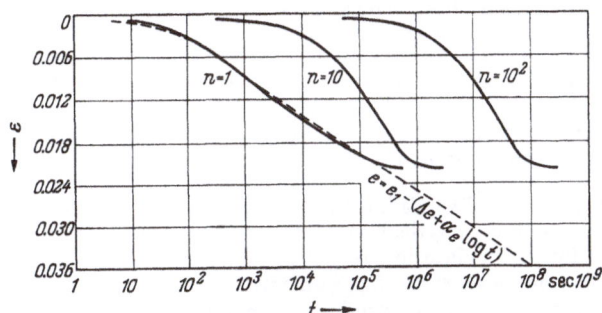

toriques, elles aussi, correspondent au modèle rhéologique de KELVIN-VOIGT généralisé. La solution de STROGANOV contient dix paramètres indépendants. Même ces dix paramètres ne prennent pas encore en considération la hétérogénéité des sols et leur anisotropie, ni celle de structure ni celle due aux états de contraintes actuels et préalables.

La solution correspondant aux relations linéaires peut être déduite de la solution de STROGANOV comme un cas spécial. En appliquant les coefficients constants du corps de KELVIN-VOIGT aux courbes expérimentales des essais oedométriques on ne peut pas exprimer d'une façon satisfaisante ni les effets importants de l'épaisseur de la couche sur la valeur maximum de la consolidation primaire ni les effets secondaires ultérieurs des couches épaisses. La figure ci-jointe montre une application de la théorie de consolidation de MM. GIBSON et Lo (1961) dont le modèle rhéologique est la connection en série d'un corps poreux saturé de HOOKE avec un corps poreux saturé de KELVIN-VOIGT. Bien que dans la première phase la courbe de consolidation s'approche bien de la courbe oedométrique observée, elle se décline trop tôt vers l'asymptote horizontale tandis que la compression observée continue dans la direction de la ligne logarithmique inclinée. Les courbes de consolidation des couches épaisses ont pratiquement l'allure des courbes de consolidation de TERZAGHI et les effets secondaires ne se font pas valoir sur la valeur finale de la compression spécifique primaire et n'interviennent pas après la dissipation des pressions intersticielles élevées.

L'application du modèle de MAXWELL aux relations des tenseurs déviateurs, et du modèle de KELVIN aux relations sphériques (ANAGNOSTI, 1963) conduit à la fin de consolidation aux états hydrostatiques qui, bien sûr, eux aussi ne correspondent pas à la réalité.

The questions I would like to ask Professor SCHIFFMAN are: 1. What are, in his opinion, the possibilities of expressing the soil properties by linear rheological models. 2. When applying non-linear models, what are our limitations in expressing the properties of real soils and in predicting the deformations of soil subjected to technical loads.

Replying the answer of Professor SCHIFFMAN:

The experimental consolidation curves which have been presented by Professor SCHIFFMAN have secondary branches depending on the logarithm of time. A theory of consolidation based on KELVIN's rheological model does not correspond to such consolidation curves.

References

ANAGNOSTI, P. (1963): Stresses, deformations and pore pressure in triaxial test obtained by a suitable rueological model. European Conference on Soil Mechanics and Foundation Engineering, Wiesbaden.

GIBSON, R. E., and K. Y. Lo (1961): A theory of consolidation for soils exhibiting secondary compression. Norges Geotekniske Institut, Publ. Nr. 41.

STROGANOV, A. S. (1963): One-dimensional deformation of soil as non-linear visco-elastic medium. European Conference on Soil Mechanics and Foundation Engineering, Wiesbaden.

Question posée par TAN TJONG-KIE: 1. In assuming a model, it is important to base it on a definite physical concept; which models have you used for the spherical and for the deviatoric stress-strain relationships?

2. How do you measure the 5 parameters?

3. According to my new insight it is not necessary to assume a definite model; I have derived theoretically that the time dependent part of the unconfined modulus of compression can be obtained experimentally by subtracting the TERZAGHI settlement from the experimental settlement curve from oedometer tests. In this manner the complete continuous spectrums of retardation-times can be obtained and on this base it is possible to make simplifications of the model.

4. Concerning Professor ŠUKLJE's question, I wish to remark that a hydrostatic state of stress has been measured in the soil around the underground tunnels in London (stiff clay), whereas settlements with constant rate have been measured in practice. So there are practical examples of my theoretical conclusion (1961, 1954) that when the deviatoric deformations will be unrestricted a hydrostatic state of stress will be created ultimately.

Question posée par R. V. WHITMAN: It has often been noted that the plot of secondary compression vs. log time is not really straight. Rather, the curve often bends downwards at the end of the record. I note that the theoretical curves also show this pattern of behaviour.

This raises a practical question. The engineer must look at field settlement results and estimate what the final settlement will be.

The question inevitably arises: can he assume that settlement will continue to follow a semi-logarithmic law, or may the rate of settlement increase at some time within the foreseeable future?

I wonder whether here is a point on which theory might help the engineer with regard to his understanding of the problem, and hence in arriving at his judgments?

Réponse de R. L. SCHIFFMAN: The question posed by Prof. IRMAY, suggests that an improvement of the solution might be obtained if the relative flow velocities were considered. Prof. IRMAY is correct in stating that primary consolidation behavior might be more fully described by this consideration. This writer understands that Dr. GIBSON has recast the consolidation problem in LAGRANGIAN coordinates and thus considers the relative flow velocities. Unfortunately, the problem of secondary consolidation is most concerned with the proper choice of a stress-strain-time relationship. The secondary effect has little or nothing to do with the flow of water since much of the deformation occurs after pore water flow has substantially ceased.

Dr. PIERRARD's question concerning the state of stress utilized in the secondary consolidation relationship reinforces the dilemma of the development of these relationships. Assuming a linear relationship, it is logical to develop the secondary consolidation relationships by the separate consideration of volumetric (spherical or isotropic triaxial) and deviatoric relationships. This was done for a variety of hypothesized cases. The dilemma arises in the development of relationships applicable for laboratory testing and for engineering use. The one-dimensional relationship is the one most used in engineering, both for experimental study, in the oedometer test, and for field predictions. Unfortunately, as the analysis shows, the stress-strain relationship developed for an oedometer test is not unique to a particular deviatoric and volumetric state. The oedometer test thus appears to be of limited usefulness in the development of stress-strain parameters to be used in two and three-dimensional analysis. In one-dimensional analysis it may, however, be of some use in the development of rates of secondary consolidation.

The development of experimental techniques to evaluate the model coefficients is beyond the scope of this particular study.

Professor ŠUKLJE has correctly questioned the applicability of the use of linear systems in studying stress-strain properties of real soils. It is certainly agreed that a non-linear, large strain, anisotropic, non-homogeneous stress-strain relationship is desirable. The question is, however, whether such a relationship can tbe properly formulated; and if formulated, whether this relationship can form the constitutative relationships for a solvable boundary value problem.

It is believed that at least three approaches to this general problem are desirable. In the first place, we should always be attempting to develop sophisticated non-linear theories. It is only in this way that the frontiers of our knowledge will be pushed back. In solving practical problems even the linear rheological theories are presently impractical. Thus the second approach is the development of empirical methods for arriving at reasonable solutions to practical problems. The third approach is the one studied in this paper. This, the middle ground of a linear theory applied to problems of secondary consolidation. This middle ground has two positive aspects. It, first of all, tends to bridge the wide gap between empirical methods and non-linear theories. Its second use is that it sometimes can be used as an approximation to the non-linear problem. A multi-parameter system of rheological models can be used to approximate a segment of non-linear behavior over a finite range of consideration.

Professor TAN has emphasized the fact that model relationships should be based upon a definite physical concept. This is quite true. The five parameter system is a reflection of first order volumetric (spherical) and deviatoric systems. In this system the volume change is represented by a three parameter model for providing an instantaneous elastic and a delayed elastic response. The deviatoric model provides a MAXWELL type behavior, providing a relaxation type of behavior. As is shown in this study, a complete linear system would have a volume change consisting of an instantaneous elastic element (spring) in series with an infinite number of delayed elastic (KELVIN units). A complete deviatoric law would have a MAXWELL unit in series with an infinite series of KELVIN units.

Prof. TAN presents some interesting data on the long time effects in secondary consolidation. It would appear that field observations tend to confirm the theory and experiments in showing the relaxation of deviatoric components during secondary consolidation. It is probably true that part of the deviatoric state of stress is dissipated during secondary consolidation. However, there must be shear stresses present at all stages of consolidation. Thus, the retarded part of deviatoric stress, will develop while the relaxing part of the deviator will dissipate.

This writer disagrees with Prof. TAN's advocacy of the use of unconfined compression and oedometer tests to determine secondary consolidation parameters. If the phenomena involves deformations in more than one coordinate direction, the use of these two tests will not determine the volumetric and deviatoric components of the stress-strain behavior. In fact, the moduli determined from these tests will be related to each other and will not be independent. If the phenomena involves one-dimensional deformation, the oedometer test, with pore pressure measurements will be as good as both tests in determining the behavior parameters. It is probable that the two tests most likely to provide useful results are the isotropic consolidation test for volume change and the torsion test of thin tubes for deviatoric behavior. Both tests should be either drained creep tests or undrained tests with pore pressure measurements.

Prof. WHITMAN's remarks concerning types of secondary consolidation deformations are most appropriate. It has recently been observed by Dr. BJERRUM that a linear logarithm of time plot is observed for constant loading. Cyclic loading shows up as a linear time plot. This observed effect has been theoretically developed by the writer, for these different types of loading.

Most soil engineers recognize the time-dependent nature of soil systems. The current practice of using the theory of elasticity in making engineering predications, on the other hand, tends to cloud the recognition of some of the effects. The soil properties are time-dependent. As such the deformation response of the soil mass will be time-dependent. Any boundary condition, such as the loading, which is time-dependent will affect the time-deformation response of the soil. Thus, in arriving at an engineering judgment, the soil engineer should not only examine the soil properties but the relationship of the boundary conditions to these soil properties, and the engineering consequences of this relationship.

Contribution de J. POTTIER: Remarques générales sur l'utilité de considérer les phénomènes physiques et physico-chimiques dans la mécanique des milieux argileux.

Les remarques que je désire présenter sont relatives aux deux communications de MM. SCHIFFMAN et MURAYAMA, et plus généralement aux études de sols argileux. Je dois dire tout d'abord que je n'ai pas de compétence personnelle en mécanique des sols et que les remarques suivantes sont largement inspirées de notre expérience de l'étude des écoulements dans les milieux poreux consolidés argileux.

Dans l'étude des milieux argileux, ou, plus généralement, des milieux finement dispersés, les effets physiques ou physico-chimiques prennent une importance particulière dans l'ensemble des propriétés du milieu, et notamment des propriétés mécaniques.

Les études de mécanique des sols s'attachent à la détermination, théorique ou expérimentale, de la loi Rhéologique. Mais souvent, elles laissent de côté la physique du matériau: en présentant seulement l'aspect mécanique, ne risquent-elles pas d'être incomplètes, et même de ne pas atteindre leur but?

Et en effet, on propose souvent des lois, ou des modèles rhéologiques complexes (éléments divers: élastique, plastique, visqueux, etc., en série ou en parallèle). Ces modèles théoriques contiennent de nombreux paramètres, dont on cherche l'ajustement par l'interprétation rhéologique de telle expérience. Cela assure-t-il vraiment leur valeur par rapport à la nature profonde du matériau? On peut se poser la question, notamment quand on voit discuter des modèles différents pour des expériences différentes (avec ou sans drainage, uniaxiale ou triaxiale).

Citons, par comparaison, la théorie cinétique des gaz. A partir des propriétés atomiques et moléculaires, on obtient, suivant le besoin telle propriété mécanique ou thermodynamique. Les modèles rhéologiques sont au contraire enfermés dans leur propre domaine: ils copient leur objet sans pouvoir le faire comprendre. Je ne suis pas moi-même spécialiste en physico-chimie, mais l'efficacité de cette discipline dans l'étude des écoulements dans les milieux argileux me persuade qu'elle offre aussi de grandes possibilités dans la mécanique de ces mêmes milieux. Je m'exprimerai dans un langage simple, voire approximatif, mais mon seul souhait est que tel chercheur s'inspire de ces remarques pour orienter ses études de mécanique.

1. *Definition du Materiau.* La précision est un souci essentiel de toute étude scientifique. Les données physico-chimiques d'un sol argileux, ou argilo-siliceux comprennent la température, la pression, le nombre des phases présentes, leur nature minéralogique, leur proportion pondérale, leur état de dispersion (forme, dimensions de particules).

Une mention particulière doit être faite pour la phase aqueuse. La teneur en eau a un rôle mécanique bien connu; mais le contenu ionique de cette eau n'est pas moins important. L'échange des ions sur les argiles est un phénomène physico-chimique majeur et les charges électriques de ces ions sont la source de forces qui réagissent sur les propriétés mécaniques. Dans une expérience de tassement avec expulsion d'eau, la composition ionique de cette eau peut donner de précieux renseignements sur les forces électriques.

Dans un autre ordre d'idées, si une expérience de mécanique suggère un effet d'orientation des feuilets argileux, il faut penser aux possibilités des outils habituels du physicien (Rayons X, par exemple).

2. *Comportement de l'Eau.* Les argiles ont la forme de plaquettes (chargées électriquement) qui retiennent sur leurs faces une certaine quantité d'eau (molécule polaire). Cette eau n'a pas les propriétés usuelles de l'eau en mécanique des fluides classique. L'eau ainsi liée aux argiles est dans un état proche de l'état solide. On comprendra ainsi (et ceci s'inspire des théories moléculaires de la

viscosité, par exemple celle d'EYRING) que la viscosité de l'eau liée soit très grande, soit 10 ou 100 poises. On est alors tenté de rapprocher les deux régimes de consolidation, primaire et secondaire, respectivement du mouvement de l'eau libre (Théorie de TERZAGHI) et de l'eau liée. D'un autre côté, pour l'influence de la température sur le comportement rhéologique l'écart de l'état d'eau liée à l'état solide est proportionnel à l'écart de température avec le point de fusion de l'eau. On retrouve ainsi la forme du résultat obtenu par S. MURAYAMA dans la fig. 8 de son étude.

Un autre point, que nous nous contenterons de signaler est que le mouvement de l'eau dans un milieu argileux peut n'être pas newtonien.

Enfin, si l'on tient compte de la présence de l'eau entre les feuillets, on comprendra qu'entre deux états successifs de compression ou de tassement, il peut y avoir expulsion d'une partie de cette eau. Ce mouvement n'est pas inclus dans la théorie de TERZAGHI pour laquelle la résistance au mouvement de l'eau est uniquement la perméabilité qui correspond à la circulation de l'eau libre. Or, l'épuisement laminaire à pression latérale constante d'une lame liquide donne une loi en fonction du temps t en $t^{-1/3}$, beaucoup plus lente que celle de TERZAGHI. Ce phénomène pourrait donc aussi expliquer la partie secondaire de la consolidation avec drainage.

3. *Forces Électriques.* Nous nous contenterons du rappel de notions très simples: les faces des plaquettes d'argile sont entourées d'ions négatifs tandis que les arêtes du pourtour sont chargées positivement. Les forces électriques peuvent se composer selon deux géométries particulières: plaquettes à faces parallèles ou contact angulaire entre la face d'une plaquette et l'arête d'une autre. Il me semble que c'est en étudiant la mécanique de tels assemblages que l'on doit rechercher la structure et la justification de modèles rhéologiques. Ce point de vue est d'ailleurs développé largement dans la communication de E. C. W. A. GEUZE.

3.3 Étude sur la Consolidation Sphérique des Sols Partiellement Saturés Soumis à la Filtration Linéaire

Par

L. Šuklje

Introduction

L'étude se rapporte à l'application directe des phénomènes observés aux éssais triaxiaux ou oedométriques soumis au drainage axial, dans le but de les adapter à la consolidation des couches cylindriques d'épaisseur uniforme quelconque et soumises à l'état de contraintes et de déformations spécifiques totales similaire à celui de l'échantillon. L'étude est basée sur la méthode semi-graphique d'isotaches qui avait été préalablement appliquée à une étude analogue de la consolidation sphérique des sols saturés (ŠUKLJE, 1957, 1963). Les isotaches y sont définies comme les lignes des pressions effectives en fonction des indices des vides correspondant aux différentes vitesses constantes de la consolidation sphérique.

Selon l'hypothèse de base de la méthode d'isotaches la vitesse du changement de l'indice des vides dépend de l'indice des vides moyen $\bar{e}$ ainsi que de la pression sphérique effective moyenne $\bar{\sigma}'_{\text{oct}}$. Cette hypothèse doit être limitée aux sols dont la déformabilité est isotrope; la dilatation (déformation sphérique due au tenseur déviatorique de contraintes) n'y est considérée non plus (ŠUKLJE, 1963). En appliquant la méthode d'isotaches aux couches indéformables latéralement (essais oedométriques), la pression axiale effective moyenne $\bar{\sigma}'_1$ et l'indice des vides moyen $\bar{e}$ sont considérés gouvernant la vitesse du tassement $\frac{\partial \bar{e}}{\partial t}$.

La présente étude est limitée aux sols non saturés dont les bulles d'air sont complètement entourées d'eau. On suppose que l'air s'échappe des vides ensemble avec de l'eau. Par conséquent les degrés de saturation S_r au début et à la fin de consolidation sont supposés égaux et le coefficient k se rapporte au débit commun de l'eau et de l'air.

Les isochrones sont supposées d'avoir la forme des paraboles du degré n quelconque (d'habitude on prend $n = 2$). On suppose la loi de DARCY valable admettant la variation du coefficient de perméabilité k avec l'indice des vides.

On tient compte de la loi de BOYLE-MARRIOTTE ainsi que de la fusibilité de l'air dans l'eau en appliquant la loi de HENRY. L'influence des contraintes superficielles de l'eau interstitielle est prise en considération.

Vitesse du Changement du Volume et la Pression Interstitielle Moyenne

Les fig. 1 et 2 respectives représentent la courbe de consolidation sphérique d'un échantillon triaxial ($\bar{e} =$ indice des vides moyen) présentée à l'échelle logarithmique (fig. 1) ou $\sqrt{t}$ (fig. 2) pour le temps t.

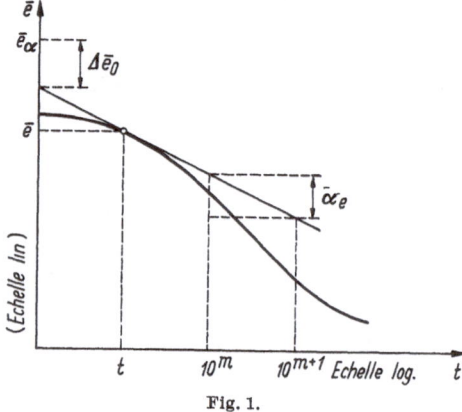

Fig. 1.

La dérivée par t de l'équation de la tangente au point quelconque $(t, \bar{e})$ est

$$\frac{\partial \bar{e}}{\partial t} = -\frac{\bar{\vartheta}_e}{N t^m}. \qquad (1)$$

Nous y avons appliqué les notifications préalables et les suivantes:

$$N = 2{,}3026,$$

(a) $m = 1,$ (fig. 1)

$$\bar{\vartheta}_e = \bar{\alpha}_e$$

si le temps est présenté à l'échelle $\log_{10} t$ (voir ŠUKLJE 1957, 1963),

$$N = 2,$$

(b) $m = \dfrac{1}{2},$ (fig. 2)

$$\bar{\vartheta}_e = \bar{\beta}_e$$

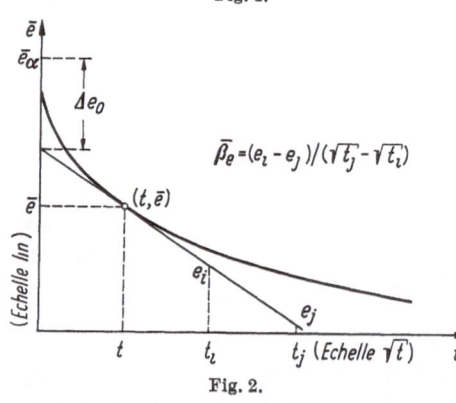

$$\bar{\beta}_e = (e_i - e_j)/(\sqrt{t_j} - \sqrt{t_i})$$

Fig. 2.

si le temps est présenté à l'échelle $\sqrt{t}$.

Supposons que les isochrones sont des paraboles du degré n (fig. 3). La vitesse du changement du volume total d'un échantillon triaxial saturé doit être égale au debit simultané aux surfaces drainantes. En supposant le drainage à la surface supérieure seulement (fig. 3) et la loi de DARCY valable, la condition ci-dessus conduit à l'expression suivante pour la pression moyenne de l'eau interstitielle $\bar{u}_w$:

$$\text{(I)} \quad \bar{u}_w = \frac{\gamma_w}{N(n+1)} \cdot \frac{\bar{\vartheta}_e'}{(1+\bar{e})} \cdot \frac{h^2}{kt^m} \tag{2}$$

$$\text{si} \quad \bar{u}_w \leq \frac{n}{n+1}\Delta\sigma,$$

$$\text{(II)} \quad \bar{u}_w = \Delta\sigma\left\{1 - \frac{\Delta\sigma}{\gamma_w} \cdot \frac{Nn}{n+1} \cdot \frac{1+\bar{e}}{\bar{\vartheta}_e'} \cdot \frac{kt^m}{h^2}\right\} \tag{3}$$

$$\text{si} \quad \bar{u}_w \geq \frac{n}{n+1}\Delta\sigma.$$

Nous y avons appliqué les notifications de la fig. 3 et préalables (cf. ŠUKLJE, 1957, 1963).

Sur la fig. 4 la courbe a représente la courbe de consolidation volumétrique de l'échantillon triaxial ou oedométrique $\bar{e} = \bar{e}(t)$ d'un sol non saturé; $\bar{e}_\alpha$ est l'indice des vides initial et S_r le degré de saturation initial. La courbe b est une courbe de consolidation apparente correspondant à la perte de l'eau et de l'air s'échappant des vides, l'air supposé incompressible. Par conséquent, la différence $\Delta\bar{e}_2$ entre les courbes a et b représente la compression des vides remplis d'air qui est due à la pression augmentée de l'air dans les vides.

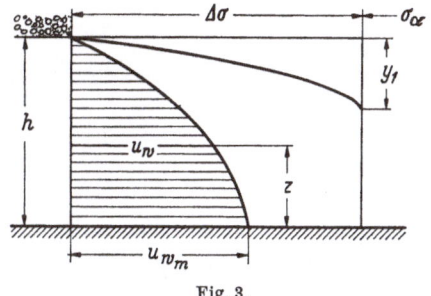

Fig. 3.

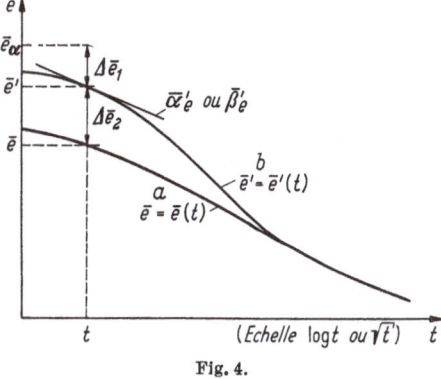

Fig. 4.

Si nous appliquons les expressions 2 et 3 aux sols non saturés, les gradients $\bar{\vartheta}_e'$ doivent se rapporter à la courbe de consolidation apparente b, tandis que les indices des vides $\bar{e}$ correspondent à la courbe a qui gouverne aussi le coefficient de perméabilité k correspondant au débit commun de l'eau et de l'air.

L'Éxpression pour la Pression Interstitielle de l'Air

Le volume de l'air interstitiel V_a' correspondant, au temps t (fig. 4), à la courbe b de consolidation apparente $\bar{e}' = \bar{e}'(t)$ supposée connue, peut être exprimé par l'équation:

$$V'_a = \bar{e}'(1 - S_r)\,V_s.\tag{4}$$

Nous y avons supposé que le degré de saturation S_r reste constant en ce qui concerne la courbe de consolidation apparente b. V_s signifie le volume de la matière solide supposée compacte.

La courbe b correspond à la supposition que la pression de l'air initiale $\bar{u}^*_{a0}$ ne change pas pendant la consolidation. En réalité, cependant, elle doit s'accorder avec la pression de l'eau interstitielle. Si nous indiquons la pression réelle moyenne de l'air interstitiel au temps t par $\bar{u}^*_a$

$$\bar{u}^*_a = \bar{u}^*_{a0} + \varDelta\bar{u}_a\tag{5}$$

et le volume correspondant de l'air par V_a, la loi de Boyle-Mariotte donne la condition:

$$V'_a\,\bar{u}^*_{a0} = V_a\,\bar{u}^*_a.\tag{6}$$

Le volume réel de l'air à temps t, V_a, peut être exprimé comme suit:

$$V_a = V'_a - \varDelta\bar{e}_2\,V_s - \bar{e}'\,V_s\,S_r\,H\varDelta\bar{u}_w\tag{7}$$

où H signifie le coefficient de Henry de fusibilité de l'air dans l'eau, et le troisième membre la quantité supplémentaire de l'air fusé dans l'eau pendant le temps t de la consolidation; l'augmentation de la pression de l'eau interstitielle $\varDelta\bar{u}_w$ peut être exprimée par l'équation:

$$\varDelta\bar{u}_w = \varDelta\bar{u}_a + \mu_t\,\varDelta\sigma_{\text{oct}}.\tag{8}$$

Le membre $\mu_t\varDelta\sigma_{\text{oct}}$ représente l'influence du changement des contraintes superficielles de l'eau interstitielle; il exprime la différence entre la pression transmise, pendant l'augmentation des pressions moyennes totales de $\varDelta\sigma_{\text{oct}}$, sur l'eau interstitielle par des surfaces des grains, et celle transmise par l'air interstitiel.

Introduisant dans l'équation (6) les expressions (4), (5) et (7), (8) on obtient l'équation

$$A\varDelta\bar{u}^2_a + B\varDelta\bar{u}_a + C = 0,\tag{9}$$

où

$$A = \bar{e}'S_r H,\tag{10}$$

$$B = \varDelta\bar{e}_2 + \bar{e}'[S_r H(\bar{u}^*_{a0} + \mu_t\,\varDelta\sigma_{\text{oct}}) - (1 - S_r)],\tag{11}$$

$$C = \bar{u}^*_{a0}(\varDelta\bar{e}_2 + \bar{e}'S_r H\mu_t\,\varDelta\sigma_{\text{oct}}).\tag{12}$$

La solution pour la pression supplémentaire moyenne de l'air interstitiel au temps t est alors:

$$\varDelta\bar{u}_a = \frac{-B \pm \sqrt{B^2 - 4AC}}{2A}.\tag{9a}$$

L'eau et l'air commencent à s'échapper quand la surpression $\bar{u}_w$ devient positive. Pendant la consolidation suivante appelée primaire les pressions interstitielles de l'eau et de l'air doivent être égales:

$$\bar{u}_w^* = \bar{u}_a^*. \tag{13}$$

En notifiant les pressions interstitielles correspondant au potentiel hydrostatique et atmosphérique à la limite drainante de la couche par $\bar{u}_0$ et les surpressions respectives de l'eau et de l'air interstitiel par $\bar{u}_w$ et $\bar{u}_a$, nous avons les relations:

$$\bar{u}_w = \bar{u}_w^* - \bar{u}_0, \tag{14}$$

$$\bar{u}_a = \bar{u}_a^* - \bar{u}_0. \tag{15}$$

La condition (13) est alors équivalente à la condition

$$\bar{u}_w = \bar{u}_a, \tag{16}$$

et l'équation (8) peut être developpée, dans ce cas, comme suit:

$$\mu_t \Delta \sigma_{\text{oct}} = \Delta \bar{u}_w - \Delta \bar{u}_a = (\bar{u}_w^* - \bar{u}_{w0}^*) - (\bar{u}_a^* - \bar{u}_{a0}^*)$$

$$= [(\bar{u}_w + \bar{u}_0) - (\bar{u}_{w0} + \bar{u}_0)] - [(\bar{u}_a + \bar{u}_0) - (\bar{u}_{a0} + \bar{u}_0)],$$

$$\mu_t \Delta \sigma_{\text{oct}} = \bar{u}_{a0} - \bar{u}_{w0}. \tag{17}$$

Nous *y* avons signalé par $\bar{u}_{a0}^*$ et $\bar{u}_{w0}^*$ les pressions absolues initiales de l'air et de l'eau interstitiel, et par $\bar{u}_{a0}$ et u_{w0} les surpressions correspondantes. La valeur (17) doit être introduite dans les expressions (11) et (12) pendant la consolidation primaire.

Détermination des Pressions Interstitielles

Avec les notifications precédentes nous pouvons écrire:

$$\bar{u}_w^* = \bar{u}_0 + \bar{u}_w = \bar{u}_0 + \bar{u}_{w0} + \Delta \bar{u}_a + \mu_t \Delta \sigma_{\text{oct}} \tag{18}$$

ce qui conduit à la relation suivante:

$$\Delta \bar{u}_a = \bar{u}_w - (\bar{u}_{w0} + \mu_t \Delta \sigma_{\text{oct}}). \tag{19}$$

Pendant la consolidation primaire la valeur $\mu_t \Delta \sigma_{\text{oct}}$ doit être substituée par l'expression (17):

$$\Delta \bar{u}_a = \bar{u}_w - \bar{u}_{a0}. \tag{20}$$

Pendant la compression des vides remplis d'air précédant la consolidation primaire la pression $\mu_t \Delta \sigma_{\text{oct}}$ peut être supposée de s'accroître de zéro jusqu'à la valeur $(\bar{u}_{a0} - \bar{u}_{w0})$ [équation (17)].

De cette manière les pressions initiales de l'eau ($\bar{u}_{w0}$) et de l'air ($\bar{u}_{a0}$ et $\bar{u}_{a0}^*$ respectivement) sont considérées d'être les paramètres gouvernant la condition (19) et influant le développement de la com-

pression de l'air dans les vides [voir équations (10), (11), (12)]. Ces
paramètres expriment l'influence de la composition minéralogique et
granulométrique des sols, de sa porosité et de sa saturation et, surtout,
des états de contraintes préalables et actuels. Les pressions interstitielles satisfaisant aussi bien les équations (2) ou (3) et (9a) que le rapport
(19), peuvent être trouvées en construisant la courbe de consolidation
apparente b (voir fig. 4) de la façon suivante:

On choisit, au temps t, le point $\bar{e}'$ et la direction ($\bar{\alpha}'_e$ ou $\bar{\beta}'_e$) de la
courbe b correspondant au point $t, \bar{e}$ de la courbe a. Les valeurs choisies
$\bar{e}'$ et $\bar{\alpha}'_e$ ou $\bar{\beta}'_e$ doivent satisfaire la condition de continuité de la courbe
b tenant compte de la construction préalable de la courbe. Les valeurs $\bar{e}$
et $\bar{\alpha}'_e$ ou $\bar{\beta}'_e$ ainsi que le coefficient de perméabilité k correspondant à
l'indice des vides $\bar{e}$ et à la saturation effective S_{rt}, doivent être introduites dans l'équation (2) ou (3); la surpression moyenne de l'eau
interstitielle $\bar{u}_w$ en résulte. La même valeur $\bar{e}'$ et la valeur $\Delta \bar{e}_2$ correspondante (voir fig. 4) doivent être introduites dans l'équation (9a) pour
trouver la surpression de l'air interstitiel $\bar{u}_a$. Les deux valeurs $\bar{u}_w$ et $\bar{u}_a$
doivent satisfaire la condition (19). Si la condition (19) n'est pas satis-
faite il faut choisir d'autres valeurs $\bar{e}$, $\bar{e}'$ et $\bar{\alpha}'_e$ ou $\bar{\beta}'_e$ et répéter tout le
procédé.

Pour faciliter le procédé l'auteur recommande la construction des
graphiques auxiliaires suivants:

a) La relation

$$\frac{\bar{\vartheta}'_e(1+\bar{e})}{k} = \frac{N(n+1)}{\gamma_w h_s^2} \bar{u}_w t^m \tag{21}$$

qui se dégage de l'équation (2), permet de construire, pour différentes
valeurs $\bar{u}_w = $ const ainsi que pour les valeurs N, n choisies et γ_w, h_s
données, les diagrammes:

$$y = \frac{\bar{\vartheta}'_e(1+\bar{e})}{k} = f(t)_{\bar{u}_w = \text{const.}} \tag{22}$$

Si les courbes de consolidation sont présentées à l'échelle logarithmique
du temps ($m=1$), on recommande de présenter la fonction (22) aux
échelles logarithmiques du temps et des ordonnées y pour obtenir des
tracés linéaires. La fig. 8 en représente un exemple.

b) L'équation (9) peut être résolue en exprimant l'indice des vides $\bar{e}'$
de la courbe de consolidation apparente b en fonction de la différence
$\Delta \bar{e}_2$:

$$\bar{e}' = D \Delta \bar{e}_2, \tag{23}$$

$$D = \left\{ \frac{1 - S_r}{1 + \dfrac{\bar{u}^*_{ao}}{\Delta \bar{u}_a}} - S_r H (\Delta \bar{u}_a + \mu_t \Delta \sigma_{\text{oct}}) \right\}^{-1}_{\Delta \bar{u}_a = \text{const}} \tag{23a}$$

La fig. 9 représente un example des droites $\bar{e}' = D\Delta\bar{e}_2$ correspondant aux différentes valeurs constantes de $\Delta\bar{u}_a$ et aux certaines valeurs choisies ou données des paramètres S_r, H, $\bar{u}_{a0}^*$ et μ_t.

Construction du Système d'Isotaches

Une fois les pressions interstitielles $\bar{u}_w$ et $\bar{u}_a$ connues, les pressions effectives sont déterminées par l'équation (BISHOP, 1961):

$$\bar{\sigma}'_{oct} = \bar{\sigma}_{oct} - \bar{u}_a + \chi(\bar{u}_a - \bar{u}_w) \tag{24}$$

qui peut être conçue aussi de la façon suivante:

$$\bar{\sigma}'_{oct} = \bar{\sigma}_{oct} - \chi\bar{u}_w - (1 - \chi)(\bar{u}_{a0} + \Delta\bar{u}_a). \tag{24a}$$

Le paramètre χ dépend du degré de saturation S_r ainsi que des propriétés physiques des sols, de sa composition et de l'état de contraintes. Dans la première approximation il peut être pris égal à S_r.

L'équation (1) permet de construire pour chaque courbe de consolidation $\bar{e} = \bar{e}(t)$ la courbe de vitesses $\dfrac{\partial \bar{e}}{\partial t} = \omega(t)$ correspondante.

Utilisant les résultats du procédé exposé ci-dessus nous construisons sur le même diagramme aussi la courbe $\bar{\sigma}'_{oct} = \bar{\sigma}'_{oct}(\bar{e})$.

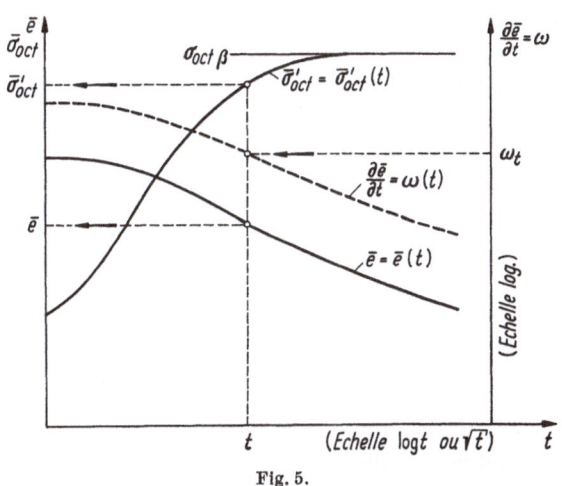

Fig. 5.

Les trois lignes coordonnées (fig. 5) permettent de déterminer pour certaines vitesses $\omega_t = \left(\dfrac{\partial \bar{e}}{\partial t}\right)_t$ les couples correspondants $(\bar{e}, \bar{\sigma}'_{oct})$. Si nous disposons des différentes lignes de consolidation $\bar{e} = \bar{e}(t)$ correspondant aux différents intervals de chargement $\Delta\sigma_{oct} = \sigma_{oct\beta} - \sigma_{oct\alpha}$, nous pouvons trouver pour chaque valeur $\omega = \text{const}$ plusieurs couples

$(\bar{e}, \bar{\sigma}'_{oct})$ qui peuvent servir pour la construction de l'isotache correspondante.

Les isotaches peuvent être construites, d'une manière analogue, à la base d'une courbe de consolidation correspondant au chargement continu, interrompu par des intervals de la charge constante.

La construction du système d'isotaches se simplifie considérablement si elle est basée seulement sur des branches secondaires des courbes de consolidation correspondant aux pressions interstitielles négligeables. C'est la phase de la consolidation où les courbes a et b (voir fig. 4) se confondent pratiquement et pour laquelle le rapport entre les pressions interstitielles de l'eau et de l'air ne doit pas être considéré davantage. Le temps où les pressions interstitielles deviennent suffisamment petites, peut être déterminé avec une précision satisfaisante de la courbe a supposant les conditions de la consolidation des sols saturés.

La pression $\bar{u}_a$ de l'air interstitiel s'accroît probablement pendant la consolidation secondaire avancée et des pressions négatives de l'eau

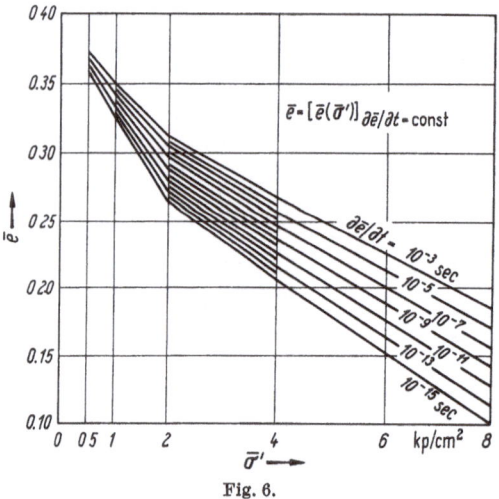

Fig. 6.

interstitielle apparaissent de nouveau; la somme $[\bar{u}_a - \chi (\bar{u}_a - \bar{u}_w)]$, cependant, reste certainement très petite s'approchant à zéro. Par conséquent, la pression effective $\bar{\sigma}'_{oct}$ [voir l'équation (24)] peut être considérée constante et égale à σ_{oct} pendant toute la consolidation secondaire.

La fig. 6 représente le système d'isotaches se dégageant des courbes de consolidation d'une craie lacustre précompactée dans l'appareil de PROCTOR et caractérisée par la limite de liquidité $w_L = 23,6\%$, l'indice de plasticité $I_P = 11,6\%$ et la composition granulométrique suivante: $17\% < 0,002$, 71% $0,002/0,06$, 12% $0,06/2$ mm.

A l'application de la méthode d'isotaches à la consolidation linéaire de l'essai oedométrique, les isotaches représentent les lignes des pressions axiales effectives moyennes en fonction des indices des vides moyens: $\bar{\sigma}'_1 = \bar{\sigma}'_1 (\bar{e})$. L'hypothèse que la vitesse du fluage dépend de la pression effective axiale, n'est pas identique à l'hypothèse qu'elle est une fonction de la pression effective sphérique parce que le rapport entre les pressions effectives axiales et latérales changent pendant la consolidation. Bien que cette hypothése différente ne soit pas tout aussi pure, elle peut quand même servir comme une hypothèse de travail utilisable.

Construction de la Courbe de Consolidation d'une Couche d'Épaisseur Quelconque

Une fois le système d'isotaches connu, la courbe de consolidation d'une couche d'épaisseur quelconque en peut être déduite, pour une pression additionnelle $\Delta\sigma_{\text{oct}} = \sigma_{\text{oct}\beta} - \sigma_{\text{oct}\alpha}$ quelconque, d'une manière analogue à celle expliquée préalablement (ŠUKLJE, 1957 et 1963) pour des sols saturés. La complication supplémentaire concerne le contrôle simultané des pressions de l'air interstitiel. Le processus est inverse à celui qui a été expliqué dans le chapitre précédent pour la construction des isotaches à la base des courbes de consolidation.

La courbe de consolidation auxiliaire b et la courbe réelle a doivent être construites simultanément. A chaque temps t la construction doit satisfaire les conditions suivantes:

1. La surpression interstitielle de l'eau $\bar{u}_w$ obtenue selon l'équation (2) ou (3) en utilisant, pour $\bar{\vartheta}'_e$, les valeurs $\bar{\alpha}'_e$ ou $\bar{\beta}'_e$ de la courbe de consolidation auxiliaire b, doit être dans le rapport exprimé par l'équation (19) avec la pression additionnelle de l'air interstitiel $\Delta\bar{u}_a$ obtenue selon l'équation (9a).

2. La vitesse de consolidation correspondant, dans le système d'isotaches, à l'indice des vides $\bar{e}$ de la courbe a au temps t ainsi qu'à la pression effective $\bar{\sigma}'_{\text{oct}}$, trouvée des valeurs $\Delta\bar{u}_a$ et $\bar{u}_w$ d'après l'èquation (24a), doit être égale à la vitesse obtenue de la courbe de consolidation a selon l'équation (1).

3. Les courbes de consolidation a et b doivent être continues.

Le degré de saturation initial S_r ainsi que l'indice des vides $\bar{e}_\alpha$ initial doivent être connus. On suppose que le degré de saturation est le même au début qu'à la fin de l'intervalle $\Delta\sigma_{\text{oct}} = \sigma_{\text{oct}\beta} - \sigma_{\text{oct}\alpha}$.

Bien entendu, les graphiques auxiliaires $y = \dfrac{\bar{\vartheta}'_e(1+\bar{e})}{k} = f(t)_{\bar{u}_w=\text{const}}$ [l'équation (22)] et $\bar{e}' = D\Delta\bar{e}_2$ [l'équation (23)] peuvent être utilisés avantageusement aussi pour la construction de la courbe de consolidation correspondant à une épaisseur quelconque de la couche. Les diagrammes $y = f(t)_{\bar{u}_w=\text{const}}$ qui ont été construits pour une certaine hauteur h_s, peuvent servir également à une autre épaisseur $h'_s = m h_s$, quelconque en adaptant l'échelle du temps; si les graphiques ont été tracés en échelle logarithmique du temps, les valeurs t correspon-

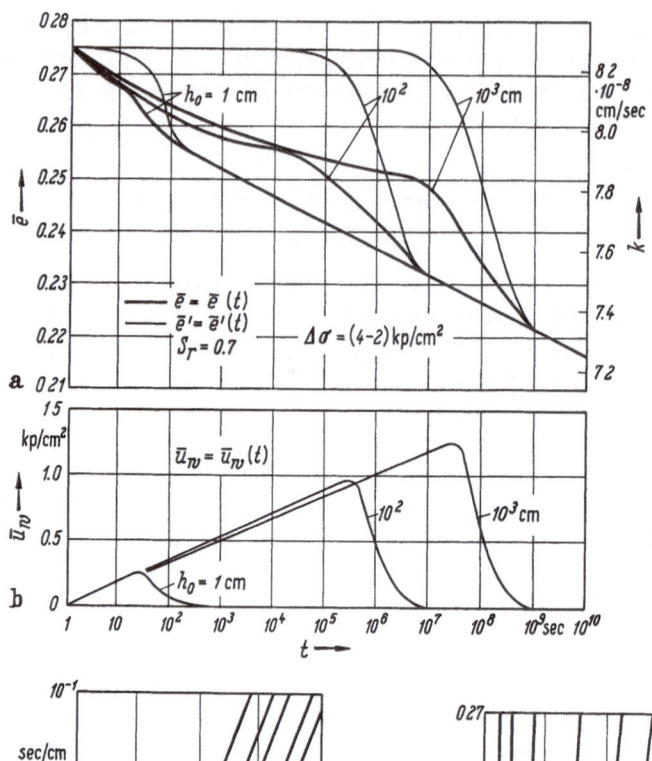

Fig. 7.

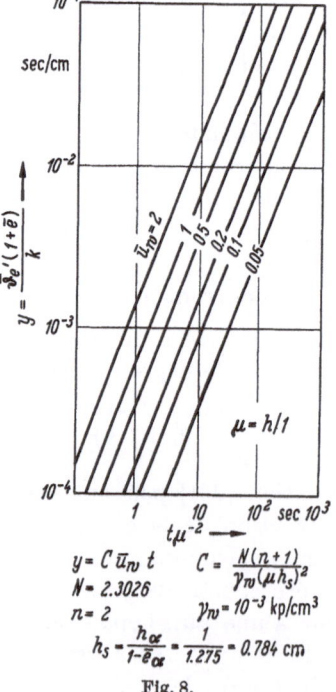

$$y = C \, \bar{u}_{rv} \, t \qquad C = \frac{N(n+1)}{\gamma_{rv}(\mu h_s)^2}$$
$$N = 2.3026$$
$$n = 2 \qquad \gamma_{rv} = 10^{-3} \text{ kp/cm}^3$$
$$h_s = \frac{h_\alpha}{1 - \bar{e}_\alpha} = \frac{1}{1.275} = 0.784 \text{ cm}$$

Fig. 8.

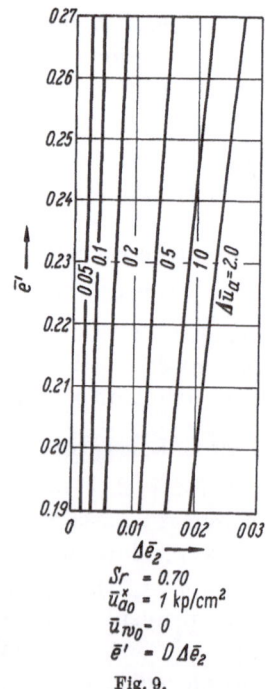

$$Sr = 0.70$$
$$\bar{u}_{a0}^x = 1 \text{ kp/cm}^2$$
$$\bar{u}_{rv_0} = 0$$
$$\bar{e}' = D \, \Delta \bar{e}_2$$

Fig. 9.

dant à la hauteur h_s doivent être multipliés avec le facteur m^2 (voir fig. 8).

Se basant sur le système d'isotaches présentées sur la fig. 6, et supposant l'indice des vides initial $\bar{e}_\alpha = 0,275$ et la pression sphérique moyenne de début $\sigma_{oct} = 2 \text{ kp/cm}^2$, nous avons appliqué la méthode expliquée ci-dessus à la construction des courbes de consolidation correspondant aux hauteurs initiales suivantes: $h = 1 \text{ cm}$, 1 m, 10 m. Les courbes résultantes $\bar{e} = \bar{e}(t)$ sont présentées, sur la fig. 7a, par les lignes épaisses tandis que les lignes minces représentent les courbes de consolidation apparentes b facilitant la construction des courbes de consolidation réelles a. Le développement des pressions moyennes de l'eau interstitielle correspondantes est montré sur la même figure sous b (fig. 7b). Les courbes présentées correspondent aux suppositions suivantes: $S_r = 0,70$, $H = 0,02 \text{ kp}^{-1} \text{ cm}^2$, $\bar{u}_{w0} = 0$, $\bar{u}_{a0}^* = 1 \text{ kp/cm}^2$.

Conclusion

Le cas traité prouve l'importante influence qui peut être exercée par l'épaisseur de la couche partiellement saturée sur le caractère et le développement de la consolidation ainsi que sur l'accroissement initial des pressions interstitielles et leur dissipation ultérieure. La porosité de départ, le degré de saturation, les pressions (souspressions) interstitielles de l'eau et de l'air initiales, la perméabilité du sol et sa compressibilité exprimé par le système des isotaches déterminent l'intensité de l'effet. Le rôle des tensions superficielles de l'eau interstitielle entourant les bulles d'air devrait être éclaircie davantage.

Références

Bishop, A. W. (1961): Discussion. Conference on Pore Pressure and Suction in Soils (London), pp. 63—66.

Šuklje, L. (1957): The analysis of the consolidation process by the isotache method. Proc. 4th Int. Conf. Soil Mech., Vol. 1, pp. 200—206.

Šuklje, L. (1963): The equivalent elastic constants of saturated soils exhibiting anisotropy and creep effects. Géotechnique **13**, pp. 291—309.

3.4 Pore-Water Pressure and Creep in One-Dimensional Compression of Silt

By

E. Schultze and J. Krause

Notations

a	Coefficient of compressibility	cm²/kg
E_s	Modulus of compressibility $= (1 + e_0)/a = v \cdot \sigma^w$	kg/cm²
e	Void ratio	1
e_0	Void ratio at the beginning of the test	1
e_1	Void ratio at the beginning of the load-increment	1
e_1'	Void ratio at the beginning of the primary compression	1
e_2'	Void ratio at the end of the primary compression	1
e_2	Void ratio at the end of the load-increment	1
Δe_0	Initial compression	1
h	Drainage length of the sample	cm
H	Height of sample	cm
J	Viscosity factor	1
k	Coefficient of permeability	cm/s
s	Degree of saturation	1
t	Time	min, s, h
T	Time factor	1
u	Pore-water pressure	kg/cm²
u_0	Pore-water pressure at the beginning of the primary compression	kg/cm²
$\bar{u}$	Pore-water pressure, averaged over the height of the sample	kg/cm²
U	Degree of consolidation	1
v	Factor in the equation of the modulus of compressibility	1
w	Exponent in the equation of the modulus of compressibility	1
w	Water content	1
P_l	Plastic limit	1
L_l	Liquid limit	1
I_p	Plasticity index	1
α	Angle	o
γ_w	Specific weight of water	kg/cm³
ε	Strain	1
η	Viscosity	h · kg/cm²
$\bar{\eta}$	Mean value of viscosity	h · kg/cm²
$\varkappa$	Creep factor	h⁻¹
σ	Total stress	kg/cm²
σ'	Effective stress	kg/cm²

$\bar{\sigma}'$	Effective stress. averaged over the height of the sample	kg/cm²
σ_{pl}	Plastic structural resistance	kg/cm²
σ_b	Bond resistance	kg/cm²
σ_v	Viscous structural resistance	kg/cm²
σ_0	Yield value, creep limit	kg/cm²
σ_{v0}	Initial viscous structural resistance	kg/cm²

Aim of the Tests

In the last few years one-dimensional compression tests on silt have been carried out in the Institut für Verkehrswasserbau, Grundbau und Bodenmechanik of the Technische Hochschule Aachen whose original purpose was to clarify the question of how big the residual pore-water pressure is at the end of the compression process. During the investigations rheologically interesting observations also came to light so that the originally intended experimental programme and its evaluation were considerably extended. As tests of this type have been up till now very uncommon, a short summary of the most important points will be given. A much more complete and full paper is under preparation.

Experimental

Large scale compression tests were carried out using diameters of 30 cm and 50 cm and heights of between 8 and 32 cm as well as small-scale preliminary tests using diameters from 5 to 15 cm and heights from 1 to 3 cm.

An artifically prepared, homogenised, water-saturated sandy silt was used as experimental soil (Fig. 1a). The soil was prepared with the help of an apparatus that was originally meant to determine the yield-behaviour of soap (Fig. 2). Certain changes were necessary in order to use the machine for this purpose.

A pressure cushion (Fig. 3) which was put under air pressure supplied by air chambers was used in order to load the samples for the large-scale tests. In this way the sample was pressed against an upper abutment.

The pore-water pressure was measured with the help of the apparatus designed by BJERRUM. Certain appliances like BJERRUM's needle, a sieve-needle and a filterstone element (Fig. 4) were examined in order to separate the pore-water pressure from the earth-pressure within the sample. The above mentioned filterstone element was proved to be most successful and as such was used during the experiment.

In the large-scale tests the course of the pore-water pressure was measured across the height of the sample in addition to settlement so that it was possible to plot the isochronous curves at various times.

The soil was set up in the liquid-limit state and with the exception of one test was loaded to 5 kg/cm² in increments of 1 kg/cm².

The relation between the coefficient of permeability and the void ratio was ascertained with the help of variable head tests in an appa-

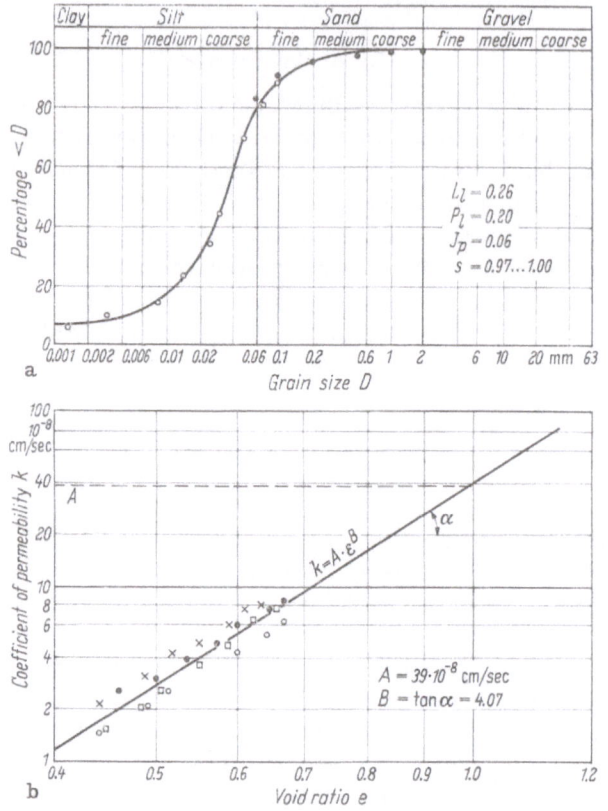

Fig. 1a and b. Characteristic lines of the investigated soil.
a) Grain-size distribution; b) Dependence of coefficient of permeability on void ratio.

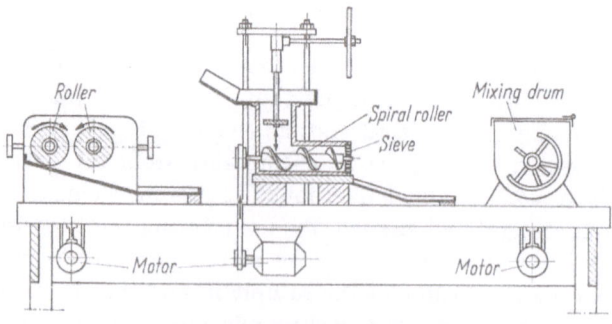

Fig. 2. Homogenizer.

ratus of the dimensions $\phi/H = 7.0/1.4$ cm. The plotting in a log-log coordinate system resulted in a straight line (Fig. 1b).

As no rheological investigations were originally planned the side friction was not measured. The results of the tests can therefore, from

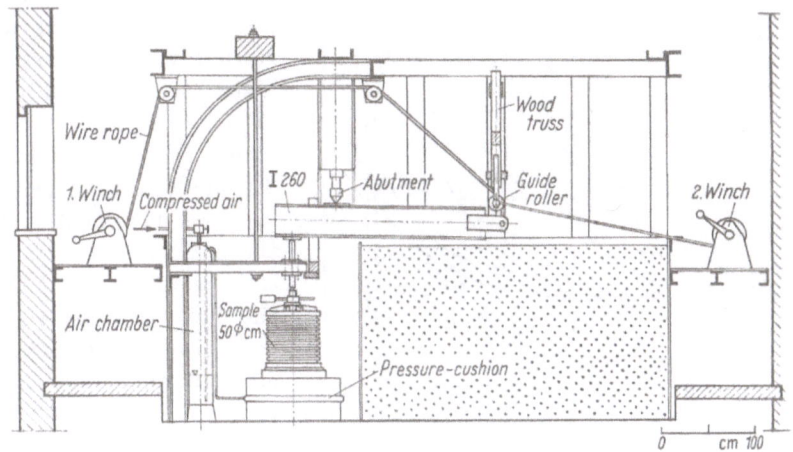

Fig. 3. Loading system.

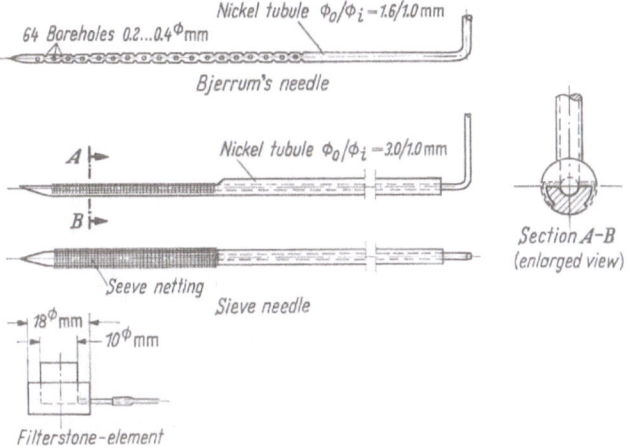

Fig. 4. Pore-water pressure measuring elements.

the rheological point of view, be only qualitatively evaluated. Quantitatively only an approximate evaluation is feasible.

Isochronous Curves

As an example of the results of the measurements, the isochronous curves for two increments, 1 to 2 kg/cm² (load increment 2), and 2 to

3 kg/cm² (load increment 3) of a sample drained on both sides are reproduced (Fig. 5). The experimental points are illustrated by means of circles.

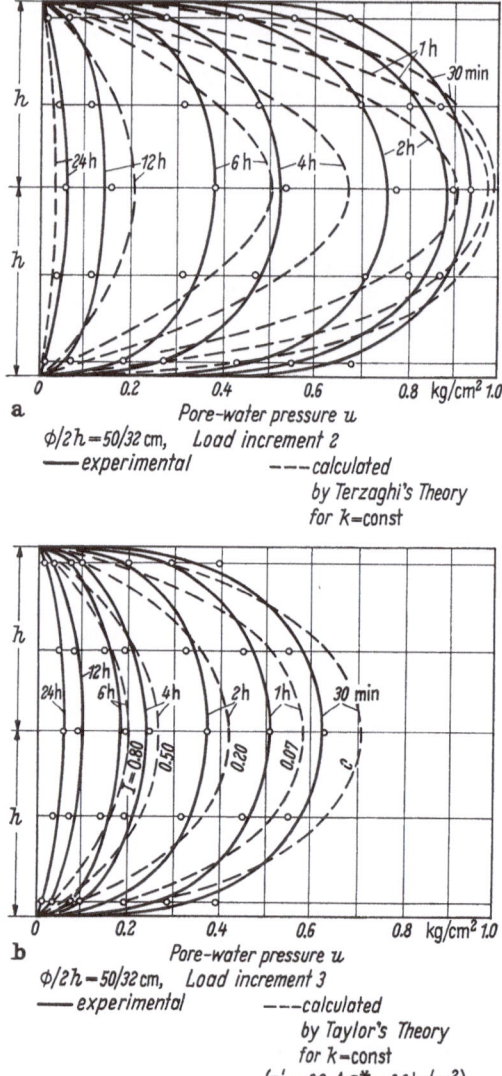

a

$\phi/2h = 50/32$ cm, Load increment 2
—— experimental   ——— calculated
by Terzaghi's Theory
for k=const

b

$\phi/2h = 50/32$ cm, Load increment 3
—— experimental   ——— calculated
by Taylor's Theory
for k=const
$(r'_p = 0.8, \Delta\sigma^* = 0.9 \text{ kg/cm}^2)$

Fig. 5a and b. Isochronous curves for the load increment 2 and 3 of the test $\phi/2h = 50/32$ cm.

For the sake of comparison the isochronous curves are plotted according to the theory of Terzaghi and Fröhlich (1936) for load increment 2 and following Taylor's theory B (1942) for load increment 3. In the latter case the viscosity coefficients J for the investigated times were respectively inserted into that porewater pressure equation of Taylor which gave the best approximation of the theoretical isochronous curves to the experimental curves.

The measured isochronous curves do not exactly agree with the theoretical curves. But the deviations are however by and large insignificant if one takes into account, on the one hand, the experimental difficulties in measuring the pore-water pressure and, on the other hand, the fact that the theories are based on simplified suppositions.

Time-Void Ratio Curve, Time-Pore-Water Pressure Curve, Time-Consolidation Curve

With the aid of the isochronous curves it is possible to separate the part of the pore-water pressure (neutral stress) from the part of the

effective stress for various times during a load increment over which
the total stress is constant. The slope of the time-pore-water pressure
averaged over the whole sample height (Fig. 8d and 9d) shows at first
a linear decrease when using the root-scale for plotting time. This
gives a curve similar in appearance to the time-void ratio and time-
settlement curves respectively (Fig. 8c and 9c). Since the first pore-
water pressure measurement was only possible after 15 minutes, due
to the response time of the pore-water pressure measuring instrument,
the beginning of the time-pore-water pressure curve had to be extra-
polated. The initial pore-water pressure reaches the value 1 kg/cm²
($= 100\%$ of the additional applied load) only during load increment 2.
During the next load increment it is appreciably lower. Similarly,
the curves show that at the end of the tests a residual pore-pressure
remains, which is only of small size and does not markedly change during
the two load increments.

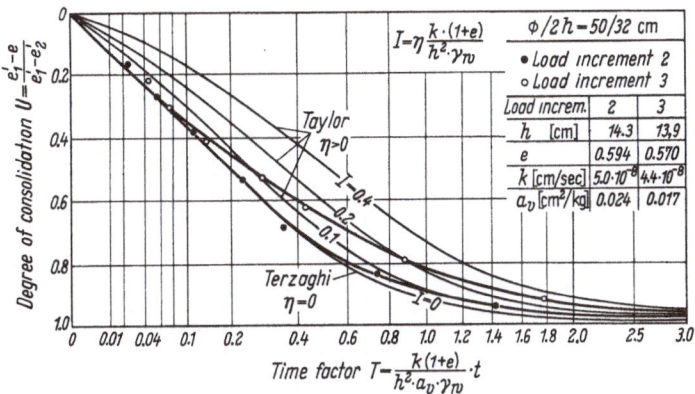

Fig. 6. Time factor-consolidation curve.

From each time-void ratio curve (Figs. 8c and 9c) a time factor-
consolidation curve can be plotted (Fig. 6). The observed curves are
compared with those obtained from the theories of TERZAGHI-FRÖHLICH
and TAYLOR. If one takes into account the fact that the coefficient of
permeability, which is from experience very unprecisely measurable,
enters linearly into the time factor then the results of the load increment 2 give a good agreement between the experimental and the theoretical curve for $J = 0$. The small deviations which become more appreciable with increasing time can be attributed to the effect of the
residual pore-water pressure which is not taken into account in the
theory of TERZAGHI and FRÖHLICH. Load increment 3 shows a marked
deviation from the theoretical curve of TERZAGHI and FRÖHLICH. In
this load increment the viscosity coefficient is obviously no longer zero.

A comparison with theoretical time-consolidation curves calculated by
Taylor for various J-values shows that the experimental curve intersects the theoretical ones and that therefore the ratio of viscosity is
not a constant but rather increases with time. This was also observed in
the comparison of the isochronous curves (Fig. 5 b) for this load increment.

Pressure-Void Ratio Curve

The pressure-void ratio curve of the test chosen as an example
(Fig. 7) exhibits a distinct curvature as is always observed with silt.
If one reproduces the pressure-void ratio curve in terms of E_s(kg/cm²)

$$E_s = \frac{d\sigma}{d\varepsilon} = \frac{d\sigma}{de}(1 + e_0) = \frac{1}{a}(1 + e_0) = v \cdot \sigma^w$$

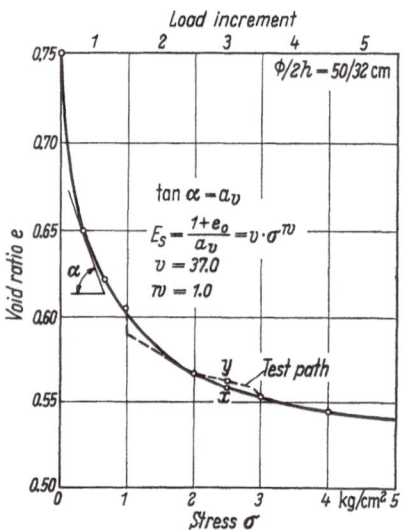

Fig. 7.   Stress-void ratio curve of the test
$\phi/2h = 50/32$ cm.

one obtains for the present case
the following values for v and w:

$$v = 37,$$

$$w = 1.$$

The pressure-void ratio curve is
then an logarithmic function,
which only occurs with undisturbed
silt during unloading and reloading, whereas for Rhein silt investigated in the Institut, an average
value of $w = 0.62$ was found for
the virgin compression (Kotzias
1963). The disturbance and artifi-
cal homogenizing of the soil before
assembly in the apparatus shows
itself in an increase in the expo-
nent.

Test Paths

The relations between pressure and void ratio *during a load increment* can be more exactly illustrated by means of the test paths. The
test paths illustrate in essence a pressure-void ratio diagram. Whereas,
however, when the plotting is done in the usual way the applied total
stresses are always equal to the effective stresses, if the time is sufficiently large to obtain complete settlement, the effective stress in the
case of the test paths can be shown to differ from the applied total stress.
Only the effective stresses averaged out over the whole sample height
can be plotted against the void ratio. Since the pore-water is slowly
pressed out of the sample during a load increment the increase in

loading does not take place suddenly, but continuously. If the size of
the average pore-water pressure at separate times is known from the
measurements, the simultaneous size of average effective stress can be
determined from it. The relation between average effective stress and
void ratio during a load increment can be built up from the simultaneously measured settlement (Figs. 8c to e and 9c to e).

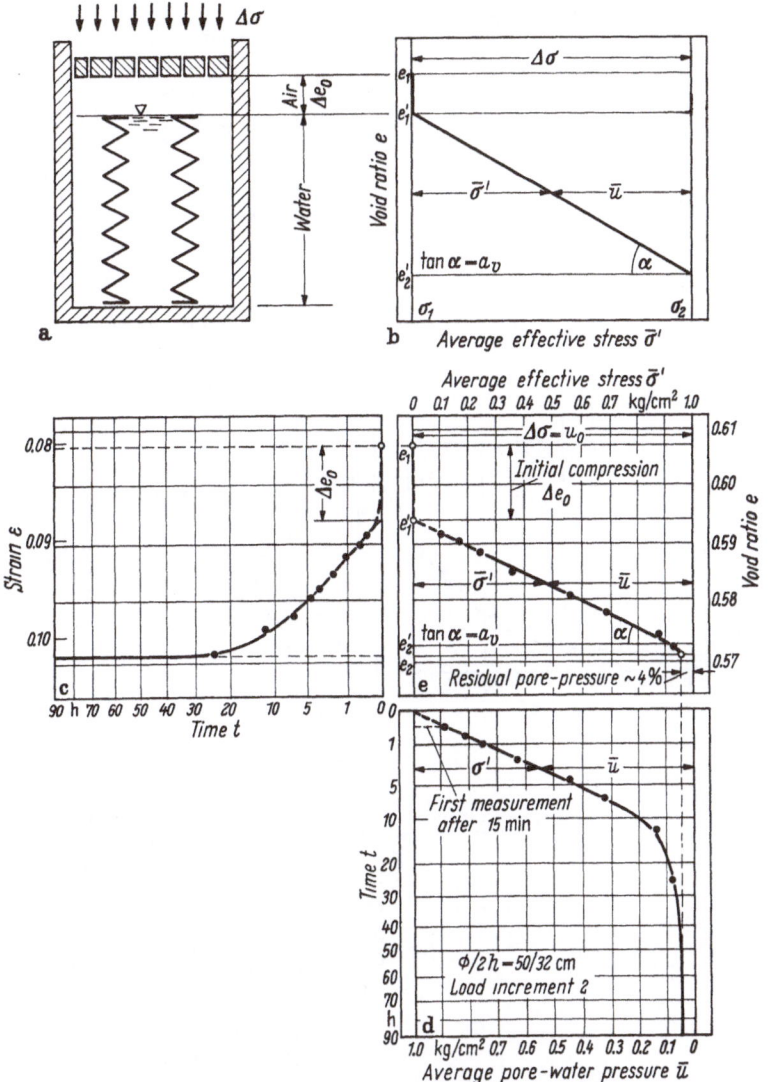

Fig. 8a—e. Comparison between theory and test results.
a) Rheological model of TERZAGHI and FRÖHLICH; b) Test path according to the Theory of TER-
ZAGHI and FRÖHLICH; c) Time-void ratio curve from the test; d) Average time-effective stress
curve from the test; e) Experimental test path.

The test paths were also inserted in the pressure-void ratio curve (Fig. 7). The different shape can be explained as follows. During a load increment of, for example, 2 to 2.5 kg/cm², the point X is reached at the end of this load increment. As compared to this, the settlement which is reached at an effective stress of also 2.5 kg/cm² in a load increment of, for example, from 2 to 3 kg/cm² at a given time, corresponds only to

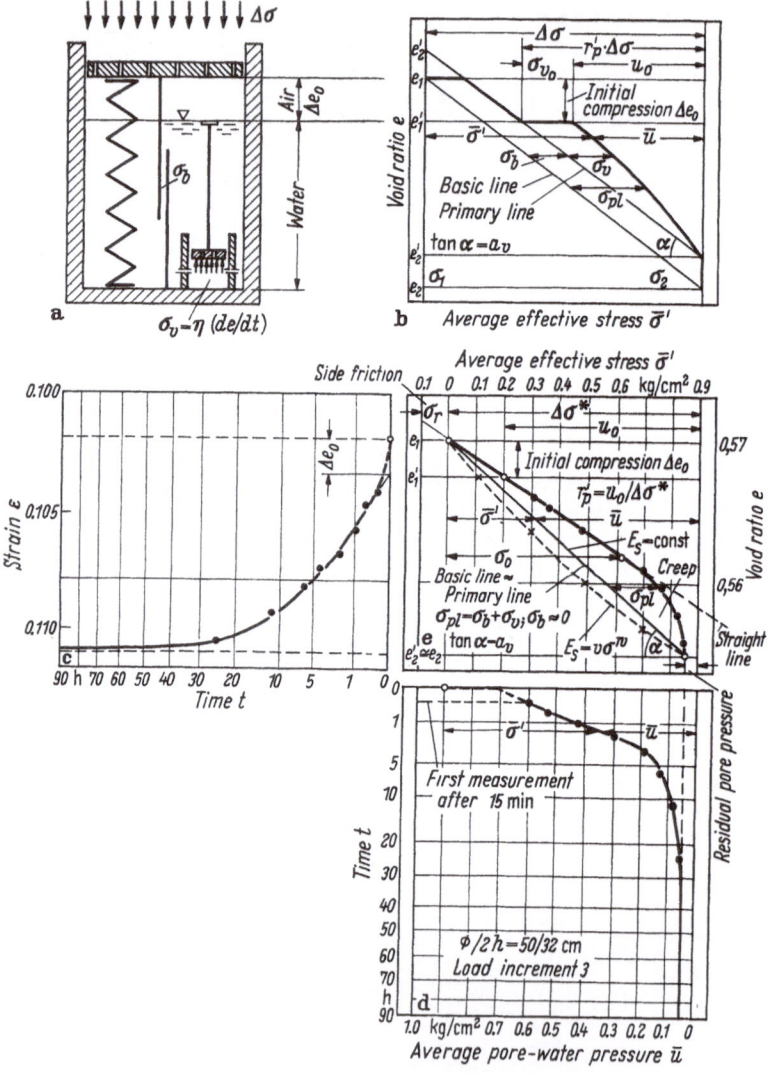

Fig. 9a—e. Comparison between theory and test results.

a) Rheological model of Taylor; b) Test path according to the theory of Taylor; c) Time-void ratio curve from the test; f) Average time-effective stress curve from the test; e) Experimental test path.

the point y. If however one interrupts the test so that this effective stress remained constant for a long time the soil would settle the distance xy under this constant stress, until finally the point x were again attained.

It was upon this consideration that TAYLOR built his theory A in 1942 which was later superseded by theory B. He called the distance $\overline{xy}$ "undeveloped secondary compression". If one considers the diagrams obtained for the two increments (Figs. 8e and 9e) the test paths can be divided into the following components.

1. *The side friction* reproduces that part of the increase in load which is absorbed by the side walls of the apparatus and therefore is not used in the compression of the sample. Since the side friction has not been measured it can only be estimated with the help of the test diagrams. In the load increment 2 the side friction is clearly small due the consistency of the sample still in the vicinity of the liquid limit (Fig. 8e). In load increment 3 it can clearly be recognised in the diagram (Fig. 9e). After this the utilised load is no more 1.0 kg/cm² but only 0.9 kg/cm².

2. *The initial compression* of the sample can be seen in that part of the settlement which has taken place in the first few seconds of the test, i.e. before the first measurement of the pore-water pressure was possible after 15 minutes. This branch of the curve appears quite clearly in both load increments. The theoretical value of the initial compression was taken from the time-void ratio diagram according to the method of TAYLOR (Figs. 8c and 9c). Here it was assumed that the initial compression occurs as a result of the momentary compression of the air still present in the sample and the apparatus in spite of the high water saturation and further that it takes place free from pore-water pressure. During load increment 2 the soil is still so soft that the initial compression may occur almost without increase in the effective stress. Compared to this, a marked rearrangement of stress on the soil skeleton is recognisable during load increment 3, corresponding to the smaller void ratio. By and large the occurrences which take place during the initial compression could only be roughly determined from the tests due to the short intervals of time which were available during observation. Further investigation of this phenomenon and its effect on the consolidation is therefore desirable.

3. *The pore-water pressure at the beginning of the consolidation settlement* is related to the initial compression. Let this be designated as initial pore-water pressure u_0. It still amounts to 100% of the applied load during load increment 2 (Fig. 8d) whereas during load increment 3 it lies appreciably below this value due to the initial compression (Fig. 9d).

4. *The relations which exist between average effective stress and void ratio immediately following initial compression* can be satisfactorily reproduced in load increment 2 by a straight line (Fig. 8e) and in load increment 3 by a curve with a linear initial branch (Fig. 9e). The shape of the test path for load increment 2 agrees with the theory of Terzaghi and Fröhlich (Fig. 8b). In load increment 3 only a rough approximation to Taylor's theory B regarding the connection between effective stress and deformation can be recognised (Fig. 9b). According to Taylor's hypothesis the excess stress which lies above the straight line joining the points at the beginning and end of the settlement during a load increment (basic line) can be attributed to the plastic structural resistance σ_{pl} of the soil. The part of the stress necessary to overcome this does not in any way hinder the dissipation of the pore-water, but is at least partly the cause of secondary creep deformations which do not necessarily need to appear only after completion of the primary settlement but rather overlap these when the material becomes somewhat more compact, as is the case in load increment 3. As hypothesis, Taylor takes the difference between the basic line and the curve to be creep (Fig. 9b). The basic line would then correspond to a purely elastic compression of the soil in the primary and the secondary region, i.e. a spring model, whereas the creep deformation would be attributed to a viscous model. In addition to the basic line there is the primary line if one splits up the structural resistance σ_{pl} into two components (Fig. 9b):

the approximately constant bond resistance σ_b which corresponds to an increase in the structural resistance due to creep in the previous load increments,

the viscous resistance σ_v which is dependent on the speed of deformation.

These concepts, introduced by Taylor into his theory B can only be indistinctly recognised in the experimentally obtained test paths. Whilst Taylor had a viscous model which was loaded right from the beginning, the test path according to the present experimental results starts off linearly — corresponding to a spring compression. The element of creep only becomes noticeable after an effective stress σ_0, which is termed the creep limit or the lower yield value. After this limiting stress the test path changes into a curved line.

The test path was obtained by joining the individual readings together. These readings are scattered in the range of the accuracy of measurement to be expected in the test. It is therefore possible that the initial portion of the curve is not linear but is slightly curved. In this case σ_0 would be zero. The accuracy with which the measurement of the pore-water pressure is possible does not suffice to explain this question.

Viscosity

According to TAYLOR's hypothesis viscosity diagrams for both load increments (Fig. 10) will be obtained from Figs. 8e and 9e. TAYLOR gives the relation between the viscous resistance and the settlement speed as follows (NEWTONian creep):

$$\sigma_v = \eta \cdot \frac{de}{dt}$$

where η is the viscosity.

According to this definition, the ratio of viscosity is given by the slope of the line joining the bond resistance, which in this case is approximately zero, and the single readings. As expected in load increment 2 in which stress and deformation progress linearly with one another (spring model) the viscosity is zero, whereas in load increment 3 it is not zero and its value varies with time. The connection between σ_v and de/dt is reproduced here by a curve.

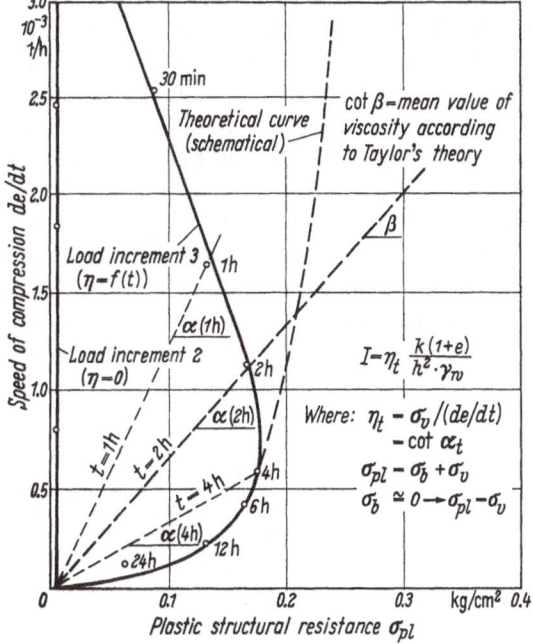

Fig. 10. Viscosity diagram.

FLORIN (1953) uses the creep factor x instead of the viscosity. This is related to the viscosity as follows

$$x = \frac{1}{a \cdot \eta}.$$

The viscosity factor J, which appears in TAYLOR's pore-water pressure equation can be calculated from the experimental value of η:

$$J = \eta \cdot \frac{k(1 + e)}{h^2 \cdot \gamma_w}.$$

TAYLOR replaces in his theory the variable viscosity by the mean viscosity $\bar\eta$.

The viscosity curve for load increment 3 is bent to the left at the top and deviates strongly from the theoretical curve represented by a dotted line. This is due to the fact that the shape of the test path postulated by Taylor (Fig. 9b) does not agree with the experimental curve (Fig. 9e) at the beginning of the consolidation.

Agreement with the Theory

In load increment 2 the material is still so soft that it essentially obeys the laws of Terzaghi-Fröhlich and Taylor respectively with $J = 0$. This was the case in all tests for all load increments with initial void ratios $e_i' > 0.59$. During load increment 3 the silt had already hardened so much that it obeyed, even if only approximately, the law of Taylor (1942) and Florin (1953) which assumes the presence of a structural resistance or creep capability of the soil ($J > 0$). According to this the viscous behaviour of the silt largely depends on its consistency. In contrast to Taylor, J is however not constant but increases with time (Figs. 5b, 6 and 10). On applying Taylor's hypothesis one obtains from the isochronous curves, the time-consolidation curve and the viscosity diagram, different relationships between the viscosity factor and time (Fig. 11).

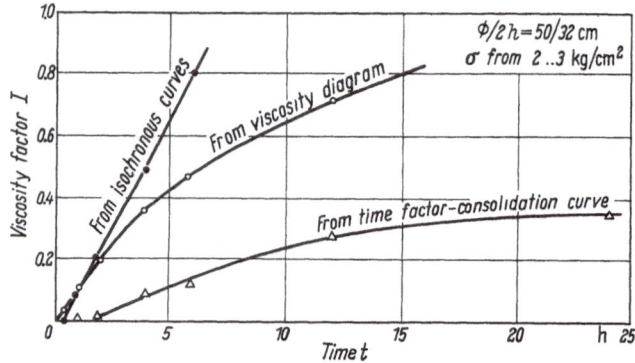

Fig. 11. Dependence of viscosity factor on time.

Rheological Model

One can conclude from the investigations that the investigated silt behaves with a soft consistency according to the rheological model of Terzaghi and Fröhlich (1936). In this model (Fig. 8a) only the springs are compressed. The relation between effective stress and strain is therefore linear. The initial compression, at which the soil skeleton takes up little or no signs of stress according to the test results, can be attributed to the fact that there exists an air gap between the piston and the spring.

The model of Taylor consists of a Kelvin model with an element of friction, which reproduces the bond resistance. In order to consider the initial compression, the length of the rod above the creep element can be shortened corresponding to the height of the air in the sample. During the initial compression the spring will be compressed by Δe_0 thereby being subjected a stress depending on its rigidity. After the initial

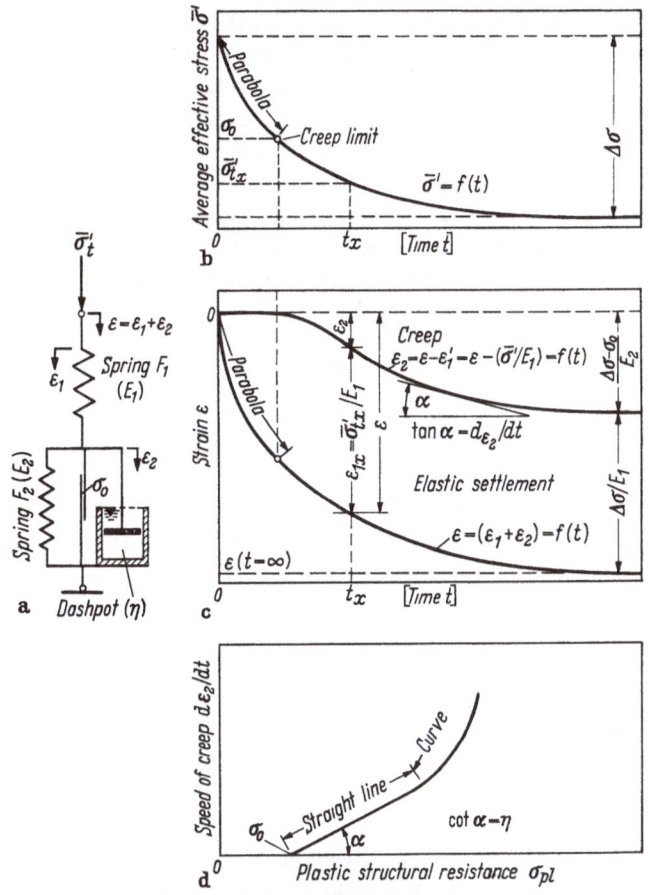

Fig. 12 a—d. Rheological model of MURAYAMA and SHIBATA.
a) Assembly of the model; b) Average effective stress-time curve; c) Relation between strain and time; d) Viscosity diagram.

compression an initial viscous resistance σ_{v0} appears in the creep element, which in turn diminishes the initial pore-water pressure. With the increase of time the spring is loaded by the effective stress. The viscous resistance of the creep element, which is a function of the compression speed, retards the settlement and makes the test path-run bend directly after the initial compression. This does not agree with the measured test paths.

On the other hand the model of Murayama and Shibata (1958, or Akai 1960) (Fig. 12a) shows a good approximation with the measured test paths. Here for effective stresses, which lie beneath the lower creep limit σ_0 only the spring F_1 is compressed and stress and settlement proceed linearly. If the effective stress exeeds σ_0 the creep element is activated. In addition to the elastic compression of the spring F_1 there now comes an elasto-plastic deformation and the test path becomes a curve. The part of the stress taken up by the spring F_2 is dependent on the speed with which the strain ε_2 of the creep piston proceeds. The line $\varepsilon_2 = f(t)$ can be determined from the measured time settlement curve (Fig. 12c). Its gradient is equal to the creep velocity $d\varepsilon_2/dt$. The creep velocity plotted against the viscous resistance of the creep element leads to a curve of viscosity (Fig. 12d), which has a similar shape to that derived by Tan Tjong-kie (1954) with the help of experiments on clay samples.

Summary

If one assumes that the behaviour of the investigated silt is typical for all silts then it can be stated that the consolidation for soils of a more compact consistency cannot be reproduced with satisfactory accuracy by any of the theories already mentioned. It is true that there is a good agreement between the measured test path and the stress-strain curve according to the model of Murayama and Shibata. But the theory belonging to the model does not consider the relation between effective stress and time in the primary range and therefore does not come into consideration for the computation of the consolidation settlements. A new theory has to be developed for the computation of the consolidation of silt with harder consistency, where according to the present test results the model of Murayama and Shibata can be taken as a basis.

References

Akai (1960): Die strukturellen Eigenschaften von Schluff. Mitteilungen aus dem Institut für Verkehrswasserbau, Grundbau und Bodenmechanik der Technischen Hochschule Aachen, Heft 22.

Florin (1953): Eindimensionales Problem der Verdichtung eines zusammendrückbaren, porösen, kriechfähigen Bodenmaterials. Mitt. d. Akad. d. Wiss. UdSSR, Abt. d. techn. Wiss., p. 797. (In russischer Sprache.)

Kotzias (1963): Die Zusammendrückbarkeit von Schluff. Mitteilungen aus dem Institut für Verkehrswasserbau, Grundbau und Bodenmechanik der Technischen Hochschule Aachen, Heft 28.

Murayama and Shibata (1958): On the Rheological Characters of Clay-Part 1. Disaster Prevention Research Inst. Kyoto Univ., 26, p. 8.

Tan Tjong-kie (1954): Onderzoeking over de rheologische Eigenschappen van Klei. Diss. Techn. Univ. Delft, p. 57.

TAYLOR (1942): Research on Consolidation of Clay. Mass. Inst. of Technology, Dep. of Civ. and Sanit. Engg., Serial No. 82.

TERZAGHI and FRÖHLICH (1936): Theorie der Setzung von Tonschichten, Leipzig/Wien: Deuticke.

Discussion

Question posée par M. BUISSON: M. SCHULTZE a-t-il pu se rendre compte des causes du changement qui se produit dans le comportement de la consolidation (loi de TERZAGHI et de M. TAYLOR). Ne serait-ce pas le fait de changement de perméabilité ? Est-ce déjà l'influence de la consolidation, secondaire (M. BIAREZ). Je pencherais vers la première raison, du fait d'observations faites jadis.

Réponse de E. SCHULTZE: There exists no separate secondary consolidation (Fig. 6). But a creep overlaps the primary consolidation (Fig. 8).

A decreasing permeability during compression would also decrease J and the time factor T (Fig. 7). The test points are than moved to the left and the $J = \text{const}$-curves have a greater slope.

Question posée par E. DE BEER: Monsieur le Professeur SCHULTZE a mesuré expérimentalement les isochrones pour un silt en consolidation à partir de teneurs en eau relativement élevées sous des accroissements de contrainte à paliers constants de 1 kg/cm².

Il a constaté qu'au début de la consolidation, la dissipation des surpressions dans l'eau se fait plus rapidement que ne correspond à la loi théorique de TERZAGHI, tandis qu'à la fin de la période de consolidation, il reste une surpression résiduelle, indiquant donc que la dissipation devient plus lente que ce que donne la loi de TERZAGHI. Il faut remarquer que près des surfaces de contact, il existe immédiatement après l'application d'un nouveau palier, des gradients hydrauliques très considérables, qui peuvent y perturber localement la structure de l'échantillon, et y donner lieu à une perméabilité accrue, ce qui peut expliquer la dissipation plus rapide des surpressions peu de temps après l'application de la charge.

La consolidation se faisant d'abord aux extrémités de l'échantillon et procédant graduellement vers le centre, ce phénomène provoque une diminution plus rapide de la perméabilité aux faces terminales qu'au centre, ce qui peut expliquer le freinage du phénomène dans le 2ème stade de la consolidation.

En fait le coefficient de perméabilité $k = f(z, t)$, ce qui explique que la loi de TERZAGHI basée sur la constance de k, ne peut au cas de fortes compressions donner la réponse rigoureusement exacte.

Il serait intéressant de faire des essais complémentaires en procédant avec des paliers plus faibles, par ex. $p = 0.25$ kg/cm², afin de contrôler si les isochrones ne sont pas influencées par la grandeur de la charge.

Réponse de E. SCHULTZE: At the two ends of the sample the slope of the isochronous curves of the tests is steeper as in the theory (Fig. 5). In the middle of the sample the inverse was observed. Perhaps that means, that the k-value of the originally homogenous sample decreases at the ends of the sample or increases in the middle. But it is not evident, what sort of perturbation has such an effect. The suggestion, to take smaller test-pressure-steps, should be followed up to bring more information.

Question posée par E. C. W. A. GEUZE:

1. Have you considered the possibility, that the deviation from the theoretical shape of the isochrones might be the result of non-uniform permeability of the clay system.

2. Have you considered the possibility, that the pore pressures as measured by the pore-pressure needle do not represent the average magnitude within the system.

Réponse de E. Schultze:

1. That is possibly the case. But it is at the moment difficult, to prove it by the present tests.

2. It seems that the measured points lie on steady curves with reasonable end-points (Figs. 5 and 6). Therefore only systematic errors concerning *all* the points are possible. The filterstone-element was very thoroughly developed and examined (Fig. 4). All measurements have the same bend. There is no further method to constate, if there are inaccuracies in the measurement of the pore pressure.

Contribution de L. Šuklje: Dans leur étude présentée au Symposium MM. E. Schultze et J. Krause ont comparé les courbes expérimentales de la consolidation linéaire des échantillons oedométriques d'un sol limoneux aux courbes théoriques correspondant à la théorie *B* de D. W. Taylor. Les suppositions de cette théorie peuvent être illustrées par le modèle rhéologique du corps de Kelvin. La partie de la pression effective proportionnelle à l'indice des vides est la somme de la « résistance statique » et du « lien » (bond), tandis que la partie proportionnelle à la vitesse de consolidation est appelée la résistance visqueuse de structure. Les suppositions mêmes de la théorie négligeant la compression secondaire qui apparaît après la consolidation primaire, ne permettent pas de faire valoir l'effet de la consolidation secondaire sur l'augmentation du coefficient de compressibilité gouvernant la consolidation primaire des couches d'épaisseurs différentes.

Cet effet a été mis en évidence par Taylor et Merchant (1942) dans une théorie précédente, ultérieurement appelée la théorie *A*, dont la supposition fondamentale est la proportionalité linéaire entre la vitesse de la consolidation secondaire et la compression secondaire non développée.

Taylor lui même a présenté une analyse critique des suppositions des deux théories et s'est rendu compte de leur imperfection. Il a indiqué la nécessité des recherches ultérieures et aujourd'hui on dispose de quelques études avancées de la consolidation linéaire. C'est pourquoi je pense qu'une analyse de la consolidation des échantillons oedométriques limitée à la comparaison des courbes de consolidation observées et de celles correspondant à la théorie *B*, ne pourrait pas éclaircir davantage le problème. Il serait désirable de confronter les observations aux autres solutions théoriques et d'en dégager des conclusions sur l'applicabilité des différentes solutions connues pour le calcul de la consolidation des couches épaisses.

Une grande difficulté qui s'oppose à l'interprétation correcte des observations expérimentales, est l'incertitude de l'extrapolation de la courbe de consolidation observée aux essais de courte durée. Dans un article publié en 1961 dans les Proceedings of the ASCE, M. Lo a confronté plusieurs essais oedométriques à la théorie de consolidation élaborée par Dr. Gibson et lui-même. Les exemples présentés conduisent à la conclusion que dans la plupart des cas la consolidation secondaire disparaît dans un délai de quelques semaines ou quelques mois. Si cela correspondait à la réalité les phénomènes visqueux porteraient sur l'interprétation des essais tandis que le calcul de la consolidation des couches épaisses serait gouverné, avec une approximation satisfaisante, seulement par les coefficients élastiques. Malheureusement des nombreuses observations de longue durée dont nous disposons ne permettent pas une généralisation de l'expérience de Lo favorable à une telle conclusion.

3.5 Correlation of Creep and Dynamic Response of a Cohesive Soil

By

Robert L. Kondner and Raymond J. Krizek

Abstract

The static and dynamic response of a cohesive soil tested in uniaxial compression is formulated in terms of a response spectra. The stress-strain-time behavior of the soil is definitely nonlinear; however, the concept of a compliance function is employed to represent the response over nine decades of time. The experimental data within this range of the time spectrum are obtained from an extensive series of transient (creep) and steady-state (vibratory) tests. The compliance function representation of the response is presented in terms of the GAUSS Error Integral. This integral gives a quantitative form of nonlinear constitutive equation for the cohesive soil and offers a unified presentation of quasi-static and dynamic response characteristics by a single phenomenological expression. Moisture content variations are taken into account by normalizing the response with respect to a strength-consistency index. The ratio of the stress level to the strength-consistency index (stress-strength parameter) is included as one of the arguments in the definition of the compliance function. This is necessary for generality, not only because of moisture content variations, but also because of the nonlinearity of the soil response. The nonlinearity is handled by reducing the response to a constant stress level which can be approximated by linear theory. The stress level considered is the special case where the applied stress vanishes. Such a condition represents the limiting magnitude of the compliance function as the applied stress approaches zero and provides a convenient datum from which the stress dependence of the compliance function is ultimately developed.

Introduction

A rigorous theoretical solution to a typical boundary value problem in soil mechanics must satisfy the equations of equilibrium (or motion), compatibility equations (or some other form of conservation equation), boundary conditions and initial conditions. The development of such a solution requires the use of some form of stress-strain-time relation for the material, and it is this aspect which specifies the solution for a particular medium. There are, at present, no such satisfactory relations available for soils, and this is, perhaps, the greatest hindrance to realistic theoretical solutions of the response of soil-foundation systems.

The formulation of a generalized stress-strain-time relationship for any given engineering material must, by its very nature, be comprehensive enough to include the entire time response spectrum of interest. Transient experiments (creep and relaxation tests) are most advantageous for measuring material behavior in the region of large time, and steady-state experiments (vibratory tests) are most useful for short time response measurements. Thus, the two types of experiments are mutually complementary and may be synthesized to phenomenologically describe response characteristics over many decades of time.

This study investigates the one-dimensional stress-strain-time behavior of a remolded clay. The concept of a compliance function is employed to represent soil behavior over approximately nine decades of the time spectrum. Experimental data within this time range are obtained from both transient (creep) and dynamic (vibratory) testing techniques and formulated in terms of the Gauss Error Integral. This integral permits the unified presentation of observed material response by a single phenomenological expression.

Material Investigated

The material investigated is a remolded clay sold commercially under the name Jordan Buff by the United Clay Mines Corporation, Trenton, New Jersey, U.S.A. The particle size distribution of the clay is given in Fig. 1 and its characteristics are as follows:

Liquid Limit	46%,	Plasticity Index	16%,
Plastic Limit	30%,	Specific Gravity	2.74.
Shrinkage Limit	20%,		

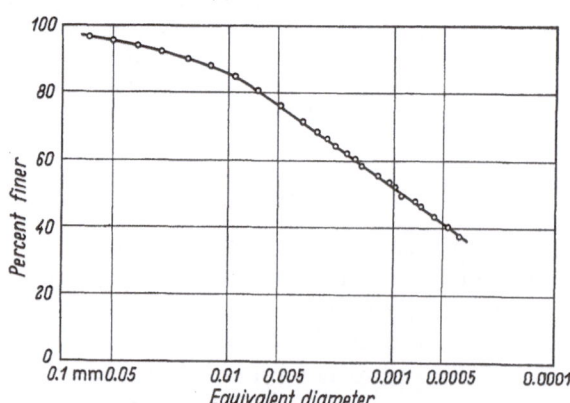

Fig. 1. Grain size distribution.

The soil specimens were prepared from a dry, powdered form by mixing with distilled water to a predetermined moisture content and passing the soil-water mixture through a "Vac-Aire" extruder. This equipment has been described in detail by MATLOCK, FENSKE and DAWSON (1951).

All tests reported herein have been performed on specimens with a length of 8.20 cm and a diameter of 3.65 cm, giving a length-to-diameter ratio of 2.25.

Creep Response

The experimental results of the creep study are given in Fig. 2 in terms of the conventional creep compliance, $J(t)$, which is the ratio of the strain at time t to the applied stress level.

This typical set of creep response curves was obtained for a particular moisture content of 32.5 ± 0.6 per cent where applied stress was

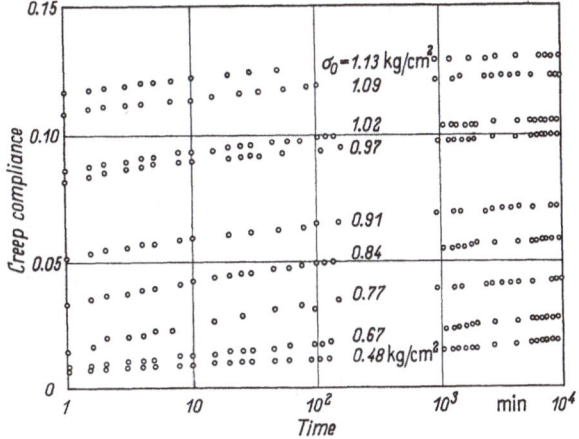

Fig. 2. Creep compliance versus time; different stress level.

varied from 0.48 to 1.13 kilograms per square centimeter. Although only four decades of time; that is, 1 to 10,000 minutes, are shown, all creep tests extended for periods of time in excess of 20,000 minutes. The response indicates a definite nonlinear behavior since $J(t)$ is not a unique curve but a function of applied stress level. Recent work by the authors indicates that effects due to small changes in moisture content can be taken into account conveniently by normalizing the response with respect to the unconfined compressive strength, q, as a consistency index. This leads to the use of a stress-strength parameter defined as the ratio of applied stress to unconfined compressive strength.

Thus, stress level and consistency must be included as arguments of the compliance function which may be defined as

$$J\left(\frac{\sigma_0}{q}, t\right) = \frac{\varepsilon(t)\,q}{\sigma_0}. \tag{1}$$

In conducting creep tests, it was noted that application of a stress level greater than approximately 68 per cent of q caused immediate failure of the specimen. Thus, the ultimate strength in creep is taken as 68 per cent of the unconfined compressive strength.

Gross (1953) notes that creep response often can be expressed as the sum of two terms; namely, a base strain associated with a particular stress level plus the time dependent response for the stress level. In terms of the stress-strength parameter this can be written

$$\varepsilon(t) = \alpha\left(\frac{\sigma_0}{q}\right) + \beta\left(\frac{\sigma_0}{q}\right)\gamma(t). \tag{2}$$

For purposes of this study, the first term will express the strain level associated with a time of one minute, which is the lower time limit for the particular form of equation to be developed, and the second term will express the time dependent response for times ranging from one minute to twenty thousand minutes, which is the upper time limit for this particular equation.

Using the explicit form for strain as a function of time and the stress-strength parameter in conjunction with the compliance function definition and creep failure criterion, the creep compliance may be written

$$J\left(\frac{\sigma_0}{q}, t\right) = \frac{0.0067}{1 - \dfrac{\sigma_0}{[\sigma_0]_{ult}}} + 0.015\,(\log t - 0.125\,\log^2 t), \tag{3}$$

$$(1 \le t \le 20{,}000),$$

where t is given in minutes. As expected, the compliance function is dependent on applied stress level.

Dynamic Response

The short time response of the clay was obtained from dynamic experiments consisting of steady-state vibratory uniaxial compression tests conducted in a specially constructed electromagnetic device.

The vibratory tests were conducted at 25 cycles per second on specimens 8.20 cm long and 3.65 cm in diameter at a moisture content of 32.4 ± 2.1 per cent subjected to a static stress of 0.54 kilograms per square centimeter. Dynamic stress amplitude, σ_D, have been normalized with respect to the unconfined compressive strength, q, to form a dynamic stress-strength parameter, $\dfrac{\sigma_D}{q}$, and results are expressed in this form.

Since the specimens were failed in the vibratory test, an unconfined compression test could not be conducted and values for q were obtained from a graph of q versus moisture content using the moisture content of the specimen. The maximum amplitude of dynamic stress to which the specimens were subjected prior to failure was found to be approximately 40 per cent of the unconfined compressive strength. Average results of the dynamic stress-strength parameter versus the dynamic strain amplitude are presented graphically in Fig. 3. These data describe a band which can be approximated within a reasonable degree of accu-

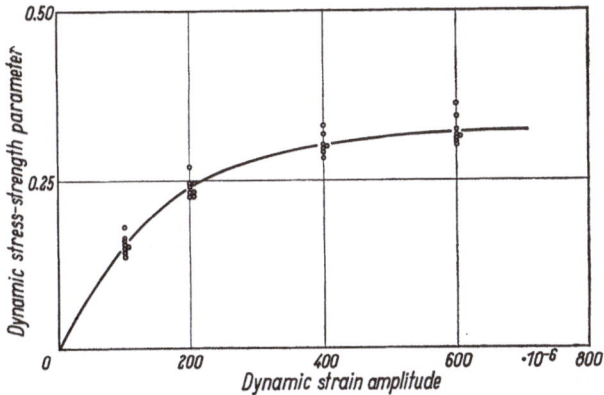

Fig. 3. Dynamic stress- strength parameter versus dynamic strain amplitude: comparison of variables.

racy by a single curve. The dynamic response of Fig. 3 is that of a non-linear material and agrees in this respect with the results of the creep study.

The vibratory experimental data can be analyzed by plotting the values taken from the curve in Fig. 3 in the form of the ratio of the dynamic strain amplitude to the dynamic stress-strength parameter versus the dynamic strain amplitude, as given in Fig. 4. Approximation of the response by a straight line, rearranging the equation in terms of the previously defined compliance function and dynamic failure criterion, and using the relation that time may be approximated by the reciprocal of the frequency, the magnitude of complex compliance becomes

$$\left| J\left(\frac{\sigma_D}{q}, t\right) \right| = \frac{0.00035}{1 - \sigma_D / [\sigma_D]_{ult}}, \quad (0.0000265 \leq t \leq 0.00053), \quad (4)$$

where t is expressed in minutes. Again, note the stress level dependence manifested by the compliance function.

The lower limit for the creep data is approximately one minute while the dynamic response has an upper time limit of approximately five ten-thousandths of a minute — corresponding to five cycles per second. There exists an intermediate region of the time spectrum where data was obtained by a special vibratory test technique for a frequency of two cycles per minute. This corresponds to a time of approximately

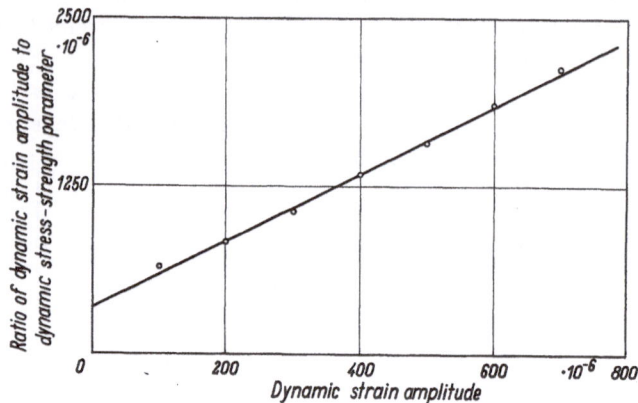

Fig. 4. Dynamic response: average values.

eight hundreths of a minute. A similar transformed hyperbolic plot can be used to obtain

$$\left| J\left(\frac{\sigma_D}{q}, t\right) \right| = \frac{0.0012}{1 - \sigma_D/[\sigma_D]_{ult}}, \quad (t = 0.0796), \tag{5}$$

where t is given in minutes.

Gauss Error Integral Representation

Eqs. (3), (4) and (5) give three expressions for the compliance function over definite limited regions of the time spectrum. It is desirable to obtain a coherent formulation of the clay behavior with a single equation at least over the region of the time spectrum covered by the experimental test data. Although the response is nonlinear, specification of a constant stress level will reduce the system to one which may be approximated by linear theory. In particular, consider the special case where the applied stress vanishes. Such a condition represents the limiting magnitude of the compliance function as the applied stress approaches zero and may be likened to the reciprocal of the initial slope of a conventional stress-strain plot. As concerned with this current analysis, this condition simply provides a convenient datum from which to develop ultimately the stress dependence of the compliance function. Subject to the specification of zero applied stress, Eqs. (3), (4 and (5) may be written

$$J(t) = 0.0067 + 0.015 \, [\log t - 0.125 \log^2 t], \quad (1 \leq t \leq 20.000), \quad (6)$$

$$J(t) = 0.00035, \qquad\qquad\qquad (0.0000265 \leq t \leq 0.00053), \quad (7)$$

$$J(t) = 0.0012, \qquad\qquad\qquad\qquad (t = 0.0796), \qquad (8)$$

where t is expressed in minutes. Values obtained from Eqs. (6), (7) and (8) are plotted in Fig. 5 as compliance function, $J(t)$, versus time, t, in minutes.

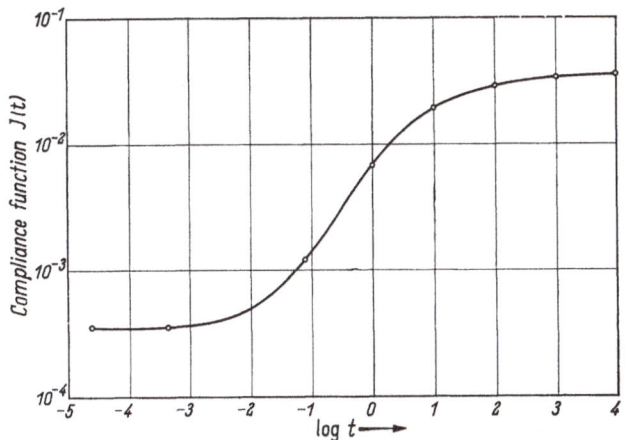

Fig. 5. Compliance function versus time: total experimental time spectrum.

Using this as a guide, the creep compliance data may be described by the GAUSS Error Integral. Such a formulation was found useful by CATSIFF and TOBOLSKY (1954) in relating the stress relaxation and dynamic properties of polyisobutylene. Also, BISCHOFF, CATSIFF and TOBOLSKY (1952) utilized this integral to unify experimental data on the stress relaxation modulus of a GR-S gum vulcanizate and polymethyl methacrylate over a series of temperatures in the transition region.

The GAUSS distribution function — or probability density function — may be written as

$$F(z) = \frac{1}{\sqrt{2\pi}} \int_{-\infty}^{k(z-\mu)} e^{-\varkappa^2/2} \, d\varkappa. \qquad (9)$$

Values for $F(z)$ may be found in almost any handbook of tables for $k = 1$ and $\mu = 0$. The form of equation given above will be utilized to describe the compliance function given in the previous graph.

For this particular study, the general variable, z, under consideration is time, t; hence, the function $F(z)$ becomes $F(t)$ and has its origin in the basic experimental data which has been expressed in the form of a compliance function, $J(t)$. In order to determine a proper form of $F(t)$ to yield a straight line when plotted versus t on probability paper, standard curvefitting techniques must be employed.

22*

For the particular study, the general variable under consideration is time, t, and the specific form obtained for $F(t)$ is plotted versus $\log t$ in Fig. 6.

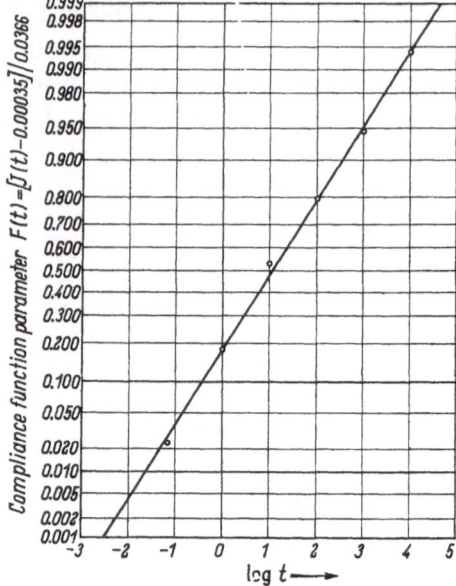

Fig. 6. Probability plot of Gaussian error integral formulation.

From this graph a slope, k, and mean, μ, are found to be 0.87 and 1.08, respectively. Hence, the special form of the Gauss Error Integral applicable in this analysis may be written

$$\frac{J(t) - 0.00035}{0.0366} = \frac{1}{\sqrt{2\pi}} \int_{-\infty}^{0.87(\log t - 1.08)} e^{-\varkappa^2/2} d\varkappa. \tag{10}$$

To handle stress dependence, algebraic manipulation of the previous equation yields this expression as representing the material behavior of the clay over approximately nine decades of time:

$$J\left(\frac{\sigma}{q}, t\right) = \frac{1}{1 - \sigma/\sigma_{ult}} \left[0.00035 + \frac{0.0366}{\sqrt{2\pi}} \right.$$
$$\left. \times \int_{-\infty}^{0.87(\log t - 1.08)} e^{-\varkappa^2/2} d\varkappa - 0.015\, H\,(\log t)\, \{\log t - 0.125 \log^2 t\} \right]$$
$$+ 0.015\, H\,(\log t)\, \{\log t - 0.125 \log^2 t\}, \tag{11}$$
$$(0.0000265 \leq t \leq 20,000),$$

where t is given in minutes. Note that this equation is a function of two variables, stress level and time, and the compliance may be represented in the time space by a family of S-shaped surves, each corresponding to a particular stress level. The particular case of the compliance function given by Eq. (11) versus the log of time for zero applied stress is shown as the solid curve in Fig. 7 while points calculated by Eqs. (6), (7) and (8) are shown by small circles. Hence, Eq. (11) may be regarded

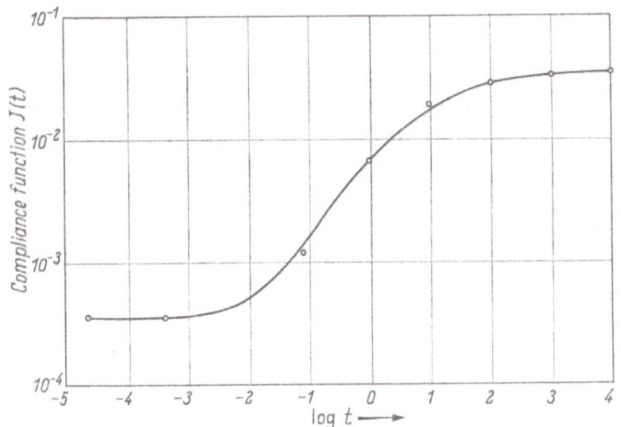

Fig. 7. GAUSSian error integral formulation of compliance function versus time: total experimental time spectrum.

as a from of constitutive equation for the particular soil investigated subject to the restrictions and limitations of the test program.

To verify that the preceding equation does reasonably represent the original experimental data, three widely separated points in the time

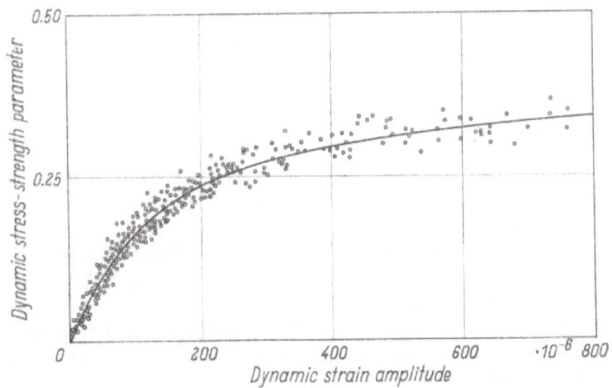

Fig. 8. Dynamic stress-strength parameter versus dynamic strain amplitude: comparison of constitutive equation with experimental results.

spectrum are investigated, these time values are 0.000106 minutes — corresponding to a vibratory oscillation of 25 cycles per second — one minute and ten thousand minutes. In Fig. 8, the solid curve represents the response predicted by the previous response equation while the data points are taken from several random tests scattered throughout the vibratory experimental program.

Conclusions

For a range of nine decades of time, the static and dynamic response of a cohesive soil tested in uniaxial compression, as given in terms of a compliance function, can be represented by a form of the Gauss Error Integral. The response obtained from both transient and dynamic experiments is definitely nonlinear. The nonlinearity is handled by reducing the response to a constant stress level which can be approximated by linear theory. The stress level considered is the special case where the applied stress vanishes. This represents a limiting condition and provides a convenient datum from which the stress dependence of the compliance function can be developed. Moisture content variations are taken into account by including the uniaxial compressive strength as a consistency index in one of the arguments of the compliance function.

Acknowledgment

The research reported herein was conducted as a phase of an investigation directed by Dr. R. L. Kondner for the U.S. Army Engineers, Waterways Experiment Station, Vicksburg, Mississippi. WES support of the study is gratefully acknowledged.

Bibliography

Bischoff, J., E. Catsiff and A. V. Tobolsky: Elastoviscous Properties of Amorphous Polymers in the Transition Region. J. Amer. Chem. Soc. **74**, 3378—3381 (1952).

Catsiff, E., and A. V. Tobolsky: Relation Between Stress Relaxation Studies and Dynamic Properties of Polyisobutylene. J. appl. Phys. **25**, 145—151 (1954).

Gross, B.: Structure of the Theory of Linear Viscoelasticity. Discussion. Proceedings of the Second International Congress on Rheology, Oxford, 1953, pp. 221—228.

Kondner, R. L.: A Non-Dimensional Approach to the Vibratory Cutting, Compaction and Penetration of Soils, Technical Report 8, Department of Mechanics, The Johns Hopkins University, 1960.

Kondner, R. L., and R. J. Krizek: A Vibratory Uniaxial Compression Device for Cohesive Soils. American Society for Testing and Materials, Proceeding **64**, 934—943 (1964).

Kondner, R. L., R. J. Krizek and H. J. Haas: Dynamic Clay Properties by Vibratory Compression. Presented at the A. S. C. E. Symposium on the Dynamic Response of Materials and Structures, San Francisco, 1963.

Kondner, R. L., and R. J. Krizek: Creep Compliance Response of a Cohesive Soil. J. Franklin Inst. **279**, 5, 366—373 (1965).

Matlock, H., C. W. Fenske, and R. F. Dawson: De-Aired Extruded Soil Specimens for Research and for Evaluation of Test Procedures. American Society for Testing and Materials, Bulletin No. 177, pp. 51—55 (1951).

3.6 Quelques Aspects de la Loi Rhéologique des Sols

Par

J. Biarez, A. Belot, J. M. Pierrard et K. Wiendieck

La loi rhéologique fait correspondre un «chemin» dans l'espace des déformations à tout chemin donné dans l'espace des contraintes, ces deux trajets étant repérés en fonction du temps. Cette loi est souvent très complexe; pour permettre le calcul, on utilise des lois partielles plus simples mais dont la validité est limitée à certains domaines ou certains chemins de l'espace des contraintes ou des déformations (cette limite peut être représentée par une surface dans un espace à six dimensions). On peut distinguer au moins les trois domaines suivants pour les matériaux qui restent «continus» pendant la déformation:

1. Domaine des déformations quasi réversibles, dont la limite dépend du mode de fabrication et de l'histoire des déformations irréversibles (fig. 1) (écrouissage isotrope ou anisotrope — «consolidation»).

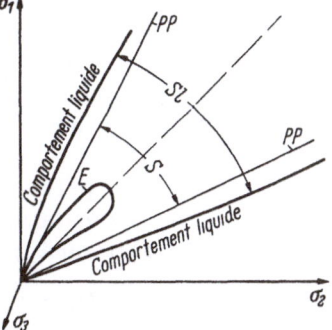

Fig. 1. Schématisation élémentaire du comportement rhéologique des sols (espace à 6 dimensions). *E* Limite du domaine de comportement reversible (avec ou sans viscosité) — limite élastique; *SL* Limite du comportement solide (dépend du chemin dans le domaine irréversible); *S* Plasticité parfaite.

2. Domaine des déformations irréversibles entre la limite précédente et la limite maximale de plasticité (pic de la courbe contrainte-déformation) qui dépend du chemin parcouru dans le domaine irréversible pour atteindre la limite. Toutefois, après avoir atteint cette limite, si l'on poursuit une déformation (lente) suffisamment grande, on obtient une limite de plasticité parfaite indépendante du mode de fabrication et de l'histoire des déformations, pourvu qu'il n'y ait pas écrasement de grains (palier final de la courbe effort-déformation).

3. Domaine du comportement liquide où un état de contrainte ne produit pas une déformation finie (Influence de la vitesse de déformation).

La mesure de la loi nécessite la reproduction des «chemins» qui seront subis en tous les points du matériau dans le problème à examiner (sauf si l'on reste dans le domaine réversible). Cette mesure correcte est pratiquement impossible; l'appareil «triaxial» permet de reproduire, pour le moment, la plus grande diversité de sollicitations, l'oedomètre, par contre, ne permet de suivre qu'un chemin souvent insuffisant (déformation monoaxiale).

Examinons quelques aspects particuliers de la loi dans les domaines réversibles et irréversibles:

A. Domaine Réversible. Viscoélasticité Linéaire [1] [2]

L'influence du temps dans la loi peut provenir du mouvement du fluide par rapport à l'ossature, ou des propriétés de l'ossature (argile). Le premier phénomène est prépondérant dans les déformations isotropes, sauf si le milieu est très dense; par contre, une déformation principalement déviatoire (faible variation de volume) sur un sol argileux mettra surtout en évidence le second phénomène. Pour mesurer les propriétés de l'ossature seule d'un "grès argileux", nous avons choisi un échantillon non saturé ($w = 1\%$) que nous avons soumis à des variations de contraintes respectivement isotropes et déviatoires.

Si la loi est élastique ou viscoélastique du 1er ordre, elle peut se décomposer en deux relations, l'une pour la partie isotrope, l'autre pour la partie déviatoire; en fait, pour les sols, un déviateur de contrainte modifie souvent le volume de l'échantillon. Cette décomposition n'est pas possible pour une viscoélasticité plus générale car, dans ce cas, la déformation finale dépend du chemin suivi par la contrainte et de la loi du temps sur ce chemin: elle sera donc différente selon qu'on applique d'abord la partie isotrope et ensuite la partie déviatoire, ou bien si l'on procède de façon inverse.

La viscoélasticité linéaire du 1er ordre s'écrit donc:

$$C_i + t_{0i}\overset{\circ}{C_1} = E_i(D_i + t_{1i}\overset{\circ}{D_i}),$$

$$C_d + t_{0d}\overset{\circ}{C_d} = E_d(D_d + t_{1d}\overset{\circ}{D_d}).$$

Le solide est défini par six coefficients qui sont des scalaires (et non des tenseurs) si la loi est isotrope. Les lettres E sont des modules d'élasticité rapportés à la forme définitive. Les t ont la dimension d'un temps; ce sont respectivement les temps de réponse des contraintes («temps de relaxation») ou des déformations («temps de fluage»). Ces coefficients peuvent être aisément obtenus avec l'appareil triaxial, à l'aide des deux essais suivants:

1. Essai isotrope ($\varepsilon_x = \varepsilon_y = \varepsilon_z$; $\sigma_x = \sigma_y = \sigma_z$,

$$\sigma_z + t_{0i}\mathring{\sigma}_z = E_i(\varepsilon_z + t_{1i}\mathring{\varepsilon}_z).$$

2. Essai déviatoire, tel que

$$\sigma_{xd} = \sigma_{yd} = -\sigma, \quad \sigma_{zd} = 2\sigma,$$

$$\sigma_{zd} + t_{0d}\,\sigma_{zd} = E_d(\varepsilon_z + t_{1d}\,\mathring{\varepsilon}_z).$$

Les essais de fluage classiques à charge constante sont délicats, car il est difficile de mesurer avec précision la déformation instantanée,

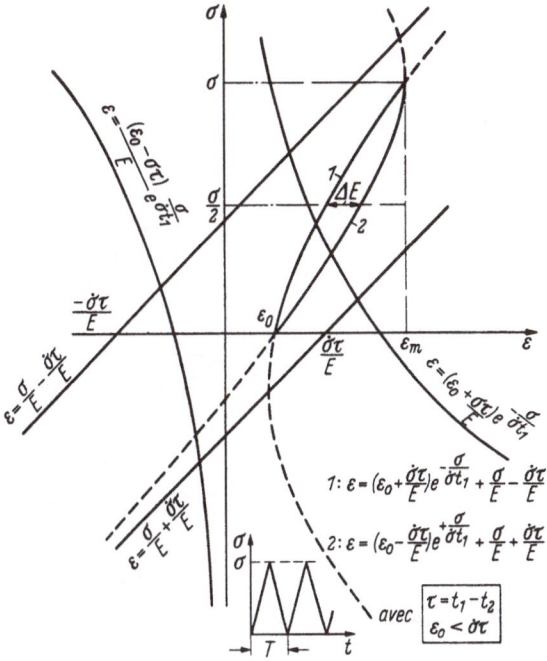

Fig. 2. Cycle limite théorique.

de même que la déformation finale qui nécessite, en outre, un temps parfois trop long, pendant lequel il faut éviter les perturbations extérieures. Nous avons préféré soumettre l'échantillon à des cycles de charge variant linéairement en fonction du temps, de période T. On obtient rapidement un cycle limite de déformation, formé par une partie de chaque courbe suivante, provenant de l'intégration des équations précédentes.

$$\varepsilon = \left[\varepsilon_0 - \frac{\mathring{\sigma}}{E}(t_1 - t_0)\right] e^{-\frac{\sigma}{\mathring{\sigma}t_1}} + \frac{\sigma}{E} + \frac{\mathring{\sigma}}{E}(t_1 - t_0) \quad \text{pour la montée,}$$

$$\varepsilon = \left[\varepsilon_0 - \frac{\mathring{\sigma}}{E}(t_1 - t_0)\right] e^{\frac{\sigma}{\mathring{\sigma}t_1}} + \frac{\sigma}{E} + \frac{\mathring{\sigma}}{E}(t_1 - t_0) \quad \text{pour la descente}$$

$$\text{(fig. 2, [2]).}$$

Il suffit alors de mesurer les déformations minimale et maximale du cycle ε_0 et ε_m, et la largeur de la boucle $\Delta\varepsilon$ pour $\frac{\sigma}{2}$.

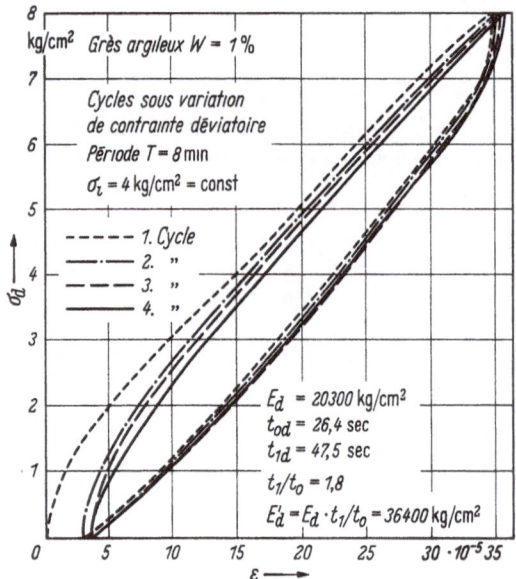

$$E = \frac{\sigma}{\varepsilon_0 + \varepsilon_m},$$

$$t_1 = \frac{T}{4X},$$

$$t_0 = t_1 - \frac{\varepsilon_0 T}{2(\varepsilon_0 + \varepsilon_m)\, th\, X},$$

$$\text{avec } x = Arg\,ch\,\frac{4 + \left(\frac{\Delta\varepsilon}{\varepsilon_0}\right)^2}{4 - \left(\frac{\Delta\varepsilon}{\varepsilon_0}\right)^2}.$$

On trouvera, sur les figs. 3 et 4, le résultat des mesures sur le grès. Le cycle isotrope présente la courbure habituelle des milieux pulvérulents (élasticité non linéaire).

Fig. 3.

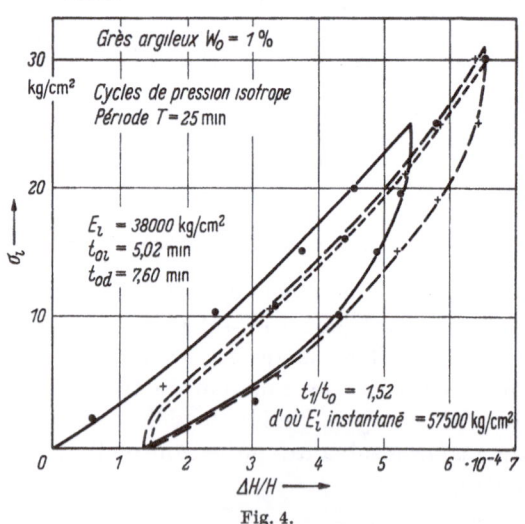

Fig. 4.

Cas Particuliers

1. Dans le cas où $t_{1i} = t_{1d}$ (temps de fluage isotrope et déviatoire égaux), on peut montrer [2] que l'essai sous contrainte monoaxiale

donne une loi différentielle linéaire entre σ_z, et ε_δ au ε_x:

$$\sigma_z + t_0 \mathring{\sigma}_z = E(\varepsilon_x + t_1 \mathring{\varepsilon}_x),$$

$$\sigma_z + t_0 \mathring{\sigma}_z = E(\varepsilon_z + t_1 \mathring{\varepsilon}_z).$$

Les coefficients E, t_0, t_1 sont des combinaisons linéaires des coefficients isotropes et déviatoires:

Déformation latérale (ε_x), déformation axiale (ε_z)

$$E = \frac{3 E_i E_d}{E_d - E_i}, \qquad E = \frac{3 E_i E_d}{2 E_i + E}$$

$$t_0 = \frac{E_d t_{0i} - E_i t_{0d}}{E_d - E_i}, \qquad t_0 = \frac{2 E_i t_{0d} + E_d t_{0i}}{2 E_i + E_d}$$

$$(\text{et} \quad t_1 = t_{1i} = t_{1d}).$$

2. Dans le cas où $t_{0i} = t_{0d}$ (c'est-à-dire temps de relaxation isotrope et déviatoire égaux), on montre que, pour un essai de déformation monoaxiale (oedomètre), la loi aura la forme précédente avec, cette fois:

Contraintes latérales:

$$E = \frac{E_i - E_d}{3},$$

$$t_1 = \frac{E_i t_{1i} - E_d t_{1d}}{E_i - E_d}.$$

Contraintes axiales:

$$E = \frac{2 E_d + E_i}{3} = (\lambda + 2\mu)$$

$$= \text{module oedométrique},$$

$$t_1 = \frac{2 E_d t_{1d} + E_i t_{1i}}{2 E_d + E_i}.$$

Dans le cas où $t_{0d} = t_{1d} = 0$, ou $t_{0i} = t_{1i} = 0$, c'est-à-dire le cas du matériau viscoélastique sous sollicitations uniquement isotropes ou déviatoires, la loi aura la forme précédente, en contrainte monoaxiale comme en déformation monoaxiale.

B. Domaine Irréversible. Plasticité

Une déformation irréversible (écrouissage) modifie la loi rhéologique du matériau car elle modifie la structure de celui-ci. Comme pour beaucoup de solides, un déviateur de déformation irréversible D_d accroît la limite élastique pour les déviateurs ultérieurs sensiblement de même orientation que D_d, et la diminue pour des déviateurs d'orientation voisine de la direction orthogonale à D_d (effet BAUSCHINGER) [3], [4], [5].

Nous avons aussi montré que ce déviateur d'écrouissage D_d crée une anisotropie qui peut s'ajouter aux précédentes si D_d est petit, et même effacer les anisotropies antérieures si D_d est assez grand. Nous avons montré que cette anisotropie mécanique est liée à l'anisotropie de l'assemblage géométrique des particules en examinant l'orientation statistique des plans tangents entre particules, si celles-ci n'ont pas une nette anisotropie de forme [3], [5].

On note, entre autres choses, qu'un déviateur de déformation suffisamment grand peut créer des assemblages de particules qui ne sont pas stables sous état de contrainte isotrope. On observe, par exemple, au cours de cycles croissants de compression monoaxiale, qu'il se forme des boucles dans la courbe contrainte — déformation, dont la largeur croît en raison des déformations irréversibles qui se créent au voisinage de l'état de contrainte isotrope (fig. 5).

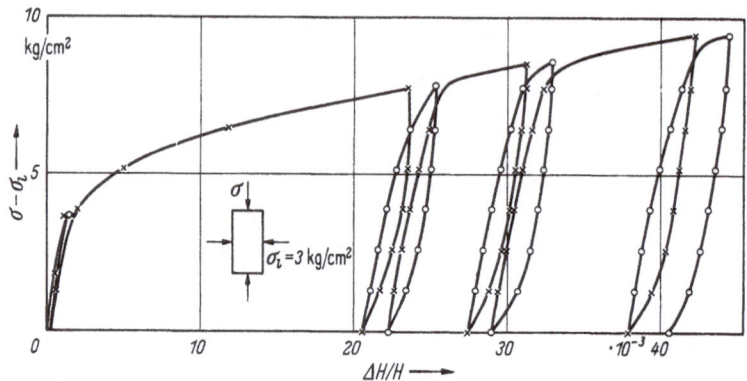

Fig. 5. Relations entre contraintes et déformations axiales. Écrouissage du sable. Échantillon de sable: $2R = 70$ mm, $H = 170$ mm.

A l'écrouissage anisotrope précédent, dû aux déviateurs de déformation, il faut ajouter, pour les sols, l'effet des déformations isotropes irréversibles qui accroissent la densité (compactage) et la limite maximale de plasticité («pic de la courbe contrainte — déformation»), mais ne modifient pas la limite de plasticité parfaite (palier final).

Les grandes déformations en plasticité parfaite, à contrainte moyenne constante, semblent se produire sans variation de volume (densité critique γ_c) (figs. 6 et 7) [3], [6]. Cette non variation de volume, correspondant au palier de plasticité parfaite, n'est visible que si la déformation est homogène; ceci nécessite la suppression du frettage aux extrémités de l'échantillon, en utilisant une fine feuille de caoutchouc posée sur graisse, ou mieux, un coussin d'air. Si l'on écrase des échantillons de densité différente en suivant un chemin analogue dans l'espace des contraintes, par exemple une compression monoaxiale, on voit que la densité critique est pratiquement indépendante de la densité initiale de l'échantillon (fig. 7). Par contre, cette densité croît fortement avec la contrainte moyenne de plasticité parfaite; elle peut être 1,5 ou 1,9 pour du sable, si la pression est 5 ou 80 kg/cm² (fig. 8). En second lieu, cette densité critique dépend du chemin parcouru dans l'espace des contraintes, elle est légérement plus forte pour une compression monoaxiale que pour un déviateur de contrainte (fig. 8); ceci peut se com-

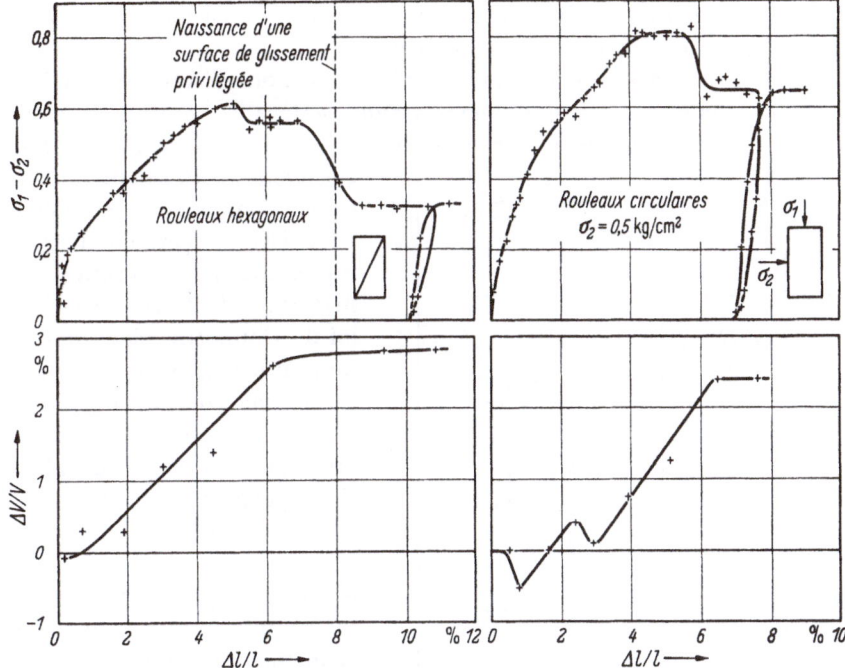

Fig. 6. Cisaillement biaxial sur rouleaux. Variation de volume. $\Delta v/v$.

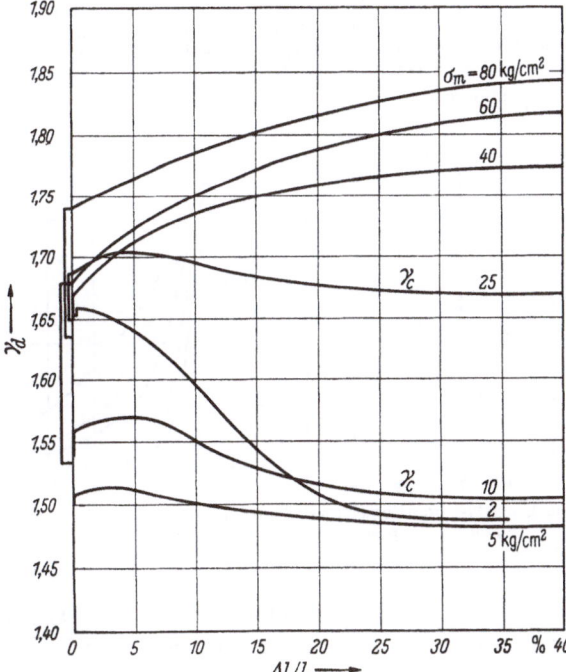

Fig. 7. Variation de densité pendant une déformation homogène. Essai triaxial. Contrainte moyenne constante σ_m; γ_d densité sèche; $\Delta l/l$ pourcentage de déformation verticale.

prendre si l'on sait que la déformation produit un écrasement des grains qui croît avec la pression moyenne (fig. 9) [6].

La loi de plasticité peut être différente si l'on effectue une déformation homogène, ou si la déformation se localise sur une surface privilégiée. Nous avons effectué des essais de compression monoaxiale sur des milieux à deux dimensions, composés de cylindres parallèles de sections diverses représentant les grains. Ce matériau obéit très bien à la loi de Coulomb. Si la déformation est homogène, on observe une courbe contrainte — déformation avec un maximum puis un palier de plasticité parfaite. Si la déformation cesse d'être homogène pour se poursuivre selon un plan privilégié, la résistance mécanique reste sensiblement la même si les cylindres sont de sections circulaires; par contre, elle diminue fortement pour des cylindres de sections hexagonales car ceux-ci s'orientent pour avoir une face

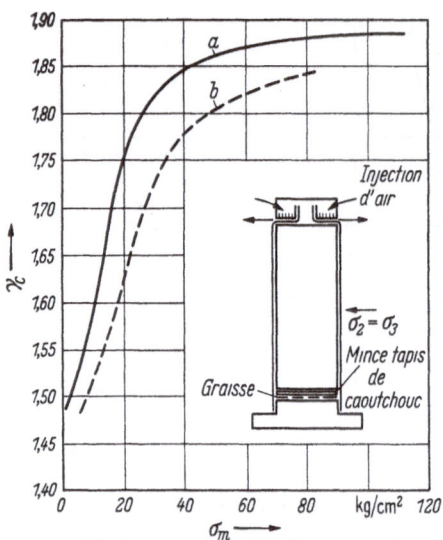

Fig. 8. Densité critique fonction de la pression moyenne. Rôle du chemin dans l'espace des contraintes. a Essais à pression latérale constante $\sigma_2 = \sigma_3$; b Essais à contrainte moyenne constante $\sigma_m = \dfrac{\sigma_1 + \sigma_2 - \sigma_3}{3}$.

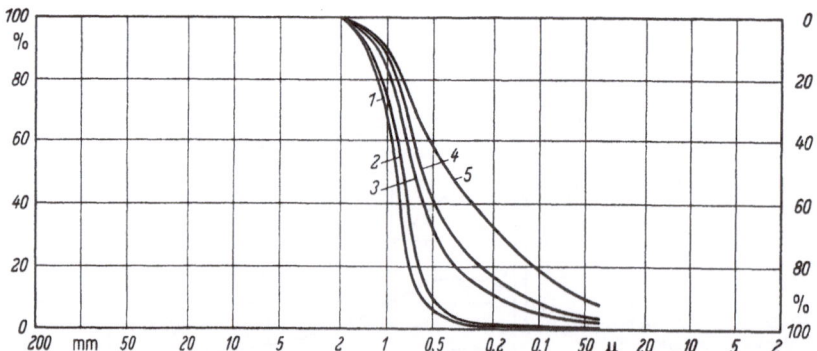

Fig. 9. Analyse granulométrique. Sables, essais triaxiaux à $\sigma_3' = $ const. 1 Avant essai et après essai à $\sigma_3 = 5$ kg/cm²; 2 Après essai à $\sigma_3 = 9$ kg/cm²; 3 Après essai à $\sigma_3 = 25$ kg/cm²; 4 Après essai à $\sigma_3 = 40$ kg/cm²; 5 Après essai à $\sigma_3 = 80$ kg/cm².

dans la surface de glissement (fig. 5) [3]. On obtient ainsi un frottement voisin de celui du métal composant les cylindres. Ce phénomène doit vraisemblablement s'observer quand les particules ont une forme nettement anisotrope dans leur majorité. Cette diminution de résistance peut contribuer à expliquer la formation de surfaces privilégiées pour la déformation. On peut rapprocher ceci de l' «angle de frottement résiduel» observé dans les argiles pour de grandes déformations qui permettent aux particules plates d'argile de s'orienter selon une surface privilégiée.

Références Bibliographiques

[1] ANGLES D'AURIAC, P.: Cours de rhéologie, Faculté des Sciences de Grenoble.

[2] PIERRARD, J. M.: Contribution à l'étude de la propagation d'ondes en milieux viscoélastiques. — Définition expérimentale de milieux rhéologiques. Thèse de Doctorat de Spécialité (3è Cycle de Mécanique des Sols), Faculté des Sciences de Grenoble, 1963.

[3] BIAREZ, J.: Contribution à l'étude des propriétés mécaniques des sols et des matériaux pulvérulents. Thèse de Doctorat ès-Sciences, Grenoble, 1961.

[4] BIAREZ, J.: Anisotropie mécanique et géométrique des milieux pulvérulents. Comptes-rendus du Quatrième Congrès International de Rhéologie, Providence (Brown University), Août 1963.

[5] BIAREZ, J., et K. WIENDIECK: Remarque sur l'élasticité et l'anisotropie des matériaux pulvérulents. Extrait des séances de l'Académie des Sciences, 9 Avril 1962, t. 254, pp. 2712 à 2714. La comparaison qualitative entre l'anisotropie mécanique et l'anisotropie de structure des milieux pulvérulents. Extrait des séances de l'Académie des Sciences, 4 Février 1963, t. 256, pp. 1217 à 1220.

[6] BELOT, A.: Remarques sur les propriétés des sols à relativement haute pression (100 kg/cm²).

Discussion

Question posée par S. IRMAY: The experimental results presented in the paper confirm a theoretical analysis of the HOHENEMER-PRAGER linear body by S. IRMAY [On the dynamic behaviour of linear rheological bodies. Quart. J. Appl. Math. Mech. 7, Pt. 4, 399—409 (1954)].

The equation of the deviator stress p may be written in several forms:
(i) That of the authors:

$$p + t_0\dot{p} = G(e + t_1\dot{e}) \tag{1}$$

t_0 = time of relaxation; t_1 = time of retardation; G = static modulus of elasticity (or asymptotic rigidity).
(ii) Or also:

$$e/t_0 + \dot{e} = p/m + \dot{p}/H \tag{2}$$

m = solid viscosity; H = dynamic modulus of elastic firmness.
(iii) introducing non-dimensional quantities: stress (P), time (θ), time-factor τ:

$$P = p/G; \quad \theta = t/t_1; \quad \tau = t_0/t_1 = G/H \tag{3}$$

we get the non-dimensional equation in $P(\theta)$ with a single parameter τ:

$$P + \tau \cdot dP/d\theta = e + de/d\theta. \tag{4}$$

This expresses a principle of similitude. Applying a periodical stress

$$p = p_1 \sin ft; \quad P = P_1 \sin \omega\theta \tag{5}$$

$f =$ frequency; $\omega = ft_1 =$ non-dimensional frequency, we get in the (p, e) plane a curve tending asymptotically towards a closed ellipse of inclination $\tan \alpha$:

$$\tan \alpha \div G \text{ for } \omega \to 0 \text{ (static loading)}, \tag{6}$$

$$\tan \alpha \div H = G/\tau \text{ for } \omega \to \infty \text{ (dynamic loading)}.$$

The power $W = p\dot{e}$ expended by the deviator stresses in the deformation of the unit volume of the body is stored by the elastic part, expended by the hysteresic and viscous parts. In the case of a periodical stress W consists of a periodical part, which vanishes in a cycle, and of a non-vanishing part:

$$W = 0{,}5(1 - \tau)\,\omega^2\,(1 + \omega^2)^{-1} \tag{7}$$

For $\tau < 1$, $W > 0$, and we have viscous dissipation.
For $\tau = 1$, $W = 0$, and there is no dissipation.
For $\tau > 1$, $W < 0$ which means one of the two:

(a) The body does not undergo any internal structural changes.
Then necessarily:

$$\tau \leq 1, \; W \geq 0, \; t_0 \leq t_1, \; G \leq H.$$

The time of relaxation does not exceed the time of retardation, the rigidity modulus G may not exceed the firmness modulus H.

(b) Often bodies, when submitted to periodical shear, are gradually destroyed and undergo internal structural and physico-chemical changes. This may be accompanied by the liberation of sufficient amounts of energy, and a "negative" dissipation may be allowed for. Then we may have:

$$\tau > 1, \; t_0 > t_1, \; G > H.$$

Such bodies may be called *endous* bodies, by analogy with endothermal chemical reactions. Some polyelectrolytes exhibit such properties.

Bodies of $\tau < 1$ may be called *exothermal* or *dissipative* bodies; while $\tau = 1$ represents *homothermal* bodies.

It is curious that $\tau = 0$ corresponds to Maxwell's firmo-viscous fluid ($\tau = 0$, $G = 0, t_1 \to \infty$), Kelvin's elastico-viscous solid ($\tau = 0$, $H \to \infty$, $t_0 = 0$) and Newton's viscous fluid. When $\tau \neq 0$, we have a solid, without any fluid properties.

It may be shown that Eq. (1) is equivalent to the following one:

$$e = g \cdot p(t) + \int^{t} F_0 \exp(-t'/t_1)\,p(t' - t)\,dt' \tag{8}$$

which consists of a perfectly elastic part and a delayed-elastic (or hysteresis part) without any fluid viscosity part.
Then:

$$\left. \begin{array}{l} g + F_0 t_1 = 1/G, \\ \tau = g/(g + F_0 t_1) \end{array} \right\} \tag{9}$$

$\tau = 1$ corresponds to $F_0 t_1 = 0$, i. e. either $F_0 = 0$ or $t_1 = 0$;

$\tau = 0$ corresponds to $g = 0$ or to $F_0 t_1 \to \infty$.

In order to take into consideration true fluid viscosity and a permanent strain, one should write:

$$d\varepsilon/dt + t_1 d^2 \varepsilon/dt^2 = m p + [g + (F_0 + m)t_1] \, dp/dt + g t_1 \, d^2 p/dt^2. \quad (10)$$

This corresponds to:

$$e = g p(t) + \int^t F_0 \exp(- t'/t_1) \, p(t' - t) \, dt' + \int^t \mu p(t') \, dt' \quad (11)$$

μ = true fluid viscosity.

It should be noted that the case $\tau > 1$ is consistent not only with deteriorating bodies, but also with energy being supplied by the work of isotropic stresses, when dissipative.

Réponse de J. M. PIERRARD: Oui, le 2e principe de la thermodynamique est bien vérifié. Nous avons bien $t_1/t_0 > 1$, ce qui exprime que l'énergie dissipée est toujours > 0 (c. à. d. toujours dissipée). On peut d'ailleurs le voir sur les formules.

De plus, nous avons laissé ces coefficients sous la forme «temps de réponse« pour la mesure de ceux-ci à partir du cycle de déformation *limite* obtenu sous des cycles de charge périodique (cf. les formules).

Question posée par J. POTTIER:

1. Les massifs utilisés (sable, grès) sont-ils complètement *saturés* en eau?

2. Pour suivre la consolidation du massif, n'est-il pas préférable d'utiliser le rapport des *vides*, ou *porosité* (nombre sans dimension) plutôt que la densité apparente?

3. Dans les montages expérimentaux, les contraintes sont assurées par des plaques planes. Celles-ci subissent-elles une force d'application fixe correspondant à une pression moyenne, ou a-t-on utilisé des dispositifs assurant une *pression locale uniforme* sur toute la plaque?

4. Un effort a-t-il été fait pour relier les résultats globaux des expériences sur sables avec une analyse des phénomènes élémentaires, comme par exemple le nombre des contacts de grains et les frottements entre grains.

Réponse de J. BIAREZ: Les échantillons de grès ne sont que partiellement saturés afin d'étudier la viscoélasticité de l'ossature.

Les calculs que nous effectuons font intervenir le poids volumique; c'est pourquoi nous utilisons souvent cette donnée. Nous nous intéressons ici, habituellement, à l'influence de la pression moyenne sur la ,,densité critique", expression que l'on utilise souvent en Mécanique des Sols; on pourrait évidemment utiliser d'autres variables fonction de celle-ci.

Les extrémités de l'échantillon reposent sur des plaques métalliques par l'intermédiaire d'une feuille de caoutchouc sur couche de graisse ou coussin d'air.

Depuis plusieurs années, nos recherches portent effectivement sur les relations entre la géométrie de l'ossature du sol et les propriétés du milieu continu fictif.

3.7 Yield Stress and Modulus of Elasticity of Soil

By

Fusayoshi Kawakami and Shoji Ogawa

Summary

This paper presents the yield stress σ_y and the modulus of elasticity E of the soils subjected to repeated stress application. Unconfined and triaxial compression tests were conducted on the compacted specimens of silty loam, clay and three kinds of clay-sand mixture by means of stress control after the application of repeated loading. The data were analyzed rheologically by choice of a simple mechanical model. Then the curves of rate of strain increasing versus stress are drawn, and bending points are found in every curve. As these points are considered as the beginning points of slip occurring in the specimen, they indicate the yield points of soils when repeated stresses are applied. From these curves, the relation between yield stress and frequency or magnitude of repeated loading and confined stress is obtained. From the ordinate of $A-C$ line in Fig. 6 ($\dot{\sigma}/E$), the modulus of elasticity of soil subjected to repeated stress is determined, then the relations between yield stress or modulus of elasticity and frequency, magnitude of repeated stress and confined stress were obtained. Finally the conclusions are drawn with regard to relation between yield stress or modulus of elasticity and frequency or magnitude of repeated stress and confined stress, and hardening effect due to repeated stress application. Importance of elastic and plastic behavior of soil when the earth structures are designed is written.

Introduction

Recently, the investigation on mechanical properties of soil subjected to the transient load or repeated load has grown more important relating to the construction of road-beds of highway etc. It is known that A. CASAGRANDE, H. B. SEED and others have conducted these investigation. Nowadays, the road-beds are designed practically on the basis of the ultimate strength or bearing capacity of material without considering the yield behavior, elastic and plastic properties of soils, which are considered to govern the properties of road-beds.

From these points of view, the writers conducted also a study on the properties of compacted soils subjected to the repeated loading.

The results of tests were analyzed rheologically and the elastic and plastic properties of soils which are subjected to repeated loading and the factors which affect the mechanical property of soil are made clear.

Rheological Model

As a rheological model of a soil, which is not ruptured, a mechanical model consisting of VOIGT, MAXWELL and BINGHAM is chosen, as shown in Fig. 1 in which E designates modulus of elasticity, η_1, η_2 etc. are coefficients of viscosity, and σ_y are yield stresses.

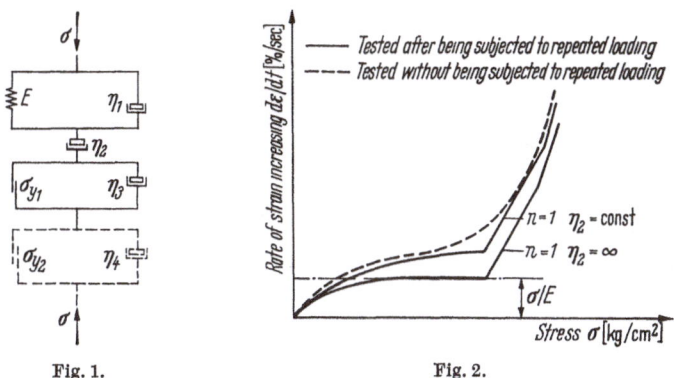

Fig. 1.   Fig. 2.

When the stress $\sigma = \dot{\sigma}t$ is applied on the model in Fig. 1, the strain is given by

$$\varepsilon = \frac{\dot{\sigma}}{E}\left(\frac{\sigma}{\dot{\sigma}} - \frac{\eta_1}{E}\right) + \frac{\eta_1\dot{\sigma}}{E^2}\, e^{-\frac{E\sigma}{\eta_1\dot{\sigma}}} + \frac{\sigma^2}{2\,\eta_2\dot{\sigma}} + \left[\frac{1}{\eta_3\dot{\sigma}(n+1)}(\sigma - \sigma_{y1})^{n+1}\right]$$
$$+ \left[\frac{1}{\eta_4\dot{\sigma}(n+1)}(\sigma - \sigma_{y2})^{n+1}\right] + \cdots \tag{1}$$

in which $\dot{\sigma} = \dfrac{d\sigma}{dt}$. Differentiating Eq. (1) by t and substituting $t = \dfrac{\sigma}{\dot{\sigma}}$, the rate of increase in strain is given by

$$\frac{d\varepsilon}{dt} = \frac{\dot{\sigma}}{E}\left(1 - e^{-\frac{\sigma E}{\eta_1\dot{\sigma}}}\right) + \frac{\sigma}{\eta_2} + \left[\frac{1}{\sigma_3}(\sigma - \sigma_{y1})^n\right] + \left[\frac{1}{\eta_4}(\sigma - \sigma_{y2})^n\right] + \cdots \tag{2}$$

Then the relations of $d\varepsilon/dt - \sigma$ are drawn as shown in Fig. 2, in which the bending points are found on the curve.

Apparatus

Two kinds of repeated loading devices, unconfined compression type and triaxial compression type, were used in the investigation.

23*

Fig. 3 shows schematically the unconfined repeated loading apparatus. The weight *3* suspended by a wire is connected to a motor *8*. The load is applied to the lever *4* periodically by the weight and transmitted to the specimen *1* through the support *6* and frame *7*. The lever is controlled to be always horizontal, and equilibrium of system of stress application was maintained.

The triaxial repeated loading apparatus, which consists of ordinary triaxial compression device and regulating device of axial stress, can

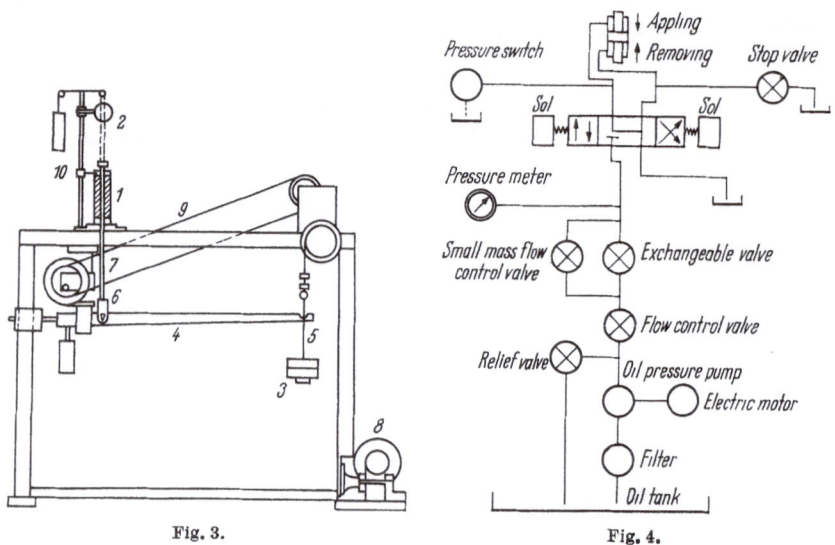

Fig. 3.   Fig. 4.

apply a repeated axial stress of a desired magnitude to a specimen which has been subjected to a constant confined stress in the triaxial cell by a controlling oil circuit, as shown in Fig. 4. By adjusting the microswitch, any desired combination of loading and release terms can be applied. By electrical signal from the time unit, the solenoid valve open to admit compressed oil, the piston is acted by oil through the flow control valve and exchangeable valve, and the axial stress is applied.

In unconfined repeated loading device, the cycles of loading term of 0.75 sec. can be applied on the specimen periodically after every interval of 0.85 sec.

The deformation of specimens is measured with the dialgage.

Test Procedure

The samples used for the investigation were silty loam (symbolized A, $W_P = 26$, $W_L = 68$), clay (symbolized B, $W_P = 33$ $W_L = 39$), and three kinds of clay-sand mixture (symbolized C, D and E, $W_P = 27$

$W_L = 34$, $W_P = 23$ $W_L = 32$ and $W_P = 19$ $W_L = 25$). The samples were prepared to have the desired water content, and compacted in a 5 cm diameter mold of a height of 12.6 cm with a Harvard Miniature Compactor. The specimens were sealed with paraffin prior to testing. Water contents, dry densities and unconfined compression strengths of specimens are shown in Table 1.

Table 1

Specimen	Type of samples	Water content (%)	Dry density (g/cm³)	Unconfined compression strength (kg/cm²)
1	A	30.2	1.43	1.64
2	A	27.3	1.45	2.37
3	A	24.3	1.45	3.39
4	A	21.8	1.45	3.94
5	A	27.3	1.50	3.04
6	A	27.3	1.48	2.82
7	A	27.3	1.43	2.02
8	B	32.0	1.40	4.52
9	C	27.8	1.45	3.60
10	D	25.5	1.47	2.52
11	E	20.5	1.64	2.35

Every kinds of specimens were applied 5000, 10000, 50000 and 100000 repetitions of five to seven grades of stresses, and some specimens were subjected to repeated stress of 5 to 1000 applications to investigate the behavior in the initial stage of repeated loading.

The unconfined compression tests were conducted by means of stress control. The rate of stress increase was $0.017 - 0.018$ kg/cm²/sec.

Test Result

1. The Relation between Stress and Strain

Some relations between stress and strain in unconfined compression tests of specimens subjected to repeated stress application are shown in Fig. 5. At the lower strain, these specimen showed rather elastic behavior and have liner relations between stress and strain.

Tangents of these relation curves got higher with the

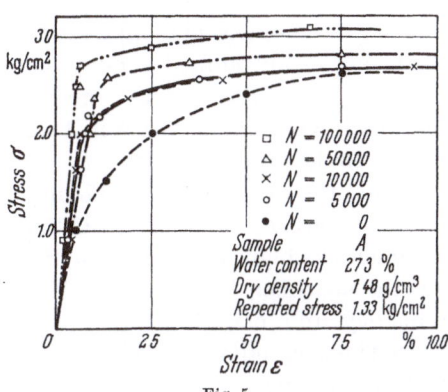

Fig. 5.

number of repetition. But at the higher strain, specimen showed plastic behavior until they ruptured. Ultimate strength increased with the frequency and the magnitude of repeated stress, but the effect of the repeated stress on the ultimate strength was insignificant. The rupture of the compressed specimens occured in the phase of brittle failure and sheared planes could be observed clearly.

Therefore it is proper that the elasto-plastic properties or yield stress should be more important than ultimate strength.

2. The Relation between Stress and the Rate of Strain Increasing

An example of the relation between the rate of increase in strain $d\varepsilon/dt$ and the axial stress σ in the unconfined compression test is shown in Fig. 6.

The outstanding bending points appeared in the curve. As these points (Point B and D) are the beginning points of slip occurring in the rheological model, they correspond to the yield stresses of the specimen. In some specimens, two or three bending points were found. It is said that this behavior of rupture in the specimen had been attributed to the step-strain phenomenon which followed the yield of the specimen.

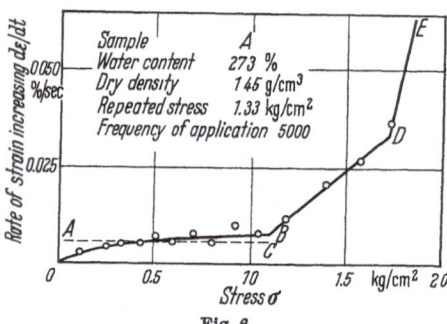

Fig. 6.

The lowest bending point corresponds to initial yield of the specimen in the rheological model. Other bending points can be considered by adding the slider-dashpot system in series.

The ordinate of $A-C$ line in initial stage of the stress versus the rate of strain increasing is expressed as $\dot{\sigma}/E$ in which $\dot{\sigma}$ is rate the of stress increasing in unconfined compression test, and the deviation of $O-B$ line from $A-C$ line and the inclination of lines $B-D$, $D-E$ etc. are due to the viscosity of the soil. From these, the value of E, η_1, η_2 etc. and σ_y could be determined.

When the frequency and the magnitude of repeated stress are small in Fig. 6, the deviation of $A-C$ line from $O-B$ line is larger and the specimen has high plasticity, but the deviation gets progressively small and the specimen has elasticity like a solid body with the increase of frequency and magnitude of stress.

In the triaxial tests, the relation between stress and the rate of strain increasing is similar to that in the unconddfined compression test and also has elasticity as long as the confined stress is low, but the effects of repetation of stress are lost as the confined stress is high.

From Fig. 6, it is easy to determine the yield stress σ_y, modulus of elasticity E and viscosity (η_1, η_2 etc.)

Consideration

1. Effect of Frequency and Magnitude of Repeated Stress Application on Yield Stress

When the magnitude of repeated stress is lower than the fatigue limit, the yield stress increases with the frequency and magnitude of stress application as shown in Fig. 7. This tendency is found on any specimen of various water contents, dry densities and soil types. The yield stress gains the ultimate value of 5000 to 100000 repetations of stress, when the repeated stress is higher, but it requires more numbers of repetation to gain the ultimate value, when the repeated stress is lower. In general, yield stress (σ_y) is given as a following function of the frequency (N).

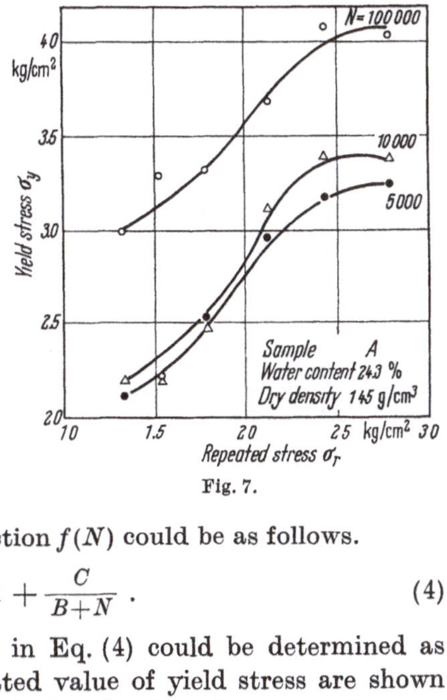

Fig. 7.

$$\sigma_y = e^{f(N)}. \qquad (3)$$

Assuming that the form of function $f(N)$ could be as follows.

$$f(N) = A + \frac{C}{B+N}. \qquad (4)$$

The value of each coefficients in Eq. (4) could be determined as shown in Table 2, and the calculated value of yield stress are shown with a dotted line in Fig. 8. The calculated values show a good agreement with the results of experiment.

Table 2

Repeated stress kg/cm²	Coefficient			σ_y, $N = 0$ kg/cm²
	A	B	C	
1.25	0.64	2.54×10^3	-1.98×10^3	0.87
1.03	0.54	2.89×10^3	-1.76×10^3	0.93
0.85	0.32	3.45×10^3	-1.58×10^3	0.87

Substituting $N = 0$, in the equation, the yield stress of specimen which has not been subjected to repeated stress can be estimated as showed in Table 2.

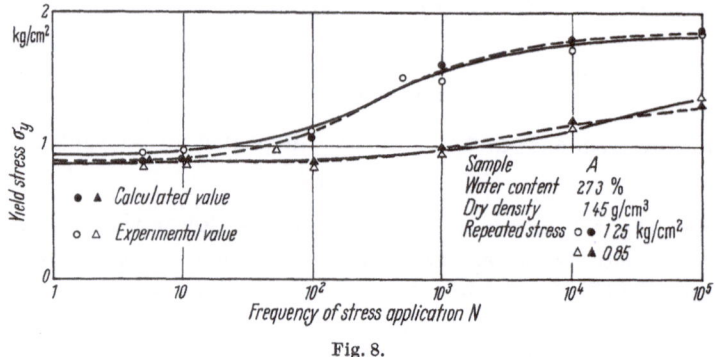

Fig. 8.

2. Effect of Frequency and Magnitude of Repeated Stress Application on Modulus of Elasticity

Fig. 9 and 10 show the relations between modulus of elasticity and magnitude or frequency of repeated stress. When the repeated stress was lower, modulus of elasticity increased with the frequency of stress application, but when the repeated stress was higher, it rather decreased. This tendency was appeared on the other kinds of specimens.

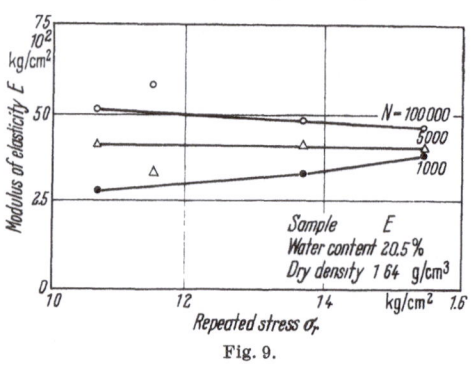

Fig. 9.

From these, it is said that there should be a critical stress to increase the modulus of elasticity, and it increases with the frequency of stress application, when the specimen is subjected to lower stress than the critical stress. But when the specimen is subjected to higher stress than that, modulus of elasticity gains the maximum va-

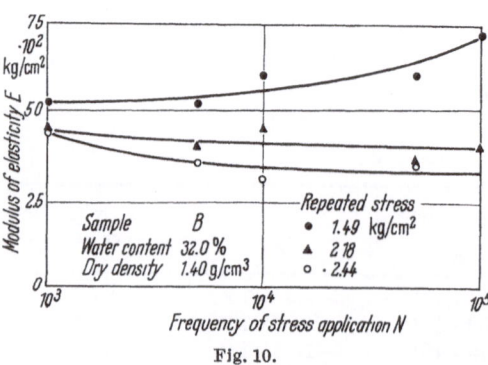

Fig. 10.

lue at a certain frequency of stress application and at the frequencies which are lower or higher than this frequency, the modulus of elasticity rather decreases.

3. Effects of Confined Stress on Yield Stress and Modulus of Elasticity

When the specimen had been subjected to the higher confined stress in the repeated loading test, the internal stress was occurred in the specimen and resisted to the repeated stress, thus it was appeared as if the apparent hardening had occurred in the specimen, and removing the confined stress, the effect of apparent hardening was lost and yield stress and modulus of elasticity in the unconfined compression test decreased with the magnitude of confined stress in the repeated loading test as shown in Figs. 11 a,

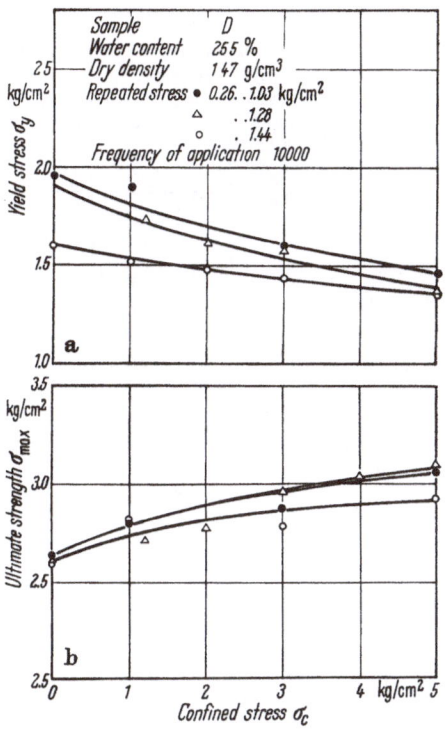

Fig. 11 a and b.

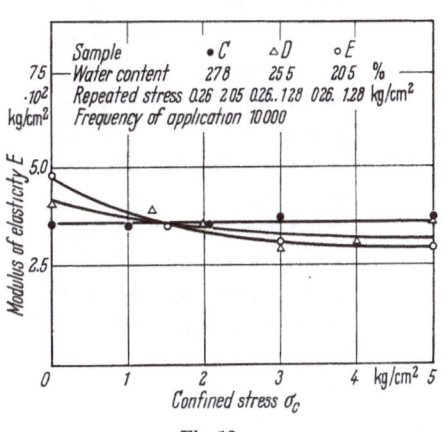

Fig. 12.

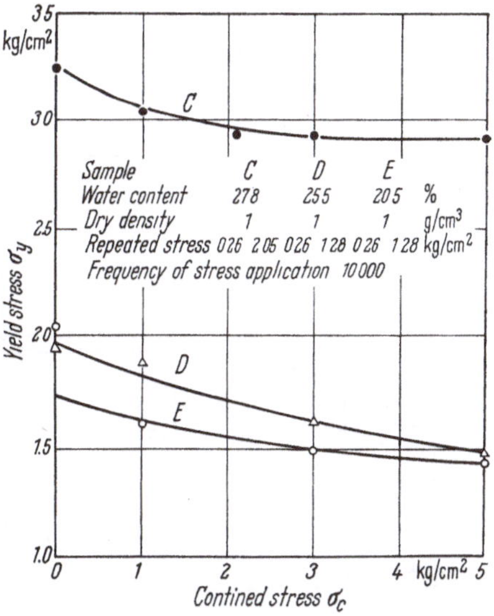

Fig. 13.

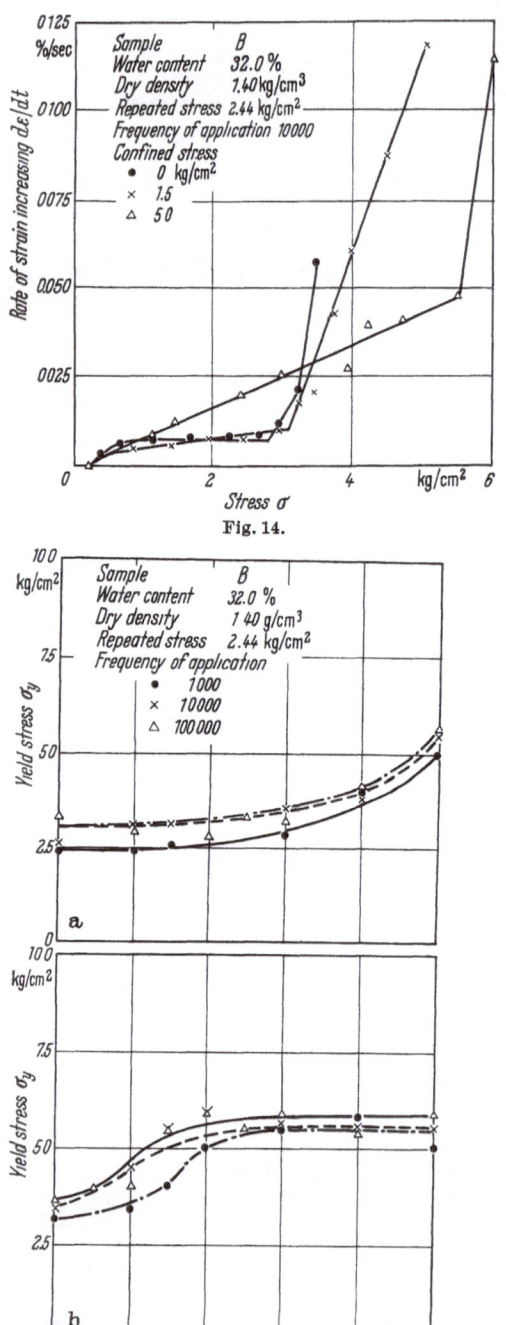

Fig. 14.

Fig. 15 a and b.

but the ultimate strength increased reversely (see Fig. 11 b).

The hardening effect of soil due to application of repeated stress was more outstanding when the confined stress was lower. This phenomenon was appeared more sensitively on the sandy soil than clayey soil. Thus the variance of yield stress and modulus of elasticity due to the confined stress in repeated loading test are higher on the sandy soil (see Figs. 12 and 13).

4. Effects of the Confined Stress in the Compression Test on the Yield Stress

When the confined stresses were lower, the specimen behaved like a elastic body by a low axial stress, but it behaved like a plastic body as the axial stress exceeded the yield stress, and culminated in failure. So the initial yield point and failure yield point could be distinguished. When the confined stress was higher, elastic and plastic behaviors could not be distinguished even at the initial stage, owing to losing the hardening effect occured by repeated stress application, and initial yield point might be coincided with the failure yield point (see Fig. 14).

The relation between the initial yield stress or failure yield stress and the confined stress in this case are shown in Fig. 15a and b and the modulus of elasticity could not be determined by rheologically.

Conclusion

Summing up the above, conclusions can be drawn as follows.

1. When the specimen was subjected to repeated stress, the elastic and plastic behaviors of specimen could be distinguished using a mechanical model as shown in Fig. 1 and the yield stress and modulus of elasticity could be determined rheologically. But when the confined stress in triaxial compression test is higher, the hardening effect due to repeated stress application is lost and it is very difficult to distinguish the elastic and plastic properties of the specimen.

2. When the magnitude of repeated stress is lower than the fatigue limit, the yield stress increases with the repeated stress, and also increases with frequency of repetition until it gains the ultimate value.

3. In general, yield stress is expressed as a function of frequency N of repeated stress and in the form as follow:

$$\sigma_y = e^{f(N)}.$$

Substituting $N = 0$, into the above equation, yield stress of a soil which has not been subjected to repeated stress application could be estimated. The relation between frequency of repeated stress and yield stress can be approximated as follow:

$$f(N) = A + \frac{C}{B+N}.$$

4. Removing the confined stress in the repeated loading test, the apparent hardening effects are lost, and yield stress and modulus of elasticity decrease with the magnitude of confined stress, but ultimate strength increases with it. The same relation was appeared on the specimen of different water contents, dry densities and other types of soil.

Effect of hardening due to the confined stress is more outstanding on the sandy soil than clayey soil.

5. The initial yield stress increases with the confined stress in triaxial compression test, and the failure yield stress also increases when the confined stress is lower, but it gets a constant value at the higher confined stress.

6. The elastic and plastic properties or yield stress of soil should be more important than ultimate strength in the compression test in the designing of the earth structures, such as road-beds.

3.8 On the Uniqueness of Yield Surfaces for Wet Clays

By

K. H. Roscoe and A. Thurairajah

Summary

The experimental evidence in support of recent theories involving the use of a unique yield surface for saturated samples of a normally consolidated remoulded clay is examined for three clays. The yield surfaces are restricted to a three dimensional space with axes of voids ratio and two observed (or corrected) stress variables. The corrections that have so far been made refer to various types of energy correction. The available data is from tests in the triaxial apparatus and in the simple shear apparatus. From the triaxial data on Kaolin it appears that different yield surfaces are obtained for each type of test (e.g. undrained or drained, etc.) but only one yield surface is obtained for undrained and drained tests in the simple shear apparatus. It is suggested that this discrepancy is due to the erroneous assumption that the dilatation in a triaxial sample is uniform. The results of partially drained tests and anisotropic consolidation tests confirm this hypothesis. It is concluded that there is probably one unique yield surface and critical state line for a clay under conditions of axial symmetry or plane strain.

1. Introduction

During the recent investigations carried out at Cambridge into the problem of stress-strain relationships of soils, and similar granular media, every effort has been made to reduce the problem to its simplest possible form and to subject any theoretical concepts to experimental test. For example attention has been confined to conditions of axial symmetry or of plane strain, thereby reducing the requisite stress and strain parameters to a minimum. It has been assumed that soils are isotropic, that dilatancy can be accounted for by changes of voids ratio e and that this parameter e together with two parameters of stress can be used to define the state of a sample. On the experimental side, testing of cohesive soils has so far been restricted to saturated remoulded samples of normally consolidated or lightly over-consolidated (wet) clays. The reason being that they tend to deform more uniformly than heavily over-consolidated (dry) clays throughout any drained test which is continued until the sample reaches the critical state. A further advantage in the use of normally consolidated clays is that they begin to yield (i.e. undergo irrecoverable deformation) as soon as they are subjected to any increase, however small, of shear stress. It has been further assumed that samples of such clays are homogeneous under all stages of test in conventional triaxial compression tests (axial symmetry) and in plane strain tests in the simple shear apparatus (hereafter referred to as the S.S.A.). This assumption is inherent in the method of measuring changes of voids ratio by observing, directly or indirectly, movements of the boundaries of the sample.

One of the main problems has been to see if it is possible to correlate the results of tests in which the samples have been subjected to various drainage facilities. The work of HVORSLEV (1937) suggested that some coordination might be achieved by defining the state of a sample in terms of the voids ratio and two observed stress variables. The question then arises, is there a unique yield (or state boundary) surface in such a space for all types of state path in triaxial compression, or in S.S.A., tests for saturated samples of a given normally consolidated remoulded clay? If not, can one be found in another space which may be determined by modifying, or adding to, the selected observed variables? If one such surface could be found in any such space then it would provide a key with which a start could be made to tackling the problem of the stress-strain relationships of soil. It was thought that the simplest key would correspond to a yield surface based upon observed rather than modified values of the stress variables. POOROOSHASB and ROSCOE (1961) have concluded from the triaxial data kindly provided by Imperial College, London that Weald Clay has one such surface. On the other hand it appears from tests carried out at Cambridge that Spestone Kaolin has a number of yield surfaces according to observed uncorrected stresses in triaxial tests but only one from S.S.A. tests. ROSCOE and POOROOSHASB (1963) assumed that the differences between these "uncorrected triaxial" yield surfaces for the Kaolin were small. Later ROSCOE, SCHOFIELD and THURAIRAJAH (1963a) assumed that a unique „triaxial" yield surface existed for the Kaolin in a similar space but in which the observed values of the deviatoric stress were modified by application of elastic and boundary energy corrections. Experimental results were quoted which suggested that the divergence between the observations and the assumptions was of the order of 5 to 10%. When these assumptions were made the problem became amenable to simple mathematics, and the soil properties could be characterised by only four fundamental constants. The resulting picture of the behaviour of the soil is, of course, strictly only applicable within the limitations that have been discussed above and may represent a gross over-simplification. As further reliable information becomes available this simple picture should be continually adapted and each change justified by an increase in the accuracy of prediction of the behaviour of soil under stress. Finally it may be possible to introduce more complicated stress patterns and soil properties, preferably one at a time, under controlled conditions.

In this paper some of the anomolies regarding the uniqueness of yield surfaces will be examined: particular attention will be given to those based upon uncorrected observed stresses in triaxial compression, and S.S.A., tests.

2. Preliminary Definitions

2.1 Stress Parameters

The stress parameters used under conditions of axial symmetry were the mean normal stress $p = \frac{1}{3}(\sigma_1' + 2\sigma_3')$, and the deviator stress $q = (\sigma_1' - \sigma_3')$, where σ_1' and σ_3' are respectively the major and minor principal compressive effective stresses. For plane strain the mean shear stress τ and the mean effective normal stress σ' on horizontal planes in the S.S.A. were used and it may be shown by a method similar to that suggested by Arthur, James and Roscoe (1964) that these stresses are within less than 2% of the mean value of the maximum shear stress τ_{max} and the mean normal stress $\frac{1}{2}(\sigma_1' + \sigma_3')$ in the S. S. A. for all the tests described below.

For the triaxial data an additional stress parameter will be considered namely q_w which is obtained from the observed deviatoric stress q by application of a boundary energy correction and an elastic energy correction (see Roscoe, Schofield and Thurairajah (1963a) and Roscoe and Schofield (1963)).

2.2 The Three Dimensional State Boundary Surfaces

When working in terms of the observed stresses the state of a sample under conditions of axial symmetry will be represented by a point in (p, q, e) space. Any change of state will be represented by a state path in this space and the projection of a state path on the (p, q) plane represents the associated stress path. In plane strain the state will be defined by a point in (σ', τ, e) space. Typical (p, q, e), (σ', τ, e) and (p, q_w, e) surfaces will be found in the references cited above.

2.3 The Two Dimensional Plot of the Yield Surface

In triaxial tests on any one of the three clays listed in Table 1 the state paths of all undrained tests are geometrically similar in (p, q, e)

space and all such paths begin on the normal consolidation line and end on the critical state line. Since for these lines q and p are linearly related to $\exp\left(\dfrac{-e}{\lambda}\right)$, where "exp" is the exponential and λ is a soil constant, the undrained yield surface will be represented by a single curve in a two dimensional diagram with $p \exp\left(\dfrac{e}{\lambda}\right)$ as abscissa and $\dfrac{q}{p}$ as ordinate. These axes are very similar to those first used by POOROOSHASB and ROSCOE (1961). In a similar manner the undrained yield surface from S.S.A. tests may be reduced to two dimensions with $\sigma' \exp\left(\dfrac{e}{\lambda}\right)$ as abscissa and $\dfrac{\tau}{\sigma'}$ as ordinate. In both types of two dimensional plot the whole of the normal consolidation line and of the critical state line will be represented by single points (see for example N and x_1 respectively in Fig. 2).

Attention has so far been confined, in this section 2.3, to state paths of undrained tests but the arguments outlined above refer also to any yield surface obtained from any series of tests with geometrically similar stress paths.

3. Materials and Modes of Sample Preparation

3.1 The Materials

The particle size distribution curves of the Weald Clay, Spestone Kaolin and Cambridge Gault Clay are shown in Fig. 1. The classification properties are shown in Table 1.

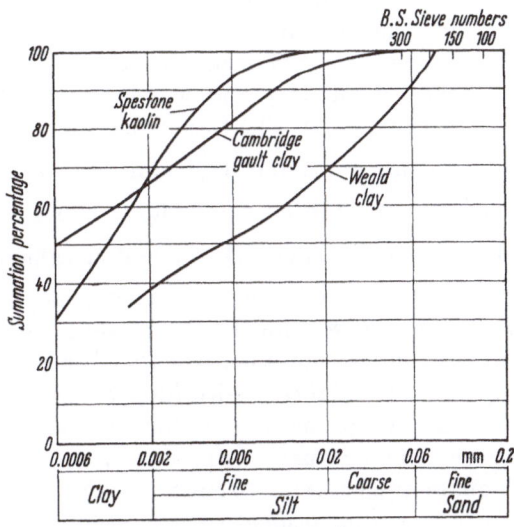

Fig. 1. Particle size distribution curves.

Table 1. *Classification Test Properties*

Material	Liquid Limit %	Plastic Limit %	Activity	Specific Gravity
Weald Clay	43	18	0.62	2.74
Spestone Kaolin	70	40	0.44	2.61
Cambridge Gault Clay	85	30	0.83	2.75

The Kaolin consisted of 99% pure kaolinite mineral.

3.2 Sample Preparation

The mode of preparation of the Weald Clay samples for triaxial tests has been described by Henkel (1956) and some of the Cambridge Gault Clay samples were prepared in a somewhat similar manner. For the latter an "undisturbed sample" at a natural moisture content of 42% was remoulded with palette knives while deaired distilled water was added until a moisture content of 55% was attained. This was the wettest consistency at which the clay could be easily handled. After mixing, a lump of the clay (about 1 c.c.) was taken on a palette knife and pressed against a metal piston in a lightly oiled brass tube of $1^1/_2$ inches diameter. The piston was withdrawn a little and another lump was pressed on top of the first. This process was repeated until a sample 3 inches long could be cut from the clay when extruded from the tube. The sample was then set up in the triaxial cell with filter drains and porous stones at the ends, and isotropically consolidated for four days prior to test.

The Kaolin samples and some of the Cambridge Gault samples were prepared in a different manner, which it was believed would considerably reduce the possibility of air inclusions in the samples. For the Kaolin a known weight of oven dry material was subjected to a vacuum of 20 inches of mercury and was gradually saturated from below by deaired distilled water. The Kaolin and water were then mixed under the same vacuum for about $1\frac{1}{2}$ hours in a special mixer (see Thurairajah (1961)) and the resulting slurry had a voids ratio of 4 (about 2.2 times the liquid limit). Triaxial test specimens $\left(1\frac{1}{2}\text{in. dia.}\right)$ were then obtained by consolidating the slurry one-dimensionally in a split tube former which was lined with a rubber sleeve 0.01 inches thick. The split tube former and liner was then removed and the samples were set up in the triaxial cells with filter paper drains and porous stones at the ends and were consolidated further under isotropic stress.

A similar procedure was used to prepare Kaolin samples for the S.S.A. but a square tube former (6 cm. × 6 cm.) was used without a sleeve lining. After one dimensional consolidation in this former to a pressure of about 10 p.s.i. less than that used in the S.S.A. test the specimen was extruded from the former, cut to the desired height, and assembled in the S.S.A. where it was further consolidated for about 15 hours prior to shear.

A few of the Cambridge Gault triaxial samples were prepared by taking an "undisturbed" sample at a natural moisture content of 42% and remoulding with palette knives while adding deaired distilled water until the moisture content was about 110% (L.L. = 85%). This slurry was then placed in the vacuum mixer and the remainder of the treatment was as described above for the Kaolin triaxial samples.

4. The Yield Surfaces

4.1 Undrained and Drained Triaxial (p, q, e) Yield Surfaces for Normally Consolidated Weald Clay

In Fig. 2 the curve Nx_1 shows the undrained (p, q, e) yield surface for normally consolidated Weald Clay samples in conventional triaxial compression tests. The points in this diagram represent results obtained in three typical undrained tests.

The curve Nx_1 from Fig. 2 is again drawn in Fig. 3 but the points in this latter diagram were obtained during three typical drained tests.

It would seem reasonable to conclude that there is one unique (p, q, e) yield surface for both drained and undrained conditions for the normally consolidated Weald Clay. The critical state line obtained from drained tests corresponds to point x_2 in Fig. 3 and this appears to be below the point x_1 for undrain-

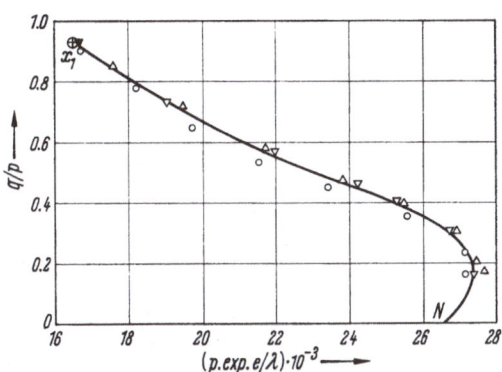

Fig. 2. (p, q, e) surface for undrained triaxial compression tests on normally consolidated Weald Clay.

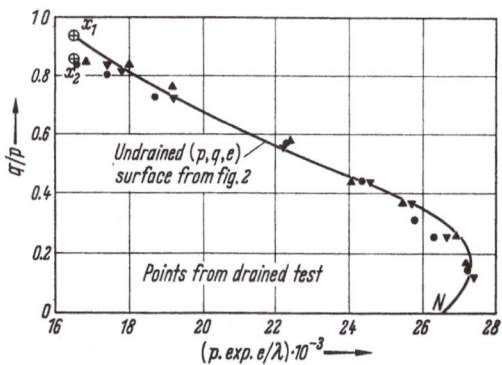

Fig. 3. Comparison of (p, q, e) surfaces for undrained and drained triaxial compression tests on normally consolidated Weald Clay.

ed tests. However the average axial strain required to attain the critical state was about 10% for the undrained tests but was more than 15% for the drained. Hence the errors in assessing the magnitude of q are likely to be larger in the drained than the undrained tests [see Roscoe, Schofield and Wroth (1959)]. It is therefore suggested that x_2 should coincide with x_1 in Fig. 3 and that there is one unique critical state line for Weald Clay samples under drained and undrained conditions.

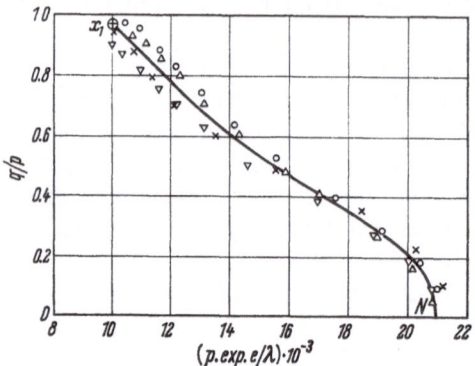

Fig. 4. (p, q, e) surface for undrained triaxial compression tests on normally consolidated Kaolin.

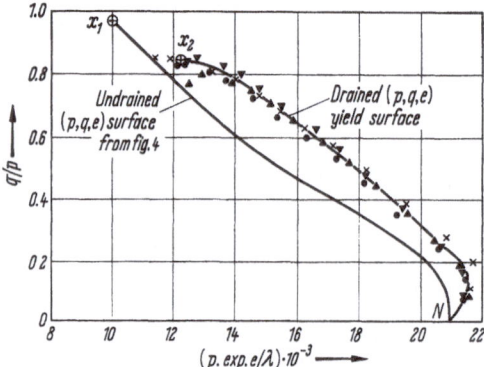

Fig. 5. Comparison of (p, q, e) surfaces for drained and undrained triaxial compression tests on normally consolidated Kaolin.

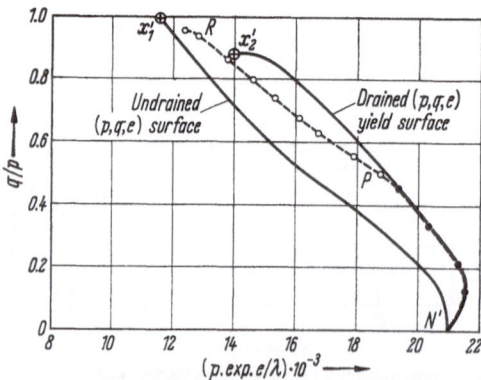

Fig. 6. Special triaxial compression test on Kaolin. First drained $(N'P)$, then undrained (PR).

4.2 Some Triaxial $[p, q, e]$ Yield Surfaces for Kaolin

The results of a variety of tests on Kaolin will now be presented which suggest that there are a large number of (p, q, e) yield surfaces for Kaolin. They will be further discussed in Section 5.

(a) Undrained and Drained — Normally Consolidated. The (p, q, e) yield surfaces obtained from four undrained and four drained triaxial tests on Kaolin are shown in Figs. 4 and 5. The four tests are typical of a series of drained and undrained tests each exceeding twenty in number, the samples ranging in initial voids ratio from 1.18 to 1.40. It is evident that the drained (p, q, e) yield surface Nx_2 and critical state line x_2 are different from the undrained (p, q, e) yield surface Nx_1 and critical state line x_1. In three dimensional (p, q, e) space the drained surface would lie outside the undrained surface, i.e. it would be further from the origin.

The results of a special test are shown in Fig. 6. In this test a sample was tested under fully drained conditions giving the curve $N'P$ but thereafter the test was completed under undrained conditions PR. The (p, q, e) yield surface for this test coincides with the drained surface up to the point P but then falls towards the undrained surface as the strain increases. It will be noted that the undrained and drained surfaces, $N'x_1'$ and $N'x_2'$ respectively in Fig. 6, do not coincide with those in Fig. 5. This is because they belong to another series of tests in which the drainage facilities during sample preparation were slightly different from that described in section 3.2. For further details see THURAIRAJAH (1961).

(b) Undrained and Drained — Samples Pre-Sheared. In Fig. 7 the curve $N'x_1'$ represents the undrained yield surface as shown in Fig. 6. The points lying on the path $N'PQRS$ refer to a special undrained test in which the axial strain was increased at a con-

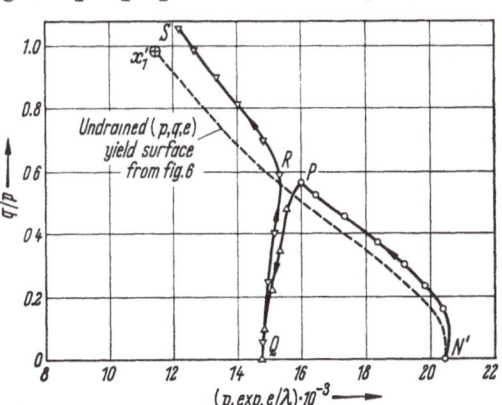

Fig. 7. Path followed by Kaolin sample (in an undrained triaxial compression test) which was sheared, deviator load removed and sheared again.

stant rate from N' to P and from Q through R to S but was decreased from P to Q. The sample was initially consolidated at 60 psi at N' and the conditions at P were: axial strain $\varepsilon_1 = 1.74\%$, excess pore pressure $u = 22.2$ psi, and $q = 26.2$ psi. While unloading from P to Q the excess pore pressure diminished and the steady value at Q was $u = 17.1$ psi. Upon reloading, the sample virtually retraced the path corresponding to PQ until at R the excess pore pressure $u = 23.9$ psi. Successive states for further loading are represented by RS which is a continuation of $N'P$. Allowing for experimental scatter the path $N'PRS$ coincides with the undrained (p, q, e) yield surface $N'x_1'$. It is evident that the sample behaved elastically over the region PQR and that little or no yield occurred during this process. This lends support to the elastic-plastic concepts of ROSCOE and SCHOFIELD (1963).

24*

The effects of unloading and reloading were further investigated in the two tests considered in Fig. 8. Both samples were subjected to a conventional drained test from N to P. The deviator stress was then removed over the path PQ at the same rate as it had been increased for the path NP. When the samples had reached the state corresponding

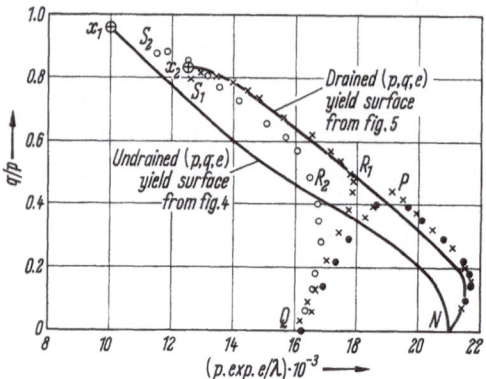

to Q they were left for about 20 hours. One sample was then subjected to a drained test and followed the path QR_1S_1, the portion R_1S_1 coinciding with the drained (p, q, e) yield surface Nx_2. The other sample was tested under undrained conditions and followed the path QR_2S_2 which lies somewhat above the undrained (p, q, e) yield surface Nx_1.

Fig. 8. Paths followed during reloading triaxial compression tests on normally consolidated Kaolin. NPQ drained, QR_1S_1 drained, QR_2S_2 undrained.

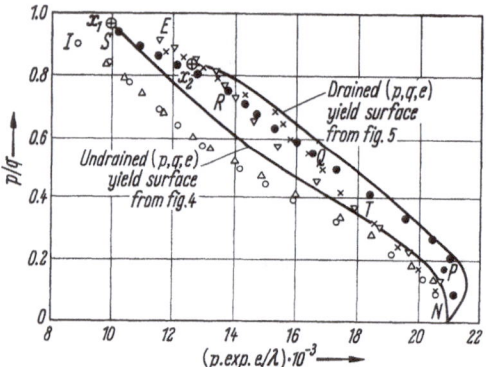

Fig. 9. Paths followed during partially drained triaxial compression tests on normally consolidated Kaolin.

(c) Partially Drained Tests — Normally Consolidated. A special device, called an injector, was used to cause controlled changes of dilatation of the sample with simultaneous measurement of the excess pore water pressure. In this way it was possible to investigate state paths of any desired shape in (p, q, e) or (τ, σ', e) space. The injector has been fully described by Thurairajah (1961) but in essence consists of nothing more than a fine tube with water in the upper end and mercury in the lower. The upper end is connected to the lead from the null device of the pore pressure measuring apparatus to the base of the sample. The lower end connects with a screw type ram full of mercury, so that by turning the screw, water may be injected into, or extracted from, the sample at any desired rate. Such tests will be called "partially drained" tests.

In Fig. 9 the curves Nx_1 and Nx_2 show the undrained and drained (p, q, e) yield surfaces taken from Figs. 4 and 5. The paths shown by points lying between NI, which are nearer the origin than curve Nx_1, refer to tests in which water was injected into the samples in such a manner that the rate of increase of volume with respect to axial strain was constant. The rate for the path nearest the origin was half that of the other.

The points represented by crosses and inverted triangles for the two paths NTE lying between the two yield surfaces Nx_1 and Nx_2 refer to two tests which were initially undrained (NT) so that a positive excess pore pressure was developed. Water was then extracted so that the rate of decrease of volume of the sample with respect to axial strain was constant for each sample but the rate for one was half that of the other. Note that both loading paths are further from the origin than the undrained surface Nx_1 and the distance is greater for the sample with the faster rate of volume change.

The path represented by the solid black circles in Fig. 9 refers to a test in which the portions NP and QR correspond to undrained conditions while during PQ the same volume of water is extracted as is injected back into the sample during RS. Notice how close P and S lie to the undrained yield surface and how the path moves away from this surface as the volume decreases but returns as the volume increases.

(d) Anisotropic Consolidation. It has been shown by ROSCOE, SCHOFIELD and THURAIRAJAH (1963a) that the results of anisotropic consolidation tests on Kaolin give straight lines when plotted on $(e, \log_e p)$ axes which are parallel to the normal consolidation and critical state lines. In the two dimensional plot of the (p, q, e) yield surface the results of each such test should be represented by a single point.

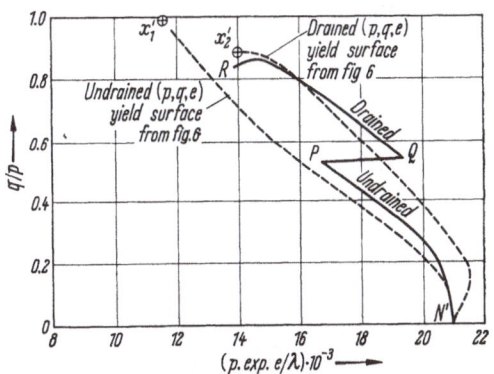

Fig. 10. Path followed during a triaxial compression test on normally consolidated Kaolin. $N'P$ undrained — PQ anisotropic consolidation — QR drained.

In Fig. 10 the results are shown of a test on a normally consolidated sample which was subjected to load under undrained conditions $N'P$ and drained conditions QR. When the sample had reached the state corresponding to P, the triaxial machine was left running at constant rate of axial strain but the cell pressure was constantly adjusted to maintain $\dfrac{q}{p}$ constant.

During this process, which lasted about 4 hours, the path PQ was traversed, instead of remaining at the point P. However the rate of average volumetric strain with respect to average axial strain $\frac{dv}{d\varepsilon_1}$ was remarkably constant [see Fig. 11 in Roscoe and Poorooshasb (1963)].

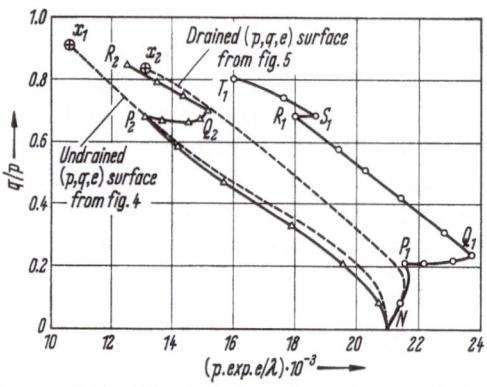

Fig. 11. Paths followed during anisotropic consolidation tests on normally consolidated Kaolin.

The paths followed during two other similar anisotropic consolidation processes are shown by P_1Q_1 and R_1S_1 in Fig. 11, but in this diagram paths NP_1, Q_1R_1 and S_1T_1 correspond to fully drained conditions. Data from a further test is represented by the path $NP_2Q_2R_2$ in Fig. 11. The portions NP_2 and Q_2R_2 correspond to undrained conditions but during P_2Q_2 water was extracted from the sample with the injector so that $\frac{dv}{d\varepsilon_1}$ was maintained constant. The constancy of $\frac{q}{p}$ during this process, which took about $3\frac{1}{2}$ hours, can be assessed by the deviation of P_2Q_2 from a straight line parallel to the abscissa axis in Fig. 11. Further discussion of these tests will be deferred until section 5.

4.3 Triaxial (p, q, e) Yield Surfaces for Normally Consolidated Cambridge Gault Clay

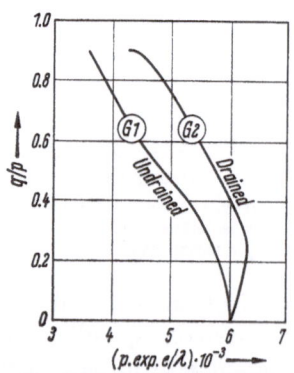

Fig. 12. (p, q, e) surfaces for triaxial compression tests on normally consolidated Cambridge Gault Clay (sample preparation as for Kaolin).

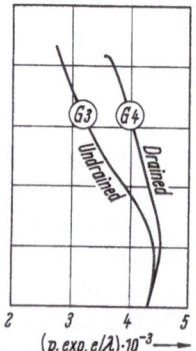

Fig. 13. (p, q, e) surfaces for triaxial compression tests in normally consolidated Cambridge Gault Clay (sample preparation as for Weald Clay).

A few triaxial tests were carried out on Cambridge Gault Clay to
see if the two modes of preparation of the samples as described in
Section 3.2 had any influence on the results. The undrained and
drained (p, q, e) yield surfaces obtained from samples prepared under
vacuum conditions similar to those employed for the Kaolin are shown
by curves $G1$ and $G2$ respectively in Fig. 12. They may be compared
with curves $G3$ and $G4$ in Fig. 13 which were obtained from samples
prepared in a similar manner to those of Weald Clay. It is suggested
that more air is likely to be present in these latter samples than the
former. Consequently the drained and undrained surfaces $G3$ and $G4$
would be expected to lie closer together, especially for small strains,
than surfaces $G1$ and $G2$. When air is present an apparent undrained
test becomes, in fact, a partially drained test.

4.4 The Yield Surface for Normally Consolidated Kaolin in the S.S.A.

The two dimensional plot of the mean (σ', τ, e) undrained yield surface as determined from five S.S.A. tests on normally consolidated samples of Kaolin is shown by the curve $A x_1$ in Fig. 14. These five tests are
typical of a series of more than twenty similar tests in which the sam-

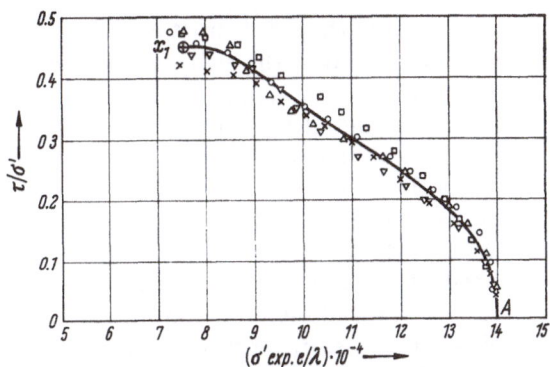

Fig. 14. (σ', τ, e) surface for undrained tests on normally consolidated Kaolin in the simple shear
apparatus.

ples covered the same range of initial voids ratio as for the triaxial
tests described in section 4.2a above. This undrained surface is again
plotted as the curve in Fig. 15 but the points shown in this diagram
refer to five typical drained tests. It is suggested that the drained and
undrained (σ', τ, e) yield surfaces, as determined from S.S.A. tests on
normally consolidated samples of Kaolin, are coincident. It can be
seen from Fig. 15 that the critical state line x_2 from the drained tests
appears to differ from the critical state line x_1 which was obtained
from undrained tests. The writers believe that x_2 should coincide

with x_1 and the observed difference may be ascribed to the increase in deviation between the actual and the assumed behaviour of the sample and the apparatus at high displacements (ΔX) of the S.S.A. At these large displacements ($\Delta X > 0.7$ inch) the assumption that the normal stress across all horizontal planes is uniform and equal to σ' may not be valid.

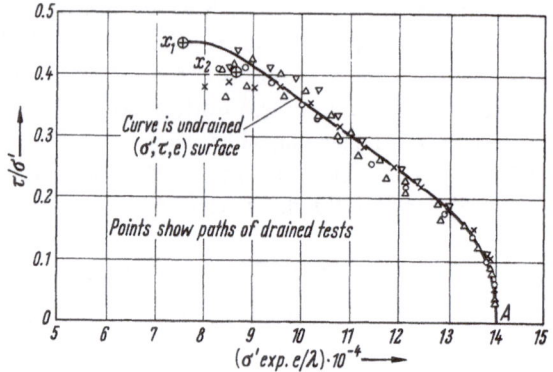

Fig.15. Comparison of (σ', τ, e) surfaces for undrained and drained tests on normally consolidated Kaolin in the simple shear apparatus.

The particular pattern of the S.S.A. used in these tests was specially designed to permit X-rays to pass through the whole of the sample in a direction normal to the plane of strain. It has been described briefly by ROSCOE (1963) and in detail by THURAIRAJAH (1961). It is interesting to note that the recorded excess pore pressure u became constant in undrained tests at $\Delta X = 0.5$ inch but the recorded volume change δV only became zero in drained tests when $0.7 < \Delta X < 0.8$ inch. This behaviour is in marked contrast to that observed in the triaxial tests. In the undrained triaxial tests on Kaolin described above, the excess pore pressure u was still increasing at an axial strain ε_1 of 25% despite the peak deviator stress q_{max} having been recorded at $\varepsilon_1 = 12\%$ approx. In the drained triaxial tests q_{max} usually occurred at a strain of about $\varepsilon_1 = 23\%$ but the recorded volume was still decreasing steadily at strains exceeding $\varepsilon_1 = 30\%$.

5. Comparison of the Triaxial and S.S.A. Data on Kaolin

The conclusions that can be drawn concerning the existence of a unique yield surface in a space relating the voids ratio and the observed values of two stress variables for the results of either S.S.A. or triaxial tests on normally consolidated samples of Kaolin are contradictory. From Fig. 15 it appears that there is one unique yield surface in (σ', τ, e) space, as determined in S.S.A. tests, at least in so far as drained and undrained conditions are concerned. In the drained tests there is some discrepancy at very high strains when the irregularity of the stress distribution in the S.S.A. becomes accentuated. On the other hand it is evident from Fig. 4 and 5 that the undrained and drained yield surfaces in (p, q, e) space, as determined from triaxial compression tests, are quite distinct. Furthermore the results of partially drained tests (Fig. 9) and anisotropic consolidation tests (Figs. 10 and 11) in the triaxial apparatus do not appear to lie on either the undrained or the drained surfaces. One possible explanation of this discrepancy in the results obtained from the S.S.A. and the triaxial apparatus may be obtained by considering the differences between the true and the measured values of the parameters that have been used for plotting the yield surfaces.

In Fig. 5 the drained (p, q, e) yield surface from triaxial tests lies further from the origin than the undrained (p, q, e) surface even at small strains when the errors in p and q are likely to be small. It is suggested that this may be due to erroneous estimation of the changes in voids ratio during the drained tests. ROSCOE, SCHOFIELD and THURAIRAJAH (1963b) showed that the axial and radial strains are not uniform along the length of a triaxial sample during a compression test (see their Figs. 14 and 15). This result has recently be confirmed by K. Z. SIRWAN using a more refined X-ray technique. The change in voids ratio estimated by measuring the volume of water traversing the boundary of a sample is always less than the actual value of the change in the portion of the sample that undergoes most deformation. Hence at any instant in a drained triaxial test on a normally consolidated saturated clay the estimated value of the magnitude of the voids ratio of the portion of the sample that is subjected to most deformation will be higher than the true value. Therefore the observed drained (p, q, e) surface will be further from the origin in Fig. 5 than the true drained surface.

It is suggested that there is much less divergence of the observed and true location of the undrained (p, q, e) yield surface. The excess pore pressure developed in the part of the sample subjected to most deformation spreads throughout the entire sample and may be recorded correctly by the standard methods, provided (i) the volume of the water and clay particles in the less deformed portions of the sample do not change under the developed excess pore pressure, and (ii) the rate of testing is sufficiently slow. Under these conditions the non-uniform deformation of samples during undrained tests will not cause much difference between the observed and true values of e and u.

The divergence of the (p, q, e) surfaces from the undrained surface in the special tests described in paragraph 4.2 may also be explained by the erroneous estimation of the voids ratio in the relevant part of the sample. For example in the tests on presheared samples shown in Fig. 8 while yielding in any drained portion of a test the results for both samples lie close to the observed drained surface Nx_2. On the other hand while yielding under undrained conditions corresponding to R_2S_2 all the observed points lay above the undrained surface Nx_1. It is suggested that this is due to the error in the apparent voids ratio change during the drained portion NPQ of the test.

In the partially drained tests, see Fig. 9, the paths NE for tests in which water was extracted lie below the fully drained surface Nx_2. This is because the rate of extraction of water was less than the rate of expulsion of water in a fully drained test: consequently the errors in the estimation of the voids ratio will be less. During the extraction process the true *reduction* of voids ratio of the portion of the sample which is subjected to most deformation will be greater than the observed average for the whole sample. Hence the observed value of the magnitude of the voids ratio will be *greater* than the true, and the observed yield surface NE lies above the undrained surface Nx_1. On the other hand when water is injected the surface NI is obtained which lies below the undrained surface Nx_1. Now the true *increase* of the voids ratio of the most deforming part of the sample will be greater than the observed average. Hence the observed magnitude of the voids ratio will now be *less* than the true and the surface NI will lie below Nx_1.

Finally since water is always expelled during anisotropic consolidation tests (see Figs. 10 and 11) the true decreases of voids ratios will always be underestimated and the longer the process is continued the further will the resultant apparent yield surface deviate from the undrained surface. Furthermore this deviation will always be in a direction away from the ordinate axes in diagrams such as Figs. 10 and 11.

The authors therefore consider that it is possible that there should be only one unique (p, q, e) yield surface and critical state line for the Kaolin as determined from triaxial compression tests with all types of drainage facilities (i.e., state paths). Furthermore they believe that this surface is most truly represented in constant strain rate tests by the surface obtained from observations in undrained tests. The rate of test should be extremely slow in the early stages to ensure that the pore pressure gradients within the sample are always of negligible magnitude.

6. The (p, q_w, e) Yield Surfaces for Kaolin

It has been assumed by Roscoe, Schofield and Thurairajah (1963a) and Roscoe and Schofield (1963) that there was one unique surface in (p, q_w, e) space for triaxial compression tests on samples of normally consolidated Kaolin with any type of uniform drainage facility. q_w was obtained from the observed deviator stress q by applying a boundary and an elastic energy correction. It was assumed that the elastic energy component due to shear was negligible in comparison with that due to volume change. From the available data it appeared that this latter assumption would not cause an error in the magnitude of q_w of more than 3.5% and in general would be much less. This error would, of course, occur when correcting observed values of q for all types of test, including drained and undrained tests. The "open" points in Fig. 16 were obtained from four undrained tests for axial strains greater than $\varepsilon_1 = 0.15\%$; the line $0x_1$ of slope $\tan^{-1} 0.9$ is the mean through these points. The "solid" points in Fig. 16 refer to four drained tests for axial strains exceeding $\varepsilon_1 = 0.5\%$ and the mean line $0x_2$ is of slope $\tan^{-1} 1.0$. Both undrained and drained (p, q_w, e) surfaces when plotted three dimensionally approximate to planes containing the e-axis. The writers believe that only one such surface should exist for undrained, partially drained, and fully

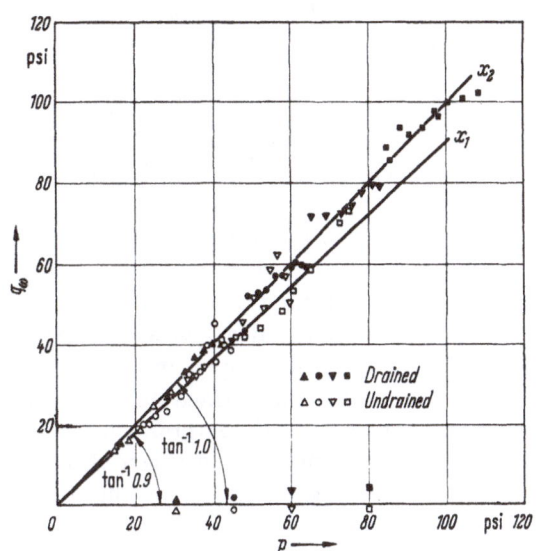

Fig. 16. Undrained $(0\,x_1)$ and drained $(0\,x_2)$ (p, q, e) surfaces for normally consolidated Kaolin in triaxial tests.

drained tests. They suggest that the undrained surface $0x_1$ at present gives the nearest approximation to this true surface. Recently C. P. WROTH has suggested that the drained surface might be made to coincide with the undrained surface if the elastic and boundary energy connections were refered to unit volume of solid material rather than to unit bulk volume of the clay. This matter is now under investigation.

7. Concluding Remarks

It is evident that before reliable conclusions can be drawn from triaxial test data either the technique must be improved so that the samples strain more uniformly or methods must be devised of measuring the voids ratio accurately in the zones of failure within the samples. The S.S.A. was originally designed with the specific object of producing uniform dilatation throughout the sample so that changes in voids ratio can be correctly evaluated from observed changes in volume of the sample boundary. In the particular model of the S.S.A. that was used to obtain the test data discussed in this paper there was a very small zone of the sample (called the dead zone, see ARTHUR, JAMES and ROSCOE (1964)) the boundaries of which were not subjected to simple shear. The latest model of the S.S.A. is designed to eliminate this zone and contains facilities for determining the complete three dimensional stress system imposed on samples in plane strain.

While waiting for this improved data it would seem reasonable to make the following tentative conclusions regarding the results presented in this paper:

(i) The reason Weald Clay appeared to have one unique surface in (p, q, e) space was due to the presence of air in the samples. If air is removed then there would probably be two independent apparent surfaces for drained and undrained triaxial tests on Weald Clay as shown by the Cambridge Gault Clay and the Kaolin.

(ii) There is one unique surface in (σ', τ, e) space for Kaolin for drained and undrained tests in the S.S.A.

(iii) If the voids ratio had been correctly measured in the most deforming part of the samples in triaxial drained and partially drained tests, then one unique surface and one critical state line would exist in (p, q, e) space for all compression tests regardless of the dilatation allowed to the samples of a given clay.

(iv) The order of the difference between the drained and undrained (p, q_w, e) surfaces for the Kaolin is less than 10%.

(v) At present undrained data give more reliable representations of a single unique yield surface and critical state line in both (p, q, e) and (p, q_w, e) space than do drained data for a given clay.

Finally the authors believe that anisotropy of the plate shaped Kaolin particles cannot be used to explain the divergence of the (p, q, e) surfaces in the triaxial test and the singularity of the (σ', τ, e) surface in the S.S.A. tests. They have carried out a few tests on samples in which the principal axes of anisotropy were 90° apart and have obtained remarkably little difference in the observed data. Further experiments are needed before firm conclusions can be drawn.

Acknowledgements

The authors would like to thank their colleague Dr. A. N. SCHOFIELD for many discussions on the subject of this paper and Mr. R. E. WARD for assistance in the experimental work. They also wish to express their appreciation for the continued interest and support given by Professor SIR JOHN BAKER, Head of the Department of Engineering, Cambridge University.

References

ARTHUR, J. R. F., R. G. JAMES and K. H. ROSCOE (1964): The determination of stress fields during plane strain of a sand mass. Géotechnique **14**, No. 4, 283—308.

HENKEL, D. J. (1956): The effect of over-consolidation on the behaviour of clays during shear. Géotechnique **6**, 139—150.

HVORSLEV, M. J. (1937): Über die Festigkeitseigenschaften gestörter bindiger Böden. (On the physical properties of remoulded cohesive soils). Ingenior-videnskabelige Skrifter. A. No. 45, 159 pp.

POOROOSHASB, H. B., and K. H. ROSCOE (1961): The correlation of the results of shear tests with varying degrees of dilatation. Proc. 5th Int. Conf. Soil. Mech., Vol.1, pp. 297—304.

ROSCOE, K. H. (1963): Contribution to the discussion session of Section 6. Proc. European Conference on Soil Mechanics and Foundation Engineering, Wiesbaden, October 1963, Vol. II, p. 129—133.

ROSCOE, K. H., and H. B. POOROOSHASB (1963): A theoretical and experimental study of strains in triaxial tests on normally consolidated clays. Géotechnique **13**, No. 1, 12—38.

ROSCOE, K. H., and A. N. SCHOFIELD (1963): Mechanical behaviour of an idealised "wet-clay". Proc. European Conf. Soil Mech. Wiesbaden, October 1963, Vol. I, p. 47—54.

ROSCOE, K. H., A. N. SCHOFIELD and A. THURAIRAJAH (1963a): Yielding of clays in states wetter than critical. Géotechnique **13**, No. 3, 211—240.

ROSCOE, K. H., A. N. SCHOFIELD and A. THURAIRAJAH (1963b): A critical appreciation of test data for selecting a yield criterion for soils. Symposium on "Laboratory Shear Testing of Soils". Ottawa, September, 1963. ASTM Spec. Tech. Pub. No. 361, p. 111—128.

ROSCOE, K. H., A. N. SCHOFIELD and C. P. WROTH (1959): Correspondence "On the yielding of soils". Géotechnique **9**, No. 2, 72—83.

THURAIRAJAH, A. (1961): Some properties of Kaolin and of sand. Ph. D. Thesis, Cambridge University.

Discussion

Contribution de E. SCHULTZE: Professor ROSCOE showed during the discussion the so-called stress-strain surfaces of kaolin and clay with the ordinates p, q and w. During the last few years a research work: HORN: Die Scherfestigkeit von Schluff. Forschungsberichte des Landes Nordrhein-Westfalen Nr. 1346, Köln/Opladen: Westdeutscher Verlag, was completed, at which a similar relationship was found out for disturbed satured silt (Fig. 1).

ROSCOE was able to determine the stress-strainsurface with the help of two experiments. On the other hand, due to the dilatation of the silt, it was possible to derive the yield surface of silt with the help of a single experiment by analysing the test paths of a consolidated undrained test (Fig. 2f). Both the true and the effective shear-parameters can be derived by laying a tangent on the upper branch of the test path between the criteria $(\sigma_1'/\sigma_3')_{max}$ and $(\sigma_1 - \sigma_3)_{max}$ after mobilizing various shearing resistances.

On comparing the results of the drained and undrained tests one came to the conclusion that the failure criterium of silt $(\sigma_1 - \sigma_3)_{max}$ is the deciding factor for the determination of the shearing parameters. The results of the experiments for this particular criterium, which is identical to the criterium $(\sigma_1'/\sigma_3')_{max}$ in the drained test, are similar for both types of tests. One arrives remarkably higher effective angles of internal friction for the criterium $(\sigma_1'/\sigma_3')_{max}$ in the undrained test. On the other hand, since the true shearing line goes through both the criteria, one can conclude that the true shearing parameters are not dependent on the choise of the criterium.

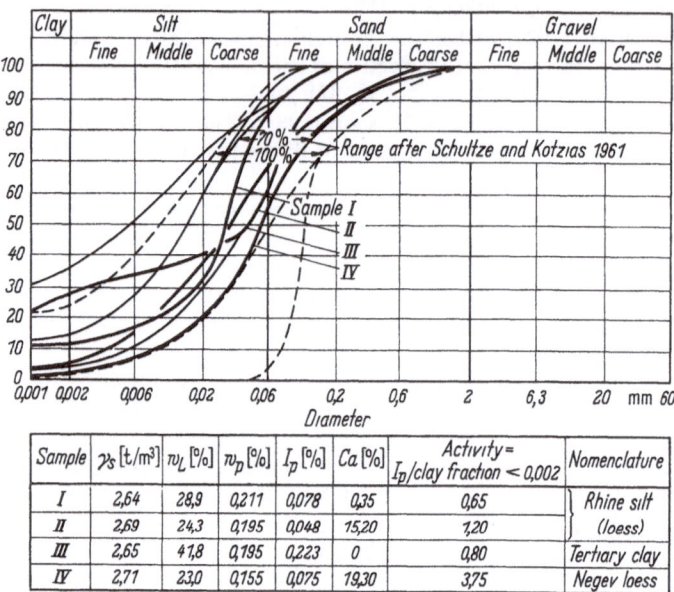

Sample	γ_s [t/m³]	w_L [%]	w_p [%]	I_p [%]	Ca [%]	Activity = I_p/clay fraction < 0,002	Nomenclature
I	2,64	28,9	0,211	0,078	0,35	0,65	} Rhine silt
II	2,69	24,3	0,195	0,048	15,20	1,20	} (loess)
III	2,55	41,8	0,195	0,223	0	0,80	Tertiary clay
IV	2,71	23,0	0,155	0,075	19,30	3,75	Negev loess

Fig. 1. Grain size distribution and soil properties of the silt.

A comprehensive view of the shearing resistance of silt can be got with the help of a three-dimensional representation of the relationship between the shearing stress, the effective normal stress in the shearing plane, and the water content (Fig. 3). It appears that all the test paths of the drained and undrained tests terminate at a single line (φ'-line, critical void ratio line) and that the upper branches of all the test paths of the undrained tests describe a yield surface in space.

The relation between the shear stress and the water content, which are necessary for the determination of the yield surface can be determined from the same experiment by measuring the change of volume caused by the consolidation that takes place during various load-increments. The curve of consolidation is a straight line in the semilogarithmic coordinate system (Fig. 2b). One plots the true cohesion c_w to its corresponding water content and draws a line parallel to the curve of consolidation through this point in the same coordinate system. In this way one derives the relationships between the true cohesion and the water content and from that the yield surface completely can be drawn.

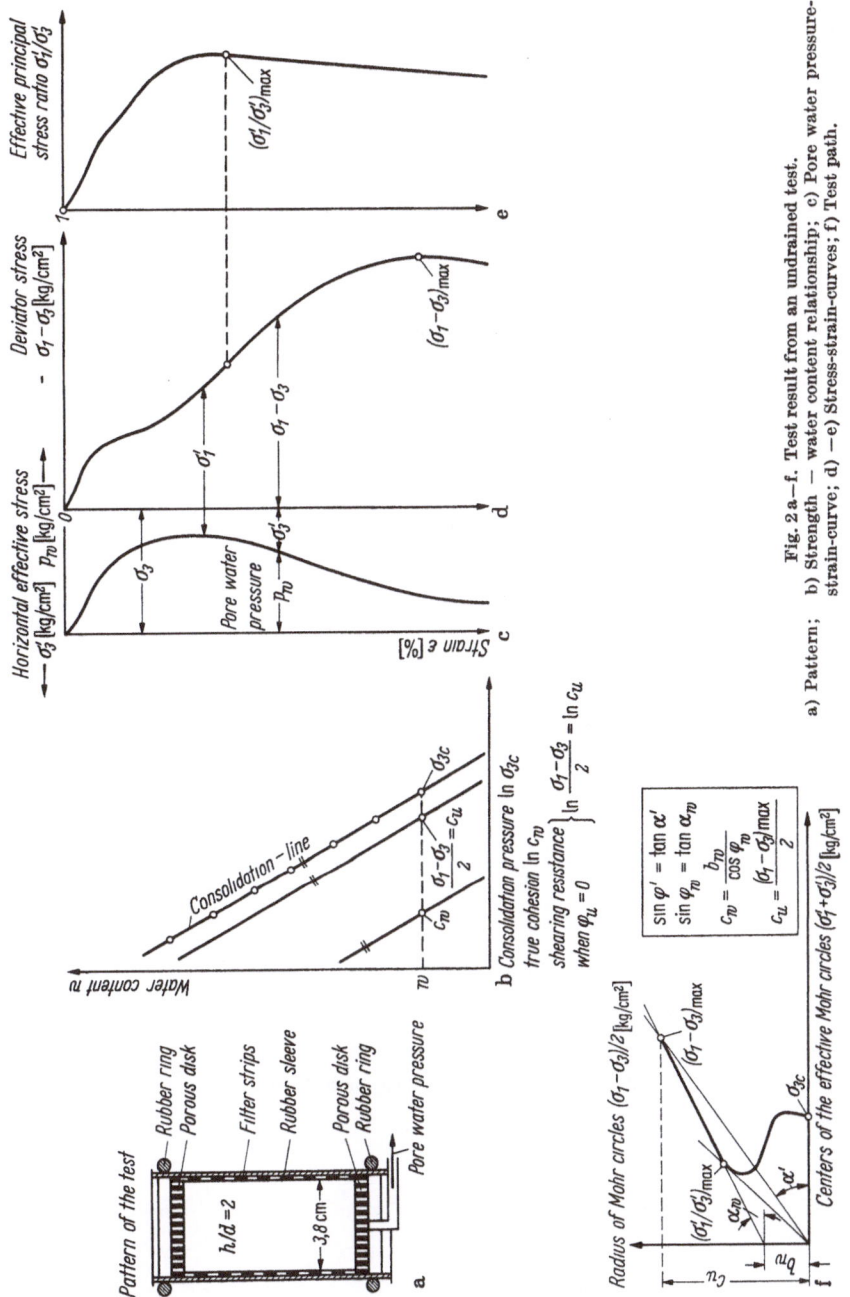

Fig. 2 a–f. Test result from an undrained test.
a) Pattern; b) Strength — water content relationship; c) Pore water pressure-strain-curve; d) –e) Stress-strain-curves; f) Test path.

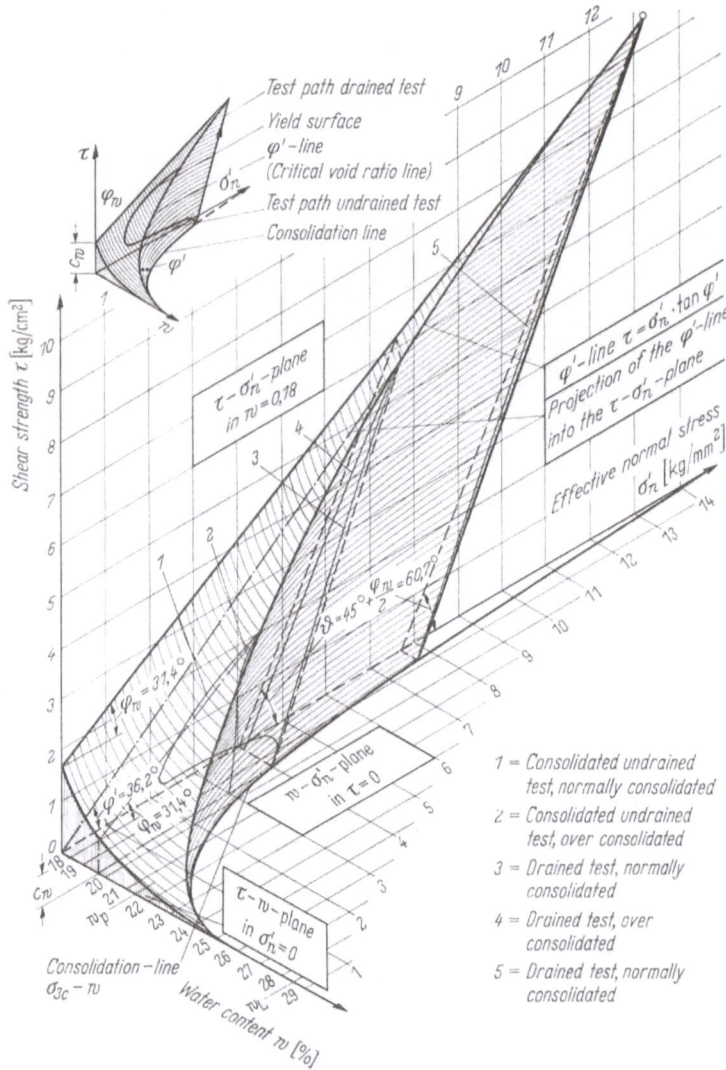

Fig. 3. Yield surface of silt in three dimensional coordinates.

3.9 L'Application de l'Essai Brésilien dans la Mécanique des Roches

Par

P. Habib, P. Morlier et D. Radenkovic

Résumé

Nous présentons ici les résultats obtenus par un procédé généralisé — les essais ayant été faits sous une pression hydrostatique variable — et nous proposons une interprétation de ces résultats fondée sur la théorie de l'équilibre limite.

1. L'Essai Brésilien Classique

Dans l'étude des propriétés mécaniques du béton, pour déterminer la résistance à la traction R_T, on utilise souvent l'essai dit brésilien: un échantillon cylindrique de dimension normalisée est écrasé sous une presse par des forces uniformément réparties le long des deux génératrices opposées (fig. 1.)

Admettant le comportement élastique jusqu'à la rupture et en négligeant la déformation au voisinage de la surface de contact entre l'éprouvette et le plateau de la presse, on peut trouver la distribution des contraintes dans le cylindre — les valeurs de σ_x, σ_y le long des deux axes de symétrie x, y sont données sur la fig. 1.

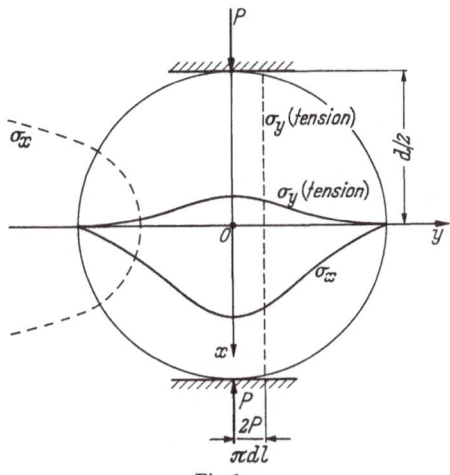

Fig. 1.

Au point 0 nous avons:

$$\sigma_x = +\frac{6P}{\pi dl}, \qquad \sigma_y = -\frac{2P}{\pi dl}$$

(P — force totale exercée, d — diamètre et l la longueur de l'échantillon; le signe positif désigne la compression). Notons qu'en principe $\sigma_y < \sigma_z < \sigma_x$ en ce point, soit en déformation plane, soit dans le cas de l'état de contraintes planes.

L'interprétation habituelle consiste à supposer que la rupture se produit lorsque σ_y atteint une valeur $R_T(b)$ proportionnelle à la résistance à la traction du matériau. La comparaison des valeurs $R_T(b)$ ainsi déterminés avec les valeurs obtenues expérimentalement par d'autres procédés (traction simple, flexion) a été faite, en particulier pour le béton, par de nombreux auteurs [1], [2], [3].

L'interprétation classique de l'essai brésilien paraît cependant un peu hâtive. En examinant la fig. 1 on voit qu'il existe au voisinage de la charge des points où les cercles de Mohr correspondants sont sécants à la courbe intrinsèque : une zone plastique doit se développer et on n'en a pas tenu compte. On verra au paragraphe suivant que cette influence est surtout sensible pour les expériences avec pression hydrostatique importante.

2. Essais sur Roches sous une Pression de Confinement Variable

Dans la plupart des cas pratiques, on a besoin de connaître le comportement des roches sous des pressions de confinement importan-

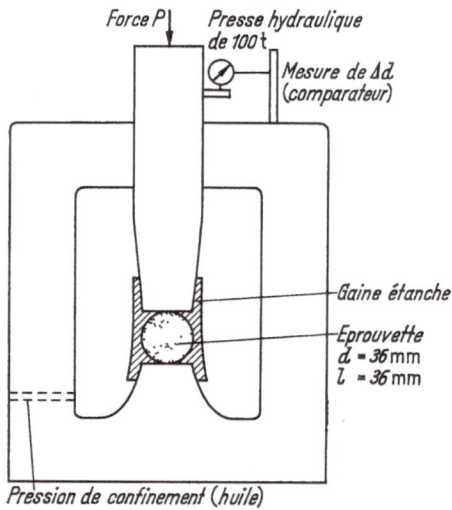

Fig. 2. Schéma de principe de l'appareil triaxial.

tes ; lorsqu'on utilise l'essai brésilien pour étudier des propriétés mécaniques des roches, il est donc assez naturel de généraliser les conditions de l'expérience en appliquant la charge diamétrale P simultanément à une pression hydrostatique q.

Pour les expériences correspondantes, nous avons utilisé un appareil triaxial représenté sur la fig. 2 : q varie entre 1 et 1000 bars : P est exercé par une presse hydraulique de 100 tonnes. Les expériences (3 essais pour $q = 1$ bar, 1 essai pour $q = 250$, 500 et 1000 bars ont été effectuées sur deux séries d'échantillons.

Nous avons etudié le *Calcaire de St-Béat* et le *Calcaire gris Ste-Anne*.

Les courbes intrinsèques correspondantes (fig. 3a et b) ont été obtenues avec le même appareil triaxial.

Le premier calcaire avait un comportement élasto-plastique prononcé dès que la pression de confinement était supérieure à 250 bars ; le deuxième calcaire gardait jusqu'à une pression de 1000 bars le comportement élastique-fragile, qu'on connait aux roches dans les conditions atmosphériques.

Les dimensions des échantillons dans les deux séries étaient $d =$ 36 mm, $l = 36$ mm. Autour de ces échantillons nous avons coulé des

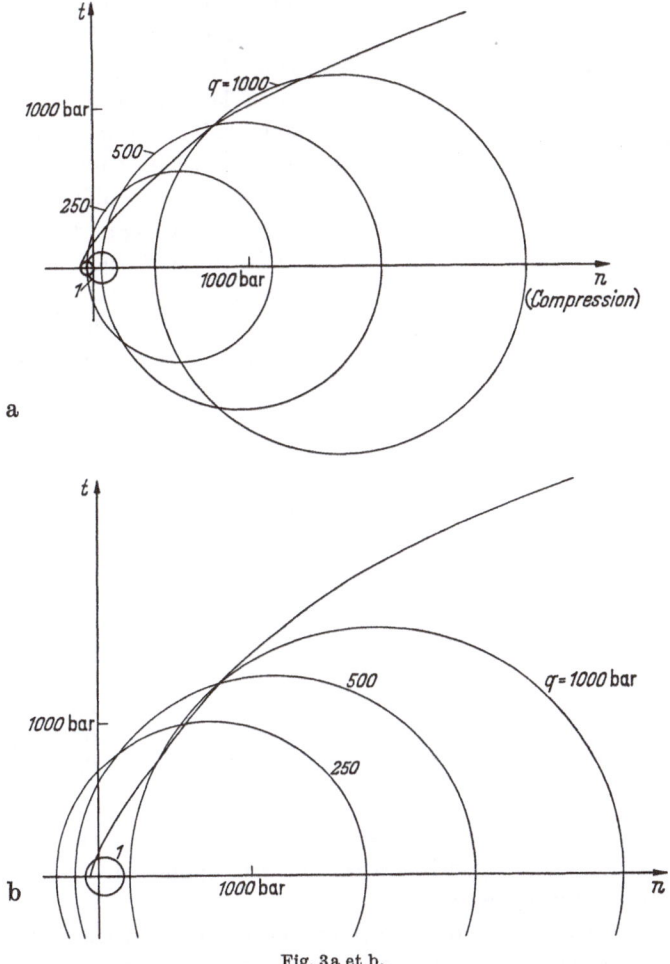

Fig. 3a et b.

gaines de caoutchouc synthétique (silastène), étanches à l'huile, par l'intermédiaire desquelles s'était exercée la pression de confinement q. Les essais étaient menés de la façon suivante: après avoir appliqué une pression hydrostatique, égale à q, sur l'échantillon, on faisait croître la force diamétrale P*. La déformation Δd du diamètre de l'éprouvette

---

* Dans le texte suivant P désigne la surcharge: $P = P' - qS$. P' étant la force axiale totale exercée par la presse et S l'aire de la section du piston.

parallèle à P était mesurée par un comparateur au 1/100 de mm. On trouve sur la fig. 4 un exemple de courbe de charge $P, \Delta d$. On y apprécie

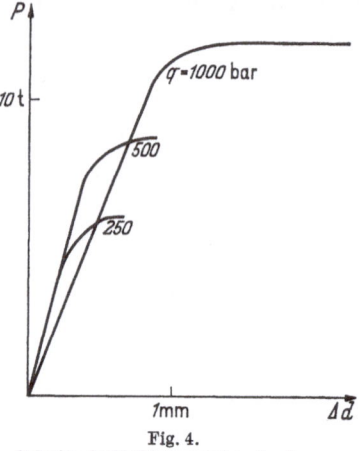

Fig. 4.
Calcaire de St-Béat: courbes de charge.

la rupture avec assez de précision (fin des courbes de charge). Les fig. 5a—d montrent l'évolution de l'aspect des échantillons brisés en fonction de la pression latérale; on voit que la largeur de la surface de contact croît avec q et la présence d'un coin plastifié sous cette surface devient de plus en plus nette.

Dans le tableau 1 nous avons donné la force P qui provoque la rupture et la largeur a (mesurée grossièrement) de la surface de contact en fonction de la pression de confinement q.

Tableau 1

	q (bars)	P (tonnes)	estimation de a (cm)
Calcaire de St-Béat	1	1	0,15 à 0,25
	250	6	0,40 à 0,60
	500	9	0,55 à 0,70
	1000	12	0,65 à 0,80
Calcaire gris Ste-Anne	1	1,4	0,10
	250	10	0,40 à 0,50
	500	13	0,70 à 0,80
	1000	16	0,80 à 0,90

En appliquant la théorie classique (§ 1) nous avons calculé les contraintes au point 0 et porté les cercles correspondants sur les courbes intrinsèques. (fig. 3a et b).

Cette interprétation de l'essai ne paraît pas satisfaisante. D'une part, même sous les pressions q importantes la surface de rupture au point 0 est verticale quoique la contrainte ne soit plus une traction mais une compression, d'autre part les cercles de Mohr calculés, qui devaient être critiques ne sont par tangents à la courbe intrinsèque — et pourtant d'autres expériences (v. par ex. [4]) montrent la validité du critère de Mohr-Caquot pour les roches. L'origine de cette divergence semble provenir de ce que dans la théorie classique de l'essai brésilien on ne tient pas compte d'une importante zone plastique qui se développe sous la surface de contact entre le cylindre et la presse. Il nous semble que ceci entraine une conception erronnée du mécanisme de la rupture; l'erreur devient surtout sensible quand on veut interpréter les expériences effectuées sous une pression hydrostatique importante.

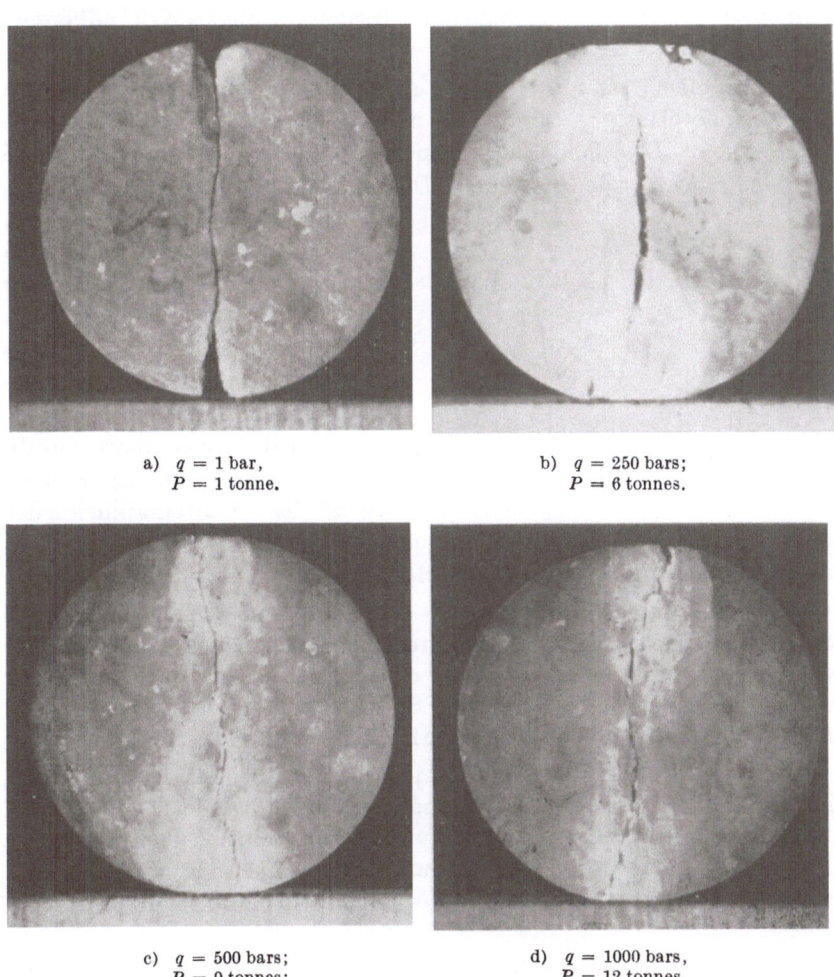

a) $q = 1$ bar,
    $P = 1$ tonne.

b) $q = 250$ bars;
    $P = 6$ tonnes.

c) $q = 500$ bars;
    $P = 9$ tonnes;

d) $q = 1000$ bars,
    $P = 12$ tonnes.

Fig. 5a–d. Aspect des échantillons brisés.

3. L'Interpretation des Essais selon la Théorie de l'Équilibre Limite

Les expériences montrent nettement qu'une zône importante de déformation plastique contenue apparaît sous la surface de contact, même en l'absence de la pression de confinement et sous des charges inférieures à celle de rupture. Nous allons supposer que la rupture se produit au moment où les deux zones plastiques partant des deux sur-

faces de contact se rencontrent au point 0 (fig. 6). Négligeant la déformation élastique, nous pouvons dire qu'à ce moment les deux croissants rigides (*AOM* et *BON*) commencent à s'écarter tandis qu'une
importante déformation plastique se produit à l'intérieur des deux
bulbes (*AOB*). (Effet de clivage entre les deux triangles *ABC* adhérents
aux surfaces de contact).

Le problème est ainsi ramené à un cas de déformation plane du
corps rigide-plastique caractérisé par une courbe intrinsèque donnée
(fig. 3a et b). Le schéma de la solution — le coin rigide *ABC* entouré
des deux faisceaux de Prandtl (*ADC* et *BCE*); partant des deux extré-

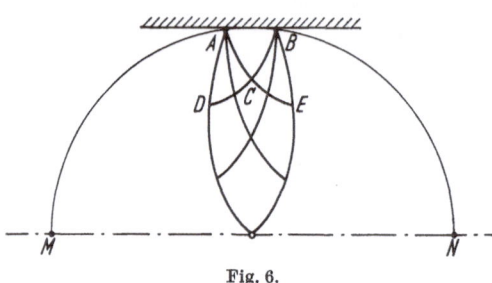

Fig. 6.

mités (*A* et *B*) de la base,
et le réseau d'interférence
DCEO- est indiqué dans
la fig. 6. Ce schéma bien
connu, a été utilisé dans
le problème de poinçonne-
ment d'une couche mince
[5]. Une comparaison du
réseau de la fig. 6 avec
les photographies des
échantillons brisés (fig. 5) indique que la conception proposée du mécanisme de la rupture est raisonnable.

Le calcul de la distribution des contraintes qui se ramène au tracé
du réseau des caractéristiques ne présente pas de difficultés; les méthodes de solution dans le cas de la courbe intrinsèque quelconque ont
été étudiées par Mandel [6] et Sokolovski [7]; en général, pour
résoudre le problème on doit recourir à l'intégration numérique.

Nous allons illustrer la théorie proposée de l'essai brésilien, en
interprétant les résultats de nos expériences avec le calcaire de *St-Béat*.
Pour ce matériau nous avons pu utiliser une solution explicite due à
Perlin [8], grâce au fait que sa courbe intrinsèque se laisse représenter
avec une très bonne approximation par la courbe théorique:

$$t = K (\sin 2\mu)^3, \quad s = 3K \left(\mu - \frac{1}{4} \sin 4\mu\right) - m \qquad (1)$$

où (v. fig. 7) s — désigne l'abscisse du centre, t est le rayon des cercles
de Mohr critiques; le paramètre μ représente l'angle de la facette
marginale, par rapport à la direction de la compression maximale; K, m
sont des constantes du matériau (pour la courbe de la fig. 7 $K = 1700$
bars et $m = 50$ bars).

Introduisant les variables $\xi = 3\mu + \varphi, \eta = 3\mu - \varphi$, φ étant l'angle
entre l'axe des x et la contrainte de compression maximale, les équations

du problème prennent la forme (canonique):

$$\frac{\partial y}{\partial \xi} = \tan(\varphi - \mu)\frac{\partial x}{\partial \xi}; \qquad \frac{\partial y}{\partial \eta} = \tan(\varphi + \mu)\frac{\partial x}{\partial \eta}, \tag{2}$$

les équations différentielles des caractéristiques étant données par

$$\frac{dy}{dx} = \tan(\varphi \mp \mu) \quad \text{pour} \quad \begin{cases} \eta = \text{const} \\ \xi = \text{const.} \end{cases} \tag{3}$$

Le changement de variables

$$X = -x\sin(\varphi + \mu)$$
$$+ y\cos(\varphi + \mu);$$

$$Y = x\cos(\varphi + \mu)$$
$$+ y\sin(\varphi + \mu)$$

$$(4)$$

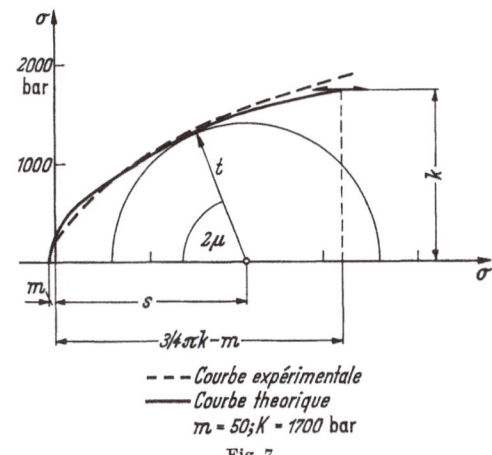

Fig. 7.
- - - Courbe expérimentale
—— Courbe theorique
$m = 50; K = 1700$ bar

ramène (2) à un système dont la solution générale peut être donnée sous une forme explicite:

$$X(\xi, \eta) = \operatorname{cosec}\frac{\xi + \eta}{3}\left[F_0(\eta) + \int_{\xi_0}^{\xi} F_1(\xi)\sin\frac{\xi + \eta}{3}d\zeta\right]$$

$$Y(\xi, \eta) = 3\frac{\partial X}{\partial \eta} \tag{5}$$

où $F_0(\eta)$ et $F_1(\zeta)$ sont deux fonctions arbitraires.

Dans un domaine où $\xi = \text{const}$, ou bien $\eta = \text{const}$, on obtient les "solutions simples" correspondantes en intégrant directement les équations des caractéristiques (3) [7].

On peut déterminer la distribution des vitesses de déformation de la même manière [8]; mais ceci demande une discussion préalable de la loi de l'écoulement que nous n'allons pas aborder ici.

Nous pouvons maintenant appliquer la solution présentée au cas particulier de l'essai brésilien (fig. 8)*. Pour simplifier le calcul, nous prenons le système des coordonnées avec l'origine à l'extrémité gauche de la surface de contact, l'axe des x orienté vers le bas.

---

* Dans l'article de Perlin, où le problème analogue du poinçonnement est traité comme exemple, les variables ξ et η sont interchangées; l'erreur d'ailleurs ne porte que sur le calcul des constantes.

Dans le domaine ABC on a l'équilibre homogène: $\varphi = 0$; $\xi_0 = \eta_0 = 3\mu_0 = \text{const}$. Donc:

$$\sigma_x = p + q = s + t; \quad \sigma_y = s - t. \tag{6}$$

(Nous admettons que la pression sur la surface de contact est due à la force diamétrale $p = P/a \cdot l$ et au confinement q séparément.)

Dans BCE on a $\xi = \xi_0$ et les lignes $\eta = \text{const}$ (seconde famille) sont les droites $y = x \tan(\varphi - \mu) + \phi(\eta)$ Introduisant les variables ξ, η et tenant compte de la condition limite $y(0) = a$, on trouve:

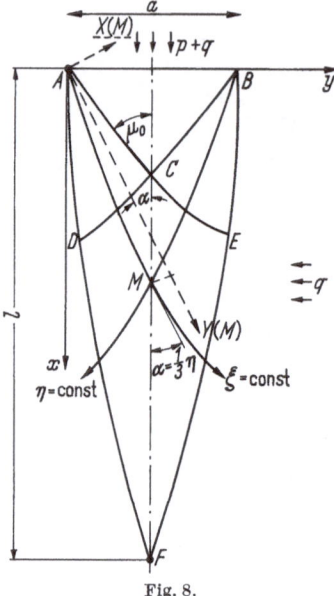

Fig. 8.

$$y - a = x \tan \frac{1}{3}(\xi_0 - 2\eta). \tag{7}$$

Les caractéristiques de la première famille sont définies par

$$\frac{dy}{dx} = \tan(\varphi + \mu) = \tan \frac{1}{3}(2\xi_0 - \eta)$$

ou bien en utilisant (7) par

$$\frac{dy}{dx} = \tan\left[\frac{1}{2}\left(\arctan \frac{y-a}{x} + \xi_0\right)\right]. \tag{8}$$

En intégrant on obtient, en forme paramétrique:

$$x(\xi_0, \eta) = C^* \frac{R}{\left[R - \dfrac{1 - \tan \dfrac{1}{2}\xi_0}{1 + \tan \dfrac{1}{2}\xi_0}\right]}, \tag{9}$$

$$y(\xi_0, \eta) = a + x \tan \frac{1}{3}(\xi_0 - 2\eta),$$

où

$$R = \left\{1 + \tan^2 \frac{1}{3}(\xi_0 - 2\eta)\right\}^{1/2} - \tan \frac{1}{3}(\xi_0 - 2\eta).$$

La constante d'intégration C^* pour la ligne CE est déterminée par la condition $x = \dfrac{a}{2} \cot \mu_0$ pour $\eta = \eta_0$.

Ensuite on a les conditions aux limites pour la zône de l'interférence le long de CE.

$$X(\xi_0, \eta) = -x(\xi_0, \eta) \sin \frac{1}{3}(2\xi_0 - \eta) + y(\xi_0, \eta) \cos \frac{1}{3}(2\xi_0 - \eta) \tag{10}$$

et le long de $CD : X(\xi, \eta_0) = 0$. Avec ceci la solution dans $CDEF$ (équation 5) prend la forme:

$$X(\xi, \eta) = \text{cosec}^2 \frac{\xi + \eta}{3} \sin^2 \frac{\xi_0 + \eta}{3} X(\xi_0, \eta). \tag{11}$$

Pour le calcul de la force de rupture nous n'avons pas besoin de connaître tout le réseau de l'interférence, mais seulement les points situés sur l'axe de symétrie. Donc, sans calculer $Y(\xi, \eta)$, nous pouvons déterminer directement (v. fig. 8):

$$(x)_{y=a/2} = \frac{a}{2} \cot \frac{\eta}{3} - X \operatorname{cosec} \frac{\eta}{3}. \quad (12)$$

À cause de la symétrie, nous avons ici $\sigma_y = s - t$ (cf. 6).

Adoptant une valeur de μ_0 nous trouvons $p + q$ correspondant selon (6). Ensuite pour les différents rapports $\dfrac{a}{b} = \dfrac{1}{n}$, c. à. d. $\dfrac{a}{R} = 2(1 + 4n^2)^{-1/2}$, nous pouvons calculer la pression de confinement q d'après la condition de l'équilibre $\int\limits_0^b \sigma_y dx = q b$. Enfin nous avons $P = p \cdot a \cdot l$ Nous obtenons ainsi la relation entre P et q, qui dépend du rapport donné $\dfrac{a}{b}$.

Pour les différentes valeurs de μ_0 nous avons déterminé ainsi un certain nombre de points conjugués $P(q)$ et $a(q)$ (fig. 9) en cherchant à aligner les deux séries de points sur les courbes expérimentales. Il faut bien dire qu'il est difficile de mesurer la largeur de la surface de contact $\left(\dfrac{a}{R}\right)$ avec une bonne

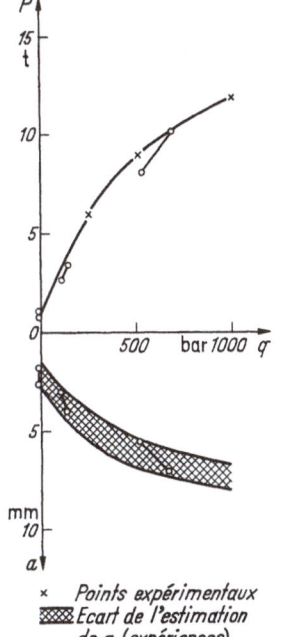

× Points expérimentaux

▨ Écart de l'estimation de a (expériences)

↗° Intervalle de variation de a/b considéré pour un μ_0 donné (calcul)

Fig. 9.

précision; pourtant une corrélation très proche des deux rapports $P(q)$ et $a(q)$ sur toute la longueur de l'intervalle considéré semble convaincante.

Conclusion

L'étude théorique et expérimentale que nous venons de présenter montre l'intérêt de l'interprétation de la rupture par la plasticité et cela sous un double point de vue.

Elle permet d'une part de définir le mécanisme de la rupture diamétrale d'un cylindre dans des conditions de charge complexe. On est assuré que la solution proposée donne des résultats plus corrects pour les essais de routine effectués à la pression atmosphérique. Inversement la comparaison de l'essai de traction, de l'essai de compression simple et de l'essai brésilien est susceptible de donner des informations sur la forme de la courbe intrinsèque dans un certain intervalle, au moyen d'essais très simples.

Mais d'autre part elle montre que l'application de la théorie de l'équilibre limite à des corps comme les roches dont la courbe intrinsèque n'est pas linéaire, donne d'excellents résultats et qu'elle peut être utilisée pour de nombreux autres problèmes.

Références

§ 1

[1] L'Hermite: Que savons nous sur la rupture du béton ? Travaux, juin 1954.

[2] Halabi: Recherches expérimentales systématiques sur la résistance à la traction des bétons. Revue des Matériaux, Janvier 1962.

[3] Narrow et Ullberg: Correlation between tensile splitting strength and flexural strength of concrete. J. Amer. Concrete Inst. Janvier 1963.

§ 2

[4] Griggs and Handin: Rock Deformation. Geological Society of America, Mémoire 79, 1958.

§ 3

[5] Prandtl: Anwendungsbeispiele zu einem Henckyschen Satz über das plastische Gleichgewicht. Z. a. M. M. 3 (1923).

[6] Mandel: Equilibres par tranches planes des solides à la limite d'écoulement, 1942.

[7] Sokolovski: Statics of Soil Media (Édition en anglais, 1960); même auteur, Sur les équations de la théorie de plasticité (en russe). Prikl. Mat. Mekh. 19 (1955).

[8] Perlin, Sur les équations de la théorie de plasticité pour une certaine condition d'écoulement. — Prikl. Mat. Mekh. 26 (1962).

Discussion

Question posée par R. Peltier: Je pense que l'interprétation «élastique» de M. Habib de l'essai brésilien doit être complétée sur certains points. Notamment il semble que des déformations plastiques se produisent sur les géneratrices de contact des éprouvettes et des plateaux de la presse. Si on tient compte de ces adaptations plastiques des zones de contact, on constate expérimentalement que les anomalies signalées par M. Habib disparaissent.

Réponse de P. Habib: Il semble que l'opposition entre l'interprétation utilisant la théorie de l'élasticité et celle qui emploie la théorie de la plasticité soit plus apparente que réelle. Il n'y a pas contradiction entre les objections faites et le texte présenté:

1. M. Peltier précise que l'interprétation élastique doit être corrigée du fait que l'appui ponctuel est une vue de l'esprit: le contact réel est pris par un méplat et ceci entraîne une redistribution des contraintes élastiques.

Dans l'interprétation que nous proposons nous disons que des contraintes extrêmement grandes apparaissent au contact (capables de marquer les plateaux de la presse) et qu'une zone de plastification contenue se forme. Il va sans dire que cette zone plastique modifie la répartition élastique dans le reste du cylindre et si nous ne l'avons pas explicité c'est que nous ne savons comment nous en servir:

en effet le calcul élasto-plastique semble très difficile et la théorie rigide-plastique donne une solution suffisante.

2. M. PELTIER précise que la rupture se produit initialement vers le milieu de l'échantillon. Ce n'est pas incompatible avec le mécanisme proposé, au contraire; en effet la contrainte moyenne dans la roche au voisinage de l'appui est extrêmement grande; les ruptures qui se produisent dans cette zone ne peuvent être fragiles ce sont des glissements sans séparation des parties adjacentes. Lorsque les déformations plastiques sont suffisamment développées pour que les deux noyaux rigides puissent se séparer par un déplacement latéral rien ne s'oppose à ce qu'une fissure se développe à partir de la partie centrale de l'éprouvette.

3. Il est clair, en vertu du principe de St-Venant, que le cercle de MOHR représentatif de l'équilibre élastique des contraintes au centre est peu modifié par un contact qui devient local au lieu d'être ponctuel. De ce fait, l'interprétation élastique, sous pression hydrostatique élevée, est insuffisante pour expliquer les écarts observés et ramener les cercles de MOHR au contact de la courbe intrinsèque lorsqu'ils sont aussi largement sécants qu'on peut le voir sur la figure 3 b.

Question posée par M. DAYRE: Vous avez insisté, lors de votre introduction, M. le Professeur, sur la difficulté inhérente à la mécanique des roches en ce qui concerne l'obtention de résultats à dispersion suffisamment faible.

C'est dans ce but que vous avez choisi, pour réaliser vos essais, un calcaire homogène autant qu'il est possible. Vous avez donc pu obtenir des résultats peu dispersés avec un nombre restreint d'essais.

Pourriez-vous nous indiquer quelle dispersion vous avez obtenue ?

A-t-elle varié avec la pression latérale ?

Réponse de P. HABIB: La dispersion décroit avec la pression moyenne. Pour les grandes pressions latérales elle est très faible, de l'ordre de 1%, c'est à dire parfaitement négligeable et il suffit de faire un seul essai pour définir la courbe intrinsèque.

En compression simple la dispersion n'est pas négligeable et il a été nécessaire de définir la rupture par la moyenne de trois essais. Elle est de l'ordre de 10% pour un matériau aussi remarquablement homogène que le calcaire de St-Beat.

En traction simple la dispersion est plus importante encore. Elle était de l'ordre de 30% pour le matériau étudié.

3.10 Phénomènes de Rupture Fragile des Roches Isotropes et Anisotropes

Par

P. M. Sirieys

Selon les états de contraintes auxquels elles sont soumises, les roches peuvent, ou non, se déformer plastiquement. Sous pressions isotropes élevées la roche fait l'objet de déformations plastiques et se présente comme un matériau élastoplastique, la limite élastique coïncide avec la limite de plasticité. Au dessous d'un seuil, la pression critique de confinement, la rupture se produit avant l'apparition des déformations plastiques, la roche a alors les caractéristiques d'un matériau élastique-fragile, la limite élastique coïncide avec la limite de rupture fragile. Dans les deux cas se superposent des déformations différées, de nature visqueuse, montrant l'influence du tenseur vitesse de déformation dans la loi rhéologique et permettant de déterminer les constantes de temps. Nous examinons les phénomènes de rupture fragile sans tenir compte du facteur temps, c'est-à-dire que nous analysons le comportement de la roche sous vitesses de déformations élevées.

La rupture fragile de tels matériaux comporte la détermination de deux éléments: d'une part l'orientation des lignes ou surfaces de rupture par rapport au tenseur des efforts compatibles avec la cinématique du phénomène, d'autre part les valeurs des composantes du tenseur de rupture.

La surface limite de rupture fragile est déterminée à l'aide d'essais sous états de contraintes variés.

Roches Isotropes

Sous état de contrainte de compression monoaxiale, avec élimination du frettage de tête, les roches sensiblement isotropes, telles que des calcaires ou des argiles cuites, se fendent selon des isostatiques, sur des coupes où la contrainte est nulle.

En compressions triaxiales, avec deux contraintes principales égales, (essais triaxiaux classiques) effectuées sur des Calcaires de Hauteville (Comblanchien) jusqu'à des pressions latérales de 50 kg/cm², la rupture s'effectue selon les isostatiques mineures (sur lesquelles la contrainte est la plus faible compression principale). L'effort de rupture

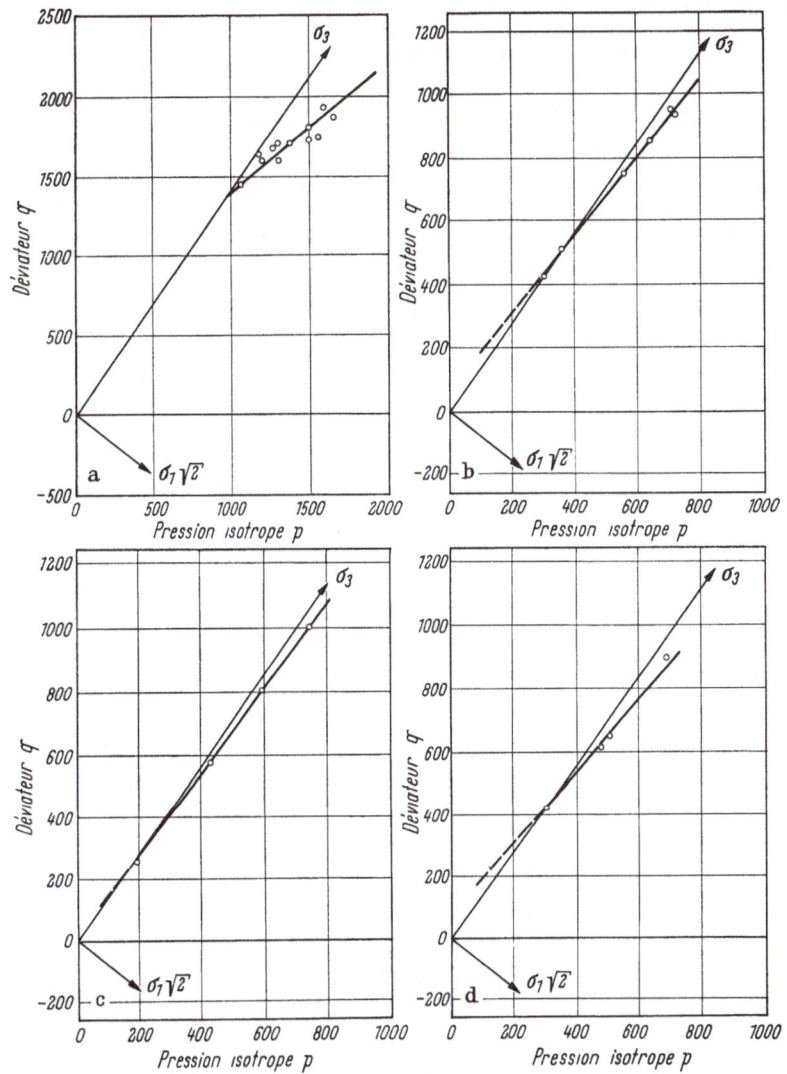

Fig. 1 a—d. Loi limite de rupture fragile de déchirement.
a) Calcaire de Hauteville; b) Calcaire Lignet; c) Laurvickite; d) Schiste cristallin.

croît avec la pression latérale, de façon sensiblement linéaire. Des résultats analogues sont obtenus avec une Laurvickite, un schiste cristallin (testé normalement à la schistosité) et le calcaire Lignet. La coupe de la surface limite de rupture par un plan $\sigma_i = \sigma_j$ (avec $i, j =$

1, 2, 3) pour ces diverses roches est une droite faisant un angle variant
de 39 à 50 degrés avec la trisectrice du trièdre principal. Les résultats
sont représentés sur la fig. 1 où sont reportées (en kg/cm²) les quantités
p et q qui sont fonctions du premier invariant du tenseur contrainte
et du second invariant du déviateur de contrainte $\left(p = \dfrac{I_1}{\sqrt{3}},\ q = \sqrt{2J'^2}\right)$.

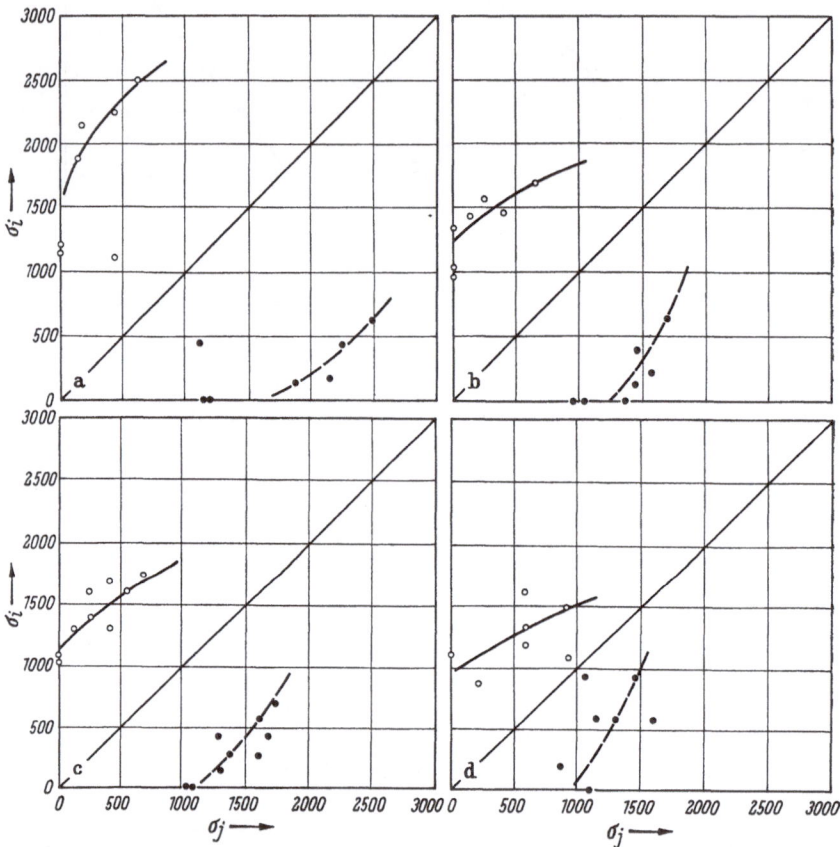

Fig. 2a—d. Loi limite de rupture fragile de déchirement (Influence de la contrainte intermédiaire).
a) Calcaire de Hauteville; b) Pseudobrèche calcaire; c) Marbre de Carrare; d) Calcaire Lignet.

En compressions triaxiales avec une contrainte principale nulle
(appelées essais biaxiaux) effectuées sur des plaques de 15 · 15 · 3 cm,
diverses roches Calcaires ont encore une surface de rupture coïnci-
dant avec une surface isostatique, perpendiculaire à la compression
nulle. En outre l'effort de rupture est fonction de la contrainte intermédiaire [1]. La coupe de la surface limite de rupture par un plan principal ($\sigma_k = 0$) est représentée par la fig. 2.

En traction indirecte, selon la technique de l'essai brésilien, des éprouvettes de béton, d'argile cuite, de calschistes sont fendues selon des diamètres parallèles à l'effort, c'est-à-dire sur des coupes où l'effort de traction est maximal.

Les essais de traction monoaxiale exécutés sur différentes roches montrent que la rupture s'effectue selon l'isostatique normale à l'effort de traction.

Enfin les expériences de flexion sur poutres de roches ou de béton sur deux appuis soumises à un moment de flexion constant dans la partie centrale [2], montrent que les fissures apparaissent dans cette zone selon les isostatiques pour lesquelles l'effort de traction est maximal (à la partie convexe de la poutre).

Si, par ailleurs, lors des essais de compression on réalise par l'intermédiaire du frettage de tête une compression non homogène, la rupture se produit par pénétration de cônes ou de pyramides de matière dans la roche elle-même. Les surfaces de rupture, d'une part, ont alors une inclinaison notable par rapport à l'effort, la contrainte de rupture, d'autre part, est multipliée par un coefficient allant de 1,5 à 4 selon la nature de la roche et l'importance du frettage.

Ces résultats expérimentaux, obtenus sous diverses sollicitations, ont en commun le fait que les ruptures apparaissent dans le domaine élastique. Ils conduisent à envisager deux types de ruptures:

1° Les ruptures isostatiques, ou ruptures d'extension qui s'effectuent selon les isostatiques mineures (en prenant la convention des compressions positives) [3]. Une ligne de rupture apparait alors comme une ligne de discontinuité du champ de vecteurs déplacement. Ces ruptures sont parfois appelées ruptures par déchirement.

La section de la surface de rupture fragile de déchirement par un plan $\sigma_i = \sigma_j$ est sensiblement linéaire. La section de cette surface par un plan $\sigma_i = 0$ montre l'influence de la contrainte intermédiaire; cette influence varie avec la structure de la roche.

2° Les ruptures de cisaillement, qui s'effectuent sur des coupes où agit la composante de cisaillement du vecteur contrainte.

La surface limite de rupture par cisaillement a encore une allure linéaire.

En définitive le seuil d'élasticité des roches se représente par trois surfaces dont les méridiennes sont tracées sur la fig. 3a: La surface de plasticité (P) d'apparition de l'écoulement plastique, la surface de rupture par extension ou déchirement (R_d) pour laquelle la surface de rupture est normale à la contrainte principale mineure, enfin la surface de rupture par caisillement (R_c) pour laquelle la rupture se produit selon des coupes où la loi du cisaillement est vérifiée.

L'équation d'une surface limite de rupture R_d ou R_c peut s'exprimer en contraintes ou en déformations (par l'intermédiaire des relations de l'élasticité). On passe d'une surface à l'autre par une dilatation cylindrique autour de la trisectrice faisant intervenir le coefficient de POISSON ν de module $k = \dfrac{1 - 2\nu}{1 + \nu}$.

La surface R_d déduite des essais précédemment décrits est assez bien représentée par un cône de révolution autour de la trisectrice (fig. 3 b).

On considère fréquemment en plasticité, que le critère d'écoulement ne dépend que de la loi élémentaire sur les coupes où apparait la première déformation plastique. Appliquée à la rupture fragile cette hypothèse conduit à deux types de lois : Celles d'une part ne faisant pas intervenir la contrainte de cisaillement décrivent les ruptures par déchirement. Ainsi a été proposée [4] la loi d'extension maximale pour le béton. Les cercles de MOHR de rupture admettent une enveloppe, mais il n'y a pas coïncidence entre les directions marginales (contact avec l'enveloppe) et l'orientation des lignes de rupture (qui affectent les isostatiques mineures).

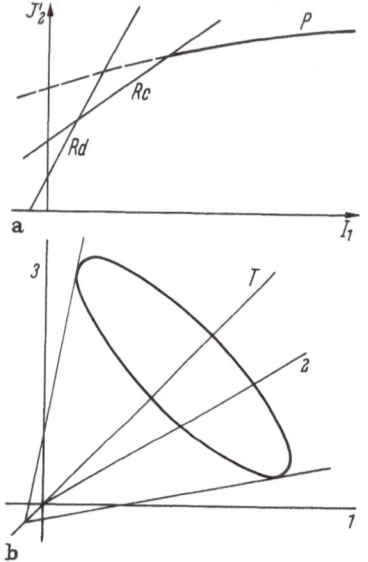

Fig. 3 a et b. a) Les trois critères de limite élastique des roches; b) Allure de la surface limite de déchirement.

Les lois élémentaires, d'autre part, dans lesquelles intervient la contrainte de cisaillement, et éventuellement la contrainte normale, sur la coupe limite (Loi de COULOMB par exemple pour les sols et de SCHMID pour les métaux) décrivent les ruptures de cisaillement. L'enveloppe des cercles de MOHR de rupture jouit dans ce cas des propriétés classiques de coïncidence entre directions marginales et lignes de rupture.

Dans tous les cas la Loi limite de rupture, ou le groupe de lois, à prendre en considération est essentiellement lié aux conditions cinématiques dépendant des données frontières du problème envisagé.

Roches Anisotropes

La structure de nombreuses roches fait apparaitre une ou plusieurs directions de plans privilégiés, c'est par exemple le cas des roches schisteuses. Elles présentent le caractère d'anisotropie discontinue. Les

joints pouvant être eux-mêmes continus (structure lamellaire ou feuilletée) ou discontinus.

Étude Expérimentale

Des schistes ardoisiers analysés en compression monoaxiale se
rompent par cisaillement le long des plans de schistosité lorsque ceux-ci
sont inclinés à 45 degrés par rapport à l'effort, par extension latérale
lorsque ces plans sont perpendiculaires à l'effort. La contrainte de
rupture a une valeur dix fois plus élevée dans le second cas que dans
le premier.

En compression triaxiale, avec une contrainte nulle sur les plans
préférentiels, ces schistes ardoisiers se fendent encore suivant la schistosité. L'effort de rupture décroit avec la pression latérale de manière
sensiblement linéaire, vérifiant ainsi la loi d'extension maximale.

Des essais de fendage, du type brésilien, effectués sur des schistes
cristallins conduisent à des ruptures d'extension ou de cisaillement
suivant l'orientation des plans de schistosité par rapport à celui de l'effort.

Un matériau idéalisé constitué par une argile cuite recimentée à
l'aide de ciment Portland selon des minces couches parallèles a été
analysé [5] en compressions monoaxiales sous diverses orientations.
Ce type d'essai permet d'éviter de fortes dispersions dans les résultats
dues à des hétérogénétités des roches in situ. Il s'applique particulièrement à l'étude des massifs rocheux faillés après injection dans les plans
des failles. Pour α (angle de la normale à la linéation avec l'effort) compris entre 0 et 40 degrés environ les ruptures obtenues sont du type
extension, les contraintes de rupture sont indépendantes de l'orientation
de l'effort. La linéation est passive, elle n'apparait pas au cours des
essais, on obtient la loi d'extension isotrope. Par contre pour α supérieur
à 40 degrés, les ruptures se produisent par glissement le long des plans
de contact brique-ciment et l'effort de rupture subit d'importantes
variations avec l'angle α. Enfin au voisinage de $\alpha = 90$ degrés on
obtient à nouveau des ruptures d'extension le long des plans préférentiels.

Diagrammes Polaires

Ces résultats sous état de contrainte unidimensionnel peuvent se
visualiser à l'aide de diagrammes polaires. Le tenseur monoaxial de
rupture est défini par ses composantes principales (r_α et 0) et son orientation α par rapport à la normale aux plans privilégiés. Il peut être
représenté par le point de coordonnées polaires (r_α, α). Une transformation simple (fig. 4a) permet à partir de ce graphique d'obtenir les

cercles de MOHR de rupture lorsque varie l'orientation du matériau par rapport à l'effort. Au point de coordonnées (r_α, α) correspond le cercle

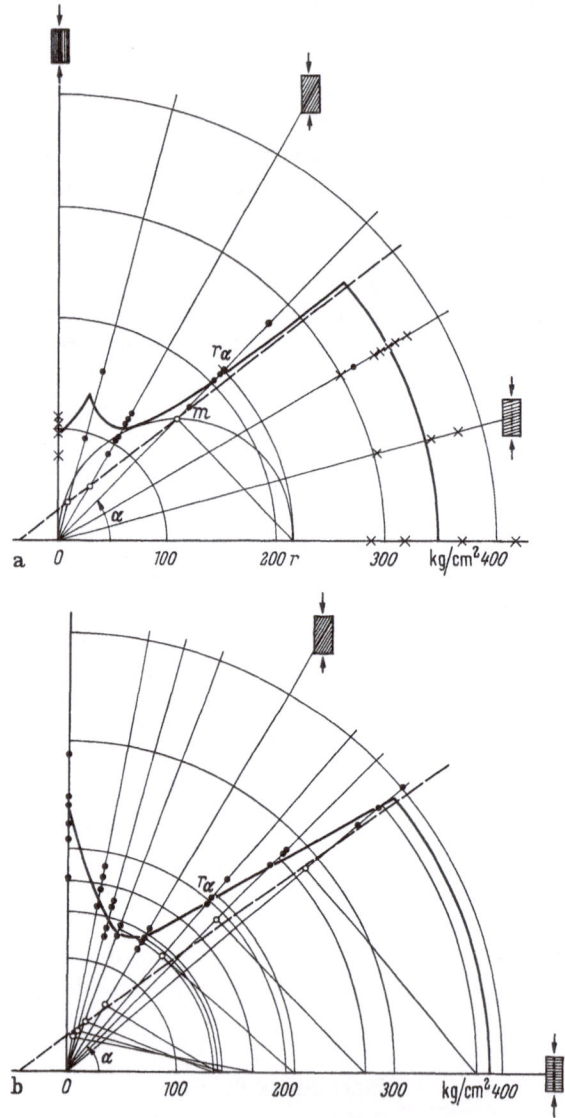

Fig. 4a et b. Rupture anisotrope d'une structure feuilletée.
a) Les trois types de rupture; b) Courbe intrinsèque de clivage.

de MOHR de rupture de diamètre Or et on lit en Om la contrainte sur la coupe privilégiée. Le lieu des points m fournit les caractéristiques de cisaillement sur les plans préférentiels.

Le matériau étudié avait pour caractéristiques de cisaillement sur les plans privilégiés une loi linéaire de la forme:

$$\tau_n = c + \sigma_n \tan \phi$$

avec $c = 27{,}5 \, \text{kg/cm}^2$ et $\phi = 37$ degrés.

Ces paramètres sont évidemment fonction du dosage en ciment dans les minces couches de feuilletage (fig. 4b).

Lois Physiques

En définitive pour de tels matériaux la résistance à la compression monoaxiale sera déterminée lorsque l'on connaitra deux lois physiques: La loi isotrope (L_i) indépendante de la schistosité et la loi valable sur les plans P de schistosité (La) qui se décompose en deux: celle de cisaillement et celle d'extension.

Si l'on ajoute une pression latérale σ_2 on peut figurer le tenseur de rupture $(r_\alpha, \sigma_2, \alpha)$ par le vecteur de coordonnées polaires (r_α, α) et tracer les courbes d'égales valeurs de σ_2. Les courbes théoriques de la résistance à la rupture ont été tracées (fig. 5) avec les lois physiques suivantes: cisaillement linéaire sur les plans P et extension maximale isotrope.

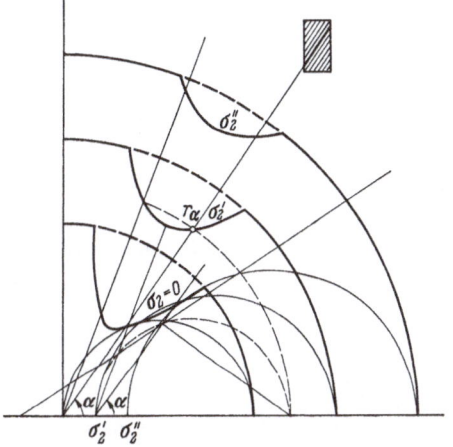

Les surfaces limites isotropes de rupture de matériaux possédant une telle structure sont donc tronquées par deux familles de surfaces (l'une d'extension, l'autre de cisaillement) déduites des deux lois élémentaires sur les plans privilégiés. Chaque

Fig. 5. Influence de la contrainte latérale sur la résistance de clivage.

famille de surfaces dépend de deux paramètres, ceux d'orientation des plans préférentiels par rapport aux axes principaux du tenseur des efforts [6]. Une loi élémentaire de cisaillement linéaire $\tau_n = a + b\sigma_n$ conduit à une famille de cônes, ou de cylindres si le coefficient b est nul. Une loi élémentaire d'extension du type $\varepsilon_n = cte$ conduit, traduite en contraintes par les équations de l'élasticité, à une famille de plans.

Les structures à joints discontinus font l'objet de lignes de rupture mixtes: leur direction est intermédiaire entre celle de l'effort et celle de la linéation. Une telle ligne, avec des schistes cristallins, est constituée d'éléments de cisaillement le long des joints et d'éléments d'extension.

26*

Enfin, la présence de plusieurs directions de joints est fréquemment rencontrée dans les massifs rocheux, c'est également le cas des monocristaux. Les surfaces limites de rupture sont alors tronquées par plusieurs familles de surface de cisaillement anisotrope. Et le plus souvent, pour de telles structures, l'ensemble des lois sur les coupes préférentielles est plus restrictif que la loi isotrope de rupture; on obtient dès lors comme seules surfaces limites de rupture fragile celles déduites des lois relatives aux surfaces de joints. Un cristal de calcite par exemple soumis à un état de compression monoaxial plan fournira, en prenant pour loi élémentaire sur les plans de clivage celle de SCHMID, un diagramme de rupture représenté par la fig. 6.

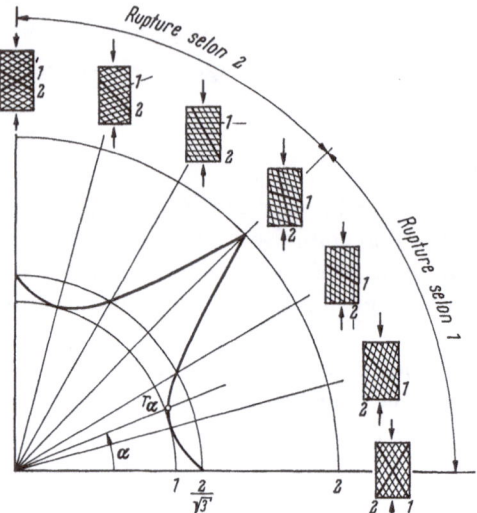

Fig. 6. Rupture des matériaux à deux plans de clivage.
(Cas théorique d'un cristal de Calcite.)

L'étude de la stabilité des massifs rocheux dépend le plus souvent uniquement des lois physiques de rupture sur les plans privilégiés que constituent la schistosité et les systèmes de joints et diaclases.

References

[1] DAYRE, M. et P. M. SIRIEYS: Phénomènes de rupture fragile et de Viscoélasticité des roches isotropes et anisotropes. Comptes Rendus de l'Académie des Sciences **259**, 1163—1166 (1964).

[2] LEBELLE, P., et J. PERCHAT: Compte-Rendu des essais — Enseignement expérimental du béton armé. Annales ITBTP, Série BBA 32, Février 1955.

[3] BRIDGMAN, P. W.: Considerations on Rupture under Triaxial Stress. Mech. Engng., p. 107—111, 1929.

[4] L'HERMITE, R.: Idées actuelles sur la technologie du béton. Documentation technique du bâtiment et des Travaux Publics, 1955.

[5] MORE, G.: Contribution à l'étude de la rupture fragile des matériaux isotropes et anisotropes. Thèse de 3° Cycle, Faculté des Sciences de Grenoble, 1964.

[6] SIRIEYS, P. M.: Lois limites des matériaux rocheux (Anisotropie discontinue). Z. angew. Math. Mech., Sonderheft, **42** (1962).

Discussion

Cf., p. 195, la citation de S. IRMAY.

3.11 On the Tensile Strength Test of Disturbed Soils

By

Hiroshi Hasegawa and **Masayuki Ikeuti**

Synopsis

So little is known as to the tensile strength of soils, that no precise quantitative analysis of some properties of soil structures is yet possible. This paper describes the results of the tests on disturbed soils by the simple tension test apparatus which the authors have devised.

1. Introduction

It is frequently observed that tensile cracks develop near the ground surface prior to the failure of slopes, and it is conceivable that these cracks may greatly influence the failure of slopes. The causes of the tensile stresses near the ground surface are explained [1] in a qualitative manner as follows (Fig. 1 refers.). Let us consider the equilibrium of a slice, ΔH high, then the weight of the slice, ΔW, will form a couple ΔWl, with the vertical component of ΔR, ΔR_v, which acts along the surface of failure as a resultant of the shearing and the normal resistances along that surface.

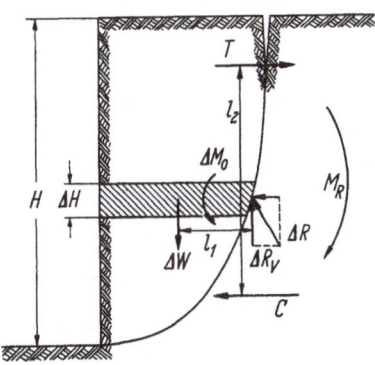

Fig. 1. The causes of tension cracks.

This couple produces an overturning moment ΔM_0. The same thing will happen in all other slices of the sliding wedge of soil. Equilibrium will therefore require the presence of a restoring moment:

$$M_R = Tl_2 = Cl_2 = \Sigma \Delta M_0.$$

This restoring moment produces a tensile stress in the wedge. Very little is known, however. About the tensile strength and the tensile strains of soils, so no precise quantitative analysis of the problem is yet possible.

Tensile strength tests are usually made on briquettes similar to those used in testing Portland cement and are all molded in a standard manner [2]. It was reported that great variations frequently occurred in determining the tensile strength. In the cement tests the stress concentration occurred at the neck of the briquette and the value of the factor of stress concentration was about 1.7.

If discontinuities exist in soil mass, the analysis, taking into consideration only the isotropic and homogeneous plastic material, does not agree with actual cases. For example, slickensided clays of the stiff-fissured type repeated trouble, and the strength at failure was found to vary between 15% and 28% of the undisturbed shear strength. It is considered the reason for this is that the unsolved problems of such as the concentrations of stress and the alternations of the distribution of the pore water pressure manifest themselves.

Strength properties of plastics are, in general, sensitive to both time and temperature. It is considered that temperature has less effect on soil than on the other materials such as high polymers. In this investigation there is no consideration of temperature, but, of course, care has been taken to control temperature in connection with the moisture content of specimens.

2. Simple Tension Test Apparatus

The tension test apparatus is grouped into two types. This grouping is dependent upon the loading direction acting on the test pieces: Vertical type and horizontal type. The defect of the horizontal type apparatus is that frictional resistance is encountered in the moving parts and the guiding parts. The authors, however, have believed that the horizontal type apparatus is usable for soil test if this defect is improved, as the soil sample containing much water has fluidity and is easier to handle horizontally than vertically. In order to decrease the frictional resistance in moving parts, by setting it level to prevent deflection occurring from the beam action of the test piece, to support the test piece exactly level, and to prevent eccentric loading, the authors have tried to test by floating the specimen upon hydrargyrum [3]. Fig. 2 shows the tension test apparatus, and controlled-stress test are examined.

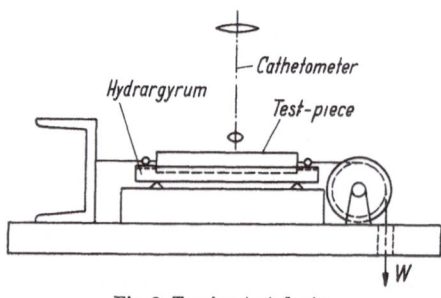

Fig. 2. Tension test device.

It is difficult to transmit the tensile load to the specimen. The authors tried to bury an object that would hook into the test piece, and under some conditions carried this out by using the eye-piece, which is so called for it resembles the eyebar, the tension member of truss.

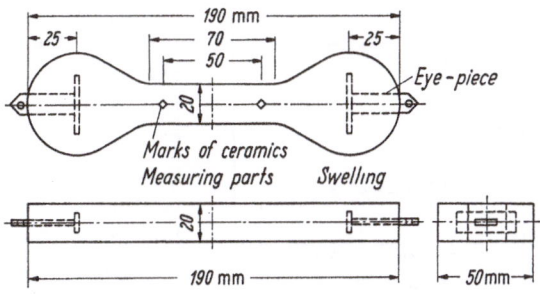

Fig. 3. Test piece and eye-pieces.

It is made of small thin steel plates and forms a cross handle. The form of the test piece is a rectangular parallelepiped and the swellings in which the eye-pieces are buried are attached at both ends. The test piece and the eye-pieces are shown in Fig. 3.

The deformation of the specimen is determined by a cathetometer, by measuring the displacement of marks. The mark is a very small grain of white ceramic and is set up along two axes: One in the loading direction and the other in the rectangular for the loading direction in the horizontal plane.

3. Materials and Method

Tensile strength tests have been performed on the disturbed soils of OBARADAI, YOKOSUKA. Table 1 shows the properties of soil, and the grain size distribution is shown in Fig. 4.

Table 1
The Properties of the Soil

Specific gravity	2.84
Plasric limit (%)	80
Liquid limit (%)	98
Plasticity index	18
Optimum moisture con-	
tent (%)	82
Shrinkage limit (%)	79

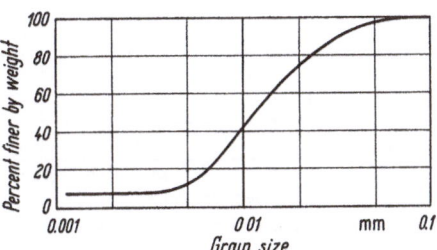

Fig. 4. Grain size distribution of specimen.

The soil was dryed in the air and for the specimen the soil that passed through sieve 74μ was used. In this experiment, soil, in the state of being on the verge of loosing stability thus approximating slope failure, of rather high moisture contents were selected. To make an established moisture content, water was added to the soil and mixed to diffuse the moisture. The moist soil was wrapped in vinylon cloth and left for one or two days. A test piece was remolded in the mold which was devised for this test, and was compacted with evenly distributed blows of a tamper. The properties of the specimen are shown in the following Figs. Fig. 5 shows the relation between densities and the moisture content w: The solid symbols show the density γ and the open symbols show the dry density γ_d Fig. 6 shows the relation between the void ratio e and the moisture content w, and Fig. 7 shows the relation between the degree of saturation S and the moisture content w. These figures show also the compaction characteristics of the soil.

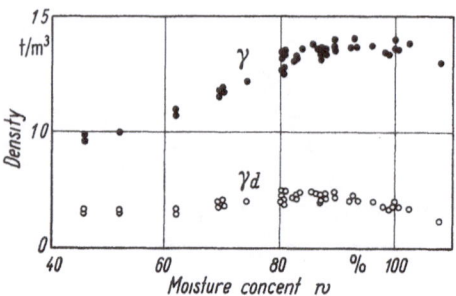

Fig. 5. Relation between the density and the moisture content.

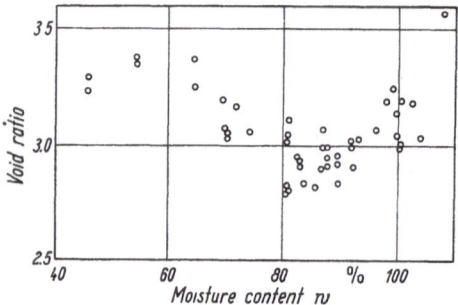

Fig. 6. Relation between the void ratio and the moisture content.

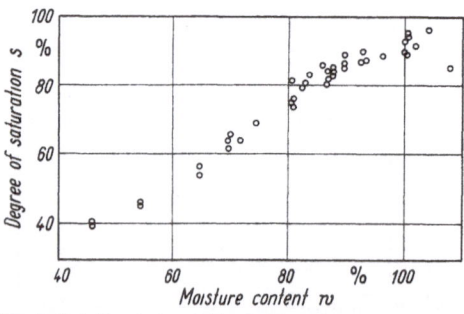

Fig. 7. Relation between the degree of saturation and the mostuire content.

Tensile stress σ and strain ε were computed by the following equations:

$$\sigma = P/A_0,$$
$$\varepsilon = (L - L)/_0L_0,$$

where

P = Tensile load (kg),

A_0 = Initial cross area of the specimen (cm),

L = Distance of the marks (mm),

L_0 = Initial distance of the marks (mm).

4. Experimental Results

The properties of remolded soils which were determined by this experiment can be summarized as follows. Fig. 8 shows the frequency distribution of the location where the tensile cracks occurred. The cracks occurred perpendicularly to the axis of loading, namely along the principal plane, and sliding ruptures were not observed. The axis of abscissa is the distance (cm) from the loading side along the axis of the test piece. The numbers of the test pieces were 63. The contact planes with the eye-pieces are 2.5 and 16.5. Under 6 and over 13 are the parts where the cross section varies, and the test pieces which broke at these points were unstatisfactory samples for the tensile strength tests. In the area from 6 to 13 the cross section of the test piece is constant, and the data taken as a result of cracks occurring in this area may be used for the tensile strength tests. The almost all test pieces, with moisture contents under $P.L.$ and some over $L.L.$ broke along the contact planes with

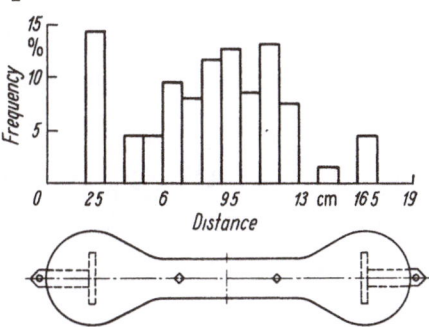

Fig. 8. Frequency distribution of the locations where the tensile cracks occurred.

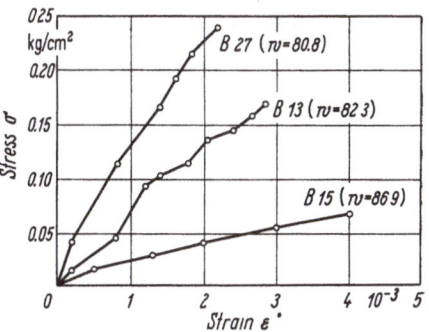

Fig. 9. Stress-strain diagram.

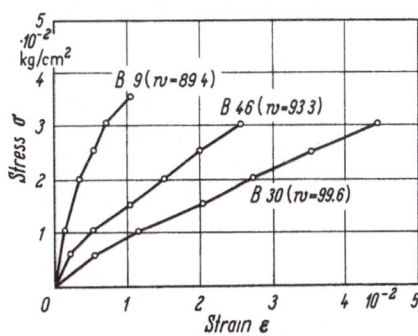

Fig. 10. Stress-strain diagram.

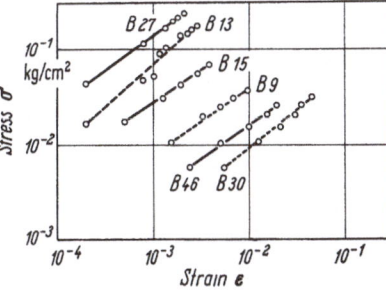

Fig. 11. Stress-strain diagram by the logarithmic scale.

the eyepieces. No test pieces, with a moisture content between $P.L.$ and $L.L.$ broke at those planes.

Typical stress strain diagrams are shown in Figs. 9 and 10, where w is the moisture content. The stress-strain relations of materials which do not follow Hooke's Law are generally expressed as follows:

$$\varepsilon = \alpha \sigma^m,$$

or

$$\sigma = \beta \varepsilon^n,$$

where α, β, m, and n are constants.

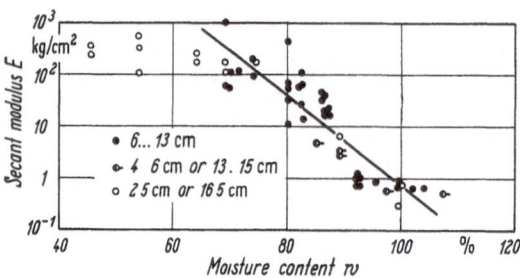

Fig. 12. Relation between the secant modulus of elasticity and the moisture content.

Fig. 11 shows the relation between stress and strain by the logarithmic scale. The values of n of samples shown in Fig. 11 very between 0.81 and 0.65.

Fig. 12 shows the relation between the secant modulus of elasticity $E = \dfrac{1}{a} \sigma^{1-m}$ at

the stress that was observed just before the ultimate strength and the moisture content w. The solid symbols are the data from the tests when the tensile cracks occurred between 6 and 13. The semi-solid symbols are under 6 and over 13, and the open symbols are at 2.5 and 13.5. The symbols other than the solid symbols mer y have the meaning of the condition of center zone when the rupture had occurred. Same symbols are used for Figs. 13 and 14. The relation between the secant modulus of elasticity E and moisture content w is shown as follows:

$$E = e^{-0.195w + 19.3}.$$

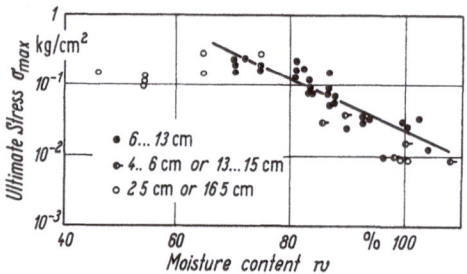

Fig. 13. Relation between the ultimate stress and the moisture content.

Fig. 13 shows the relation between the ultimate stress $\sigma_{\max}$ and the moisture content w and the relation is shown as:

$$\sigma_{\max} = e^{-0.0831w + 4.58}.$$

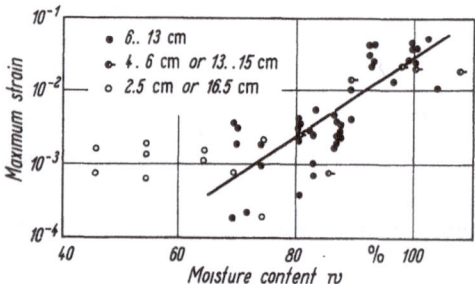

Fig. 14. Relation between the maximum strain and the moisture content.

Fig. 14 shows the relation between the maximum strain $\varepsilon_{\max}$ and the moisture content w and the relation is shown as:

$$\varepsilon_{\max} = e^{0.122w - 15.8}.$$

Fig. 15 shows the stress-strain relations of the repeated loading test that the cracks had occured in the tensile strength test. The curve shown by the open symbols is that for the sample B + 42 (the moisture content is 92.3%). The crack occurred at $\sigma_{\max}$ 3.10×10^{-2} kg/cm² and $\varepsilon_{\max}$ 4.15×10^{-2}. The tensile strength test was completed at this point, and upon unloading the strain became to 2.14×10^{-2}. This was considered as the permanent set. The result of loading for the cracked test piece was σ_c 1.63×10^{-2} kg/cm² and ε_c 4.05×10^{-2}. The ratio of $\sigma_c/\sigma_{\max}$ was 0.44 and $\varepsilon_c/\varepsilon_{\max}$ was 0.949. The frequency distribution of the ratio of the decrease of strenght $\sigma_c/\sigma_{\max}$ is shown in Fig. 16. The reciprocal number of this ratio may be considered as the factor of stress concentration by cracks. The ratio of strain $\varepsilon_c/\varepsilon_{\max}$ varied between 0.949 and 1.204 and the arithmetical mean was 1.034.

Fig. 17 shows the relation between strain and time under constant stress. The solid symbols show the result of constant stress test ($\sigma = 6.81 \times 10^{-3}$ kg/cm²) on the

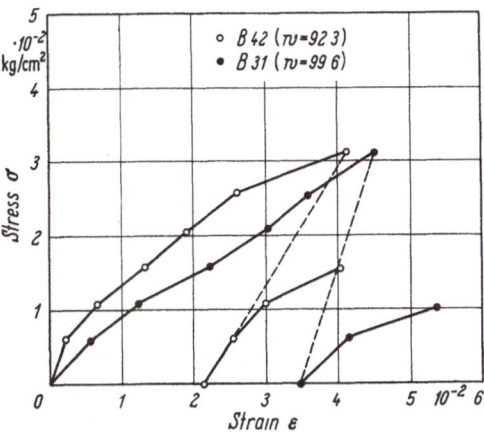

Fig. 15. Stress-strain diagram of the repeated loading for the cracked test piece.

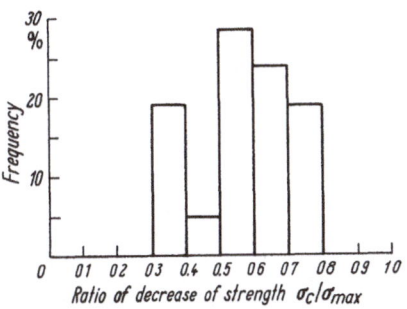

Fig. 16. The frequency distribution of the ratio of the decrease of strength.

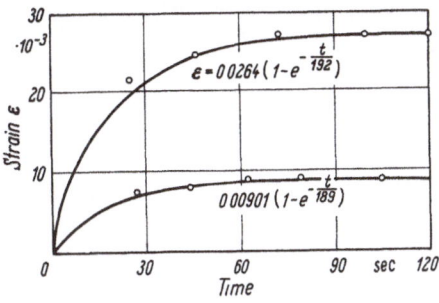

Fig. 17. Relation between strain and time under constant stress.

specimen of the moisture content of 103.0%, and the open symbols show the result of the repeated test on the same test piece 16 minutes after the first test.

5. Conclusions

The findings presented in this paper can be summarized as follows:

1. In the tension test of disturbed soils, this apparatus is workable. There remains, as yet, many points to be developed. The eye-piece is considered an effective means of transmitting the tensile loads to the specimen: under some conditions.

2. The rupture occurred along the principal plane, and sliding rupture by the shearing stress was not observed. The maximum strain was 5.5×10^{-2}, and it may be said that the soil is a brittle material.

3. The moisture content influenced the tensile properties, and as the moisture content varied from 80% to 100%, the maximum stress decreased about one-fifth, the maximum strain increased about ten times, and the secant modulus of elasticity decreased about one-fiftieth.

4. The maximum factor of stress concentration by the cracks was about 3, and in the analysis of the stability of the slope, one must take deliberate notice of the depth to which the tension cracks extend.

5. The relation between time and strain under constant stress was shown as $\varepsilon = c_1 (1 - e^{-t/c_2})$.

The authors wish to express their gratitude to Professor Yasuo Mashima for his valuable advice in the course of the work. Thanks are also due to Dr. Yoshinori Ohira for all possible help in providing the laboratory and equipment.

Bibliography

[1] Tschebotarioff, G. P.: Soil Mechanics, Foundations, and Earth Structures, 1957.
[2] Hogentogler, C. A.: Engineering Properties of Soil, 1937.
[3] Hasegawa, H., and M. Ikeuti: Some Studies on the Tensile Strength Test of Soils. In Japanese, 1961, Annual Meeting of J. S. S. M. F. E.

Rapport Général
Relatif à la 3ᵉ Sous-Section

J. Biarez

Il est une propriété du sol dont on a peu discuté au cours de ce Congrès, je veux parler de la pression d'eau interstitielle; mais, s'il a semblé inutile d'en parler, c'est que la mesure et le rôle de cette pression sont aujourd'hui bien connus, en particulier grâce aux travaux d'un grand absent, je veux nommer BISHOP. En fait, des liaisons intermoléculaires de l'eau adsorbée, à la submersion des barrages, l'eau a été notre premier souci.

Après le fluide interstitiel, notre attention se porte sur les grains auxquels nous associons l'eau adsorbée. Nous avons ainsi une ossature dont nous discutons les propriétés à l'échelle du grain. Mais, l'étude, à cette échelle, aura son intérêt prodigieusement accru, si nous pouvons faire l'hypothèse d'un milieu continu fictif. Nous disposons alors des concepts, de la formulation et des nombreux résultats de la Mécanique des Milieux Continus. La loi rhéologique trop complexe sera divisée en lois partielles simples et continues, acceptables dans des domaines à définir de l'espace des contraintes ou des déformations par exemple. Ainsi, va s'engager un dialogue entre l'étude à l'échelle du grain et l'étude à l'échelle de l'échantillon supposé continu.

Pour les milieux pulvérulents, WINTERKORN, KÉZDI, HAYTHORNTHWAITE, nous ont indiqué tout l'intérêt du calcul pour passer des propriétés d'une échelle à l'autre. Par ailleurs, WIENDIECK, en effectuant des déformations aussi homogènes que possible, montre le lien entre l'anisotropie des propriétés mécaniques et l'anisotropie de la géométrie statistique de l'assemblage. Celle-ci possède une orientation constante par rapport au tenseur de déformation et contrainte en plasticité parfaite; en outre, dans ce dernier domaine, la variation de volume semble nulle, mais la densité critique croît avec la pression moyenne. Dans un appareil de cisaillement, ROSCOE examine avec précision, aux rayons X, la variation de densité dans l'espace de l'échantillon et dans le temps. En dernier lieu, KERISEL cite des essais où la déformation dépend du chemin parcouru dans l'espace des contraintes.

Pour les milieux argileux, l'influence du temps devient primordiale. La loi rhéologique devra faire correspondre un chemin dans l'espace des déformations à tout chemin dans l'espace des contraintes, les deux étant repérés en fonction du temps.

Les lois rhéologiques étant très compliquées, il est judicieux de choisir les états de contraintes et les chemins de contraintes les plus simples et, comme l'a suggéré TAN TJONG-KIE, c'est de cette manière seulement qu'il est commode d'étudier la validité de la loi de superposition. Quand le matériau n'obéit pas à une loi linéaire, il en résulte un problème très complexe et on sera réduit à simuler, au laboratoire, des chemins voisins de ceux suivis par le sol en place. La meilleure simulation n'étant pas toujours obtenue à l'aide d'essais courants du type oedomètre.

Poursuivant la voie qu'il avait ouverte il y a plus de quinze ans avec GEUZE, TAN TJONG-KIE, s'appuyant sur les observations de ROSENQUIST au microscope électronique, explique le comportement visqueux et élastique de l'argile soumise à un cisaillement; il en déduit aussi que l'ossature doit être viscoélastique sous état de contrainte isotrope, que la perméabilité est une fonction du temps, et souligne l'importance de l'écrouissage isotrope dont il donne l'influence sur la variation de perméabilité et de compressibilité. S'appuyant enfin sur des essais dynamiques, l'auteur propose une méthode de calcul.

TAN TJONG-KIE et, par ailleurs, MURAYAMA, présentent des essais permettant de connaître les tenseurs de contrainte de rupture, pour des déformations infiniment lentes, en mesurant la déformation pendant des temps égaux, sous des charges constantes diverses. Or, cette vitesse de déformation augmente beaucoup plus rapidement à partir de certaines valeurs du tenseur de contrainte. On obtient ainsi facilement la limite de rupture qui est vraisemblablement la limite des comportements solide et liquide. Cette méthode simple et rapide semble directement utilisable par l'ingénieur pour l'étude de l'équilibre limite des ouvrages à long terme. Un autre aspect de l'équilibre limite est présenté par SKEMPTON qui montre que l'utilisation de l'angle de frottement "résiduel" permet d'expliquer, d'une manière satisfaisante, des glissements de terrain récents.

A l'intérieur du domaine solide, semble exister un domaine réversible, plus ou moins net selon les auteurs, dont la limite dépend du mode de fabrication, de l'histoire des déformations irréversibles et de l'orientation du tenseur utilisé dans l'expérience. L'existence de ce domaine réversible est nettement confirmée par BJERRUM, au cours d'essais de charge in situ où l'auteur estime que le calcul en élasticité est parfois suffisant.

Associer ces divers comportements à des modèles rhéologiques peut être tentant, mais GEUZE nous met en garde contre ces modèles à une dimension pour un espace à trois dimensions; il insiste sur l'étude indispensable des propriétés à l'échelle du grain. Toutefois, nos connaissances dans ce domaine ont à faire de grands progrès. SKEMPTON rappelle l'orientation privilégiée des particules à la fin du cisaillement; FUJIMOTO utilise la loi de EYRING pour les liaisons intermoléculaires dans la couche adsorbée, mais, faute de connaissances suffisantes à cette échelle, on s'aidera des observations sur le milieu continu pour faire des hypothèses sur le comportement de l'ossature. On peut supposer que le bord d'une particule d'argile possède une liaison réversible, pour les faibles sollicitations, avec la face d'une autre particule, puis se déplace sur cette face et, enfin, le lien se rompt. MURAYAMA estime que ces trois types de déplacements peuvent être simultanés, mais en nombre relatif, variable en fonction de la grandeur de la déformation et du temps.

BJERRUM complète ces études en indiquant que la loi de déformation en fonction du temps est modifiée si le sol in situ subit des cycles de chargement.

PIERRARD, en supposant la viscoélasticité linéaire du premier ordre pour un grès presque saturé, montre l'intérêt des cycles de déformations pour mesurer les coefficients intervenant dans la loi. IRMAY indique que les résultats expérimentaux confirment les calculs qu'il a effectués il y a plusieurs années.

Les propriétés de l'ossature qui produisent la consolidation secondaire doivent, en fait, se superposer à l'écoulement du fluide interstitiel. Dans ce cadre, SCHULTZE montre qu'au cours de la consolidation dans l'oedomètre, la pression interstitielle mesurée est de 10 à 20% plus faible au centre que ne l'indiquent les théories de TAYLOR, TERZAGHI et FRÖHLICH; elle est, par contre, plus forte

sur les bords. Le tassement en fonction du temps est en meilleur accord avec
la théorie pour les sols peu denses que pour les autres. Diverses explications
sont proposées par l'auteur; en outre, DE BEER suggère l'influence de la perméabilité qui varie dans l'espace et le temps. On doit ici mentionner l'intérêt de la
méthode de calcul de ŠUKLJE qui permet de passer directement des mesures
faites dans l'oedomètre au tassement de la couche. Le temps nous manque
malheureusement pour insister sur l'importante étude de ROSCOE sur les argiles,
en relation avec le potentiel plastique.

En dernier lieu, un troisième type de sol est enfin examiné par HABIB, MORLIER et RADEN-
KOVIC; il s'agit de roches soumises à des essais du type brésilien. Après avoir
montré les imperfections du calcul en élasticité, les auteurs proposent une solution
plastique utilisant la courbe intrinsèque non linéaire de Perlin, qui permet une
solution explicite. Les calculs sont en bon accord avec l'expérience, pour des
pressions de confinement allant jusqu'à 1.000 bars. M. PELTIER estime qu'en
tenant compte de la déformation de l'échantillon aux points de contact, la solution
élastique donne de bons résultats aux faibles pressions.

En dernier lieu, pour le domaine des ruptures fragiles, SIRIEYS détermine la
surface limite de rupture fragile et montre que les surfaces de rupture sont
souvent des isostatiques pour des échantillons isotropes. En cas d'anisotropie
discontinue, l'auteur et ses collaborateurs, DAYRE, GUEROULT et MORE, observent
l'influence de l'orientation de l'anisotropie, avec la sollicitation extérieure et
proposent une méthode de calcul.

Ce rappel trop rapide des exposés sur les mesures rhéologiques au laboratoire
ne doit pas nous faire oublier que le programme prévoyait l'étude des mesures in
situ. Nous regrettons très vivement l'absence d'études sur ce sujet car, à quoi
peuvent servir nos travaux si nous ne savons pas prélever un échantillon non
remanié d'argile sensible ou de gravier, ou si nous ne savons pas mesurer les
propriétés mécaniques de la roche qui soient satisfaisantes à l'échelle qui nous
intéresse.

4.1 Vérification Expérimentale des Méthodes de Calcul des Rideaux d'Ancrage de Palplanches Simplement Ancrées

Par

J. Verdeyen et **J. Nuyens**

1. Énoncé de Problème

On considère le cas d'un rideau de palplanches simplement appuyées en tête dont l'appui est retenu par un rideau d'ancrage simple. Cette construction se compose des parties constitutives suivantes (fig. 1):

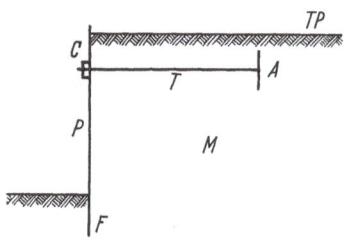

Fig. 1. Parties constitutions d'un rideau de palplanches simplement appuyé.

un rideau de palplanches P soutenant un massif de terre M et dont la fiche F est enfoncée dans le sol. Cette fiche réalise un encastrement constituant un premier appui pour le rideau, si on considère celui-ci comme une poutre. Un second appui est constitué par le chaînage C, retenu par les tirants T, lesquels sont eux-mêmes ancrés dans le sol par le rideau d'ancrage A.

Ce rideau peut être entièrement situé à une certaine distance sous le niveau du terre-plein TP, ou peut y affleurer.

Considéré sous l'angle de l'art de l'ingénieur, le problème à résoudre peut être énoncé de la manière suivante:

Déterminer quelles sont les dimensions et la position qu'il faut donner à un rideau d'ancrage pour qu'il puisse résister à une force donnée.

Il existe dans la littérature un certain nombre de méthodes générales plus ou moins rationnelles pouvant s'appliquer à la résolution du problème énoncé ci-dessus ou de méthodes particulières qui y sont spécialement adaptées.

L'objet de la communication est de faire connaître les résultats d'essais sur modèles effectués au laboratoire de mécanique des sols de l'Université Libre de Bruxelles. Ces résultats ont permis de faire un choix parmi les méthodes particulières relatives aux ancrages courts et d'adopter une méthode générale applicable au problème des ancrages longs.

Rappelons que l'on parle d'ancrage long lorsque le système d'ancrage est suffisamment éloigné du rideau de palplanches pour que les massifs de terre influencés par le rideau de soutènement d'une part et le rideau d'ancrage d'autre part, ne s'interpénètrent pas; dans ce cas, le système d'ancrage peut être calculé indépendamment du rideau de soutènement.

Dans l'autre cas, les massifs influencés par les deux systèmes s'interpénètrent; il y a une influence réciproque et le système d'ancrage doit être calculé compte tenu de sa position par rapport au rideau de palplanches. On a, dans ce cas, un ancrage court.

2. Classification des Méthodes de Calcul

On sait que l'on classe habituellement les méthodes de calcul de pression des terres comme appartenant à quatre groupes:

les méthodes aux extrêmes;

les méthodes basées sur la théorie de la plasticité, c'est-à-dire sur les états d'équilibre limites;

les méthodes basées sur la théorie de l'élasticité;

les méthodes empiriques basées, dans la plupart des cas, sur des résultats expérimentaux uniquement.

Chacun de ces groupes se caractérise comme suit:

Les méthodes aux extrêmes sont caractérisées par le fait qu'elles font appel à une seule condition d'équilibre puis en exprimant que cette condition correspond à un extrêmum (poussée ou butée). De plus, il est nécessaire pour que la condition d'équilibre puisse être expérimentée sans ambiguïté, que la forme de la ligne de glissement soit telle que les contraintes inconnues n'y apparaissent pas.

Dans la méthode de RENDULIC, cette condition est obtenue en choisissant comme ligne de glissement une spirale logarithmique d'angle égal à l'angle de frottement du sol. De cette manière, l'ensemble des contraintes agissant sur les facettes qui constituent la ligne de glissement passe par le pôle de la spirale. Dans le cas particulier d'un sol purement cohérent, la spirale logarithmique se réduit à un cercle. C'est le cas de la méthode de FELLENIUS. Dans le cas particulier d'un pôle situé à l'infini, la spirale logarithmique devient une droite. C'est le cas de la méthode de COULOMB.

Les méthodes de la plasticité permettent en principe de déterminer à n'importe quel point les contraintes au moyen des équations d'équilibre d'un élément de sol et de l'expression des critères de rupture. On sait que ces méthodes théoriquement exactes conduisent à des systèmes d'équations aux dérivées partielles non intégrables dans le cas général. Parmi ces méthodes, il en existe de plusieurs types. Les méthodes des conditions aux limites supposent que la ligne de glissement rencontre le terre-plein et l'ouvrage de soutènement à des angles statiquement corrects, de manière à ce que des relations entre les contraintes aux limites puissent être déterminées. Les autres méthodes de la plasticité étudient l'équilibre de tout le massif situé au-dessus de la ligne de glissement. Certaines, comme celle de Sokolowski ou de Caquot donnent la résolution numériquement approchée correspondant à l'état d'équilibre limite. D'autres, comme celles de Drucker, Hodge, et Prager, supposent que la pression des terres est comprise entre deux valeurs données respectivement par un champ de contraintes statiquement admissible et un champ de vitesse cinématiquement admissible. Dans ce dernier cas, la forme de la ligne de glissement doit être une spirale logarithmique. La méthode générale de Brinch Hansen appartient aussi aux méthodes de la plasticité puisqu'elle fait appel à des figures de rupture utilisant une condition aux limites particulière qui rend possible le calcul à partir des conditions d'équilibre. Cette méthode utilise les équations de Kötter pour calculer les variations de contraintes le long d'une ligne de rupture.

Dans le même ordre d'idée, Bent Hansen a remplacé dans le cas de la rupture linéaire la ligne de glissement par une zone étroite plastique. Si l'on suppose que le prisme élastique de glissement est incompressible, les lignes limites de cette zone sont des lignes d'élongation nulle et les directions des rayons vecteurs sont leur conjuguées. Il en déduit que la zone de glissement est bordée de spirales logarithmiques.

Nous citons les méthodes de l'élasticité pour mémoire car elles sont peu adaptées aux problèmes de la pression des terres.

En ce qui concerne le quatrième groupe des méthodes empiriques aucune de ces méthodes courantes ne convient particulièrement aux problèmes des ancrages longs mais certaines méthodes concernent les ancrages courts, telles la méthode de Kranz, la méthode particulière de Brinch Hansen adaptée aux ancrages courts, et la méthode de Jelinek.

La méthode de Kranz suppose que la ligne de glissement est une droite joignant le pied du rideau de soutènement au pied du rideau d'ancrage, et suppose également que l'on connaît la résultante des pressions agissant sur le rideau de soutènement. La sécurité est exprimée par rapport à la force disponible pour l'ancrage.

La méthode de Brinch Hansen suppose que la ligne de glissement est une spirale logarithmique joignant le pied du rideau de soutènement au pied du rideau d'ancrage, mais ne fait aucune hypothèse quant à la résultante des pressions agissant sur le rideau de soutènement. La sécurité est exprimée par rapport aux paramètres de la rupture.

La méthode de JELINEK est une combinaison de ces méthodes qui applique le calcul statique de la méthode de KRANZ, à une courbe de glissement en forme de spirale logarithmique. La sécurité trouvée est plus faible que celle trouvée par la méthode de KRANZ mais est sans relation directe avec celle que l'on trouve par la méthode de BRINCH HANSEN.

3. Description du Modèle Expérimental

Le problème posé étant un problème plan de déformation de par sa symétrie de translation, on a utilisé pour le résoudre l'appareil de TAYLOR-SCHNEEBELI. Cet appareil comporte un bâti *1* destiné à contenir un massif bidimensionnel de rouleaux *2*. Les essais sont menés généralement comme suit (fig. 2a et b):

Le rideau de palplanches *7* est accroché à son chaînage *8* lequel est maintenu par un support amovible *13*. Le massif *2* est ensuite mis en

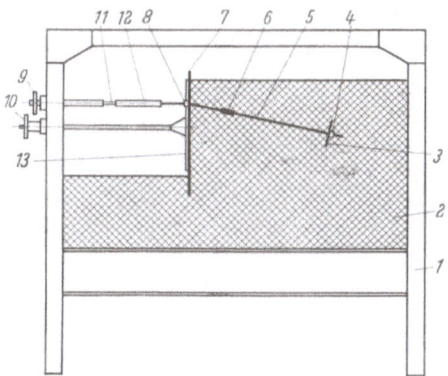

Fig. 2a. Appareil de TAYLOR-SCHNEEBELI.

place, ainsi que l'ancrage *3* et son chaînage *4*. Des deux côtés du massif, les deux chaînages sont réunis par des tirants *5*. On commence la première phase proprement dite de l'essai en libérant le rideau de l'appui de mise en place *13* en actionnant la commande *10*. On photographie le mouvement des rouleaux en vue de déterminer quelle est la zone de massif qui intervient dans cette première phase. Lorsque la stabilisation s'est produite, on prépare la deuxième phase de l'essai en accrochant le chaînage *8* du rideau de palplanches au dispositif de traction *12* à l'aide de la commande *9*. On photographie le mouvement, lors de la rupture.

Le modèle de sol est constitué d'un mélange à poids égaux de rouleaux d'aluminium de 5 et de 8 mm de diamètre qui présentent l'avantage, lorsqu'ils sont bien mélangés au départ, et mis en place soigneusement, de donner une disposition tout à fait quelconque sans direction et sans zone privilégiée. Les caractéristiques du modèle sont les suivantes :

poids spécifique apparent du massif $\qquad \gamma = 2\,T/m^3$,
angle de frottement du massif $\qquad \varphi = 22°$,
angle de frottement maximum sol sur paroi $\quad \delta = 19°$.

Fig. 2b. Appareil de Taylor-Schneebeli.

4. Résultats des Essais sur Ancrages Longs

Les essais de traction sur des rideaux d'ancrages longs ont permis d'arriver aux conclusions suivantes relatives au rideau représenté à la fig. 3 :

a) Il ne faut pas limiter la résistance opposée à l'ancrage à l'intégrale des pressions du cas particulier de Rankine, sur la hauteur h du rideau d'ancrage (méthode classique) ;

b) si on prend l'intégrale des pressions du cas particulier de Rankine, sur toute la hauteur H allant du bas de l'ancrage au terre-plein (méthode classique modifiée), on se rapproche plus de la réalité, surtout si $\dfrac{H}{h}$ n'est pas trop élevé. Ceci a été confirmé par Terzaghi qui estime que la méthode classique modifiée est valable pour des rideaux d'ancrage si $\frac{H}{h} < 2$. Cette propriété est clairement mise en évidence par les

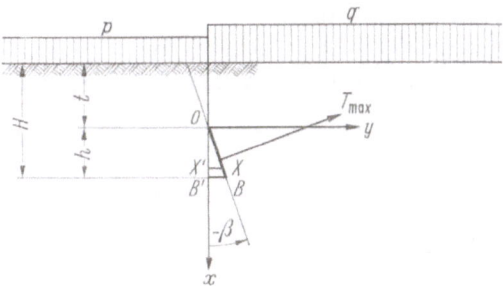

Fig. 3. Rideau d'ancrage long.

photographies obtenues dans les essais sur modèles à l'appareil TAYLOR-SCHNEEBELI (fig. 4).

c) Les photographies montrent que l'on est en présence d'équilibres curvilignes. Cependant, si on calcule la résistance d'un rideau d'ancrage

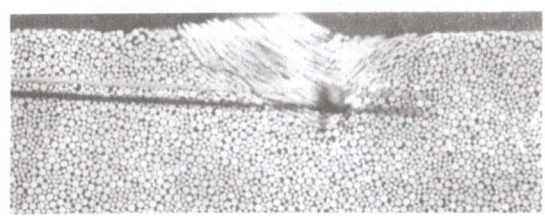

Fig. 4. Ancrage long à faible profondeur.

par une méthode supposant un équilibre curviligne sur toute la hauteur, par exemple la méthode de CAQUOT, on trouve généralement une résistance appréciablement trop élevée. Cette différence s'explique par l'existence au-dessus de l'ancrage d'une zone immobile mise en évidence

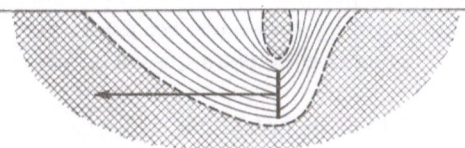

Fig. 5. Figure de rupture d'un ancrage long.

par les photographies, et stylisée à la fig. 5. Les photographies montrent également que cette zone immobile tend à disparaître pour les valeurs de $\frac{H}{h}$ faibles mais non nulles. C'est ce qui explique la concordance de la méthode classique modifiée pour $\frac{H}{h} < 2$.

d) Les photographies d'essais (figures 4 et 6) montrent que les figures de rupture qui normalement devraient être rectilignes dans la partie supérieure sont curvilignes sur toute la hauteur. Cela peut s'expliquer par le fait que lorsque le mouvement tend à se produire, la porosité tend à augmenter et ce phénomène se produit d'autant plus facilement que la pression due au poids du sol est faible, c'est-à-dire

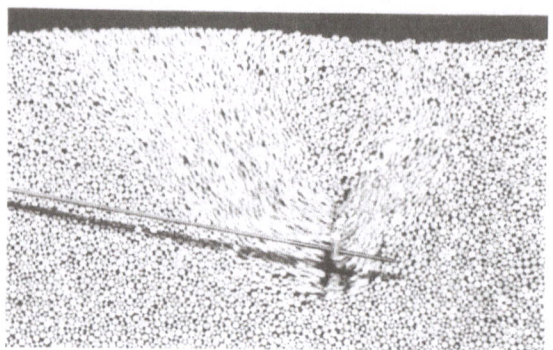

Fig. 6. Ancrage long profondeur moyenne.

qu'on est près de la surface. Il en résulte que l'angle de frottement diminue lorsqu'on remonte vers la surface et que, par conséquent, l'angle que fait la ligne de glissement avec l'horizontale, c'est-à-dire $\left(\frac{\pi}{4} - \frac{\varphi}{2}\right)$ augmente.

e) Un rideau d'ancrage incliné sollicité par une force perpendiculaire a pratiquement la même résistance qu'un ancrage vertical si l'inclinaison par rapport à la verticale est inférieure à 10°.

f) En conclusion des alinéas qui précèdent, on voit que si l'on veut mettre au point une méthode de calcul permettant de prédéterminer avec plus de précision la résistance à la traction d'ancrages longs, il est nécessaire de tenir compte des faits principaux suivants:

1° les figures de rupture sont curvilignes à la partie inférieure;

2° il existe une zone immobile au-dessus de l'ancrage.

5. Méthode de Calcul des Ancrages Longs

Soit la figure de rupture représentée à la fig. 7. La zone en amont de l'ancrage est en état d'équilibre curviligne de butée, la zone en aval, en état d'équilibre curviligne de poussée, et la zone intermédiaire reste immobile. Cette constatation expérimentale a permis de faire l'hypothèse que la résistance maximum à la traction d'un ancrage long peut être calculée en déterminant les poussées et les butées par la méthode de Sokolovski suivant le schéma représenté à la fig. 7. Les calculs de la méthode de SOKOLOVSKI à l'ordinateur ont permis l'établissement d'abaques. Ces abaques permettent de déterminer très rapidement la valeur de la résistance à la traction des rideaux d'ancrage longs.

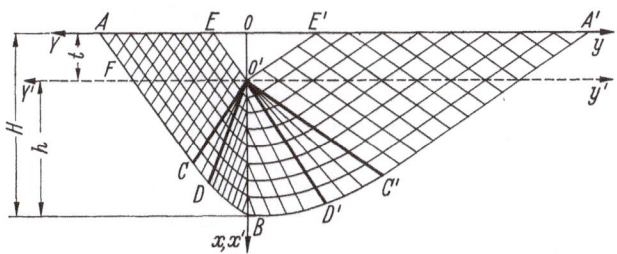

Fig. 7. Application de la méthode de SOKOLOWSKI an calcul des ancrages longs.

D'après la comparaison de quelques résultats numériques quantitatifs qui ont pu être déterminés lors des essais, il semble que l'angle δ de frottement sol sur paroi ne soit pas entièrement mobilisé lors du mouvement des rideaux d'ancrage.

Le problème à résoudre est le suivant:

Quelle est la charge limite que l'on peut exercer perpendiculairement à l'ancrage OB enterré dans un massif de caractéristiques φ, c, γ? Le bord supérieur O de l'ancrage est surmonté d'une hauteur de terre t. Le terre-plein horizontal est sollicité par une surcharge uniforme de valeur q d'un côté de l'ancrage dans le sens de la butée et de valeur p de l'autre côté, dans le sens de la poussée. Soit l'axe des x vertical ayant pour origine O et l'axe des y horizontal ayant la même origine.

Résolution

En vertu de la remarque 4e. ci-dessus, on limite la résolution au cas du rideau vertical.
On pose:

$$k_1 = p + \gamma t + \frac{c}{\tan \varphi},$$

$$k_2 = q + \gamma t + \frac{c}{\tan \varphi}.$$

On considère le long de OB sur l'axe des x la coordonnée réduite d'un point quelconque X, telle que:

$$x_1 = \frac{\gamma}{k_1} OX$$

pour le calcul des poussées et

$$x_2 = \frac{\gamma}{k_2} OX$$

pour le calcul des butées.

On cherche pour les valeurs désirées de l'angle de frottement terre sur terre φ et de l'angle de frottement terre sur mur δ dans l'abaque

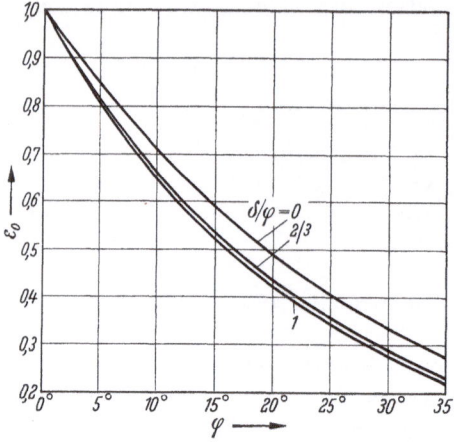

Fig. 8. Abaque No. 1.

$n° 1$ (fig. 8) la valeur de ε. La poussée perpendiculaire à l'ancrage correspondant au point X est donnée par :

$$\sigma_1 = k\,\varepsilon\,(1 + x_1) - \frac{c}{\tan\varphi}.$$

On cherche pour les valeurs désirées de φ et δ, dans l'abaque $n° 2$ (fig. 9), la valeur de ω. La butée perpendiculaire à l'ancrage correspondant au point X est donnée par :

$$\sigma_2 = k\,\omega\,(1 + x_2) - \frac{c}{\tan\varphi}.$$

La traction maximum $T_{\max}$ est donnée par la résultante des butées σ_2 diminuée de la résultante des poussées σ_1.

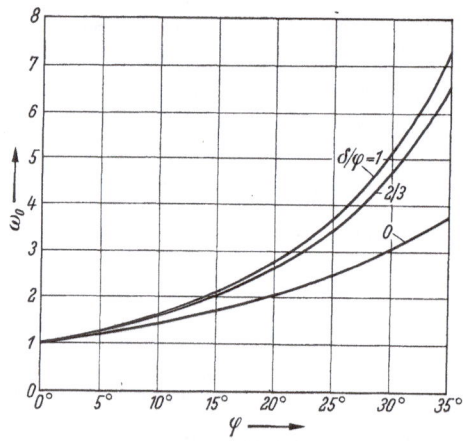

Fig. 9. Abaque No. 2.

6. Résultats des Essais sur Ancrages Courts

Les essais de rideaux d'ancrages courts dont la sécurité est nettement supérieure à un ont montré les phénomènes suivants :

Au début du mouvement de la construction, les figures de glissement qui apparaissaient se rapprochent de la figure que l'on obtiendrait par interpénétration d'une zone de poussée sur le rideau de soutènement et d'une zone de butée résistant au rideau d'ancrage. Lorsque la déformation augmente, la zone intéressée par le mouvement augmente et la courbe limite de cette zone qui relie le bas du rideau d'ancrage approximativement au point de l'appui libre du rideau de soutènement devient plus régulière et convexe vers le haut. Une des figures les plus typiques et les plus régulières obtenue par photographie du mouvement correspond à un arc de spirale logarithmique d'angle $\varphi = 22°$ (fig. 10).

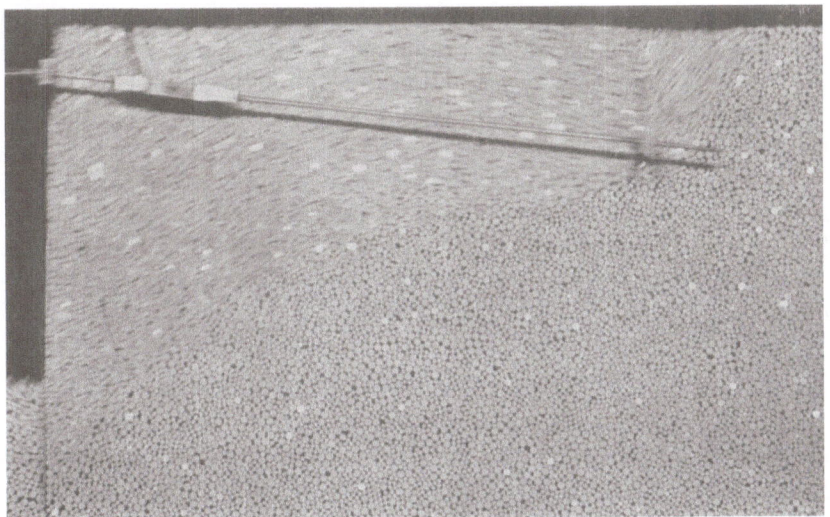

Fig. 10. Figure de rupture d'un ancrage court.

Cette coïncidence vient à l'appui de la méthode aux extrêmes de RENDULIC ainsi que des conclusions théoriques déduites des théories de la plasticité de DRUCKER, HODGE et PRAGER. Citons ici aussi les méthodes de BRINCH HANSEN et de JELINEK qui se basent également sur l'utilisation d'arc de spirale logarithmique pour la résolution du cas des ancrages courts. Une comparaison entre la méthode empirique de KRANZ qui suppose une ligne de glissement droite et une méthode analogue qui supposerait une ligne de glissement en forme d'arc de spirale logarithmique montre que la traction admissible la plus faible est obtenue avec la dernière méthode.

On peut en outre déduire de ces essais un ensemble de remarques d'ordre pratique:

a) Lorsqu'on utilise un ancrage court ou un ancrage long relativement déformable, il est dangereux de calculer le rideau de soutènement avec une fiche en appui libre (free earth support).

b) Si on prend une fiche légèrement supérieure à la fiche en appui libre et si la position de l'ancrage est telle que la sécurité est proche de l'unité, les ruptures se produisent d'après le schéma décrit ci dessus.

c) On a intérêt à ne pas placer les ancrages trop bas si cela donne lieu au placement d'un tirant fort incliné sur l'horizontale pour les raisons suivantes:

1° Si le sol est meuble aux environs du rideau de soutènement, le tirant peut tourner autour de l'ancrage sans que celui-ci ne bouge. Ceci a pour effet un déplacement du rideau de soutènement qui peut reporter un accroissement considérable de pression sur la fiche et provoquer la ruine de la construction par effet de bêche.

2° Si le sol est résistant, la flexion dans les tirants peut être considérable.

Lorsque la sécurité de la construction se rapproche de 1, la ligne de glissement curviligne se rapproche d'autant plus de la ligne droite. On peut déduire de ceci que l'hypothèse de Kranz suivant laquelle la ligne de rupture est une droite lorsque la sécurité vaut l'unité est justifiée.

7. Méthode de Calcul des Ancrages Courts

En conclusion des remarques qui précèdent, il nous a été possible de nous rendre compte que la spirale logarithmique d'angle égal à l'angle de frottement convient particulièrement bien comme forme de ligne de la glissement. La méthode de calcul la plus facile basée sur cette hypothèse est une variante de la méthode de Jelinek exposée ci-dessous (fig. 11).

1. On dessine sur un calque le rideau de soutènement avec la position souhaitée de l'ancrage.

2. On trace sur un papier la spirale logarithmique d'angle φ.

3. On fait circuler le calque sur le papier en maintenant sur la spirale le pied G du rideau de soutènement et le pied I du rideau d'ancrage.

4. On trace pour chaque position de la spirale la position correspondante de son pôle P.

5. On trace le lieu des pôles P.

6. On calcule le poids W_1 du massif AGJ, GJ étant la droite inclinée de $\frac{\pi}{4} - \frac{\varphi}{2}$ avec GA (équilibre de Rankine). On calcule la poussée $E_1 = \frac{(AG)^2}{2} \gamma \tan^2\left(\frac{\pi}{4} - \frac{\varphi}{2}\right)$ d'où on déduit $E_2 = \frac{E_1}{\cos \delta}$. La composition de E_2 et de W_1 donne le vecteur Q_1.

7. On met en place Q_1 et son alignement.

8. On calcule le poids W_2 du massif $JGIL$ limité par un arc de spirale GI choisi arbitrairement. La composition de Q_1 et de W_2 donne

le vecteur Q_2. On met en place son alignement q_2. Celui-ci intercepte l'alignement de T en un point K correspondant à l'arc de spirale choisi.

9. On en déduit une position probable du pôle P en traçant la tangente KP au lieu des pôles.

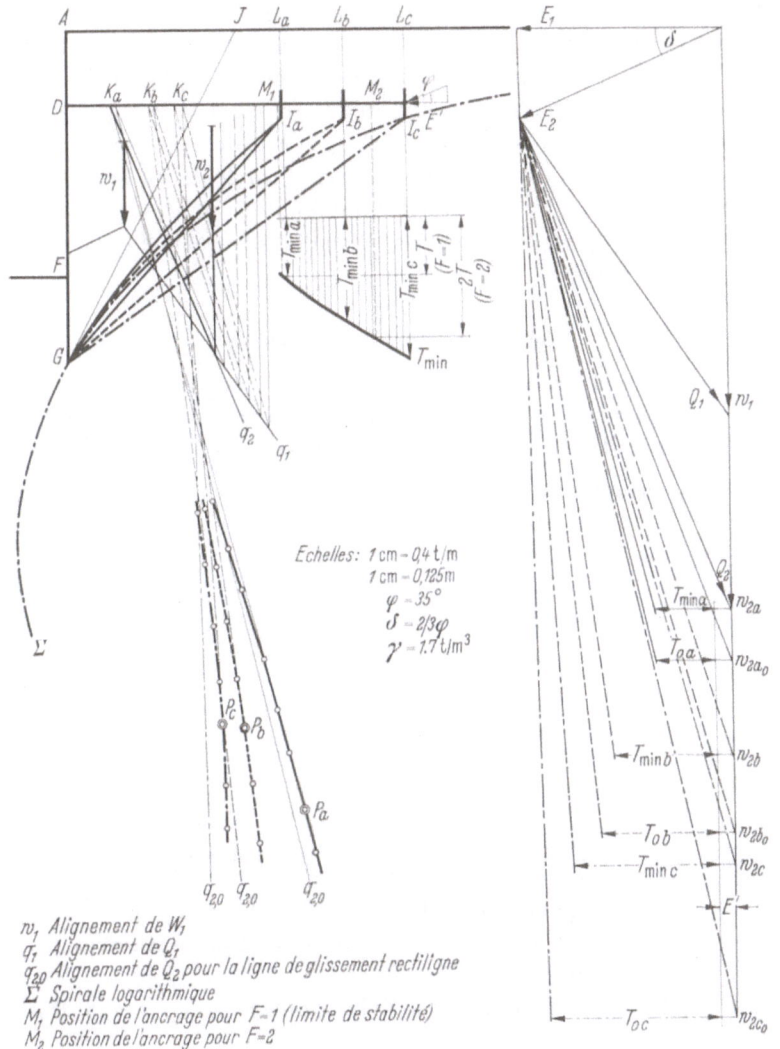

Fig. 11. Méthode de calcul d'un ancrage court.

10. On trace l'arc de spirale correspondant et on calcule les nouvelles valeurs de W_2 et Q_2 auxquelles correspond une nouvelle position du point K.

11. On corrige la position du pôle P en traçant la tangente au lieu des pôles par K.

12. On trace l'arc de spirale correspondant.

13. On recommence les tracés repris au 10 ci-dessus, ce qui donne la position définitive du point K avec une approximation suffisante.

14. On mène sur l'épure des vecteurs la parallèle à KP, ce qui permet de déterminer la valeur correspondante de l'effort T dans le tirant.

15. On soustrait la valeur de la poussée E' derrière la plaque d'ancrage.

La méthode qui vient d'être exposée se base sur le fait que le point K se déplace très peu lorsqu'on change d'arc de spirale. Comme la variation de W_2 est aussi assez faible, c'est la tangente au lieu des pôles qui est la plus inclinée sur la verticale qui donne la plus petite valeur de T. A titre de comparaison on a représenté les valeurs de T_0 données par la méthode de Kranz.

Résistance due à la Cohésion

Remarquons la propriété suivante. Quelle que soit la forme de la courbe Σ joignant le pied G du rideau de soutènement au pied I de l'ancrage, la résultante R_c des vecteurs cohésions $c\,ds$ est représentée à l'échelle $\dfrac{1}{nc}$ par le vecteur GI (fig. 12), $\dfrac{1}{n}$ étant l'échelle des longueurs.

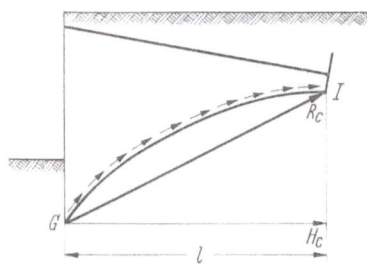

Fig. 12. Calcul de la résistance due à la cohésion.

La composante horizontale H_c de R_c est représentée à la même échelle par la distance horizontale l entre le rideau de soutènement et le pied de l'ancrage.

Le supplément $\varDelta T_c$ de résistance horizontale de l'ancrage, dû à la cohésion est donc donné par

$$\varDelta T_c = l \cdot c$$

Il résulte de ce qui précède que dans un sol purement cohésif, il n'est pas nécessaire de faire une hypothèse sur la forme de la ligne de glissement.

4.2 Équilibre Limite de «Fondations» en Milieu Pulvérulent à deux Dimensions

Par

J. Biarez, M. Boucherie, L. M. Boucraut, J. Haeringer, D. Martin, B. Montel, R. Nègre, P. Stutz et K. Wiendieck

Les recherches que nous avons effectuées sur l'équilibre limite de fondations rigides en milieu pulvérulent à deux dimensions avaient pour but, d'une part de préciser les lois et les schémas de calculs par comparaison avec l'expérience et, d'autre part de donner des solutions au moins partielles, à des problèmes pratiques.

Pour faciliter la comparaison avec le calcul, nous avons choisi un milieu à deux dimensions formé par l'empilage de cylindres parallèles de sections diverses. Ce modèle avait, en outre, l'avantage d'avoir une loi de plasticité parfaite très proche de celle de Coulomb. Nous avons examiné des fondations rectangulaires verticales soumises à une force verticale centrée, orientée vers le bas ou le haut, puis à une force inclinée excentrée. En second lieu, nous avons soumis des rectangles verticaux minces à une translation horizontale ou à une force horizontale; et, enfin, nous avons examiné le cas où la fondation était circulaire et soumise à une sollicitation analogue à une roue de véhicule.

I. Fondation Soumise à une Translation Verticale vers le Bas
[1], [2], [3], [4].

De nombreux essais d'enfoncement de fondation rectangulaire à base horizontale (B) ont été effectués pour mieux connaître l'aspect statique et cinématique du phénomène; ces résultats ont été comparés à divers calculs de plasticité afin d'examiner l'intérêt de certaines lois cinématiques et de certains schémas de solutions.

A. Aspect Statique

L'expérience montre l'existence d'un coin sensiblement rigide, sous la fondation, dont la dimension croît avec la surcharge latérale uniforme γD (fig. 1). Un premier calcul suppose ce coin limité par deux droites CO et CO', inclinées de β (ici négatif) sur la verticale. La force d'enfoncement Q de la fondation peut être obtenue en écrivant l'équilibre limite des dièdres COE et $CO'E'$, en tenant compte du poids

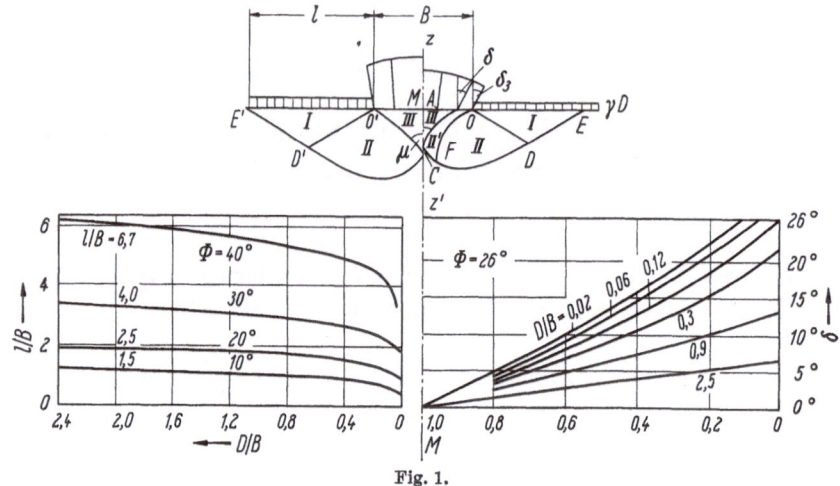

Fig. 1.

du triangle $COO' = \dfrac{\gamma B^2}{4 \tan \beta}$. Le calcul montre que l'on fait une faible erreur en additionnant la solution d'un dièdre non pesant subissant une surcharge γD et d'un dièdre pesant (coefficient de butée b) [2].

On peut ainsi écrire :

$$\frac{Q}{\gamma B^2 L} = \frac{1}{4}\, b\, \frac{\sin(\phi - \beta)}{\sin^2 \beta}$$
$$+ \frac{D}{B} \tan\left(\frac{\pi}{4} + \frac{\phi}{2}\right) \frac{\sin(\phi - \beta)}{-\sin \beta}\, e^{2\left(\frac{\pi}{4} + \frac{\phi}{2} - \beta\right)\tan\phi} + \frac{1}{4 \tan \beta}.$$

On remarque que $\dfrac{Q}{\gamma B^2 L}$ présente un minimum en fonction de β, et que cette valeur particulière de $|\beta|$ décroît avec la surcharge latérale γD de $\dfrac{\pi}{2} - \phi$ à $\dfrac{\pi}{4} - \dfrac{\phi}{2}$; ceci est en bon accord avec l'accroissement des dimensions du coin rigide que l'on observe. Par ailleurs, l'expérience montre que la force d'enfoncement Q oscille autour d'une moyenne inférieure de 10 à 20% au résultat du calcul; par contre, les valeurs maximales en sont très proches (fig. 2a). Pour simplifier le calcul pratique, on peut admettre une valeur de β indépendante de la profondeur. Divers auteurs ont choisi $\beta_1 = -\left(\dfrac{\pi}{2} - \phi\right)$ et $\beta_2 = -\left(\dfrac{\pi}{4} - \dfrac{\phi}{2}\right)$. Pour être dans le sens de la sécurité, on peut prendre la première valeur (β_1) pour obtenir la minimum du premier terme :

$$\frac{1}{2}\, N_\gamma = \frac{b}{4 \cos^2 \phi}$$

et, simultanément, la deuxième valeur (β_2) pour obtenir le minimum du second terme:

$$\frac{D}{B} N_q = \frac{D}{B} e^{\pi \tan \phi} \tan^2 \left(\frac{\pi}{4} + \frac{\phi}{2} \right) \qquad \text{(fig. 2b)}.$$

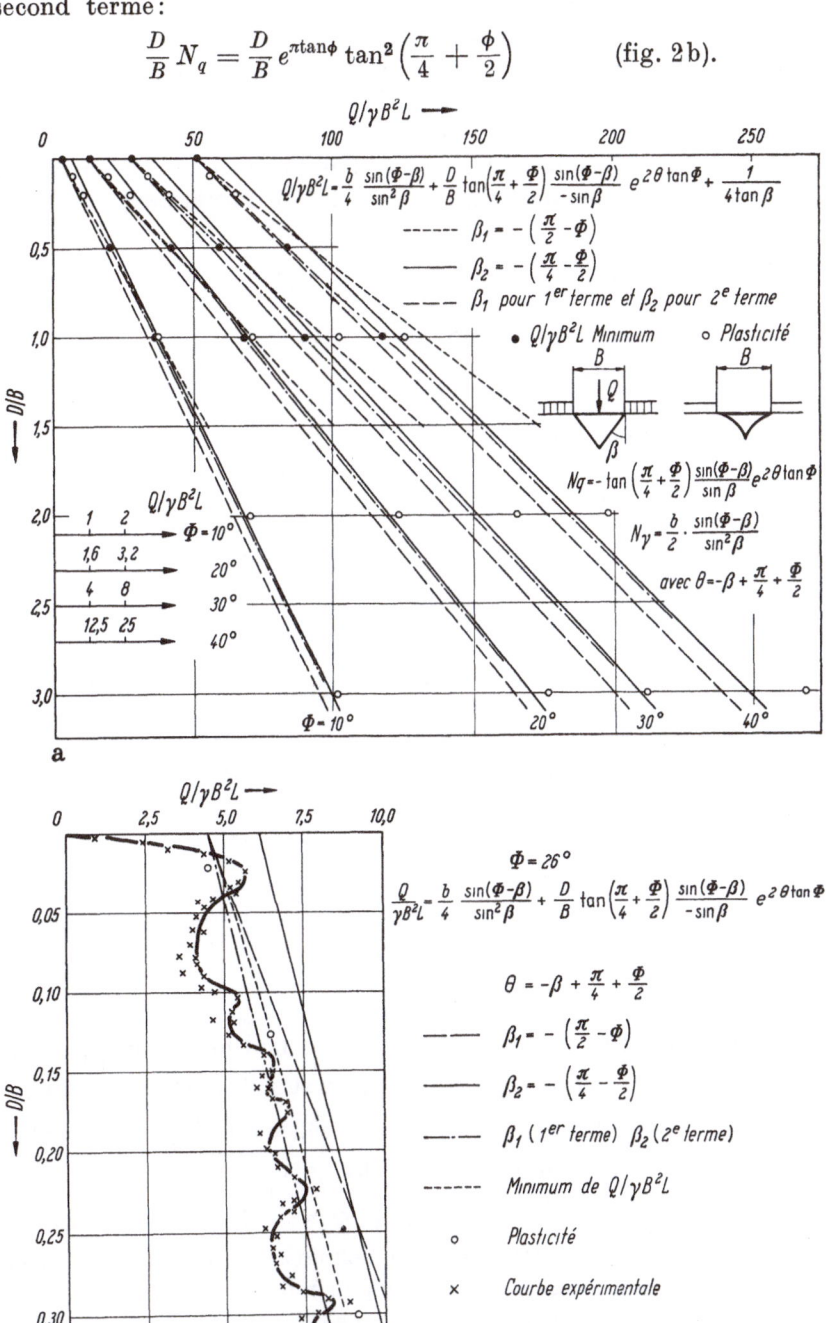

Fig. 2a et b.

La méthode précédente semble fournir une approximation suffisante pour l'ingénieur; elle a toutefois l'inconvénient de supposer un point singulier en C, et de ne fournir aucune indication sur la distribution des contraintes sous la base de la fondation; nous [2], [4] avons donc examiné une solution statiquement correcte où le coin «rigide» est limité par deux bicaractéristiques faisant partie de l'éventail issu de O (deuxième zone de PRANDTL — Solution LUNDGREN); ces deux

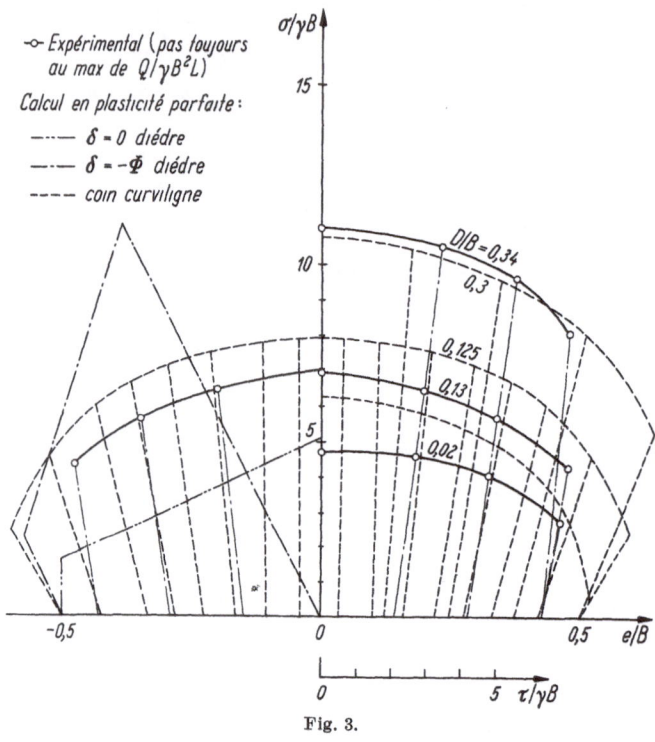

Fig. 3.

lignes sont choisies de telle manière qu'elles fassent l'angle $\left(\dfrac{\pi}{4} - \dfrac{\phi}{2}\right)$ avec la verticale CM pour éviter le point singulier. On obtient ainsi une force portante Q très voisine de la précédente $\left(\dfrac{Q}{\gamma B^2 L} \text{ minimum}\right)$ et un coin COO' croissant avec la surcharge γD. Mais, pour de très faibles valeurs de γD $\left(\dfrac{D}{B} = 0{,}125 \text{ pour } \phi = 26°\right)$, la bicaractéristique limite CO devient tangente à OO' et l'inclinaison δ_3 de la contrainte en O sur OO' devient égale à ϕ. Pour de plus faibles surcharges, nous [4] limitons donc la zone à la bicaractéristique tangente OF et nous admettons le frottement totalement mobilisé sous une partie OA de la base de la fondation. Le calcul de la zone II' se fait donc à partir des données sur OF et $\delta = \phi$ sur OA.

Pour connaître la distribution des contraintes sous la fondation, nous avons [4] supposé le coin CMO (ou CMA) en plasticité. Les

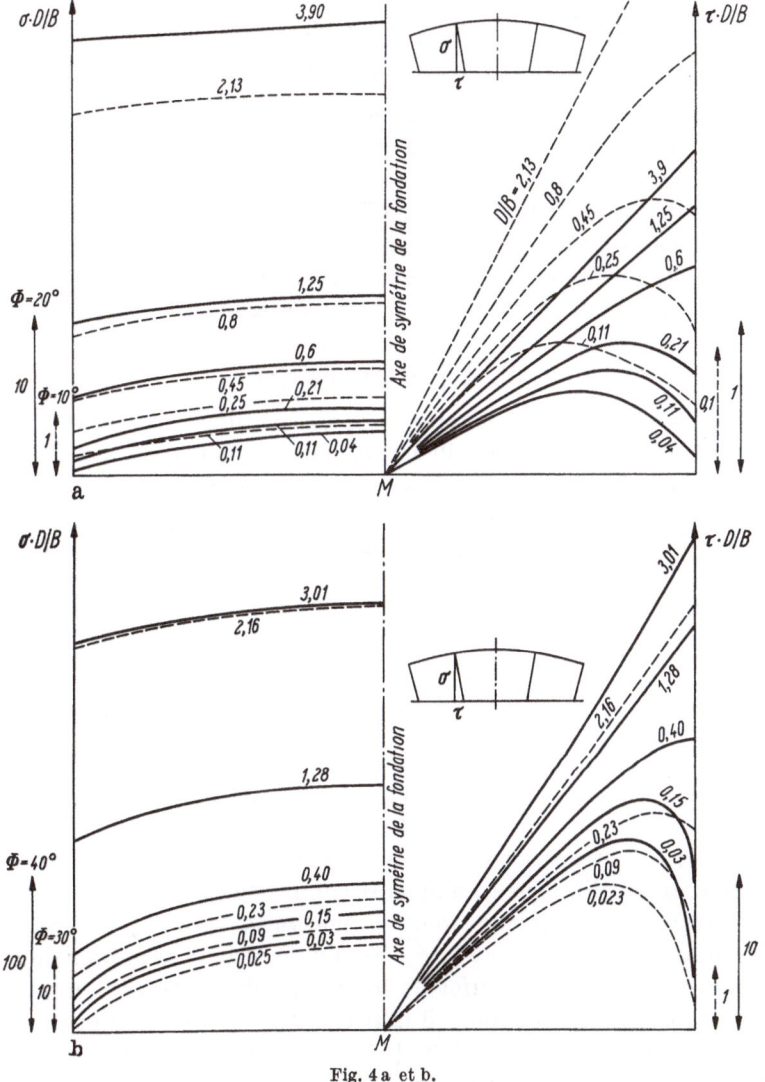

Fig. 4 a et b.

contraintes sont connues sur CO, de même que l'orientation des contraintes sur CM; en effet, CM est un plan principal mineur puisque nous avons choisi $OCM = \left(\dfrac{\pi}{4} - \dfrac{\phi}{2}\right)$ et un point C sans singularité pour cette zone. On obtient ainsi une distribution des contraintes sur la base de la fondation, analogue aux résultats expérimentaux [3]—[4] (fig. 3). L'hypothèse de l'élasticité dans le coin rigide donne une distribution différente [5]. Sur la fig. 3, les composantes normales expérimentales sont plus faibles que celles données par le calcul, car la distribution des contraintes n'a pas été mesurée pour une valeur maximale de $\dfrac{Q}{\gamma B^2 L}$. L'expérience, comme le calcul, montre que les contraintes sont moins inclinées quand la surcharge croît (fig. 1, 3). La mesure des composantes normales et tangentielles a été faite à l'aide de dynamomètre à friction, dont le principe avait été utilisé pour les composantes normales dès 1930, à l'Université d'Iowa.

Si la fondation n'est pas rugueuse comme précedemment, la bicaractéristique limite OI est choisie de telle manière que δ_3 soit inférieur ou égal à l'angle de frottement sol — fondation.

B. Aspect Cinématique

Nous avons examiné diverses lois cinématiques, en particulier le potentiel plastique et, par ailleurs, la loi de non variation de volume avec identité des directions principales des tenseurs de contraintes et

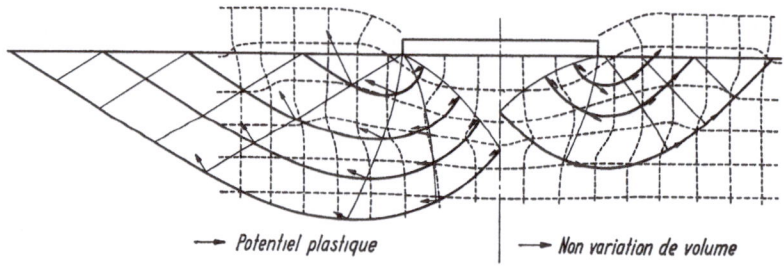

⟶ *Potentiel plastique*   ⟶ *Non variation de volume*

Fig. 5.

déformations. Cette seconde loi semble donner un meilleur accord avec l'expérience (fig. 5). Elle est d'ailleurs confirmée par les essais biaxiaux sans frettage de tête, qui montrent une non variation de volume pendant les grandes déformations [2], [6]. Toutefois, il faut se garder de conclusions trop hâtives; en effet, la loi rigide-plastique que suppose le calcul est loin d'être vérifiée; il faut une déformation non négligeable pour entrer en plasticité parfaite [11], et il existe une couche dans cet état, au long de CDE, qui perturbe les trajectoires observées dans les zones I et II. En outre, la courbe contrainte-déformation présente un maximum avant le palier final; les observations ne doivent donc être faites qu'après un certain enfoncement de la fondation pour atteindre ce palier.

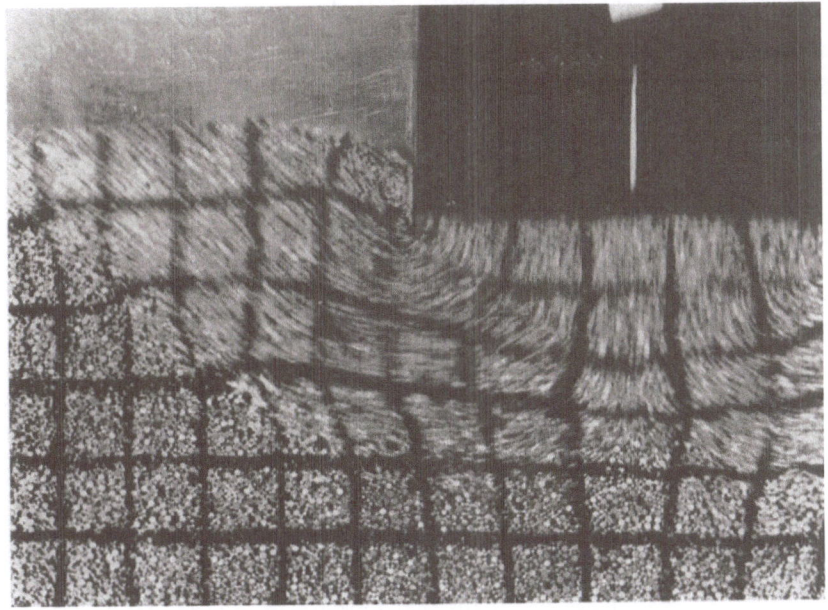

Fig. 6.

II. Fondation Soumise à une Translation Verticale vers le Haut
[2], [8], [9]

Si l'on poursuit l'enfoncement de la fondation précédente, en mesurant la force de frottement sur les parois verticales, on observe que cette force est nettement inférieure à la valeur obtenue en supposant les dièdres latéraux en plasticité parfaite avec OS en butée ($OST-O'S'T'$) [2]. Par contre, si l'on soulève la fondation, le frottement latéral est très voisin du calcul de butée avec la nouvelle inclinaison des contraintes (fig. 7). Ceci est en accord avec les observations et calculs cinématiques qui montrent l'existence d'une zone très mince en plasticité, le long de la fondation lors du soulèvement, tandis que, pour l'enfoncement, l'expérience montre encore une zone mince alors que le calcul nécessiterait un volume important. L'entrée en plasticité du dièdre latéral nécessite donc des déplacements de la fondation nettement différents selon l'orientation des contraintes.

Le soulèvement d'une fondation réelle de section circulaire nécessite un calcul de révolution que nous n'avons pas achevé. L'étude de nombreuses expériences de laboratoire et de chantier semble montrer qu'un calcul approché satisfaisant peut être fait, si $\phi = 0$, en utilisant le calcul plan précédent. Si l'angle de frottement est nettement différent de zéro, on peut utiliser le calcul empirique suivant: on admet

28*

qu'un tronc de cône de sol, de demi-angle au sommet $\alpha = -\dfrac{\phi}{8}$, est associé à la fondation; on suppose que la distribution des contraintes sur une génératrice du cône obéit à l'équation de Kötter et que la

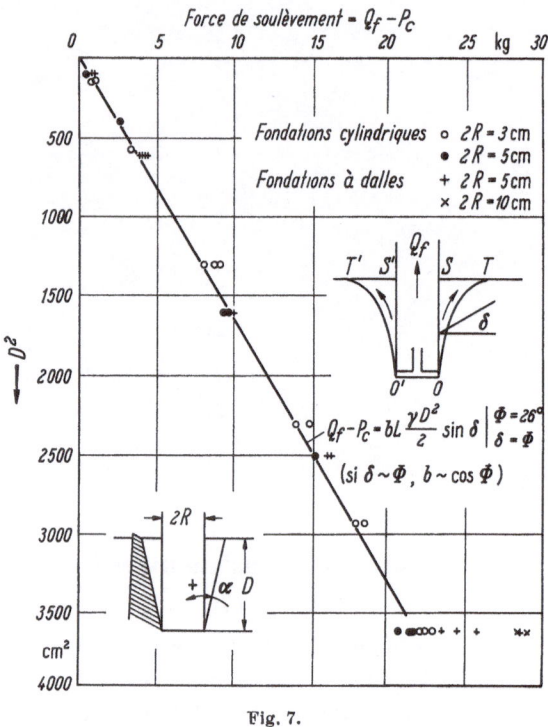

Fig. 7.

contrainte à l'origine est donnée par la solution du dièdre non pesant de Prandtl; on obtient ainsi:

$$Q_f = P_c + P_T + Q_\phi + Q_c,$$

Q_f force de soulèvement de la fondation de rayon R et profondeur D,
P_c poids de la fondation cylindrique,
P_T poids du tronc de cône de sol de poids spécifique γ,

$$P_T = \gamma \pi R^2 D \left(-\frac{D}{R} \tan \alpha + \frac{1}{3} \left(\frac{D}{R} \right)^2 \tan^2 \alpha \right),$$

Q_ϕ force due au frottement ϕ,

$$Q_\phi = \gamma \pi R D^2 \frac{\sin 2(\phi + \alpha)}{2 \cos^2 \alpha} \left(1 - \frac{D}{R} \frac{\tan \alpha}{3} \right),$$

Q_c force due à la cohésion c,

$$Q_c = \frac{2\pi R D c}{\tan \phi}\left(1 - \frac{D}{R}\frac{\tan \alpha}{2}\right)\left(b'\cos\phi(\tan\phi + \tan\alpha) - \tan\alpha\right),$$

$$b' = \tan\left(\frac{\pi}{4} + \frac{\phi}{2}\right)\frac{\cos n - \sin\phi\cos m}{\cos n + \sin\phi\cos m},$$

$$m = -\frac{\pi}{4} + \frac{\phi}{2} + \alpha, \qquad \sin n = \sin\phi\sin m.$$

Si $\alpha = 0$

$$Q_f - P_c = \gamma\pi R^2 D\left(\frac{D}{R}\sin\phi\cos\phi + 2\frac{c}{\gamma R}b'\cos\phi\right).$$

Pour un poids donné de fondation, il semble préférable de réaliser des fondations profondes de petit diamètre.

III. Fondation Soumise à une Force Inclinée Excentrée
(faible profondeur) [2], [11], [12]

Nous avons vu précédemment une solution pour une fondation soumise à deux surcharges égales γD de chaque côté; on obtient une solution symétrique en limitant le coin sous la fondation par deux bicaractéristiques appartenant respectivement aux réseaux en éventail issus de O et O'. Si l'on choisissait deux lignes non symétriques de ces éventails, se coupant sous l'angle $\left(\frac{\pi}{2} - \phi\right)$, on obtiendrait deux ten-

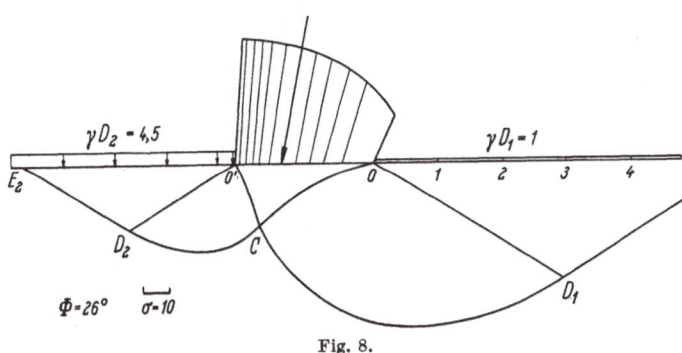

Fig. 8.

seurs de contraintes différents au point d'intersection C. Pour obtenir le même état de contrainte en C, il faudrait supposer une surcharge différente de chaque côté de la fondation, comme l'a montré STUTZ (fig. 8).

L'expérience confirme ce résultat; en effet, si l'on enfonce une fondation avec une force inclinée et surcharges latérales égales, l'observation montre que les grandes déformations ne se produisent que d'un côté de la fondation. L'hypothèse de plasticité parfaite peut se concevoir de ce côté, mais de l'autre, il faudrait tenir compte d'une loi plus complète faisant intervenir les zones de faible déformation. En l'attente d'une théorie plus complète, nous utiliserons une formule empirique permettant d'extrapoler les résultats expérimentaux (fig. 9).

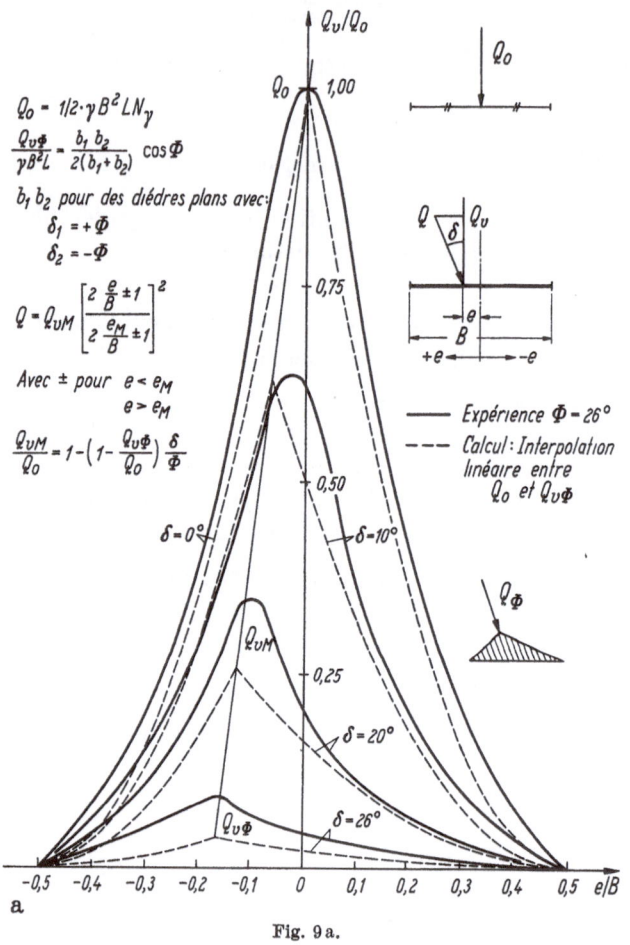

Fig. 9 a.

Si l'on exerce une force d'inclinaison constante δ mais d'excentricité variable e, l'expérience montre que la force de rupture est nulle si elle est appliquée aux bords de la fondation, et passe par une valeur maximale pour une excentricité particulière e_M qui semble voisine de $\frac{e_M}{B} = \frac{1}{6} \frac{\delta}{\phi}$ ou mieux: $\frac{e_M}{B} = \left(0,13 \frac{\delta}{\phi} + 0,03\right) \frac{\delta}{\phi}$. Pour une excentricité nettement différente, la fondation se soulève localement.

Pour $\delta = \phi$, les contraintes sont toutes parallèles; on peut donc calculer la distribution des contraintes du côté des grandes déformations, avec les formules habituelles de dièdre en plasticité parfaite : de l'autre côté, les contraintes sont inférieures à celles de butée, toutefois, un tel calcul donne une force proche de la valeur maximale observée, dont la composante verticale peut alors s'écrire :

$$\frac{Q_{v\phi}}{\gamma B^2 L} = \frac{b_1 b_2}{2(b_1 + b_2)} \cos \phi,$$

b_1 et b_2 : coefficients de butée.

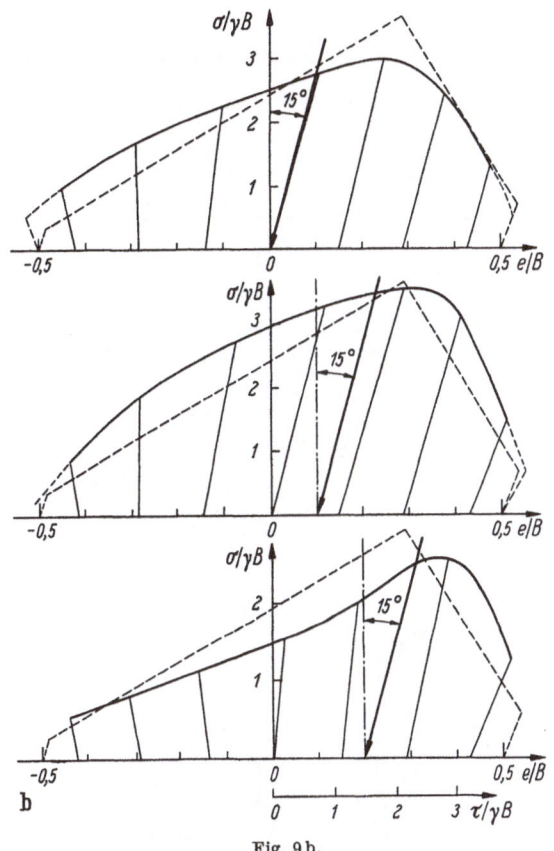

Fig. 9 b.

Pour des inclinaisons différentes de $\delta = \phi$, un calcul analogue avec deux dièdres en butée donne une valeur trop faible pour des contraintes toutes parallèles. Or, l'on connait la force portante avec inclinaison nulle (Q_0), et l'on peut admettre une variation linéaire de Q_{vM} entre Q_0 et $Q_{v\phi}$:

$$\frac{Q_0 - Q_{vM}}{Q_0 - Q_{v\phi}} = \frac{\delta}{\phi}.$$

Le rôle de l'excentricité peut être obtenu en généralisant l'ancienne loi empirique qui consistait à admettre que, pour une force verticale excentrée, on pouvait appliquer le calcul usuel à une fondation de largeur fictive, telle que la force soit centrée. Il faut écrire ici:

$$\frac{Q_v}{Q_{vM}} = \left(\frac{0,5\,B \pm e}{0,5\,B \pm e_M}\right)^2 \qquad \text{(fig. 9)}.$$

Cette méthode a permis d'obtenir d'utiles indications sur le comportement d'une roue circulant sur un milieu à deux dimensions.

IV. Translation d'Écran Vertical Mince [10]

Soit une fondation rectangulaire verticale, de faible largeur B et d'épaisseur L. Soit h sa hauteur et D la distance entre la surface du

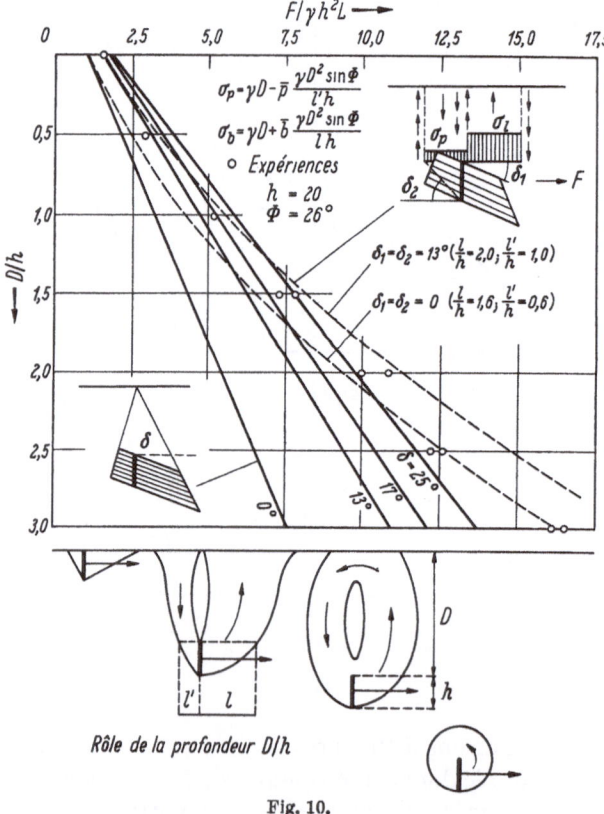

Fig. 10.

sol et la partie supérieure de la fondation. L'observation des déformations montre que la solution doit totalement changer pour différentes profondeurs D (fig. 10).

A. Écran en Surface $D = 0$ (fig. 10)

a) Si l'écran est soumis à une translation horizontale avec $D = 0$, il se forme, devant et derrière, des volumes en plasticité parfaite, que l'on peut calculer avec la méthode des dièdres, d'où les coefficients de butée b et poussée p. La force nécessitée par la translation est appliquée aux 2/3 de la hauteur, en milieu pulvérulent; sa projection horizontale est:

$$F = \frac{1}{2} \gamma h^2 L (b - p) \cos \delta.$$

L'expérience est en bon accord avec cette formule, en prenant pour δ la valeur maximale du frottement sol — écran.

b) Si l'écran est soumis à une force horizontale, située de telle manière que l'écran demeure vertical, celui-ci a tendance à se soulever; les contraintes ne sont plus inclinées de la valeur maximale. On ignore ces inclinaisons δ_1 et δ_2, mais l'on sait qu'elles sont inférieures à ϕ et doivent satisfaire les trois équations d'équilibre,

$$\frac{1}{2} \gamma h^2 L (b \cos \delta_1 - p \cos \delta_2) = F$$

et

$$\frac{1}{2} \gamma h^2 L (b \sin \delta_1 - p \sin \delta_2) = P - R \sim 0.$$

$P = $ poids de l'écran,
$R = $ force appliquée par le sol sous sa base.
La valeur de $P - R$ est souvent négligeable.

La troisième équation consiste à écrire que la force est au tiers de la hauteur, en supposant δ_1 et δ_2 constants.

La valeur de p étant petite par rapport à b, la seconde équation montre que si δ_2 peut varier entre $+\phi$ et $-\phi$, la variation de δ_1 sera beaucoup plus petite, et l'erreur due à l'ignorance de l'inclinaison des contraintes sera faible; par ailleurs, le mouvement relatif de l'écran par rapport au sol invite à choisir $\delta_2 = \phi$, ce qui donne une bonne correspondance avec la force mesurée (fig. 10).

B. Écran à Faible Profondeur $\left(\frac{D}{h} < 6\right)$ (fig. 10).

A très faible profondeur, on peut appliquer, avec un certain succès, la formule précédente en supposant que les zones de RANKINE s'étendent jusqu'à la surface libre, mais, dès que la profondeur croît légèrement, on voit que la zone en plasticité n'est pas aussi large. Au-dessus des zones en butée et en poussée, on observe des colonnes verticales soumises à un déplacement vertical. Il n'est pas impossible de trouver une solution plastique rigoureuse en accord avec les mesures, mais un calcul approché donne des résultats satisfaisants. Il suffit d'admettre que les dièdres appuyés sur l'écran subissent, à leur partie supérieure, une surcharge uniforme égale au poids des colonnes précédentes, plus ou moins le frottement de butée sur les faces latérales ($\bar{b} = \bar{p} = \cos \phi$

L'équation de projection sur l'horizontale s'écrit:

$$F = \frac{1}{2} \gamma h^2 L (b \cos \delta_1 - p \cos \delta_2) + \gamma D h L (b' \cos \delta_1 - p' \cos \delta_2)$$
$$+ \gamma D^2 h L \sin \phi \left(\frac{b' \bar{b}}{l} \cos \delta_1 + \frac{p' \bar{p} \cos \delta_2}{l'} \right),$$

b' et p' sont les coefficients de butée et poussée des dièdres non pesants.

$$\left. \begin{array}{c} b' \\ p' \end{array} \right\} = \frac{\cos \delta \pm \sin \phi \cos \gamma}{1 \mp \sin \phi} e^{\pm 2\theta \tan\phi}, \quad \sin \gamma = \frac{\sin \delta}{\sin \phi}, \quad 2 = -(\gamma + \delta).$$

l et l' sont les largeurs des colonnes précédentes; en première approximation:

$$l = h \tan \left(\frac{\pi}{4} + \frac{\phi}{2} \right), \qquad l' = h \tan \left(\frac{\pi}{4} - \frac{\phi}{2} \right).$$

La seconde équation de projection permet de calculer les limites étroites de δ_1, sachant que δ_2 est compris entre $+\phi$ et $-\phi$. En prenant $\delta_2 = \delta_1 = 0$, l'erreur est petite.

$$P - R_1 + R_2 = \frac{1}{2} \gamma h^2 L (b \sin \delta_1 - p \sin \delta_2)$$
$$+ \gamma D h L (b' \sin \delta_1 - p' \sin \delta_2)$$
$$+ \gamma D^2 h L \sin \phi \left(\frac{b' \bar{b}}{l} \sin \delta_1 + \frac{p' p}{l'} \sin \delta_2 \right).$$

La troisième équation donne la position X de la force comprise entre la moitié et les deux tiers de la hauteur.

$$F \cdot X = \frac{1}{3} \gamma h^3 L (b \cos \delta_1 - p \cos \delta_2)$$
$$+ \frac{1}{2} \gamma h^2 D L (b' \cos \delta_1 - p' \cos \delta_2)$$
$$+ \frac{1}{2} \gamma h^2 D^2 L \sin \phi \left(\frac{b' \bar{b}}{l} \cos \delta_1 + \frac{p' \bar{p}}{l'} \cos \delta_2 \right).$$

A trois dimensions, la force mesurée est environ deux fois plus grande ($\phi = 30°$). Une formule approchée du même type, mais tenant compte d'une largeur fictive plus grande de la fondation, semble donner des résultats satisfaisants en l'attente d'un calcul à trois dimensions [10].

C. Écran Profond $\dfrac{D}{h} > 6$ (fig. 11)

Dans ce cas, le sol refoulé par l'écran n'atteint plus la surface libre, mais revient derrière l'écran; les trajectoires sont sensiblement circulaires. On peut effectuer un calcul approché en écrivant la rotation d'un volume de sol de rayon r inconnu; on choisit le minimum de F en fonction de r. Cette valeur de r étant voisine de h, on obtient un bon accord avec l'expérience en écrivant:

$$F = 4\pi\gamma DhL \tan \phi.$$

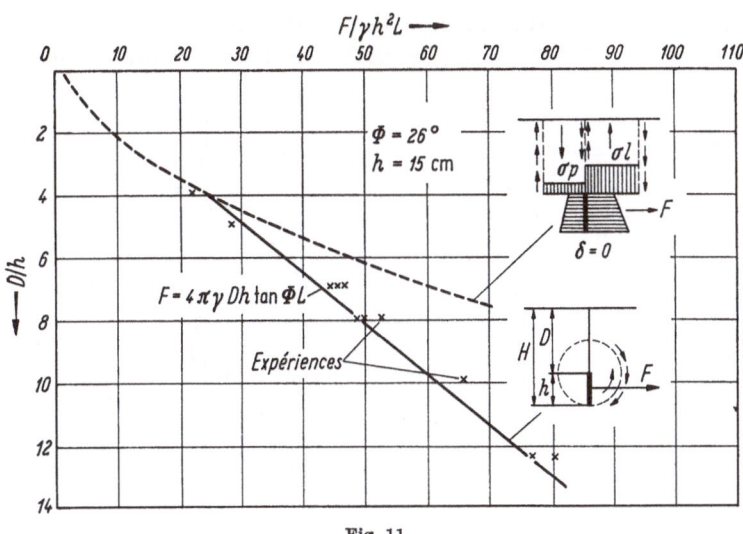

Fig. 11.

On a supposé que la moyenne des composantes normales sur le cercle est égale à γD; ceci permet de satisfaire les trois équations d'équilibre.

V. Écran Vertical Mince Soumis à un Moment

Soit un écran vertical soumis à un moment et situé dans un milieu semi-indéfini limité par un plan horizontal. Nous avons montré qu'il existait, à la partie inférieure de l'écran, une zone circulaire de sol qui restait associée à la fondation pendant sa rotation, tandis qu'à la partie supérieure, on pouvait admettre que les forces étaient en poussée (p) et en butée (b) [13], [14] avec une inclinaison δ.

Pour calculer l'équilibre limite (fig. 12), nous avons admis que les contraintes, au long du cercle, avaient pour moyenne des composantes normales γX, donc le moment nécessaire à la rotation du cercle est:

$2\pi R^2 \gamma X L \tan \phi$. Le moment appliqué à l'écran pour produire l'équilibre limite peut s'écrire :

$$\frac{M}{L} = 2\pi R^2 \gamma X \tan \phi + (b - p)\,\frac{\gamma (X - R)^2}{2}\,\cos \delta \left(\frac{X}{3} + R\right).$$

En admettant que le rayon du cercle correspond au minimum de ce moment, on obtient un bon accord avec les observations statiques et cinématiques [2], [15].

Nous indiquons, sur la fig. 12, les contraintes mesurées sur modèle [10].

On note que la distribution des contraintes ne présente pas de discontinuité au centre de la zone circulaire, contrairement aux

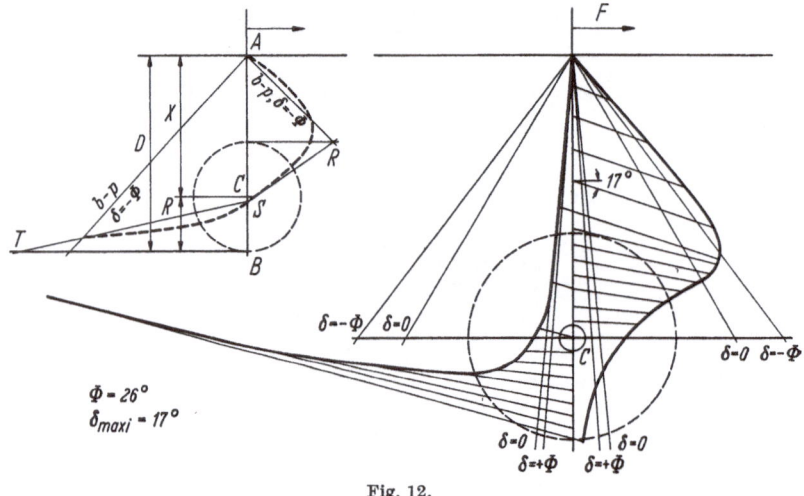

Fig. 12.

résultats de calculs supposant la plasticité dans cette zone (Un résultat opposé a été obtenu précédemment pour la zone apparemment rigide sous la base des fondations). L'inclinaison des contraintes est voisine de la valeur maximale, sauf dans la zone centrale du cercle. Un ordre de grandeur de la distribution des contraintes sur l'écran est obtenu en écrivant l'équilibre de la fondation et en supposant, comme répartition des contraintes supposées parallèles, la ligne polygonale $ARST$ avec, pour inconnues, BS et BT.

Cette méthode peut encore être utilisée avec succès si le sol est à un niveau différent de chaque coté de l'écran. Dans le cas particulier où le moment est nul, on obtient la grandeur de la dénivelée qui produit seule la rupture (palplanches sans ancrage).

VI. Conclusion

Les exemples précédents montrent les méthodes utilisées et les difficultés rencontrées dans l'étude des problèmes de mécanique des sols posés à l'ingénieur, même après de très importantes simplifications des données réelles. La plasticité parfaite dans un milieu à «deux dimensions» donne d'utiles indications si le sol obéit à cette «loi partielle» dans un volume important situé autour de la fondation; sinon, il faut tenir compte d'une loi plus complète. A titre d'exemple, nous avons donné quelques solutions pratiques semi-empiriques, qui peuvent être facilement compliquées, mais qui sont couramment utilisées.

Références Bibliographiques

[1] BUREL, M.: Étude expérimentale de la force portante des fondations par analogie avec des rouleaux. Thèse de Doctorat de Spécialité (3 è Cycle de Mécanique des Sols), Faculté des Sciences de Grenoble, 1960.

[2] BIAREZ, J.: Contribution à l'étude des propriétés mécaniques des sols et des matériaux pulvérulents. Thèse de Doctorat ès-Sciences, Grenoble, 1961.

[3] TURPIN, J.: Contribution à l'étude de la répartition des contraintes sous une fondation plane rugueuse en milieu bidimensionnel pulvérulent. Thèse de Doctorat de Spécialité (3è Cycle de Mécanique des Sols), Faculté des Sciences de Grenoble, 1963.

[4a] STUTZ, P.: Contribution à l'étude de la loi de déformation plastique des sols. Thèse de Doctorat de Spécialité (3è Cycle de Mécanique des Sols), Faculté des Sciences de Grenoble, 1963.

[4b] STUTZ, P.: Répartition des contraintes sous une fondation dans l'hypothèse de la plasticité parfaite. Compte-rendu des séances de l'Académie des Sciences **259**, 729—732 (1964).

[5] GORBUNOV-POSSADOV, M., et R. V. SEREBRJANYI: Calcul des ouvrages sur appui élastique. Comptes-rendus du 5è Congrès International de Mécanique des Sols et des Travaux de Fondations, Paris, 1961.

[6] BELOT, A.: Remarques sur les propriétés des sols à relativement haute pression (100 kg/cm²). Thèse de Doctorat de Spécialité (3è Cycle de Mécanique des Sols), Faculté des Sciences de Grenoble, 1964.

[7] LUNDGREN, H.: Determination by theory of plasticity of the bearing capacity of continuous footings on sand. Comptes-rendus du 3è Congrès International de Mécanique des Sols et des Travaux de Fondations, Zürich, 1953.

[8] MONTEL, B.: Contribution à l'étude des fondations sollicitées à l'arrachement. Phénomène plan, milieux pulvérulents. Thèse de Doctorat de Spécialité (3è Cycle de Mécanique des Sols), Faculté des Sciences de Grenoble, 1963.

[9] MARTIN, D.: Fondations profondes sollicitées à l'arrachement en milieu cohérent tridimensionnel. Thèse de Doctorat de Spécialité (3è Cycle de Mécanique des Sols), Faculté des Sciences de Grenoble, 1963.

[10] BOUCRAUT, L. M.: Mesures des composantes normales des contraintes le long d'un modèle de palplanches rigides. Thèse de Doctorat de Spécialité (3è Cycle de Mécanique des Sols), Faculté des Sciences de Grenoble, 1963.
—: Equilibre limite d'un milieu pulvérulent à deux dimensions, sollicité par un écran rigide. Thèse de Docteur-Ingénieur, Faculté des Sciences de Grenoble, 1964.

[*11*] Picchiottino, J.: Force portante des fondations — Charges à inclinaisons et excentricités variables — Étude expérimentale par analogie. Thèse de Doctorat de Spécialité (3è Cycle de Mécanique des Sols), Faculté des Sciences de Grenoble, 1962.

[*12*] Haeringer, J.: Contribution à l'étude de la force portante des fondations soumises à une force inclinée excentrée en milieu pulvérulent à deux dimensions. Thèse de Doctorat de Spécialité (3è Cycle de Mécanique de Sols), Faculté des Sciences de Grenoble, 1964.

[*13*] Biarez, J.: Remarques sur la cinématique des massifs enterrés. Comptes-rendus de la conférence de Bruxelles sur les problèmes de pression des terres, 1958.

[*14*] Capelle, J. F.: Contribution à l'étude des massifs en rotation. Thèse de Doctorat de Spécialité (3è Cycle de Mécanique des Sols), Faculté des Sciences de Grenoble, 1960.

[*15*] Wiendieck, K.: Sur la sécurité des palplanches non ancrées. Bautechnik, Novembre 1962.

4.3 Compressibility of a Certain Volcanic Clay

By

Yasunori Koizumi and Kojiro Ito

Abstract

The settlement of buildings resting on a volcanic clay, known as Kanto loam, shows particular behavior which is quite different from that of sedimentary saturated clay. The authors have been observing for about eight years the actual settlement of buildings built on the Kanto loam. The Kanto loam is a very porous, unsaturated clay. The compression of this clay consists mainly of the initial compression (immediate compression) and the following creep, the consolidation process being practically ignored. This paper deals mainly with an experimental approach to the compressibility of the Kanto loam. Long term compression tests were made and rate of the creep was observed. It has been found from these tests that both the volumetric creep and the shearing creep exist and they proceed straightly against the logarithm of the elapsed time. Comparison has been also made between the observed and calculated settlement, and the result shows a fairly satisfactory agreement.

General Description of the Kanto Loam

The greater part of the upperland of the Kanto district including Tokyo is covered with brown or yellowish brown clays with a thickness of several or more than 10 meters. These brown clays were brought at the diluvial epoch by eruptions of volcanoes locating around the Kanto Plain. In Tokyo area the stratum of the Kanto loam is generally less than 10 meters in thickness and often contains a thin layer of fractions of pumice, the lower part being more clayey. Below the Kanto loam follows a dense sandy formation (Fig. 1). Most of the buildings less than five stories high in these areas are safely supported directly on the Kanto loam.

The engineering properties of the Kanto loam are quite different from those of ordinary sedimentary clays. X-ray studies of the clay

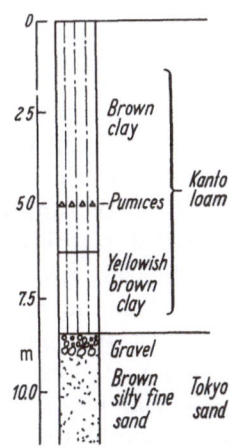

Fig. 1. Typical boring log in the upper land of Tokyo.

[1] have shown that it is composed mainly of amorphous allophane and in its lower part existence of hydrated halloysite is evident.

An example of the index properties of this clay is shown in Table 1. The samples were taken at the site of the Building Research Institute, whence samples for main tests described below were also taken. It is noticeable that it is very porous and highly plastic in a remolded state.

Table 1. *Index Properties of the Kanto Loam*

Specific gravity	2.83 g/cm³
Unit weight	1.19 g/cm³
Water content	120%
Void patio	3.92
Degree of Saturation	85%
Sand content	12%
Silt content	30%
Clay content	58%
Liquid limit	136.0%
Plastic limit	81.1%
Plastic index	54.9
Unconfined compression Strength (undisturbed)	2.4 kg/cm
Unconfined compression Strength (remolding)	0.25 kg/cm
Sensitivity	9.6
Preloading stress	2.2 kg/cm

According to the studies of Misono and others [2], the Kanto loam consists of aggregations of 0.02 to 0.06 mm in size which can not be disintergrated into a smaller size by usual mechanical agitations. However, when the clay is dispersed by means of a chemical treatment, the clay content is found more than 50 per cent.

They have been overconsolidated with a preloading stress of $2 \cdot 5$ kg/cm² which might have been caused by partial drying or some chemical process. There exist many hair cracks and pin holes, which are predominat in the upper part. In most cases they are unsaturated. The degree of saturation ranges usually from 60 to 95%. Therefore, the consolidation process may be ignored practically in the compression of the Kanto loam. Fig. 2 shows a result of the isotropic consolidation test with the undisturbed Kanto loam. Since the total pressure is carried by the intergranular stress soon after loading, the progressive settlement following the initial settlement is due to the dislocation of soil grains or the creep; when a load is applied on the Kanto loam, the initial compression (elastic compression) occurs first and the creep

follows.
$$\varepsilon(t) = {}_0\varepsilon + {}_c\varepsilon(t),$$

${}_0\varepsilon$: initial strain,

${}_c\varepsilon$: strain due to creep.

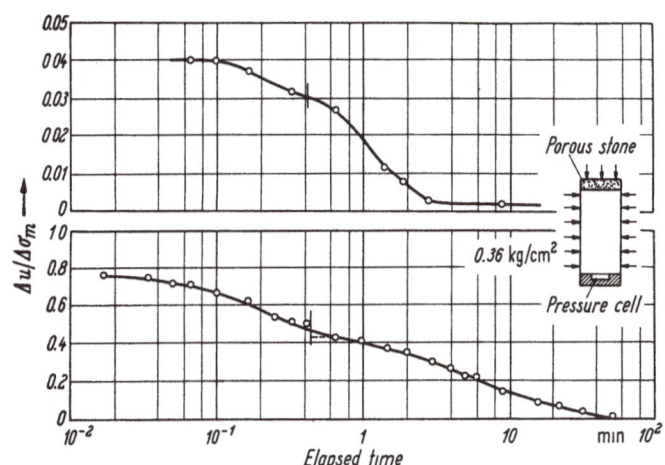

Fig. 2. Isotropic consolidation test.

Initial Settlement

In spite of the existence of many cracks and pin holes, the Kanto loam is considered to be macroscopically homogenous soil from its aeolian origin. Fig. 3 shows the comparison of vertical and horizontal strains caused by isotropic pressures. The specimens were 14 cm in height and 7 cm in diameter, and all-round pressure was applied. The lateral deformation was measured directly with an accuracy of 0.001mm by means of a specially devised instrument [3] as shown in Fig. 4. The result shows that the soil is quite isotropic in this sense.

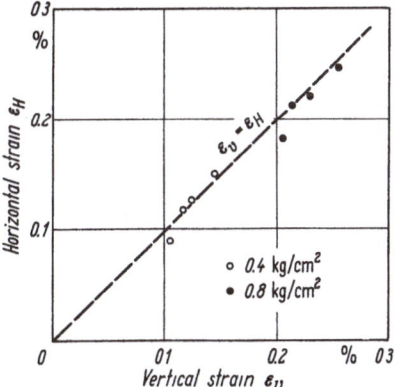

Fig. 3. Comparison between vertical and horizontal strains under isotropic pressure.

The results of compression tests show that compressive strains of the clay are approximately proportional to the stresses lower than the half of the ultimate strength.

Based on the results mentioned above, the initial settlement of a footing resting on the Kanto loam may be estimated by using the theory of elasticity, if the elastic constants are given.

According to many compression tests on the Kanto loam, Young's modulus is roughly proportional to the compressive strength and given by

$$E = 145q_c \tag{1}$$

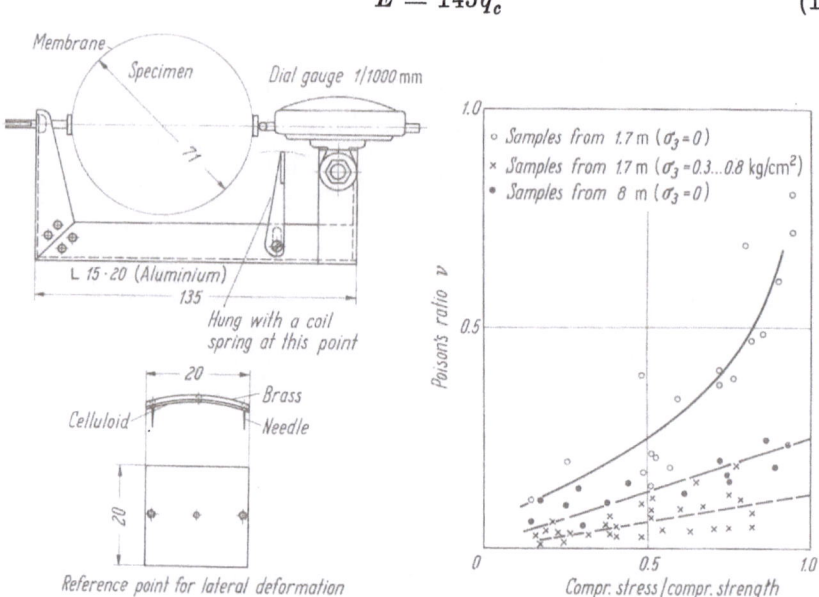

Fig. 4.　Device for measurement of lateral deformation.

Fig. 5.　Relationship between Young's modulus and compressive stress.

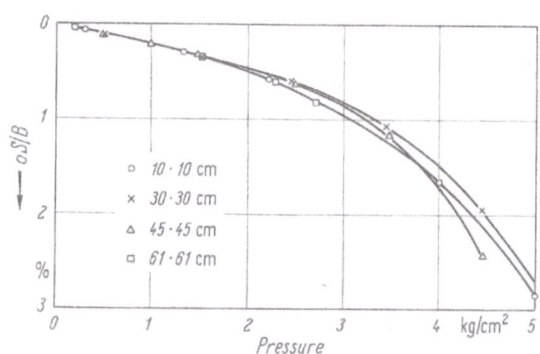

Fig. 6. Plate bearing test on the Kanto loam (load ∼ initial settlement).

where q_c is the compressive strength ($\sigma_3 = 0.5\ \text{kg/cm}^2$). The value of Young's modulus ranges mostly from 120 to 380 kg/cm².

Poisson's ratios measured are plotted against compr. stress/compr. strength in Fig. 5. Provided that a lateral deformation is restrained to a certain degree, Poisson's ratio ranges from 0.1 to 0.2. These values agree well with the results obtained from the wave velocity measurements in the Kanto loam [4].

Fig. 6 shows the results of the plate bearing tests on the Kanto loam with various sizes of the loading plate. Initial settlements were taken from readings at 10 minutes after loading. It will be found that the settlement is proportional to the width or diameter of the loading plates within a wide range of the loading intensity. The initial YOUNG's modulus was estimated as 315 kg/cm², assuming $\nu = 0.15$. The ultimate bearing capacity was about 6 kg/cm², regardless of the loading area.

Creep Settlement

A long term plate bearing test was conducted on the Kanto loam. A cube of practically undisturbed Kanto loam was taken out of a test pit and enclosed in a steel box, the surface of which was sealed with grease and rubber sheets. The gaps between the soil and the wall were filled with gypsum. The test was performed in a humid room at a constant temperature. The diameter of the loading plate was 5 cm and the intensity of the load was 1.5 kg/cm². The settlement of the plate was observed for about 4 months.

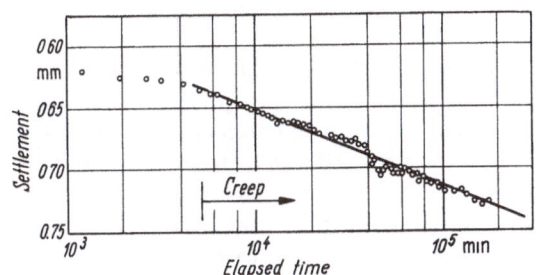

Fig. 7. Creep measurement on plate bearing test.

The result is shown in Fig. 7. A considerable initial settlement as much as 0.6 mm occurred suddently after loading. The settlement proceeded so gradually for several days and then followed a straight line but with a steeper slope when plotted against the elapsed time on semi-logarithmic paper. According to observations for about 8 years of actual settlements of buildings founded on the Kanto loam, the slope $\dfrac{d}{d(\log t)} S$ at the straight portion on the S-log t curve does not change its value for the period. We will call it the rate of logarithmic creep settlement and designate it as $\bar{S}$.

In order to investigate the laws govering the creep settlement of the Kanto loam, we made long term compression tests such as uncon-

29*

fined compression tests, isotropic compression tests and anisotropic compression tests varying the stress condition. The triaxial compression apparatus without a pressure cell was used for the anisotropic compression tests to faciliating the measurement of the lateral displacement. The schema of the apparatus is shown in Fig. 8. The lateral pressure is

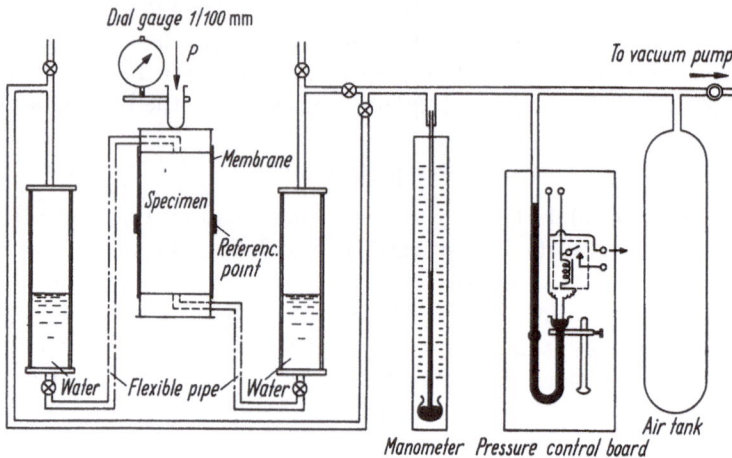

Fig. 8. Triaxial apparatus without a pressure cell.

mobilized by applying the negative pressure in the specimen enclosed with a rubber membrane. In order to prevent evaporation of water in the specimen, a vaccum pump is connected through two water tanks to the ends of the specimen. All the tests were performed at a constant temperature. The specimens, 7 cm in diameter and 14 cm in height, were prepared from block samples taken from the same test pit as the soil block used for the plate bearing test. The average index properties of the samples have already shown in Table 1.

Test A: Observation of Creep under Isotropic Stress. When a specimen is subjected to an isotropic pressure, the strain will be caused only by the volumetric compression since there exists no shear stress. In this case

$$\varepsilon_1 = \varepsilon_3 = \frac{1}{3}\varepsilon_v,$$

ε_1 : axial strain,
ε_3 : lateral strain,
ε_v : volumetric strain

The tests were conducted for about 3 months under a constant isotropic pressure and the axial strain was observed. The lateral strain was also observed on some speciments. The pore pressure was measured on one sample and it was found that it had disappeared within a few minutes after loading.

An illustrative example is shown in Fig. 9. First the initial compressive strain of about 25×10^{-4} occurred, then the strain increased some

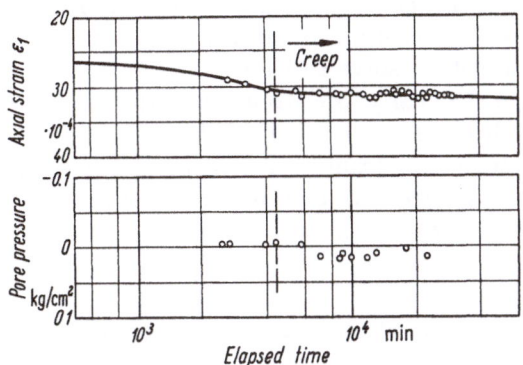

Fig. 9. Creep under isotropic stress.

amount for several days and even after that it proceeded straightly with a very small tangent with respect to the logarithm of the elapsed time.

Test B: Observation of Creep under Deviator Stress. Under an anisotropic pressure the deformation of specimen is due to both the compressive and shear strains, since the shear stress exists as well as the compressive stress. The volumetric strain is expressed as

$$\varepsilon_v = \varepsilon_1 + 2\varepsilon_3 \qquad (2)$$

and the shear strain is proportional to

$$\varepsilon_t = \varepsilon_1 - \varepsilon_3. \qquad (3)$$

Long term unconfined compression tests were made with the specimens sealed with a rubber membrane, and the axial and lateral strains were observed for about 4 months. In Fig. 10 is shown an illus-

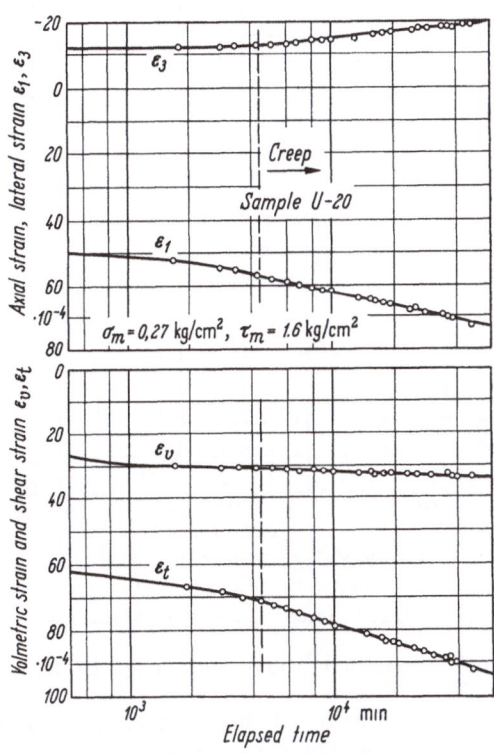

Fig. 10. Creep under deviator stress.

trating example in which the mean principal stress σ_m was 0.27 kg/cm²
and the maximum shear stress τ_m 1.6 kg/cm².

ε_1: After a very gentle increase over a period of several days
presented by a slightly upward curvature, it went on straightly with
a steeper slope with respect to log t.

ε_3: ε_3 was of the nearly same trend as ε_1, the slope of ε_3 after several
days being slightly less than a half of that of ε_1.

ε_v: The volumetric strain increased so slowly even after several
days, with the same trend as seen in the previous figure.

ε_t: The shear strain proceeded with a steeper slope after several
days.

Test C: Observation of Creep under Various Stress Conditions. Creep
tests were conducted on the Kanto loam using the triaxial apparatus
as shown in Fig. 8, where
the stress condition was
changed in three steps. An
illustrating result is shown
in Fig. 11. Firstly, an isotropic pressure 0.4 kg/cm²
was applied. At this stage
no shear stress exists and
only a small amount of
volumetric creep was ob-
served. Then, the axial
load was applied. At this
stage shear creep corres-
ponding to the shear stress
occurred. Finally, the volumetric stress was dou-
bled, keeping the shear
stress unchanged. While the
shear creep did not change
its trend, the rate of the
volumetric creep increased
according to the increase
of the mean principal
stress.

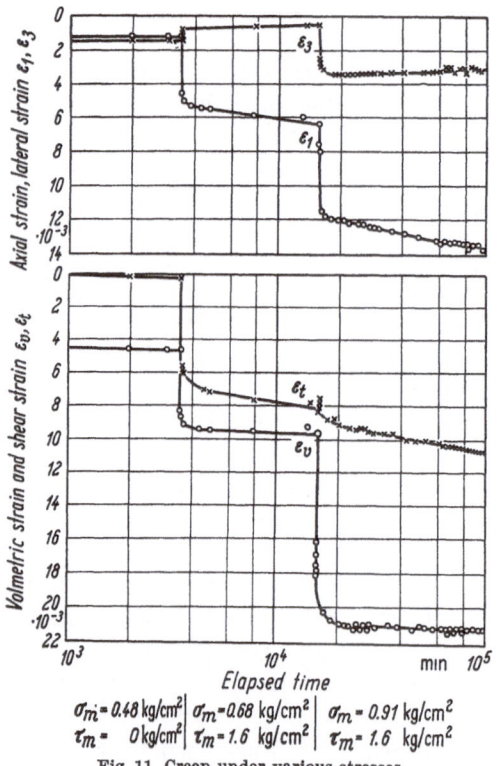

| $\sigma_m = 0.48$ kg/cm² | $\sigma_m = 0.68$ kg/cm² | $\sigma_m = 0.91$ kg/cm² |
| $\tau_m = 0$ kg/cm² | $\tau_m = 1.6$ kg/cm² | $\tau_m = 1.6$ kg/cm² |

Fig. 11. Creep under various stresses.

The experimental re-
sults described above may lead to the following conclusion.

1. The strain of the Kanto loam proceeds through two stages after
initial strain. The earlier stage lasts for several days after loading where

the volumetric strain is pronounced. Total strain during this stage, $_c\varepsilon(t_1)$ is roughly proportional to the initial strain as shown in Fig. 12. They appear to be a phenomenon somewhat analogous to the elastic aftereffect. Since, at all events, it occupies only a minor part of the whole strain which is considered to occur for a long time, we will put it out of consideration for the time being. In the following stage the shear strain predominates and the strains proceed straightly for many years on semi-logarithmic paper. We will define, in this paper, the progressive strain in this stage as the creep.

2. Possibly the volumetric creep, as well as the shear creep may exist. The former may be caused by the flow of soil particles into voids.

From measurements of the creep rates under various stresses, it has been found that the yield stress necessary to cause creep is very low in this clay, and the rates of the logarithmic volumetric creep $\bar{\varepsilon}_v$ and shear creep $\bar{\varepsilon}_t$ are nearly proportional to the mean principal stress and the deviator stress, respectively, as shown in Fig. 13. Therefore, they will be given by the following formula,

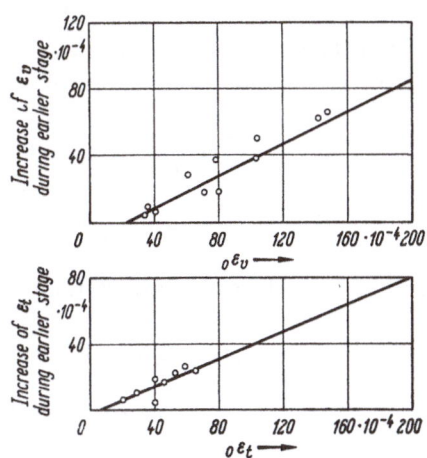

Fig. 12. Relationship between progressive strain in earlier stage (elastic after-effect) and initial strain.

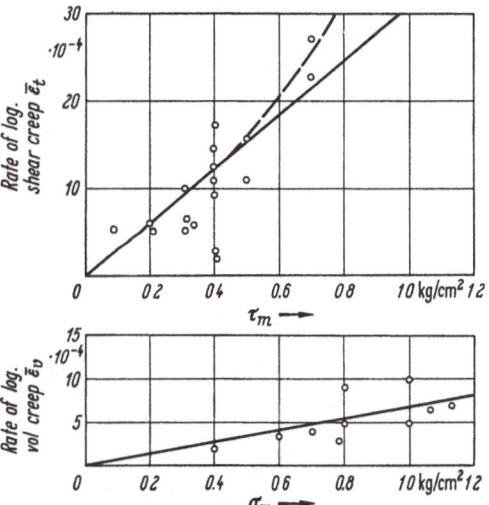

Fig. 13. Relationship between logarithmic creep and stress.

$$\bar{\varepsilon}_v = \frac{d}{d(\log t)}\varepsilon_v = 3\beta_1\sigma_m \tag{4}$$

$$\bar{\varepsilon}_t = \frac{d}{d(\log t)}\varepsilon_t = 6\beta_2\tau_m, \tag{5}$$

where

$$\sigma_m = \frac{1}{3}(\sigma_1 + 2\sigma_3), \quad \tau_m = \frac{1}{2}(\sigma_1 - \sigma_3) \tag{6}$$

and β_1 and β_2 are called the coefficients of logarithmic volumetric creep and shearing creep, respectively.

From the above, using Eqs. (2), (3) and (6) the rates of the logarithmic axial creep and radial creep will be given by

$$\bar{\varepsilon}_1 = \frac{d}{d(\log t)}\,\varepsilon_1 = \beta_1\sigma_m + 4\beta_2\tau_m, \tag{7}$$

$$\bar{\varepsilon}_3 = \frac{d}{d(\log t)}\,\varepsilon_3 = \beta_1\sigma_m - 2\beta_2\tau_m. \tag{8}$$

The observed values of β_1 and β_2 range, with a few exception, as belows:

$$\beta_1 = 1.5 \text{ to } 4.2 \times 10^{-4} \text{ cm}^2/\text{kg},$$

$$\beta_2 = 3.5 \text{ to } 7.5 \times 10^{-4} \text{ cm}^2/\text{kg}.$$

Creep Settlement of a Loaded Area

For simplicity in the analysis, only points on the axis of symmetry below a foundation will be considered.

If at the points under consideration σ_1 and σ_3 are the vertical and horizontal stresses respectively, then after application of load they are

$$\sigma_1 = \gamma Z + \Delta\sigma_1,$$

$$\sigma_3 = K\gamma Z + \Delta\sigma_3,$$

where γ is the unit weight of the soil and K is the earth pressure coefficient at rest.

Using Eqs. (6), (7) and (8), the rate of the logarithmic creep settlement of the footing resting on the Kanto loam of thickness H is given by

$$\bar{S} = \frac{d}{d(\log t)}\,S = \beta_1 \int_0^H \sigma_m\,dZ + 4\beta_2 \int_0^H \tau_m\,dZ,$$

$$= \frac{\beta_1}{3} \int_0^H \{(1 + 2K)\gamma Z + (\Delta\sigma_1 + 2\Delta\sigma_3)\}\,dZ \tag{9}$$

$$+ 2\beta_2 \int_0^H \{(1 - K)\gamma Z + (\Delta\sigma_1 + \Delta\sigma_3)\}\,dZ.$$

If the thickness of the clay is moderately large, compared with the footing width, Eq. (9) will be reduced to

$$S = \frac{1}{6}\beta_1 \left[(1 + 2K)\,\gamma H_e^2 + (1 + 2\lambda)\,H_e q\right]$$
$$+ \beta_2 [(1 - K)\,\gamma H_e^2 + (1 - \lambda)\,H_e q]. \tag{10}$$

$$H_e = 2 \int_0^H \frac{\sigma_1}{q}\,dZ,$$

$$\lambda = \frac{\int_0^H \Delta\sigma_3\,dZ}{\int_0^H \Delta\sigma_1\,dZ}.$$

The first term of Eq. (10) expresses the effect due to the volumetric creep and the second term due to the shear creep. H_e and λ depend only on the geometrical figure of footing and are calculated by Boussinesq's formula. In the case of a rigid circular footing of diameter B,

$$H_e = 1.57\,B, \quad \lambda = 0.08 \quad (\text{assuming} \quad \nu = 0.15).$$

The observed rate of logarithmic creep settlement in the plate bearing test shown in Fig. 7 was 0.0058 cm per log cycle of time and the calculated one was 0.0065 cm per log cycle of time, assuming $K = 0.3$. The agreement is fairly satisfactory.

Observed and Calculated Settlement of a Building on the Kanto Loam

The authors have made observations for about 8 years of the settlements of several buildings resting on the Kanto loam.

They show that every settlement has proceeded straightly against log t within the observation period.

Here, the observation result of a 4-story apartment house, 56 m long and 5.5 m wide, will be shown.

The building was constructed in 1954 at Hitotsugi-matchi, Minato-ku, Tokyo. The site is overlaid, below a fill of 1 m thick, by the Kanto loam up to a depth of 7 to 8 m below ground surface. The shear strength of this Kanto loam is lower than that of the usual ones and ranges from 0.3 to 0.5 kg/cm². A dense fine sand formation follows below the Kanto loam (Fig. 14).

The building is supported by individual footings with an average load intensity of 6 t/m^2 at a depth of 1.2 m below the surface.

Ten reference points were fixed at the footins with nearly equal spacings, and a bench mark reaching to the sand formation was constructed.

Observed settlements during the construction are plotted against the acting foundation pressures in Fig. 15. It is seen that the settlement

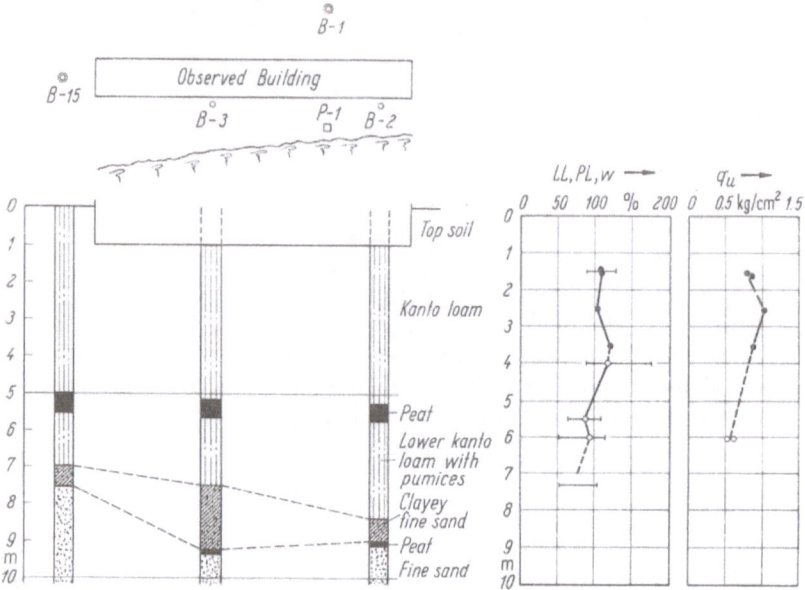

Fig. 14. Soil profile and soil properties.

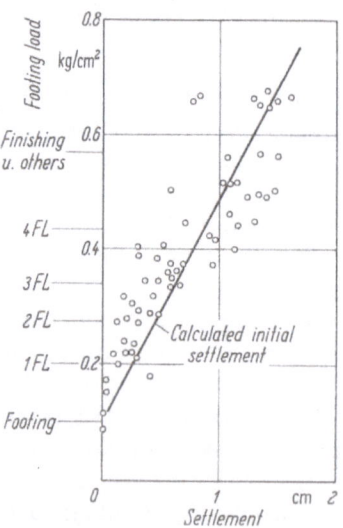

Fig. 15. Settlement during construction.

increases proportionally with an increase of the foundation pressure. The settlement during the construction consists mainly of initial settlement. The straight line in the figure represents the initial settlement calculated using Eq. (1).

The progression of the settlement is shown against the logarithm of the elapsed time in Fig. 16. The mean of the rate of the logarithmic creep settlement is 1.6 cm per log cycle of time. The calculated rate is 1.32 cm per log cycle of time which is seen to be good agreement with the observations.

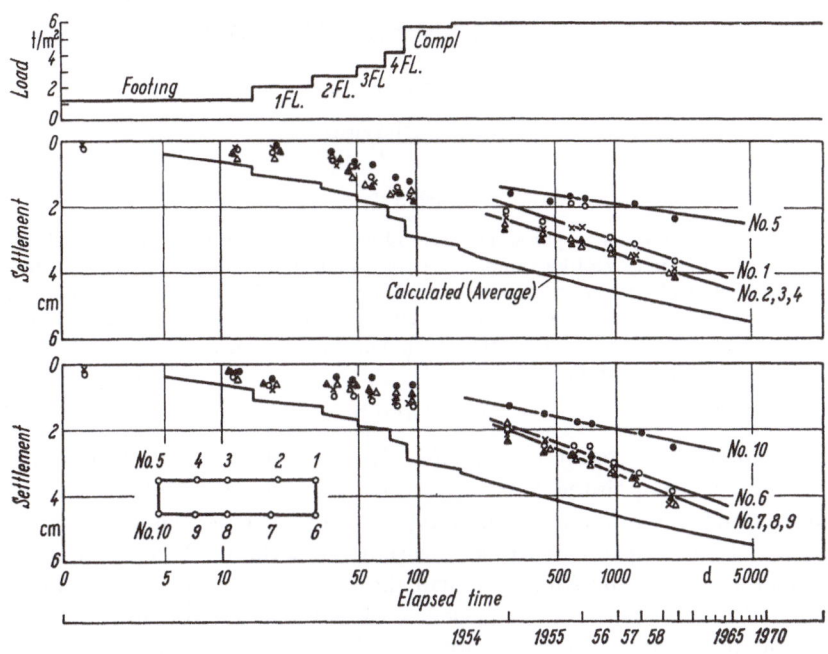

Fig. 16. Observed and calculated settlement.

References

[1] Sudo, T.: Clay minerals, 1953, Iwanami, Tokyo.

[2] Misono, S., and S. Sudo: Studies on the micro-structures of soil, Report of Agri. Institute, 1958.

[3] Koizumi, Y., and K. Ito: Device for measuring the lateral strain in the compression test of unsaturated clays, J. of Soil and Foundation, No. 42 (1962).

[4] Iida, K.: Behavior of soil due to a change of water content, Research of Foundation, No. 50 (1942).

4.4 Secondary Settlements of Structures Subjected to Large Variations in Live Load

By

Laurits Bjerrum

Abstract

In the paper are compared the secondary settlements of normal buildings with those of structures like tanks and silos which have a large variation in live load. It is shown by three examples that large variations in load on the foundation soil will result in increased secondary settlements. The three structures considered thus showed secondary settlements which continued at an unchanged rate over the period of observation.

Introduction

In his closing remark to section VI at the Rotterdam Conference 1948, TERZAGHI pointed out that the rate of secondary settlements of some structures remains practically constant over periods of decades, whereas that of others decrease so rapidly that the settlements plotted against a logarithmic time scale will show a straight line. In connection with a study of the relation between computed and observed settlements of structures (BJERRUM, 1964) the author came across some settlement records which proved the general validity of TERZAGHI's remarks. This study was therefore extended to include also a review of the secondary settlements, and it is the result of this study which is presented in this paper.

Review of Settlement Observations

The above mentioned review of the relation between computed and observed settlements was based on a collection of settlement records from about 60 buildings, not including pile foundations. Only one third of these observations were, however, carried far enough to form a reliable basis for a study of the secondary settlements. A further number of structures had to be excluded as the observations of the secondary settlements could be influenced by general subsidence phenomena. From a detailed study of the settlement-time curves of the remaining buildings it became clear that the structures could be separated into two groups according to their behaviour during the secondary phase of the settlement process.

The first group includes the majority of the structures, and their behaviour therefore represents what will be called a "normal" type of secondary settlements. The characteristic feature of the normal type of secondary settlement is that the rate of settlement is decreasing with time. If plotted against a logarithmic time scale, the secondary settlements of structures with a normal behaviour will in general lead to an approximately straight line. Examples of this type of secondary settlements have been published for instance by BUISMAN (1936) and ZEEVAERT (1958). For comparison the settlement records of two structures with a normal secondary behaviour have been included in this paper. The two structures are the Waterloo Bridge in London which is shown on Fig. 4 (COOLING and GIBSON, 1955) and Sinai Hospital in Toronto shown on Fig. 5 (CRAWFORD and BURN, 1962).

The second group of structures are those showing an "abnormal" secondary behaviour. By this it is understood that the secondary settlements will not show a regular decrease in rate with time. The most prominent examples of this group thus showed a constant rate of secondary settlements and three of these will be described in detail below. There were, however, also in this group some structures representing transition forms between a "normal" and an "abnormal" behaviour.

From a comparison between the structures showing different types of secondary behaviour it became immediately evident that there was an essential difference between the structures in the two groups in the way in which the load on their foundation varied. The normal group included normal buildings, bridge piers and embankments, i. e. structures where the live load forms only a small fraction of the total load and the variation in stresses applied to the foundation is small. The second group included such types of structures as oil tanks and silos where the live load forms an essential part of the total load and where the foundation is subjected to large variations in stresses. In this group there were also found examples on tall structures like free-standing chimneys and towers which show no variation in live load, but where a variation in load on foundation soil could be a result of oscillations caused by wind action.

It became clear that the secondary settlements of a structure are influenced by a variation in loads on the soil. If the variations in load are small, the rate of secondary settlement will decrease with time, and in general the settlement curve will plot as a straight line in a logarithmic time scale. Any variation in live load seems to increase the secondary settlements and if the variation is large — more than 50% of the maximum load — as in silos and tanks, the rate of secondary settlement may remain unchanged over at least one or two decades after the primary settlements have ceased.

The above finding seems to hold good in a variety of different types of soil. There are, however, no case records which indicate their validity in soft compressible clays, the reason being that the available settlement records hardly exceed the period of primary consolidation.

The above finding is illustrated below by three case-records describing structures which have a large variation in live load and for comparison there are also included the settlement records of two structures built but on similar types of soils, where the variation in live load is small.

Oiltank in Drammen

In the spring of 1953 the Norwegian Geotechnical Institute was consulted in connection with the foundation of an 8000 m³ oil tank on

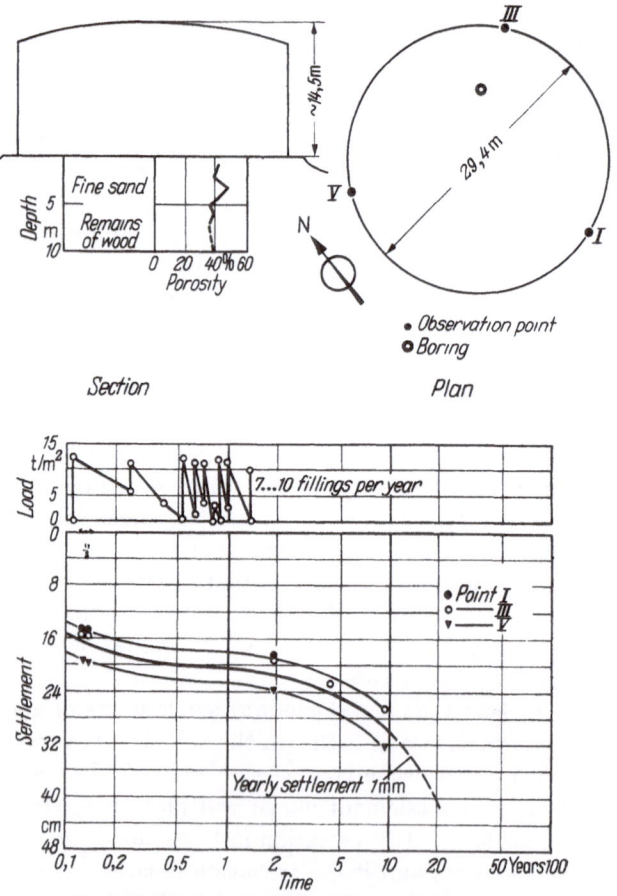

Fig. 1. Settlement record for oil tank in Drammen.

an island at the outlet of the river of Drammen, Norway (KUMMENEJE, 1955). The tank was built in 1953 and the settlements have been followed by regular measurements over a period of 10 years.

The subsoil conditions were explored by borings and soundings. To a depth of about 20 metres a fairly homogeneous loosely deposited postglacial sand was found. The porosity of the sand varied between 35 and 43 per cent. Embedded in the sand were found layers of sand containing remains of wood.

The tank is a steel tank with a diameter of 29.4 metres. It is placed directly on the ground resting on a well-compacted gravel base, and with the walls of the tank resting on a small reinforced concrete footing wall. The additional load on the ground is for full tank $16.5 t/m^2$.

During the first test filling of the tank in the spring of 1954, settlement observations were made at six points around the concrete ring. As the load increased an immediate average settlement of 13 cm was observed, which in the subsequent three days increased to 16 cm. By unloading the mean settlement was reduced by $1-2$ cm and by reloading an increase of $2-3$ cm was measured.

In the following years the tank has been emptied and filled $7-10$ times every year. The settlements have continued at a rate of 1 cm per year, see Fig. 1, and today they amount to about 31 cm.

It should be mentioned that the settlements plotted on Fig. 1 are independent of eventual regional subsidence occuring as they represent the settlements of the tank relative to the surrounding terrain.

It is of interest to compare the settlements of the oil tank in Drammen with the settlements of a grain silo built on a fine sand in Rijeka which has been described by NONVEILLER (1963). Also for this structure it was observed that the settlements increased considerably as a result of variations in the live load.

Grain Silo, Moss

The city of Moss is located at the east side of the Oslofjord about 60 kilometres south of Oslo. Part of the city is built up on a heavily overconsolidated glacial till (ground moraine) overlain by a sand layer with varying thickness. The glacial till has proved to form an excellent foundation for even the heaviest type of structures.

In the harbour area the Moss grain mill has for instance built three grain silos, all of them resting on the glacial till. Silo No. 1 was built in 1939 and the settlement of this silo has been followed up to the present.

At the site where silo No. 1 is built, the glacial till extends to great depth, exceeding 30 metres. The till is surprisingly uniform. The water content changes only little with depth, varying between 8 and

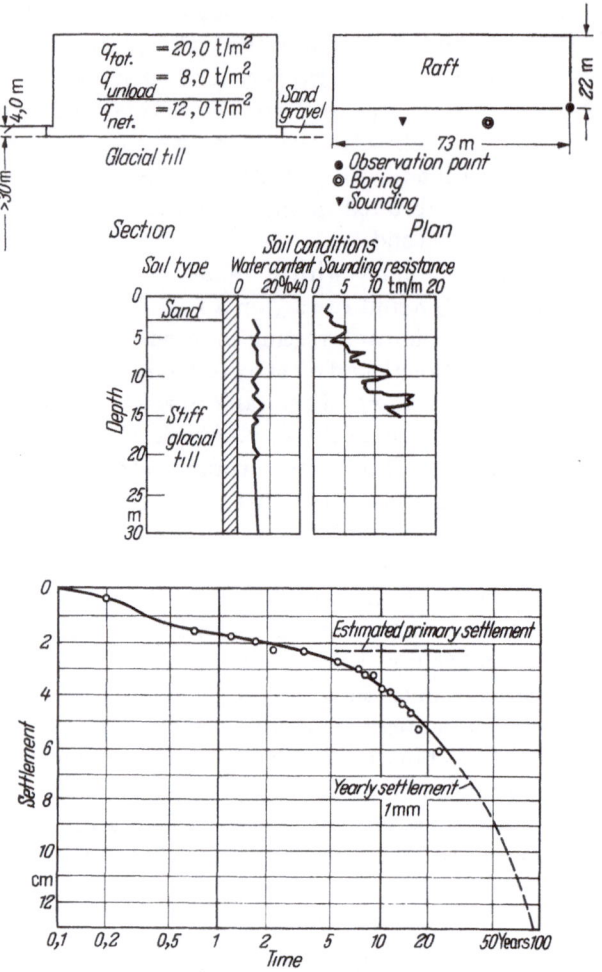

Fig. 2. Settlement record for grain silo I, Moss.

12%, see Fig. 2. The unit weight of the till is varying from 2.15 to 2.35 t/m³. At the site the till is covered by about 4 metres of sand and gravel. The top layer of the glacial till is locally softened somewhat; the water content of this softer top layer is as high as 13—14%.

The silo is founded on a continuous reinforced concrete slab poured on the surface of the glacial till, i. e. at a depth of about 4 metres below the surface. The dimensions of the silo are given on Fig. 2. The total load on the ground is for maximum filling of the bins 20 t/m². About 50% of this load is live load. The maximum net load is 12 t/m², of which the major part is live load. As in all grain silos the live load is varying over the year from a maximum to a minimum. In addition there are considerable changes in the distribution of the load due to shifting of the grain within the bins.

In Fig. 2 are shown the settlements of one of the observation points. As seen from the figure, the settlements of the silo increased as the load was applied. The primary settlements are supposed to be about 2.2 cm and they were finished a couple of years after the construction. Since then the settlements have, however, continued at a constant rate. Over a period of 23 years the settlements have increased by about 0.1 cm per year. The total settlement of the considered point amounts today to 6 cm of which two thirds have appeared during the secondary phase of the settlement process.

The settlement curves for the other observation points are similar in shape to the curve shown on Fig. 2, only the total magnitude of the settlements being different. The settlements over the silo vary from 4.6 to 10.4 cm, the higher values being the result of a compression of the softer top layer found locally.

For comparison Fig. 5 shows the settlement records for the Mt. Sinai Hospital in Toronto, Canada. This building rests on a glacial till rather similar to the soil forming the foundation for the silo in Moss, but in contrary to the silo the variations in live load of the hospital are very small. As seen from the settlement observations the settlement behaviour of the hospital is in all respect "normal". The secondary settlements are small compared with the primary settlements and their rate is diminishing with time. The data and settlement observations for the Mt. Sinai Hospital have been published by CRAWFORD and BURN (1962).

San Jacinto Monument, Texas

On april 21st, 1836, General Sam Houston defeated a Mexican army at a site located about 20 miles southeast of Houston, near the San Jacinto River. This event was commemorated 100 years later by the construction of the San Jacinto Monument, a 174 metre high reinforced concrete obelisk. The settlements of this monument have been carefully followed and in several papers DAWSON (1947 and 1948) has described the monument and its settlements. Due to the courtesy of Professor DAWSON, the author has received the most recent records of the settlements, so that the settlement curve plotted in Fig. 3 shows the results of 21 years observations.

The monument rests on a 36-metre thick deposit of the stiff-fissured Beamont clay. The geotechnical data of the clay are summarized in Fig. 3.

The monument is a 174-metre high reinforced concrete obelisk which rests on a monolithic concrete base, 38 metres square, with a thickness which varies from 4.5 m in the middle to 1.5 m along the edge. The weight of the monument causes an additional load on the clay of 14.5 t/m².

The settlements of the monument have been followed carefully since the foundation was poured. The measurements were made by

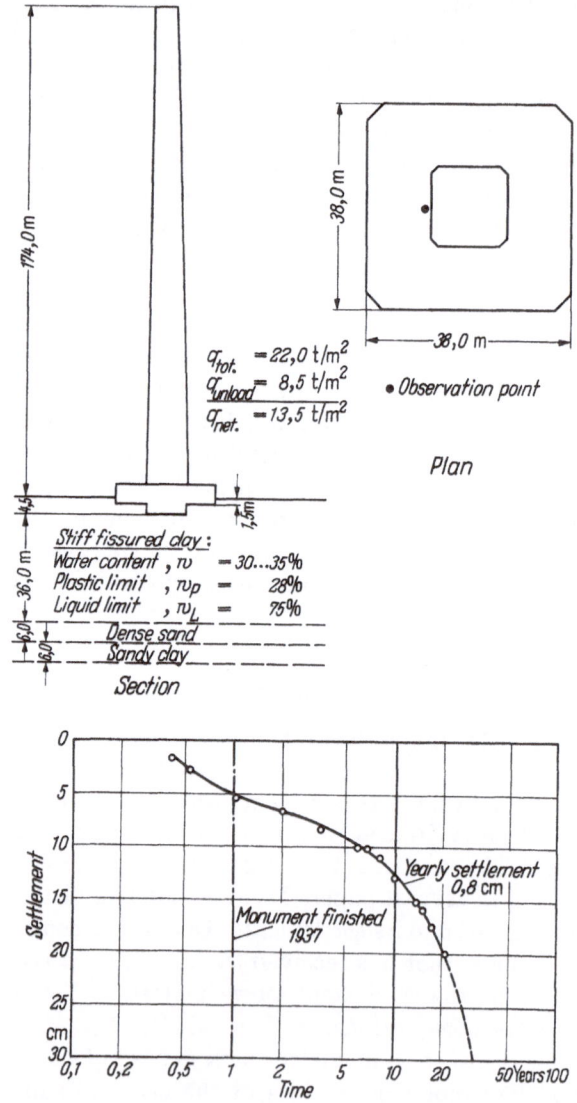

Fig. 3. Settlement record for San Jacinto Monument, Texas. (After DAWSON, 1947.)

precise   levelling, the bench marks being established at a depth of
about 5 metres in some distance from the monument. The settlements
thus represent the movement of the monument relative to the surrounding terrain.

During the construction of the monument the settlements increased
as the load was applied and at the end of the construction they amount-
ed to 5 cm.

In the following years the rate of settlements decreased regularly
and about three years after the monument was finished, the monument showed a tendency to come to equilibrium with a primary settlement of about 7—8 cm. Instead of showing the normal secondary
type of settlements, the settlements continued however, at a constant
rate. Over a period of about 15 years the rate of settlements has been
nearly constant at 8 mm per year. The settlement curve is, however,
not absolutely smooth, but shows some irregularities.

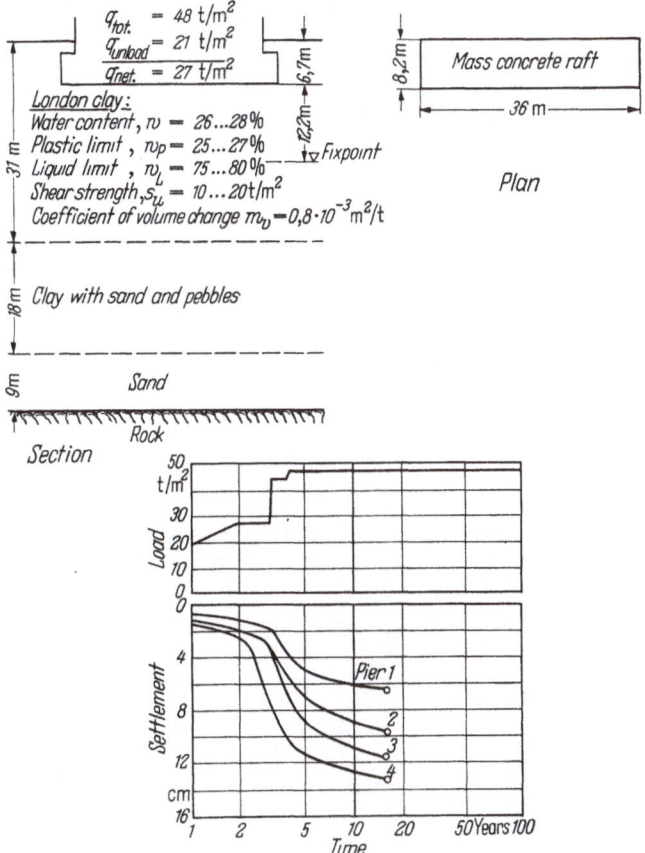

Fig. 4. Settlement record for Waterloo Bridge, London. (After Cooling and Gibson, 1955.)

This monument is located in an area which intermittantly is subjected to tropical storms. As the structure is a 174 m high free-standing obelisk placed on a free area, it is subjected to vibrations and dynamic forces, causing oscillations perpendicular to the direction of the wind. From a study of the irregularities in the settlement curve, DAWSON (1947) was able to correlate them with the occurrence of special high wind velocities. He could therefore draw the tentative conclusion that the abnormal type of secondary settlements observed is a result of the alternating loads to which the foundation clay intermittantly is subjected.

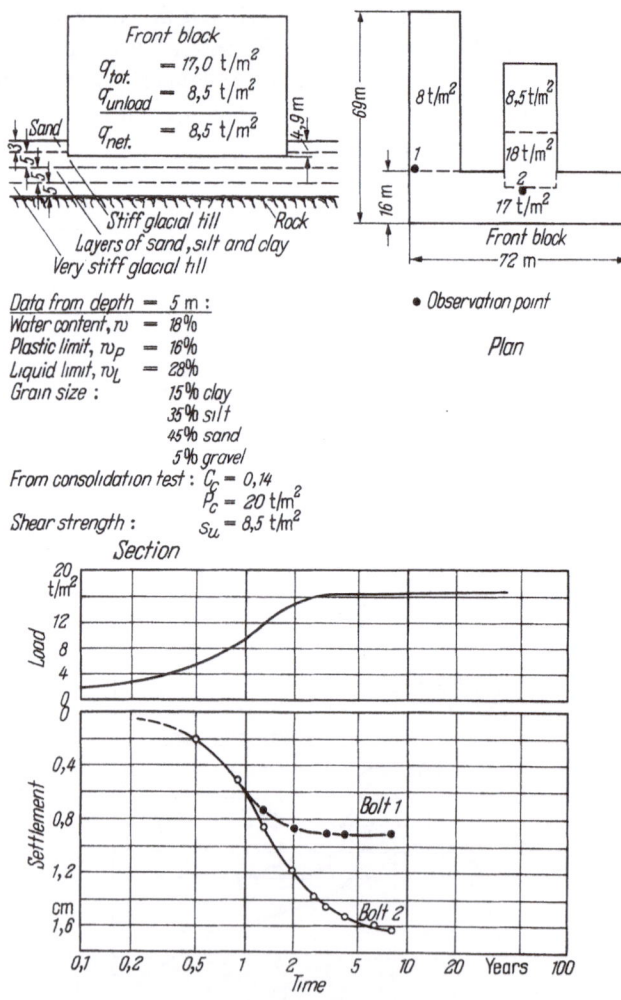

Fig. 5. Settlement record for Sinai Hospital, Toronto. (After CRAWFORD and BURN, 1962.)

It is of interest to compare the settlements of the San Jacinto Monument with the settlements of the Waterloo Bridge, which are shown on Fig. 4. The piers of the Waterloo Bridge are resting on the well-known London clay, which in several respects is very similar to the Beamont clay. The alternations in load of the piers of the Waterloo Bridge are very small compared with the dead load. As seen from Fig. 4, the secondary settlements of the piers of the Waterloo Bridge are what is called normal, showing a regular decrease in rate with time.

Concluding Remarks

It is of interest to notice that two of the structures described above, the San Jacinto Monument and the Silo in Moss, are resting on soils belonging to the group of soils which are known not to show significant secondary settlements compared with the primary settlements for normal static loads. In spite of this the two structures with varying live loads have experienced secondary settlements which after two decades are of the order of two to three times the estimated primary settlement. For the two structures where the variations in live load are known — the tank in Drammen and the silo in Moss — it is evident that the number of cycles per year of the live load is relatively small. It is therefore most likely that the abnormal behaviour of these structures are to a higher degree due to the magnitude of the variation in load than due to their frequence.

From the described review of available settlement records it is possible to draw a conclusion of some practical importance: Large secondary settlements can be expected of structures with large variations in live load and this is the case even in soils where the secondary settlements in general are assumed to be small. For such structures the secondary settlements may continue at a rate which shows no measurable reduction in time even two or three decades after the primary settlements are finished.

It is not possible from the available information to draw any conclusions concerning the mechanism of the large secondary settlements observed. Any suggestion has therefore necessarily to be of a purely speculative nature. Below it is attempted to analyze the observed phenomena in the light of a general mechanistic picture of what happens when a footing is loaded (BJERRUM, 1963).

The secondary — as well as the primary — settlement is a consequence of an adjustment of the particle arrangement to comply with the new set of stresses. It results from relative movements bending and crushing of particles. Each time a failure occurs at a contact point it is like a chain reaction followed by internal movements and deformations. The rate at which this readjustment of the structure occurs is influenced by viscous phenomena at the contact points and a certain time is therefore required before all particles have found a stable position and the settlement ceases. It is easily understood that if the shear stresses in the contact points between the particles vary as a result of a variation in the load on the foundation, this will cause an acceleration and increase of the creep phenomena which otherwise would have led to a secondary settlement occurring at a diminishing rate.

In this process we can distinguish between two different components of the secondary settlement. The first component is due to a volume reduction of the soil column located beneath the footing. The second component is due to the loss of ground below the footing resulting from lateral yield.

The volume change component is dominating when the width of the footing is large compared with the thickness of the compressible soil layers. In very compressible soils — like peat — this component is probably forming the major contribution to the secondary settlements independent of the width/depth ratio. Even in such cases where the secondary settlements are accellerated by variation of the load, one would expect that the volume change component would show a diminishing rate with time, as for instance observed in oedometer tests with cyclic loading (Schultze, 1961).

The relative importance of the lateral yield component is greatest when the shear stresses approach the shear strength of the foundation soil and in addition it depends on the type of soil. There are no obvious reasons for believing that the rate of the yield component should diminish with time, at least not if the shear stresses exceed a certain fraction of the shear strength of the foundation soil and there in addition is a cyclic variation of the shear stresses.

Based on this general mechanistic picture it seems reasonable to suggest that secondary settlements occurring at a constant rate are associated with a lateral yield of the foundation soil. This explanation was for instance proposed by Terzaghi in his closing discussion of Section VI at the Rotterdam Conference (Terzaghi, 1948).

In this connection it is relevant to mention the results of a drained triaxial test on a very stiff-fissured clay, an undisturbed sample of the Lillebelt clay. The sample was initially consolidated under the effective pressures $\sigma_1' = 2.62$ kg/cm^2 and $\sigma_3' = 2.12$ kg/cm^2. The value of σ_1' was then changed regularly by a cyclic loading of the sample. With a frequency of 15 minutes the load was increased over a period of about 8 minutes by 0.1 kg/cm^2, or 0.04 σ_1. The cyclic loading caused additional compression of the sample, which continued even after 4000 applications of the cyclic load. What is most interesting is, however, that during the period of cyclic loading a regular and continuous swelling of the sample took place. This means that the vertical compression of the sample caused by the variation in load must exclusively have been a result of a lateral yield. This finding thus seems to confirm TERZAGHI's original suggestion.

References

BJERRUM, L. (1963): Opening address. European Conference on Soil Mechanics and Foundation Engineering, Wiesbaden. Proceedings, Vol. 2, pp. 3—5.

BJERRUM, L. (1964): Relasjon mellom målte og beregnede setninger av byggverk på leire og sand; NGF-foredraget 1964. Oslo. 92 pp.

BUISMAN, A. S. K. (1936): Results of long duration settlement tests. 1st International Conference on Soil Mechanics and Foundation Engineering, Cambridge, Mass. Proceedings, Vol. I, pp. 103—106.

COOLING, L. F., and R. E. GIBSON (1955): Settlement studies on structures in England. Institution of Civil Engineers. Conference on the correlation between calculated and observed stresses and displacements in structures. Papers, Vol. I, pp. 295—317.

CRAWFORD, C. B., and K. N. BURN (1962): Settlement studies on the Mt. Sinai Hospital, Toronto. Eng. J. 45, 12, pp. 31—37. (National Research Council, Canada. Division of Building Research. Research paper 178.)

DAWSON, R. F., and W. E. SIMPSON (1948): Settlement records on structures in the Texas Gulf Coast area. 2nd International Conference on Soil Mechanics and Foundation Engineering, Rotterdam. Proceedings, Vol. 5, pp. 125—129.

DAWSON, R. F. (1947): Settlement studies on the San Jacinto monument. 7th Texas Conference on Soil Mechanics and Foundation Engineering, Austin, Texas. Proceedings.

GIBSON, R. E., and K. Y. LO (1961): A theory of consolidation for soils exhibiting secondary compression. (Norges geotekniske institutt. Publikasjon 41, pp. 1—16).

KUMMENEJE, O. (1955): Fundamentering av oljetank i Drammen. Bygg, Vol. 3, No. 9, pp. 239—243 (Norges geotekniske institutt. Publikasjon 12).

MacDONALD, D. H., and A. W. SKEMPTON (1955): A survey of comparisons between calculated and observed settlements of structures on clay. Institution of Civil Engineers. Conference on the correlation between calculated and observed stresses and displacements in structures. Papers, Vol. I, pp. 318—337.

NONVEILLER, E. (1963): Settlement of a grain silo on fine sand. European Conference on Soil Mechanics and Foundation Engineering, Wiesbaden. Proceedings, Vol. I, pp. 285—294.

SCHULTZE, E., and G. COESFELD (1961): Elastic properties of ballast. 5th International Conference on Soil Mechanics and Foundation Engineering, Paris. Proceedings, Vol. I, pp. 323—327.

TERZAGHI, K. (1948): Closing discussion on foundation pressure and settlements of buildings on footings and rafts. 2nd International Conference on Soil Mechanics and Foundation Engineering, Rotterdam. Proceedings, Vol. VI, Section VI, p. 118.

ZEEVAERT, L. (1958): Consolidacion de la arcilla dé la ciudad de Mexico. American Society for Testing Materials. Special Technical Publication, No. 232, pp. 18—27.

4.5 Les Barrages en Terre et la Mécanique des Sols

Par

Pierre Londe

Introduction

Cette communication pourra paraître hors de propos dans le cadre de débats aussi savants que ceux d'un colloque de l'IUTAM. Elle n'aborde, en effet, le fond d'aucun des grands problèmes qui se posent aux chercheurs et ne prétend nullement proposer des solutions ou des découvertes. Son objet est tout autre.

Il s'agit de présenter, à l'occasion d'une réunion qui groupe des spécialistes de tous les pays du monde — professeurs, savants, physiciens, mathématiciens, chercheurs de mécanique des sols — les réflexions du constructeur de barrages. L'auteur précise ici qu'il se fait l'écho d'une expérience collective, et parle au nom d'une équipe ayant travaillé en intime collaboration pendant plus de quinze ans.

La communication se divise en trois parties. 1. Une analyse critique des moyens dont dispose actuellement l'Ingénieur pour faire le projet d'un barrage en terre, d'où il ressort que ces moyens sont encore loin d'être parfaits. 2. Les principes directeurs qui doivent guider l'Ingénieur dans l'état actuel de ses connaissances. 3. Les différents domaines dans lesquels l'auteur suggère que les principaux efforts de progrès soient accomplis.

En attirant l'attention des chercheurs sur ses besoins les plus urgents le praticien espère non pas infléchir leurs travaux dans un sens étroitement utilitaire, mais plutôt leur offrir un cadre d'action concertée. Si elle y parvient, même partiellement, cette contribution aura atteint son but.

I. Analyse Critique des Moyens

Dans l'état actuel de la technique le projeteur de barrages en terre a trois outils principaux à sa disposition:

l'observation des précédents,
les principes de la mécanique des sols,
les calculs.

Cette classification, certainement trop schématique, mériterait d'être nuancée. Elle est toutefois adoptée ici pour les besoins de l'exposé et aussi pour souligner, dès maintenant, l'intérêt relatif que l'auteur attache à ces trois outils, énumérés dans l'ordre d'efficacité décroissante.

1. L'Observation des Précédents

a) Les Ruptures. C'est de l'observation critique des réussites et surtout des ruptures, mineures ou catastrophiques, qu'au cours des siècles se sont lentement dégagées les Règles de l'Art.

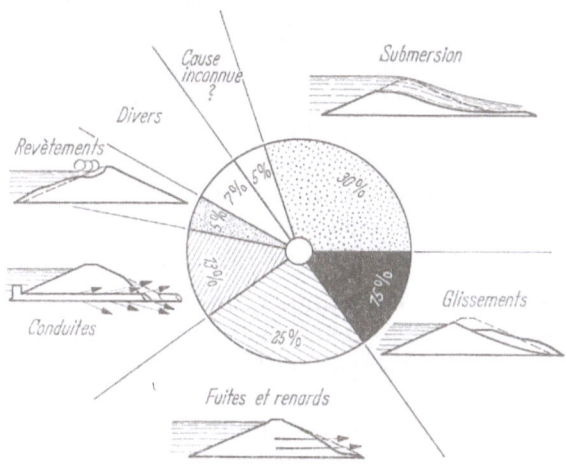

Fig. 1.

Une statistique de 1953 (MIDDLEBROOKS) portant sur deux cents barrages en terre ayant subi des désordres graves — la rupture complète pour bon nombre d'entre eux — peut se résumer par la fig. 1. Les données de cette statistique couvrent plus de cent ans d'observations au cours desquels les Ingénieurs ont appris, souvent à leurs dépens, à construire des déversoirs plus grands, à organiser les remblais en zones pour résister à l'érosion interne, à se soucier des propriétés mécaniques des matériaux et des conditions de mise en place, à agir sur les pressions interstitielles, etc...

Des progrès considérables ont ainsi été faits, et les accidents sont de plus en plus rares bien que les barrages soient plus nombreux et plus hauts. Mais des ruptures récentes prouvent que la perfection n'est pas encore atteinte et qu'il reste à tirer de nouvelles leçons de l'observation attentive des précédents.

L'interprétation des ruptures est souvent si difficile que beaucoup d'entre elles restent controversées ou au moins conjecturales. La statistique de la fig. 1 ne doit pas faire illusion à cet égard; elle est plus un tableau des effets que des causes.

Quand un ouvrage est détruit et qu'aucun témoin ne peut décrire, de façon objective, la succession des événements, la reconstitution de l'enchaînement des mécanismes destructeurs et l'isolement de la cause initiale sont des opérations douteuses. On peut faire des hypothèses plausibles, mais qui restent seulement plausibles. En retenir une plutôt que les autres est une démarche parfois nécessaire, mais il faut garder présent à l'esprit sa nature d'hypothèse de travail.

Vouloir ramener les causes à un défaut unique est rarement justifié. Un exemple est celui de la rupture du barrage de Baldwin Hills, en Décembre 1963. Il a fallu qu'une faille joue d'une vingtaine de centimètres pour ouvrir une voie d'eau dans le masque, mais il a fallu *aussi* que le matériau de la fondation soit très érodable pour que l'ouvrage périsse en quelques heures. A l'opposé, le tremblement de terre de 1930 déclenchait une fuite grave, avec entraînement de matière solide, dans le barrage de Chatsworth n° 2 (Californie). Deux jours plus tard l'eau sortait claire et le débit était dix fois plus faible. Le massif, autocolmatable, s'était adapté à la fissuration.

b) Les Barrages Existants. L'autre panneau du dyptique des précédents, et fort heureusement de beaucoup le plus nombreux, est celui des barrages qui se comportent bien. Mais il n'est pas sans danger. S'inspirer d'un précédent dont on ne sait pas *tout* est peut être rééditer une erreur latente. Si un glissement ne peut pas passer inaperçu, la fissuration d'un noyau peut demeurer cachée, pour ne citer qu'un exemple. Combien de noyaux d'étanchéité sont fissurés sans qu'on le sache ? Combien de remblais ne doivent de tenir qu'à une circonstance fortuite, par exemple colmatage ou inversement filtration naturelle ? Nul ne peut répondre. C'est que trop peu de barrages sont munis des appareils de mesure qui seuls permettraient de connaître leur véritable comportement. Trop peu ont leur histoire écrite depuis le commencement, et il est rare qu'on connaisse suffisamment bien la nature des matériaux utilisés et leurs conditions de mise en place, plus rare encore qu'on connaisse la fondation.

2. Les Principes de la Mécanique des Sols

La Mécanique des sols offre son analyse scientifique des phénomènes qui permet de comprendre le comportement des matériaux et par conséquent de progresser. Mais les modèles physiques, expérimentaux et mathématiques de la mécanique des sols sont toujours simplifiés, soit par commodité, soit par ignorance, par rapport à la complexité des conditions naturelles qu'ils prétendent représenter.

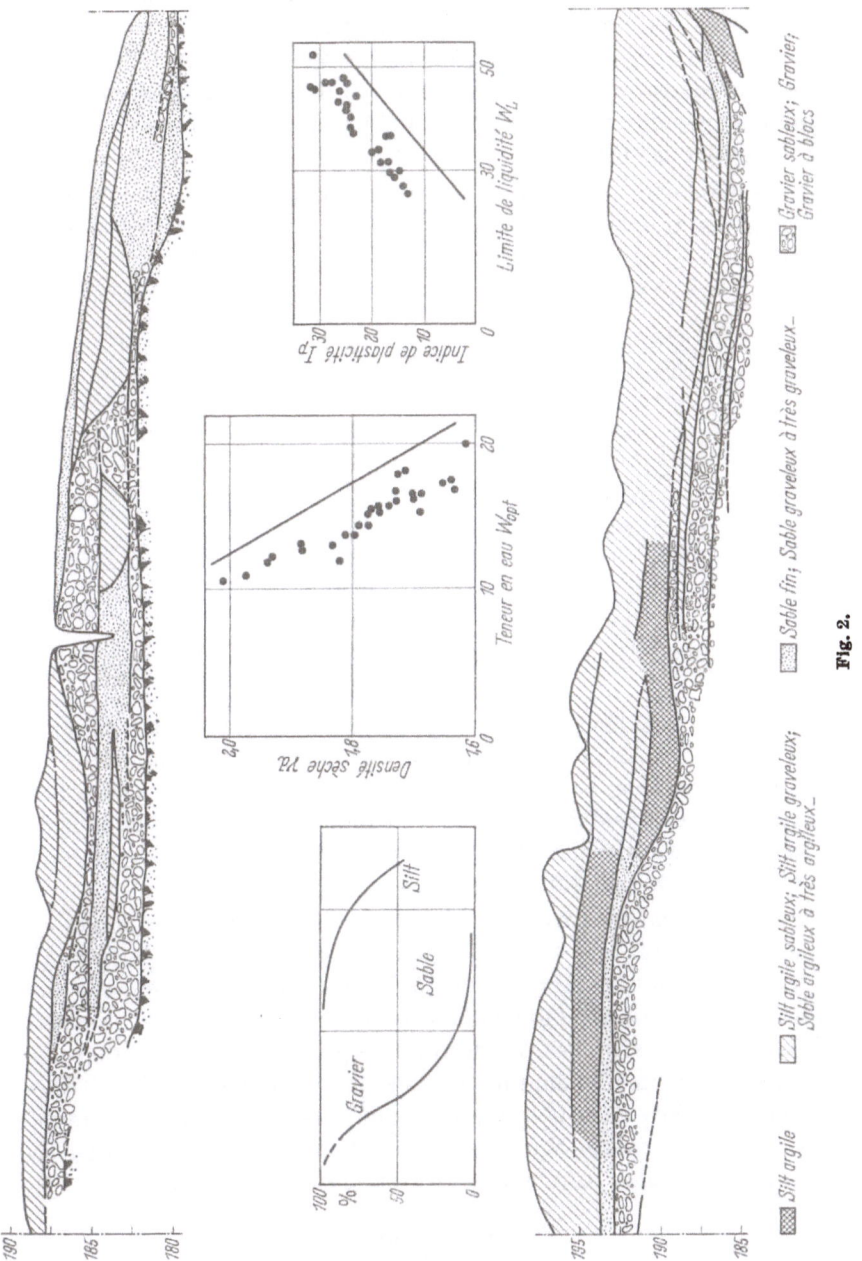

Fig. 2.

Silt argile

Silt argile sableux; Silt argile graveleux;
Sable argileux à très argileux.

Silt argile

Sable fin; Sable graveleux à très graveleux.

Gravier sableux; Gravier;
Gravier à blocs

a) L'Hétérogénéité des Matériaux. Une zone d'emprunts formée de lentilles de sols divers ne se laisse pas «moyenner» aisément d'autant plus que l'analyse statistique de cette moyenne est fonction du mode d'extraction. La fig. 2 donne un exemple concret de cette difficulté. Il n'est que de se reporter aux graphiques donnant la dispersion des granulométries, des limites d'ATTERBERG et des Optimums PROCTOR pour apprécier la difficulté.

b) La Géologie. Que penser d'autre part, pour reprendre le mot de TERZAGHI, du «détail géologique mineur» qui par sa présence peut

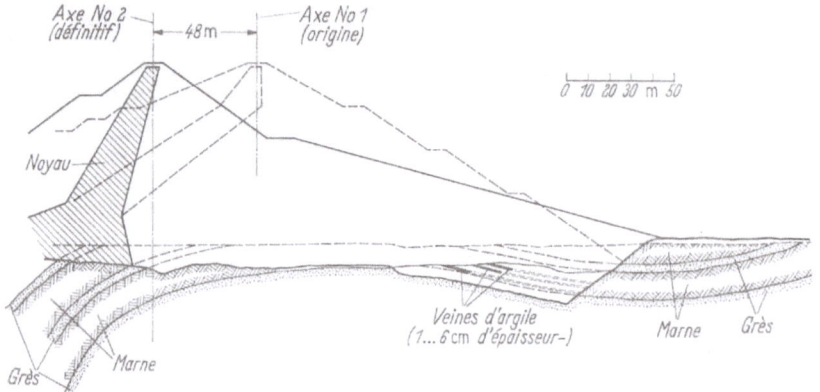

Fig. 3.

bouleverser les conclusions de l'analyse faite avant qu'on ne l'ait remarqué? La fig. 3 montre l'importante modification apportée, en cours de construction, au profil d'un grand barrage en terre et enrochements, à la suite de la découverte fortuite dans les fondations de minces couches continues d'argile plastique, qui avaient échappé à l'examen géologique et à une centaine de sondages.

c) Les Phénomènes Inconnus. On ne doit appliquer les principes et les lois de la Mécanique des Sols qu'avec des hypothèses largement encadrantes, et en gardant présent à l'esprit qu'on raisonne sur un nombre limité de critères qui ne définissent certainement pas complètement les données naturelles. La découverte progressive de phénomènes nouveaux — comme la mesure erronée de la pression interstitielle lorsque l'air pénètre dans la cellule (BISHOP), ou la sensibilité de certains sols à se fissurer même sous de faibles allongements (LEONARDS et NARAIN) — conduit à penser qu'il en reste encore d'inconnus.

3. Les Calculs

Les calculs font immédiatement penser aux études de stabilité. En fait d'autres calculs peuvent être non moins importants quant à la sécurité bien que n'intervenant pas directement dans les calculs de stabilité. Ce sont le calcul des débits et des pressions d'écoulement, le calcul des tassements notamment si l'ouvrage comporte des structures rigides, le calcul des poussées sur des murs de pied ou de bajoyers, etc... Les calculs de stabilité donnent un moyen de classer les différents ouvrages quant à leur sécurité vis-à-vis du glissement et jouent de ce fait un rôle à part.

a) Les Approximations des Calculs de Stabilité. L'apparente rigueur des calculs ne doit pas faire perdre de vue les approximations souvent grossières introduites aux différents niveaux:

On résoud l'hyperstatisme des forces en présence faute de pouvoir accéder aux véritables relations contrainte-déformation, par des hypothèses, variables selon les méthodes, mais toujours arbitraires (direction donnée des forces entre tranches, par exemple). L'importance de l'hyperstatisme apparaît sur la fig. 4, où l'on voit schématiquement que selon qu'une zone a des tassements supérieurs à l'autre, l'aire des efforts résistants mobilisables, qui mesure indirectement la sécurité, varie très appréciablement.

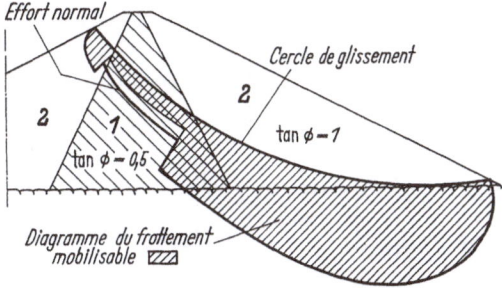

Tassement de 1 > tassement de 2

On ne tient pas compte des relations entre la déformation et la résistance au cisaillement, qui varient fortement selon les matériaux et font que l'analyse d'un remblai

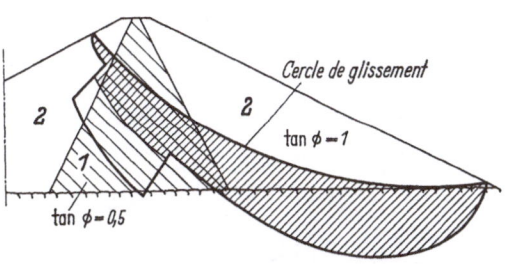

Tassement de 2 > tassement de 1

Fig. 4.

non homogène devrait se faire par un calcul en rupture progressive, malheureusement inabordable.

On ne tient pas compte des forces dans la troisième dimension, qui doivent pourtant jouer un rôle important dans certains cas, notamment dans les remblais en vallée étroite ou en vallée à profil irrégulier. On sait que la contrainte intermédiaire peut jouer un rôle appréciable sur les phénomènes de rupture. Ne pas en tenir compte est à priori incorrect sans qu'on sache d'ailleurs estimer la grandeur de l'erreur ainsi commise.

Enfin et surtout on utilise des valeurs numériques données par des essais de laboratoire qui peuvent ne pas être représentatifs: dispersion de l'échantillonnage, remaniement, effet d'échelle des dimensions et du temps, mode opératoire, histoire du matériau, etc...

Certaines méthodes de calcul ont pu être justifiées par l'analyse à postériori de glissements naturels. Mais sauf les cas exceptionnels de terrains homogènes, les auteurs de ces vérifications sont alors nécessairement guidés dans le choix des paramètres de résistance par la connaissance du glissement.

b) Les Remblais Argileux et les Remblais Sableux. D'une enquête sur 65 barrages homogènes dont 14 avaient été le siège de glissements de talus, SHERARD a tiré une très intéressante conclusion. La fig. 5 donne espace la distribution de ces barrages en fonction de la granulométrie moyenne de leur matériau. On constate que *tous* les talus en matériaux très fins (D 50 inférieur à 6 microns) se sont rompus, alors *qu'aucun* talus en matériau sableux n'a souffert, malgré des pentes parfois très raides et des modes de construction peu recommandables. Il semble qu'on touche là une notion fondamentale qui n'est pourtant pas considérée par les calculs habituels: la sécurité d'un massif argileux est beaucoup plus difficile à assurer que celle d'un massif sableux.

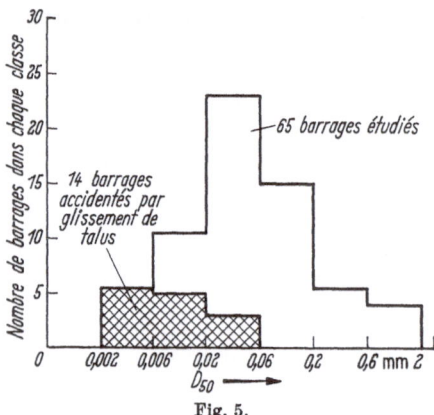

Fig. 5.

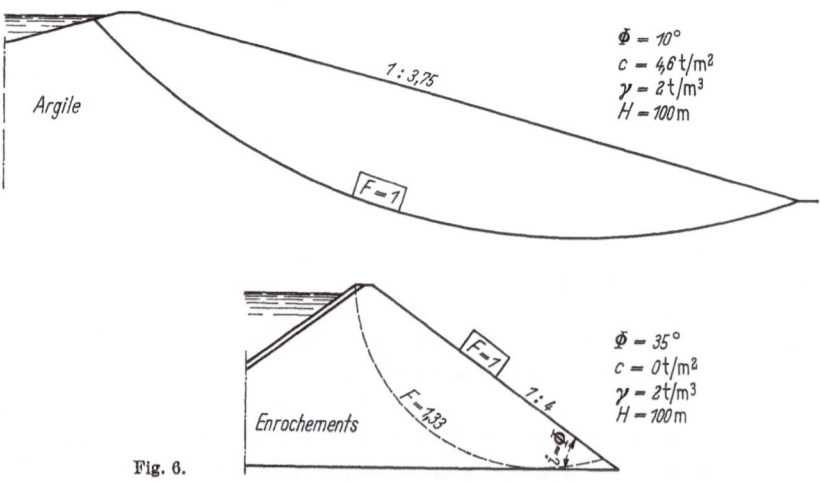

Fig. 6.

Cela tient d'abord à la difficulté d'estimation de la résistance au cisaillement des argiles. Pour les faibles charges on risque la surestimation de cette résistance si l'essai de laboratoire laisse naître une cohésion apparente, ou si le matériau a tendance à gonfler; c'est pourquoi les petits barrages sont à cet égard plus menacés que les barrages très hauts. Au contraire la plupart des alluvions sableuses ont des angles de frottement d'environ 37°, se drainent facilement, et sont construites avec des talus plus plats que 1,4/1 (= cot 37°).

Mais une autre distinction s'impose. Dans un massif argileux qui aurait un coefficient de sécurité égal à l'unité la rupture imminente est une *rupture profonde*, menaçant le remblai dans la masse, alors qu'avec le même coefficient de sécurité limite un massif d'enrochements ne risque que de laisser quelques cailloux rouler à *sa surface* (fig. 6.) Il apparaît donc que les notions habituelles d'après lesquelles on définit la sécurité ne traduisent pas nécessairement des risques comparables.

II. Principes Directeurs

a) Penser à l'Improbable. C'est avec ces outils encore imparfaits que le projeteur doit construire. Il ne peut pas attendre, et chaque année apparaissent de nouveaux barrages, de plus en plus hauts, en matériaux de plus en plus difficiles et sur des fondations de plus en plus complexes.

Ces extrapolations sont possibles à condition de mettre en oeuvre la totalité de nos connaissances et de savoir en apprécier les limites. L'enseignement du passé est clair; l'énorme majorité des accidents est venue de l'ignorance ou de la sous-estimation d'un phénomène et non d'une insuffisance de calcul. C'est ainsi que le projeteur a appris à donner à l'ensemble barrage-fondation une structure organique qui le mette à l'abri des manifestations destructrices de l'eau, érosion interne, sous-pression, pression interstitielle. Les meilleurs projets sont ceux où non seulement les phénomènes connus seront maitrisés mais également où l'improbable pourra se produire sans dommage.

Dans un projet récent les calculs ne justifiaient pas l'installation de puits filtrants à l'aval de la digue, du moins si l'on admettait les coefficients de perméabilité les plus probables de la fondation et du remblai. Mais ces hypothèses ne pouvaient être tenues pour certaines, malgré un grand nombre de mesures, eu égard à la longueur d'une dizaine de kilomètres du barrage. Aussi pour parer à l'imprévu, il a été décidé de disposer un puits filtrant tous les 25 m. Des piézomètres tous les 100 m, permettront de suivre l'évolution des sous-pressions.

Dans un autre cas, pour un barrage fondé sur un calcaire karstique, il est apparu prudent de disposer un épais filtre aval pour le cas où un conduit karstique déboucherait sous le remblai, caché par les alluvions, après un long détour l'ayant fait échapper aux reconnaissances.

Le calcul de ces organes peut être fait pour en justifier les dimensions, mais ce n'est pas par le calcul qu'on est amené à les prévoir. C'est bien plutôt par une discipline intellectuelle qui consiste à imaginer l'improbable. Si la « prime d'assurance » n'est pas chère *il faut* la payer sans hésiter. Si elle est jugée trop chère il faut pousser les reconnaissances; on justifie alors aisément le prix de la prime.

b) Estimer les Risques. Une autre considération essentielle est l'estimation correcte des imperfections de nos connaissances. Certaines sont bien mieux établies que d'autres. Puisqu'il est plus difficile de mesurer et prévoir la résistance au cisaillement d'une argile que d'un sable *il faut* admettre pour le massif argileux des coefficients de sécurité plus larges que pour le massif sableux. Si le matériau est mal connu, *il faut* aussi être plus prudent.

Selon la nature du risque les précautions à prendre doivent varier. Il est surprenant de voir fixer dans certains règlements une valeur absolue du coefficient de sécurité, sans que soient évoqués les paramètres fondamentaux du choix qui devraient être : population à l'aval, volume du réservoir, fréquence des périodes de réservoir plein, durée de la vidange par les vannes de fond, qualité du contrôle de l'exécution, nature des matériaux, hauteur de l'ouvrage, nombre de précédents, etc...

c) Améliorer nos Connaissances. La difficulté à imaginer l'imprévu est déjà grande pour des ouvrages ressemblant par la taille et les matériaux à ceux déjà construits. Elle est bien plus grande s'il faut extrapoler. Pour aborder ces nouveaux problèmes le projeteur doit perfectionner, par tous les moyens, ses connaissances et ses méthodes. Il ne peut le faire seul et il se tourne avec un intérêt croissant vers les découvertes du mathématicien, du physicien, du chercheur de laboratoire, du constructeur d'appareils de mesures. Les progrès seront comme ailleurs le fruit d'un *travail d'équipe*, surtout si chaque spécialité donne le meilleur d'elle-même, c'est-à-dire ne se disperse pas. Il n'est sûrement pas souhaitable qu'un bureau d'études se risque dans la recherche fondamentale pour laquelle il n'est pas préparé, ou qu'on laisse la conception du projet à un pur calculateur qui n'en n'apercevra qu'un aspect limité. C'est par les échanges de leurs connaissances que les diverses disciplines contribueront le plus efficacement au progrès technique. D'où la valeur d'un congrès comme celui-ci.

III. Progrès Nécessaires

Aux prises avec ses propres difficultés dans sa tâche de tous les jours, le praticien peut se proposer un programme de travail et y convier ses collègues des diverses spécialités, de façon à combler au plus vite les lacunes les plus gênantes.

Les problèmes sont classés ici dans un ordre qui, aux yeux de l'auteur, représente à peu près l'urgence croissante.

1. Les Calculs

a) **Les Méthodes Classiques, leurs Avantages, leurs Limites.** Dans le domaine du calcul des talus non homogènes il est certainement nécessaire de mettre au point des méthodes moins simplistes que les méthodes généralement utilisées.

Il est vrai que les méthodes les plus employées ont le mérite, malgré leur rusticité, d'être *éprouvées*, c'est à dire confirmées par la bonne tenue d'un grand nombre d'ouvrages. Malgré toutes les critiques qu'on peut faire à la Méthode Suédoise Standard, il faut bien reconnaître qu'elle a servi d'unique calcul à plusieurs centaines de barrages, et parmi eux, les plus grands.

Mais il ne faut pas oublier que les méthodes les mieux éprouvées peuvent être mises en défaut lors d'une extrapolation nouvelle, soit dans le type du barrage, soit dans sa hauteur. Les calculs trop approchés sont tout au plus valables pour juger du sens de variation de la sécurité entre deux variantes voisines d'un même projet. Il est donc indispensable de chercher à étendre le domaine de validité des calculs de stabilité par l'introduction d'hypothèses mécaniques correctes.

b) **L'Hyperstatisme des Données.** Toute amélioration aux calculs de stabilité comportera une part d'arbitraire non réductible tant qu'on ne fera pas intervenir les relations effort-déformation des matériaux, seules capables de résoudre l'hyperstatisme.

Les améliorations ne porteront d'ici là que sur l'introduction de systèmes de forces compatibles, mais nullement certaines.

Dans ces conditions on peut se demander s'il n'est pas indispensable de déterminer, parmi l'infinité de distributions à priori possibles, celle qui donne le plus petit coefficient de sécurité. Il semble que l'on accèderait ainsi à une notion plus exacte de la sécurité surtout si le choix des distributions pouvait être guidé par une analyse critique des déformations comparées des différentes zones du remblai, dans l'esprit du schéma de la fig. 5.

c) **Les Machines Électroniques.** Si les raffinements de calcul ont longtemps rebuté le projeteur de barrages c'est qu'ils ne semblaient pas justifiés au regard des hypothèses grossières nécessairement faites sur les données numériques de la résistance au cisaillement ou des pressions interstitielles. Plus grave, ils risquaient de faire perdre le sens des réalités physiques. Aujourd'hui, grâce aux machines électroniques, il en va tout autrement. On peut aborder des calculs extrêmement complexes dans des temps raisonnables. On peut les répéter vingt, trente ou cinquante fois s'il le faut, en gardant l'esprit libre.

On peut aussi, et c'est là l'avantage majeur des calculs électroniques, faire varier largement les données numériques et déduire de

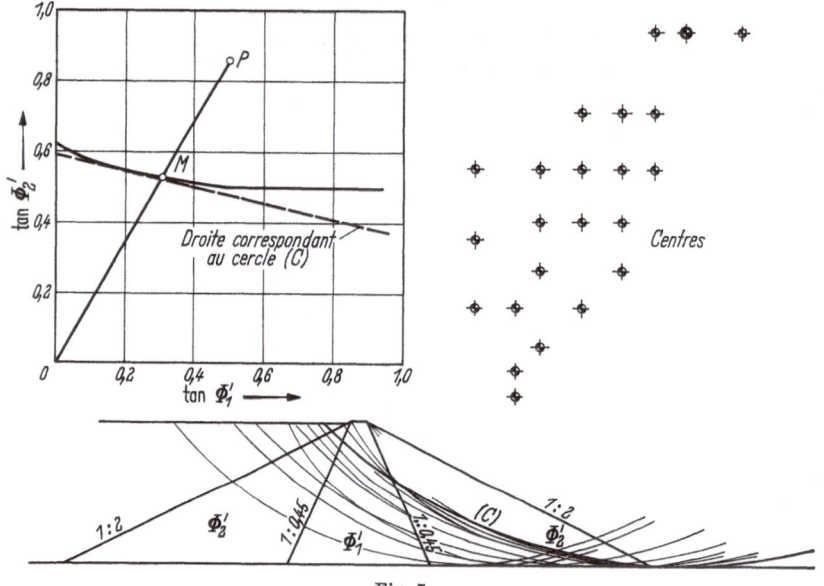

Fig. 7.

l'effet de ces variations *le poids* des différents paramètres. Cette indication est capitale pour l'estimation correcte de la sécurité. La fig. 7 donne un exemple simple de ce qu'on veut dire par le poids des paramètres. On voit immédiatement par la position du point P, donnée par le laboratoire, que l'hypothèse faite sur le frottement de la zone 2 a une importance déterminante quant à la sécurité alors que celle faite sur la zone 1 est presque sans influence. Une étude de ce genre est précieuse pour guider les recherches et apprécier correctement les éléments fondamentaux de la stabilité d'un remblai. Elle permet de plus une analyse statistique de l'influence des résultats du laboratoire sur la sécurité.

Dans les cas complexes les calculs à la machine électronique sont seuls capables de mener à bien une telle approche. Il est probable que les ordinateurs modifieront profondément notre philosophie du calcul, en le transformant en outil de recherche et de conception, alors qu'il n'apportait encore hier qu'une justification à posteriori et souvent illusoire de projets conçus d'après des règles semi-empiriques.

d) Les Coefficients de Sécurité. La mise au point d'une nouvelle méthode pose ipso facto la question des coefficients de sécurité à adopter. En effet, les valeurs usuellement acceptées n'ont pas une origine rigoureuse. La valeur du coefficient de sécurité dépend d'une définition conventionnelle et de la méthode de calcul. La valeur minimale acceptable se précise petit à petit à la lumière des échecs et des réussites comme un équilibre humainement acceptable entre économie et sécurité. Si l'on ne veut pas s'exposer à recommencer l'aventure de nos prédécesseurs il n'est qu'un moyen: calculer avec toute nouvelle méthode un grand nombre d'ouvrages existants dont les matériaux et la fondation seraient connus. La décision finale sera d'ailleurs difficile à prendre car il est évident que la nouvelle méthode ne classera pas la sécurité des ouvrages dans le même ordre que la méthode ancienne.

L'introduction d'une nouvelle méthode pose donc de sérieux problèmes et appelle une extrême prudence. Il faut se garder, en particulier, de la comparer aux méthodes éprouvées sur la base de quelques applications isolées.

2. La Mécanique des Sols

Dans le domaine de la Mécanique des Sols la nécessité d'approfondir nos connaissances se fait encore plus sentir.

a) La Résistance au Cisaillement. Si l'on fait le décompte des différentes conceptions ou définitions relatives au couple de paramètres Φ et C des sols argileux on en trouve aisément une douzaine, ce qui n'est pas étonnant puisqu'on sait qu'il n'existe pas de caractéristique intrinsèque de la résistance d'une argile. La résistance au cisaillement dépend de la déformation (elle-même dépendant de l'appareil d'essai), de la vitesse de la déformation, du mode de compactage, du temps, de l'histoire des contraintes et certainement de facteurs encore inconnus. La durée d'application des charges joue un rôle d'autant plus difficile à évaluer que l'échelle de temps n'est pas du tout la même au laboratoire et dans le vie du barrage. Devant la complexité des phénomènes il y a intérêt à les dissocier et c'est pourquoi les études en contraintes intergranulaires, de préférence aux contraintes totales, semblent recommandables. C'est en particulier, moyennant la mesure des pressions interstitielles et des déformations in situ, la seule manière de s'assurer de la stabilité d'ouvrages sans précédent, soit par la taille soit par les matériaux. Il ne faut pas oublier cependant l'extrême difficulté qu'il y a à mesurer correctement les paramètres A et B de la pression interstitielle à l'appareil triaxial. Il faut mettre en oeuvre des techniques qui ne sont pas à la portée des laboratoires et des opérateurs courants.

b) L'Influence du Temps. Des recherches urgentes concernent l'influence du temps sur l'évolution des caractéristiques mécaniques. Les études de SKEMPTON ont attaqué ce problème mais on est encore bien souvent désarmé. Un moyen d'étudier *l'affaiblissement* possible de fondations d'argile surconsolidée, très plastique et fissurée (par exemple ce qu' en climat tropical on appelle «terre noire à coton», $W_L = 100$ à 150% $W_P = 25$ à 30%) est de transformer les échantillons en boue (slurry test) avant de les cisailler au laboratoire. De nombreuses recherches sont encore nécessaires avant de pouvoir prédire le comportement à long terme de ces sols.

Dans le même domaine *l'augmentation* de la résistance de certains remblais avec le temps mérite d'être étudiée avec soin. LEONARDS a donné au Congrès de Wiesbaden (1963) d'intéressants résultats oedométriques faisant apparaître un gain de «portance» de près de 40% sur des échantillons maintenus sous charge pendant un an. Au même Congrès DENISOV communiquait des mesures russes sur le sable d'un remblai hydraulique, faites par les moyens les plus variés: chargement de plaque en fond de puits, vane test, pénétromètres dynamiques. Toutes les mesures, à indice des vides constant, mettent en évidence 'une cohésion apparaissant dès les premiers jours et croissant encore après cinq ans.

c) La Fissuration. Très importante aussi est l'étude de la fissuration des massifs sous le jeu des extensions. Les études de SHERARD, de LEONARDS et d'autres chercheurs, montrent qu'il y a là un problème capital pour la sécurité des barrages en terre. La récente fissuration du noyau du barrage de Shek Pik à Hong-Kong découverte lors d'une campagne d'injection de la fondation, témoigne de la réalité du risque. Il s'agit pourtant d'un sol normalement plastique, qui à priori ne devait pas donner de souci ($W_L = 28\%$ à 58%, $I_P = 7\%$ à 25%), d'autant plus qu'il fut compacté à plus de 2% au-dessus de l'optimum Proctor Standard. Il y a peut être là une illustration d'une sensibilité particulière à la fissuration des sols résiduels. Ce cas est, de plus, un exemple assez rare de publication objective et rapide, par les auteurs du projet, d'un incident intéressant tous leurs confrères. Ils méritent d'en être remerciés.

Les extensions dans un remblai peuvent se produire, outre par ses propres tassements, pour de multiples causes qui échappent encore au

calcul: déformations différentielles de la fondation, points durs, tremblements de terre. La résistance des matériaux à ces extensions est sans
doute une propriété fondamentale qui appelle des études approfondies.

D'après les recherches les plus récentes (NARAIN) il ne semble pas
qu'on puisse prévoir la fissuration à partir de la plasticité au sens des
limites d'ATTERBERG. Quant
à l'effet du temps sur la
sensibilité à la fissuration il
semble également indépen-
dant des critères habituels
de plasticité, comme le montre la fig. 8. Il se peut que
pour d'autres matériaux
cette résistance diminue
avec le temps. Deux faits
paraissent toutefois acquis;
1. plus l'énergie de compactage est grande plus le sol
est « fragile » à la fissuration;
2. il faut éviter de compac-
ter du côté sec de l'optimum
standard.

La thèse de SHERARD
avait déjà bien établi ce
dernier résultat, par la seule
observation d'ouvrages existants et ANDRÉ COYNE vou-
lait que les barrages aient
un « coeur tendre » pour reprendre sa propre expres-
sion. Mais on est encore loin
de pouvoir donner des règles
précises et sûres permettant
d'entrer dans le quantitatif.
Des essais de laboratoire simples sont nécessaires.

Les fissures pouvant se
transformer en renard, la

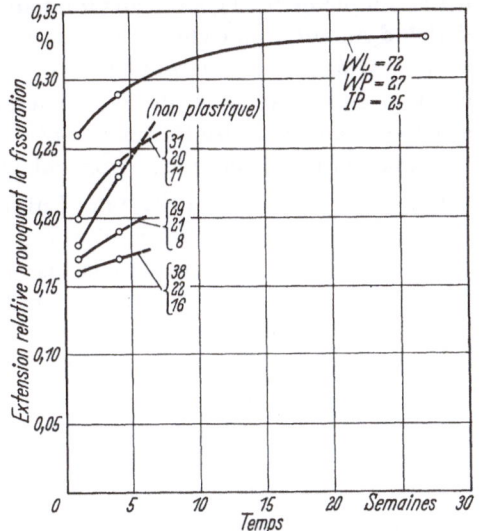

Fig. 8.

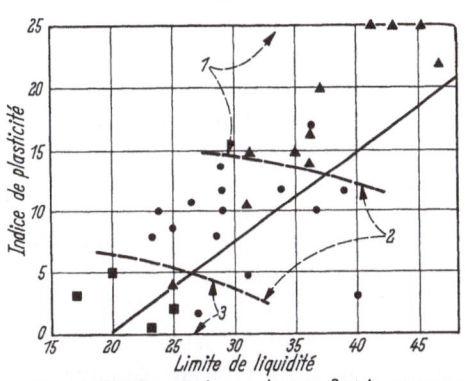

▲ et zone 1: *Résistance forte*   ● et zone 2: *Résistance moyenne*
■ et zone 3: *Résistance faible*

Fig. 9.

résistance des matériaux à l'érosion est une autre propriété importante qui justifierait la mise au point *d'essais nouveaux de laboratoire*.
La fig. 9 donne une première tentative de classement des sols, par
SHERARD, d'après l'observation statistique des précédents, quant à
leur résistance à l'érosion.

d) Les Tremblements de Terre. Les effets des tremblements de terre sur les barrages en terre sont encore fort mal connus. Un vaste champ d'étude est ouvert. Il semble en tout état de cause que la pratique actuelle qui consiste à combiner dans les calculs de stabilité les forces de poids avec des forces horizontales d'accélération 0,05 g, 0,10 g ou même 0,20 g soit sans fondement réel. Ce faisant on confère à l'effet dynamique du choc une nature statique qu'il n'a pas. D'autre part la résistance au cisaillement sous charge oscillante est très mal connue. Beaucoup plus que les glissements (il n'en existe aucun précédent) il faut craindre les tassements, la fissuration et les phénomènes de liquéfaction, tous encore non calculables.

e) Les Essais d'Identification. Enfin il est éminemment souhaitable que les essais d'identification, c'est-à-dire ceux qui permettent les comparaisons d'un site à l'autre et la dissémination de l'expérience, soient parfaitement normalisés. C'est ainsi que les résultats de l'essai Proctor varient très sensiblement avec la méthode utilisée pour faire varier la teneur en eau. Dans un cas récent on a vu sur des argiles tropicales l'optimum changer systématiquement de 17% à 22% selon qu'on faisait l'essai immédiatement ou après avoir laissé l'échantillon au repos pendant 48 heures. Il semble que les limites de liquidité augmentent aussi avec le temps de repos.

Dans une communication au Congrès de 1961 Said Youssef, Sabry et El Ramli ont donné des abaissements de 20% sur les limites de liquidité, de 10% sur les limites de plasticité et de 5% sur les Optimums Proctor de sols argileux pour une augmentation de température ambiante de 20 °C, qui peut fort bien exister entre un laboratoire tropical et un laboratoire européen.

Il est probable que le pH de l'eau utilisée peut aussi modifier les résultats.

On n'a pas encore tiré tout ce qu'on peut attendre des limites d'Atterberg. Malgré leurs définitions conventionnelles ces limites traduisent des propriétés fondamentales des sols comme en témoignent les nombreuses corrélations déjà établies avec d'autres paramètres mécaniques importants. C'est ainsi que Seed, Woodward and Lundgren ont pu s'en servir avec succès pour prévoir les propriétés de gonflement des argiles. Il est très souhaitable que d'autres corrélations simples de ce genre soient dégagées par des enquêtes statistiques bien faites. Le projeteur serait ainsi mieux armé, dans les domaines où il ne sait pas encore mesurer les caractéristiques physiques correspondantes.

3. L'Observation des Précédents

a) Importance des Mesures In-Situ. Comme par le passé, et peut-être plus encore que par le passé, l'observation des précédents reste la référence essentielle et inévitable. Elle sera grandement revalorisée par la pratique systématique de mesures in situ. Il est indispensable, pour progresser, de connaître le plus de choses possibles sur les barrages existants qui sont des essais grandeur nature. S'il est vrai qu'on ne peut pas mesurer sur un barrage tout ce qu'on mesure au laboratoire, il n'en reste pas moins qu'on peut faire beaucoup plus qu'on ne le fait en général. Ces mesures sont non seulement utiles à l'acquisition de connaissances scientifiques nouvelles, mais indispensables pour chaque ouvrage dès qu'il prend par sa taille ou les particularités de ses matériaux et de sa fondation, le caractère d'une extrapolation. Il est très regrettable de voir des ouvrages actuellement en construction, avec des hauteurs de l'ordre de 150 mètres, complètement démunis d'instruments de mesure.

b) Mesures de Déformation. *Les mesures topographiques* sont précieuses et réduisent par leur caractère de valeurs globales, les difficultés d'interprétation attachées aux mesures ponctuelles. Une des plus simples et des plus utiles est la mesure des variations de longueur entre plots scellés sur le couronnement. Si ces plots sont assez voisins (10 ou 20 m) on peut déceler les tendances à l'extension et être averti du risque de fissuration verticale. On peut utiliser le parapet pour cette mesure.

Les mesures faites sur le couronnement du barrage de Serre-Ponçon, où la brisure de l'implantation, commandée par la topographie, pouvait laisser craindre des extensions ont montré que, loin d'être en extension dans la région de l'angle rentrant, le remblai était en compression. La fig. 10 le montre clairement. Les déformations horizontales ont été imposées par la forme de la vallée et la nature des fondations beaucoup plus que par l'implantation de la crête.

Les mesures par *tubes de tassements* verticaux sont au point, bien qu'on puisse encore attendre une amélioration de la précision. Elles tendent à être remplacées par des mesures en tubes horizontaux, beaucoup moins gênants à installer sur le chantier de compactage.

Les mesures par *inclinomètres* font des progrès rapides, plusieurs appareils ingénieux ayant fait leur apparition récemment. Un point essentiel pour la qualité de ces mesures est la perfection du guidage, si l'on veut pouvoir tirer de l'appareil des renseignements utilisables pour l'étude scientifique du comportement des remblais, et pas seulement des indications ayant la nature d'un système d'alerte.

c) Mesures de Pression Interstitielle. Les mesures de pression interstitielle sont encore douteuses. Les études de BISHOP sur cette question sont capitales. Plus que tous les autres appareils actuellement utilisés la cellule de pression interstitielle risque de fausser localement le phénomène et, par conséquent, sa mesure: injection forcée d'eau dans le terrain ou au contraire envahissement de la cellule par de l'air.

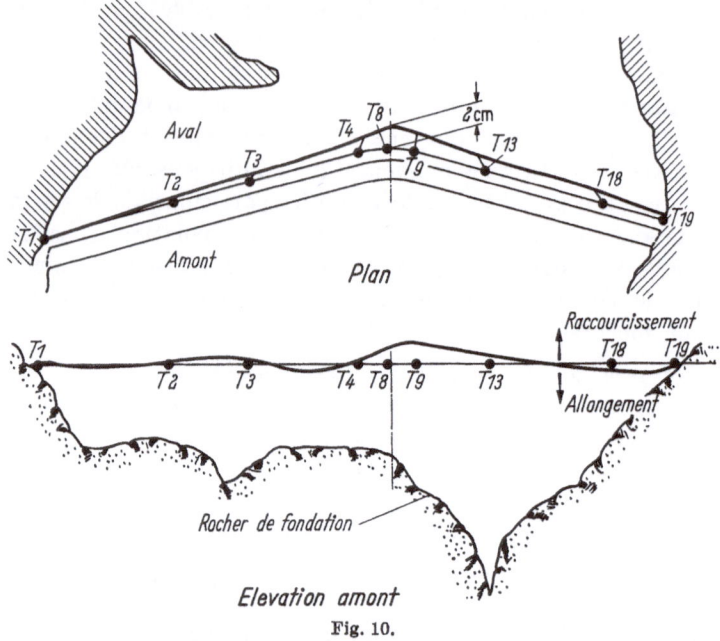

Fig. 10.

d) Mesures de Déformations Relatives. Enfin il serait très utile de pouvoir connaître les contraintes in-situ. Leur évolution donnerait des indications précieuses sur l'évolution de la résistance au cisaillement du sol et permettrait peut-être de déceler la fissuration. Mais dans ce domaine on en est encore au stade de la recherche. Il semble même que la mesure *directe* des contraintes soit impossible. Tous les essais tentés jusqu'à maintenant sont restés décevants car les appareils, aussi ingénieux soient-ils, modifient localement par leur présence le champ des contraintes. Le seul paramètre que l'on puisse mesurer avec certitude est la *déformation relative* du massif. Il faut, bien entendu, utiliser des extensomètres sans raideur propre, traduisant fidèlement les déformations du sol. Moyennant des « bases » de mesure assez longues les hétérogénéités locales seraient sans conséquences. Les Ingénieurs Mexicains ont mis en place au barrage d'El Infernillo des extensomètres ayant des bases de mesure de 2 et 4 m. Connaissant les déformations relatives et les pressions interstitielles on doit pouvoir déduire les contraintes par des essais de laboratoire. Ces mesures renseigneraient d'autre part sur les risques de fissuration.

e) Programme de Mesures In-Situ. Pour terminer il convient de souligner l'importance d'une observation attentive et scientifique des ouvrages. C'est la référence indispensable aussi bien aux recherches du laboratoire, qu'au théoricien ou au constructeur. C'est là que se situe l'objet même de notre science ou de notre art, selon le point de vue, dans son échelle réelle de dimensions et de temps.

Mais il ne faut pas perdre de vue qu'un programme de mesures in-situ, pour être utile, doit être complet. Ce n'est que d'un ensemble étendu qu'on peut tirer une interprétation sûre.

Il ne faut pas oublier non plus que ces mesures ne sont pas seulement destinées à vérifier des idées à priori, mais aussi et surtout à déceler l'imprévu et découvrir l'inconnu.

Ce sont alors les meilleurs garants de la sécurité, et les meilleurs facteurs de progrès.

IV. Conclusions

Au terme de cette revue des problèmes dont la solution doit conduire à des projets meilleurs ont peut énumérer les sujets d'étude qui semblent offrir les champs de recherche les plus féconds.

1. Calculs de stabilité, sur des hypothèses mécaniques correctes, avec meilleure appréciation de l'hyperstatisme, et utilisation des machines électroniques.

2. Méthodes de laboratoire et de calcul pour la prévision des déformations des remblais en fonction de leur nature et de leur forme.

3. Précision des critères et des modes opératoires des essais de cisaillement, et étude de l'effet d'échelle.

4. Normalisation plus stricte des essais d'identification.

5. Etude de l'influence du temps sur l'évolution des paramètres mécaniques des remblais.

6. Analyse de la fissuration et recherche de critères de laboratoire permettant la prévision.

7. Définition d'un essai d'érodabilité des fissures ouvertes.

8. Etude des effets des tremblements de terre sur les remblais et leurs fondations.

9. Amélioration de l'appareillage, et généralisation de l'utilisation des mesures in-situ: pressions interstitielles, déformations totales, déformations relatives.

Discussion

Question posée par J. KRAVTCHENKO: J'ai écouté le bel exposé de M. LONDE avec un intérêt passionné. J'ai constaté — du moins j'ai cru pouvoir tirer cette conclusion — que le projeteur d'un ouvrage ne fait qu'exceptionnellement confiance au calcul basé sur les équations dites «exactes». Je voudrais que M. LONDE précise la dernière partie de sa conférence:
1. Croit-il à la rhéologie de COULOMB ?

2. Dans quels cas les solutions disponibles dans la littérature peuvent être utilisées dans l'Art de l'Ingénieur ?

3. Quelles sont les théories mathématiques à développer par priorité en vue des applications ?

Réponse de P. LONDE: Monsieur le Professeur KRAVTCHENKO vient de nous dire ce qu'il peut faire, en tant que mathématicien, et Monsieur le Professeur BIAREZ, en tant qu'expérimentateur, pour le calcul des ouvrages en terre. Je me propose ici, à la demande d'ailleurs du Professeur KRAVTCHENKO, de compléter ces exposés en précisant ce que l'Ingénieur constructeur attend du calcul. J'ai d'ailleurs échangé à ce sujet quelques idées avec des confrères ingénieurs au cours de ces jours derniers.

J'ai pu laisser croire, surtout d'après la discussion qui a suivi mon exposé de la semaine dernière, que je niais l'intérêt des calculs théoriques. Or il n'en est rien.

En fait, je n'ai pas assez souligné le cas dans lequel je me plaçais, à savoir:

a) barrages en terre et leurs fondations,
b) calculs de stabilité.

Il reste que les calculs théoriques dont nous avons eu de beaux exemples dans ce Symposium sont extrêmement intéressants, car ils augmentent petit à petit nos moyens de prévision des comportements des sols.

Les théories sont précieuses en ce qu'elles nous permettent de mieux comprendre les phénomènes et ne serait-ce qu'à ce titre elles augmentent l'arsenal de l'ingénieur qui, comme je l'ai déjà dit, doit utiliser tous les moyens de perfectionner ses connaissances.

Ce que j'ai voulu dire est ceci.

Il semble qu'on soit encore loin d'avoir une méthode de calcul de stabilité des barrages en terre pouvant remplacer avantageusement les méthodes approchées actuelles. Rappelons que les méthodes actuelles sont essentiellement basées sur des calculs à la rupture le long de lignes arbitraires en y appliquant la loi de Coulomb à partir de répartitions d'efforts données par les hypothèses mêmes de la méthode. Ces hypothèses ne peuvent être levées qu'en introduisant les fonctions contrainte-déformation des différents matériaux.

Le calcul se termine par un nombre dit coefficient de sécurité, lui-même conventionnel, et qui n'a d'intérêt que dans la mesure où il est confirmé statistiquement.

Une nouvelle théorie de calcul de stabilité devra donc, comme j'y ai déjà insisté, établir une nouvelle statistique des coefficients de sécurité.

Que devra-t-elle prendre en compte pour être un véritable progrès ?

Le programme que je propose commence par des conditions minimales (1 et 2), se poursuit par une condition qui serait un progrès essentiel (3) et s'achève par un souhait (4) qui ferait de la méthode un outil excellent.

1. *L'hétérogénéité* du remblai et de la fondation (sous forme d'une suite discontinue de zones homogènes). γ, φ et c sont donc des fonctions discontinues des coordonnées.

2. La présence *de deux phases*, se traduisant par la superposition d'un champ de contrainte intergranulaire, et d'un champ de contrainte du fluide.

3. *Les relations effort-déformation*, distinctes selon les zones (et selon les phases solides et fluides dans chaque zone).

4. *Les paramètres rhéologiques* si l'on veut connaître les déformations en fonction du temps, et l'évolution du champ des contraintes.

Noter que 1 et 2 sont prises en compte dans les méthodes actuelles, et que 3 veut dire qu'on traite le problème en élastoplasticité.

Enfin, dernière condition à satisfaire: il faut ne pas oublier qu'une méthode ne peut être utilisée efficacement par l'ingénieur pour dessiner un projet que si elle est *rapide*, et si le calcul peut être aisément refait en prenant des valeurs différentes des paramètres physiques. On n'insistera jamais assez sur l'absolue nécessité qu'il y a à connaître l'influence des données et des hypothéses sur le résultat d'un calcul.

Les programmes électroniques actuels permettent de calculer la stabilité de 100 cercles, pour un remblai de forme quelconque, avec un nombre pratiquement illimité de zones distinctes et une distribution de pression interstitielle quelconque, en moins d'une journée et à moins de 200 F.

Bien que m'étant borné jusqu'ici au calcul des barrages en terre, il suffit de remarquer que si une nouvelle méthode peut s'accommoder de surcharges quelconques et aborder les trois dimensions elle deviendrait d'un usage quasi universel en mécanique des sols de l'ingénieur.

Pour terminer, je pense qu'on ne pourra avoir confiance en une théorie qu'après confirmation avec la nature. La plupart des théories sont confrontées avec le laboratoire seulement. C'est déjà essentiel. Mais il ne faut pas oublier le grand saut qu'il faut encore faire du laboratoire aux conditions naturelles.

C'est pourquoi il est si important d'installer des appareils de mesure dans les ouvrages.

Contribution de J. BRINCH HANSEN:

Some Remarks on Empirical Stress-Strain Relationships. At this conference I have been amazed to see the sharp division between right and left — and neither of them having very much concern for the man in the middle.

At the right we have the pure theoreticians with their wonderful mathematical theories for materials that do not exist — or at least have very little similarity to actual soils.

At the left we have the pure empirists, providing us with immense amounts of laboratory test data and foundation case records. I do not deny the value of this, but I regret that it is only very rarely crystalized into more general rules.

Finally, in the middle we have the great majority: the ordinary soils and foundation engineers, who are to do the daily work, computing settlements, bearing capacities and what not. To them, neither the right nor the left people are of much practical assistance.

Tensors and plastic potentials are — even if they were describing the behaviour of soils correctly — much too complicated for them. And as regards the enormous amount of test data and case records, they cannot be expected to know more than a tiny fraction of it.

No, what the ordinary designing engineer need, is a simple mathematical formulation of approximate empirical rules, derived from relevant and reliable test data and case records. I shall not deny, that much work of this kind has already been done, but in my opinion far too little in view of the needs for such simple rules, and the existence of abundant material for them.

In order to explain what I mean, I shall as an example take the deformations of sand. Most of us have a considerable amount of data from triaxial tests, oedometer test, isotropic consolidation test, etc., but how many have tried to express the results of such tests in a common formula ?

I know that it is difficult, because I have tried it myself and have not yet succeeded, but at least I know something about a workable form of such an empirical formula.

The following applies to the sand, we use for model tests in Copenhagen, and only in the case of first loading and increasing stresses. Also, it applies only to the case of axial symmetry, as we have in all the mentioned tests.

In a purely empirical way I have arrived at the following formulas for the principal strains:

$$\varepsilon_1 = \sqrt[3]{\frac{\sigma_m}{M}}\, e^{1,5}\left[1 - \frac{1,2\,(\sigma_1 - \sigma_3)^2}{1,1\,\sin^2\varphi\,(\sigma_1 + \sigma_3)^2 - (\sigma_1 - \sigma_3)^2}\,(1 - 4,3\,e^{1,25})\right] - \sqrt[3]{\frac{\sigma_c}{M}}\, e^{1,5},$$

$$\varepsilon_2 = \varepsilon_3 = \sqrt[3]{\frac{\sigma_m}{M}}\, e^{1,5}\left[1 - \frac{1,2\,(\sigma_1 - \sigma_3)^2}{1,1\,\sin^2\varphi\,(\sigma_1 + \sigma_3)^2 - (\sigma_1 - \sigma_3)^2}\right] - \sqrt[3]{\frac{\sigma_c}{M}}\, e^{1,5}.$$

e is the void ratio and φ the friction angle. σ_m is the mean normal pressure and σ_c the consolidation pressure in the triaxial test. M is a deformation modulus, which for our sand is $5 \cdot 10^7 t/m^2$ independent of the void ratio.

Please notice that for all 3 kinds of tests the equations represent non-linear relationships between stresses and strains.

For the isotropic consolidation test we have $\sigma_1 = \sigma_3$ and get thus simply the first term, as $\sigma_{c'} = 0$: $\varepsilon_1 = \sqrt[3]{\frac{\sigma_m}{M}}\, e^{1,5}$.

For the oedometer test we shall also put $\sigma_c = 0$ and by putting $\varepsilon_3 = 0$ we get an expression for the at-rest factor $K_0 = \sigma_3/\sigma_1$. The first equation gives then a simple expression for the compression:

$$\varepsilon_1 = \sqrt[3]{\frac{\sigma_m}{M}}\, 4,3\, e^{2,75}.$$

So far the formula is in quite good agreement with our test results. For the triaxial tests the full formula should be used, but here the agreement is, unfortunately, only good for the dense state, and I do not yet know why the loose sand does not follow it.

Therefore, the formulas above are not quite perfect yet, and I show them here only in order to give an idea of the approximate form, such formulas in my opinion must have in order to be both approximately correct and usable in practice. And I can see no reason why it should not be possible to produce similar formulas for clay, although here the time will enter as a further variable.

If such formulas can be developed, they will have several useful applications.

First, they should in principle enable a calculation of settlements, which will automatically include both the volumetric and the shear effects, which has never been possible before.

Second, by putting the volume change $\varepsilon_1 + 2\varepsilon_3$ equal to zero, a non-linear pore pressure equation may be obtained which, in contrast to existing pore pressure equations will be valid in the whole range until actual failure.

I hope that you see the immense possibilities of empirical stress-strain-time relationships, and I further hope that some of you will take up this work. If the "right" people with the mathematical abilities would cooperate with the "left" people with the empirical data, it should not be too difficult.

4.6 Mesures in Situ de la Masse Volumique et de l'Humidité des Sols par Radioactivité

Par

Bernard Wack

Les rayons gamma et les neutrons sont les seuls rayonnements radioactifs actuellement utilisables pour l'étude des sols, puisqu'ils sont assez pénétrants. Comme, de plus, les sources radioactives disponibles sont très peu encombrantes, les mesures in situ sont possibles. L'utilisation de ces deux types de rayonnement se complète très bien car elle permet de connaître deux grandeurs physiques fondamentales d'un sol: la masse volumique sèche et la teneur en eau pondérale. En effet, l'étude de la propagation du rayonnement gamma permet la mesure de la masse volumique totale:

$$\gamma = \gamma_d \cdot (1 + w),$$

et celle d'un faisceau de neutrons, la mesure de la masse volumique apparente d'eau:

$$w_a' = \gamma_d \cdot (w + E).$$

Si l'on admet que l'équivalent-eau est nul ou négligeable, on obtient la mesure de l'humidité volumique:

$$w' = \gamma_d \cdot w.$$

Deux principes de mesures sont utilisables. En laboratoire, où il s'agit d'effectuer des mesures sur échantillons, on utilise toujours le principe de la transmission directe: la mesure porte sur l'atténuation d'un faisceau du rayonnement, après la traversée de l'échantillon. La loi d'atténuation exponentielle est applicable et il est possible de bien collimater le faisceau, ce qui permet des mesures aussi fines que l'on veut. L'utilisation de ce principe de mesures sur le chantier présente une difficulté pratique puisqu'il faut avoir accès à deux faces du sol. Or, en général, une seule face de celui-ci est accessible, qu'il s'agisse de la surface des couches mises en place pour constituer une digue ou une route, ou de la surface d'un sondage de reconnaissance. C'est alors la méthode par rétrodiffusion qui semble la mieux adaptée: la mesure porte sur le rayonnement diffusé par le sol et renvoyé dans une direction proche de celle d'émission.

Mais, la méthode par rétrodiffusion présente deux graves inconvénients. En premier lieu, la composition chimique influe sur la mesure. Dans le cas de la mesure de la masse volumique, cette influence se traduit par la position de la courbe d'étalonnage: nous avons ainsi observé des différences atteignant 10% de la masse volumique, alors que la précision de la mesure est de l'ordre du pour-cent. Dans le cas de la mesure de l'humidité, il semble possible de tenir compte de l'influence de la composition chimique par le terme correctif d'équivalent-eau E. Encore faut-il connaître sa valeur pour les sols rencontrés si l'on veut effectuer la correction.

Le deuxième inconvénient des mesures par rétrodiffusion concerne le volume de la mesure. D'une part, celui-ci varie soit avec la masse volumique, soit avec l'humidité; et, d'autre part, la mesure ne donne

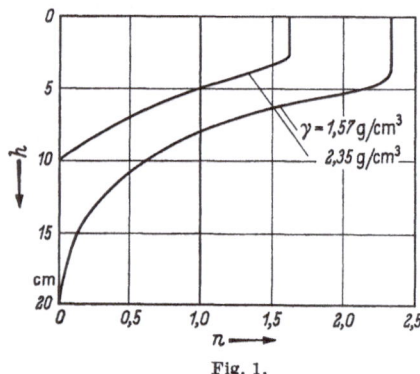 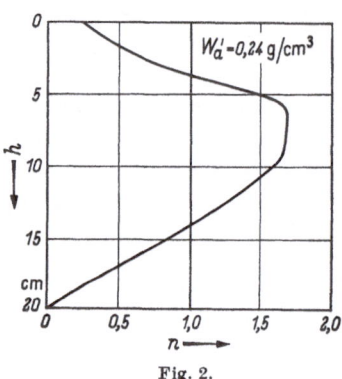

Fig. 1.   Fig. 2.

pas une valeur moyenne de la masse volumique ou de l'humidité, mais une valeur pondérée. Ainsi, dans le cas de la mesure de surface de la masse volumique, l'épaisseur «vue» diminue de 20 cm à 10 cm lorsque la masse volumique totale augmente de 1,57 à 2,35 g/cm³. La pondération de la mesure est mise en évidence sur la fig. 1 qui montre la répartition, en fonction de la profondeur h, de la quantité de rayonnement $n(h)$ qui atteint le détecteur, tel que:

$$n(h) = \frac{dN}{dh} \cdot \frac{H}{N_0},$$

N_0 étant le taux de comptage pour l'épaisseur maximale mesurée H. On constate que la plus grande partie du rayonnement se propage près de la surface. Le résultat de la mesure sera donc influencé d'une façon prépondérante par le sol de cette zone.

Pour la mesure de l'humidité, le volume de la mesure varie aussi. On admet que celui-ci est un hémisphère de rayon R, et, si l'on suppose que le volume est inversement proportionnel à la masse volumique apparente d'eau, on trouve expérimentalement que R obéit à la formule:

$$R = \frac{12}{\sqrt[3]{\gamma_d \cdot (w + E)}}.$$

La courbe de répartition du rayonnement (fig. 2) indique une influence prépondérante de la zone centrale.

Comme, la plupart du temps, on est amené à effectuer des mesures avec les rayons gamma et les neutrons au même emplacement, les courbes de pondération, différentes dans les deux cas, nous indiquent l'existence d'une erreur supplémentaire si le sol est hétérogène.

Nous avons essayé d'utiliser les jauges par rétrodiffusion pour le contrôle du compactage d'une digue en terre; la granulométrie du sol mis en place était comprise entre 0 et 150 mm. Il s'est avéré que la méthode n'est pas utilisable à une échelle industrielle dans ce cas; en effet, pour contrôler toute l'épaisseur d'une couche, il faut effectuer des mesures en profondeur et donc constituer des plateformes. La préparation de celles-ci est possible, mais est très délicate et ne peut se concevoir à une grande échelle.

Les essais que nous avons effectués ont permis, malgré tout, de se faire une idée de la mise en place du sol. Celui-ci était compacté par un rouleau vibrant de 8 tonnes. D'une part, ce mode de compactage produisait une croûte très dense en surface, due au broyage du sol par le cylindre du rouleau vibrant. Le degré de saturation de cette croûte variait avec le nombre de passages: il était de 70, 90 et 96% pour respectivement 2, 4 et 6 passages. D'autre part, la masse volumique sèche variait à l'intérieur des couches avec un gradient approximativement constant (fig. 3). La valeur moyenne était de 4,4% par 10 cm pour les couches de 40 cm et de 2,8% par 10 cm pour les couches de 80 cm. Il n'a pas été possible de trouver un sens de variation de ce gradient avec le nombre de passages du rouleau vibrant.

Nous avons, de même, obtenu le profil de l'humidité en fonction de la profondeur (fig. 3). En raison d'un séchage rapide en surface, les profils mesurés sont pratiquement identiques et indépendants des conditions de la mise en place des couches. En surface, la teneur en eau est de 4% environ et atteint une valeur approximativement constante, comprise entre 6 et 8%, à partir de 60 cm de profondeur.

Les mesures par radioactivité ont l'avantage d'être très rapides. Nous avons profité de cette qualité pour étudier statistiquement la dispersion de la masse volumique avec la grandeur du volume de la mesure. Chaque mesure était la moyenne de 16 mesures élémentaires faites, avec 4 orientations différentes, aux 4 angles d'un carré. Une mesure élémentaire correspond à un volume de 6 l environ et la mesure moyenne à un volume de 64 l; compte tenu de la granulométrie, ce

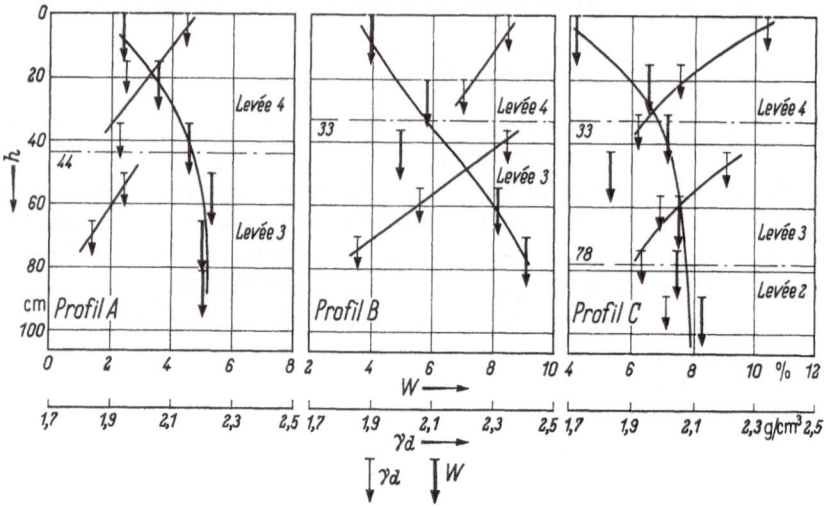

Fig. 3. Profils de densité et d'humidité.

dernier volume contient environ 5 blocs de la plus grande dimension. La dispersion sera caractérisée par le double de l'écart type.

Pour le volume élémentaire, la dispersion est de ± 11,7% (population de 1.300 valeurs). Pour le groupement des 4 mesures d'orientation différente en un même point, la dispersion est encore de ± 8,8% (population de 320 valeurs). Par contre, pour la moyenne des 4 mesures élémentaires de même orientation dans chacun des angles, la dispersion tombe à ± 3,1%. Ces dispersions sont nettement comprises entre les valeurs extrêmes possibles. La valeur maximale est celle d'un gros élément, soit 2,70 g/cm³. La valeur minimale peut être prise êgale à celle du mortier inférieur à 5 mm; si l'on considère une masse volumique moyenne de 2,15 g/cm³, la valeur minimale est égale à 1,80 g/cm³. L'intervalle extrême est donc: −16%, +25%.

Pour les mesures de l'humidité, la dispersion est bien plus importante: les mesures dans chacun des angles sont dispersées de ± 18% par rapport à la valeur moyenne. Ce grand intervalle est dû aux variations de teneur en eau entre les gros éléments et le mortier. Ainsi, pour une teneur en eau pondérale de l'ensemble du matériau de 6%, le mortier inférieur à 5 mm est à une teneur en eau de 12%, les gros éléments ayant, dans tous les cas, une teneur en eau de 1%.

Les appareils de mesure par rétrodiffusion sont beaucoup plus adaptés aux matériaux à granulométrie fine. En effet, la préparation de l'emplacement de mesure, qu'il s'agisse d'une plate-forme ou d'un sondage, est aisée et ne perturbe pas le matériau restant en place. En particulier, ces appareils présentent un grand intérêt pour les mesures

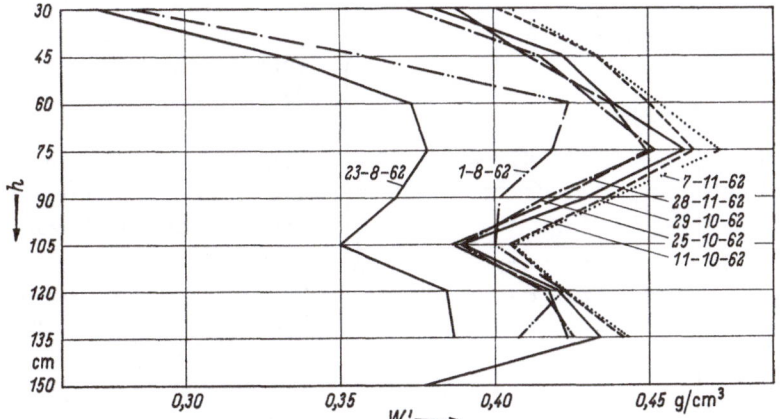

Fig. 4. Profils hydriques — Sondage No. 1.

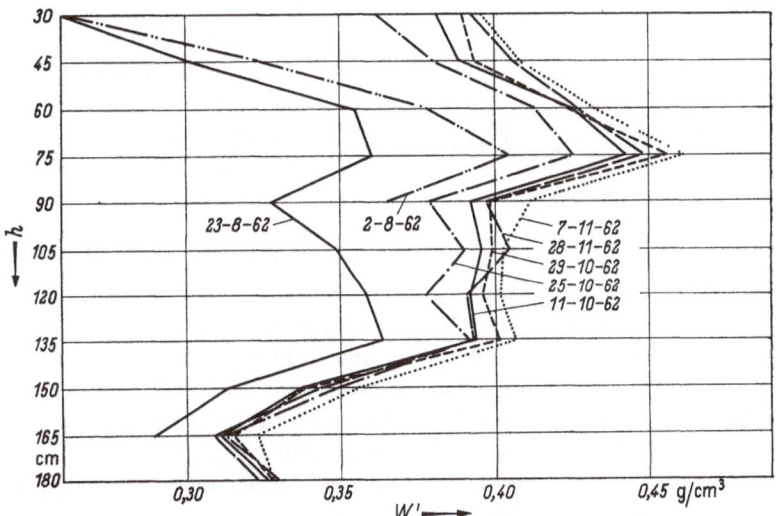

Fig. 5. Profils hydriques — Sondage No. 2.

de profils hydriques: la mesure n'étant pas destructive, il est possible, en un même emplacement, d'étudier la variation dans le temps de ces profils.

L'ensemble des profils des fig. 4 et 5 correspond à une période de fin de dessèchement suivie d'une réhumidification; ces profils se déplacent en gardant la même forme, le déplacement étant nettement plus important dans la   partie supérieure.

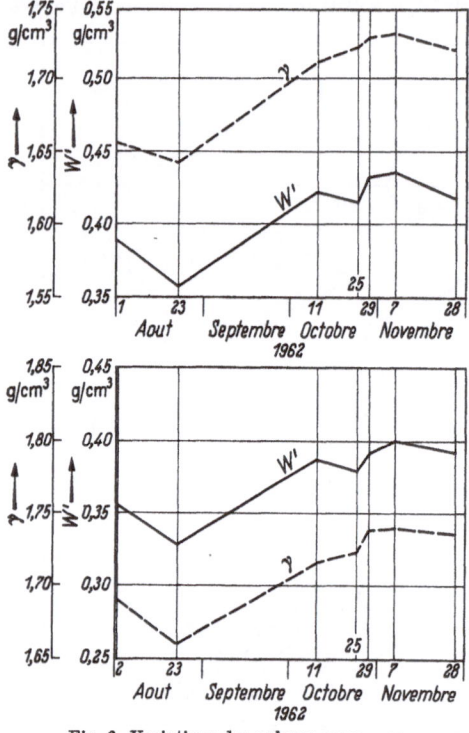

Fig. 6. Variations des valeurs moyennes.

La qualité des mesures obtenues a été confirmée par la constance de la masse volumique sèche $\gamma_d = \gamma - w'$. Ainsi, pour les valeurs moyennes des deux sondages, pendant une période de 4 mois, la masse volumique sèche obtenue par la combinaison des deux types de mesures est restée constante, le résidu quadratique moyen étant de 0,90% pour un des sondages, et de 0,47% pour l'autre (fig. 6).

La fig. 7 représente les variations de γ et w' pour le niveau 75 où elles sont les plus importantes. On constate que la constance de γ_d est un peu moins bonne, les résidus quadratiques moyens étant de 0,62 et 1,00%.

En raison des inconvénients des mesures par rétrodiffusion énumérés ci-dessus, nous nous sommes orientés vers les mesures par transmission directe pour l'utilisation sur le chantier. La mesure de la masse volumique totale, par transmission de rayons gamma, a déjà été bien étudiée et a donné lieu à quelques résultats [1]. Par contre la mesure de l'humidité a encore été peu étudiée [2], [3]; à l'aide d'un premier montage expérimental en laboratoire nous avons pu nous rendre compte des possibilités de la méthode.

Les résultats obtenus sont représentés avec une bonne approximation par une loi logarithmique d'atténuation:

$$\log N = - k \cdot L \cdot w'_a + cte,$$

N étant le taux de comptage, k une constante et L l'épaisseur de sol interceptée sur l'axe source-détecteur. Les résultats directs des mesures obtenues en changeant la position de la source, toutes choses étant égales par ailleurs, sont même excellents puisque 95% des droites $(\log N, L)$ ont un coefficient de corrélation supérieur à 0,997. Les résultats calculés, avec L constant et w_a' variable (18 échantillons), sont un peu moins bons, les coefficients de corrélation n'étant supérieurs qu'à 0,989.

A titre d'exemple, nous donnons (fig. 8) la courbe d'étalonnage obtenue pour une épaisseur de sol interceptée de 35 cm. L'équivalent-eau a été calculé à partir des résultats expérimentaux; on a obtenu $E =$ 8,09%. En première approximation, nous avons ajusté une droite avec les points expérimentaux en axes : Log N, w_a'; le coefficient de corrélation est égal à 0,990 et l'écart quadratique moyen de w_a' est égal à $\pm 0,0076\,\mathrm{g/cm^3}$, ce qui correspond à une erreur relative moyenne de $\pm 2,8\%$. Mais, en ajustant un polynôme du second degré, l'écart quadratique moyen est ramené à une valeu r correspondant à une erreur relative de $\pm 2,1\%$.

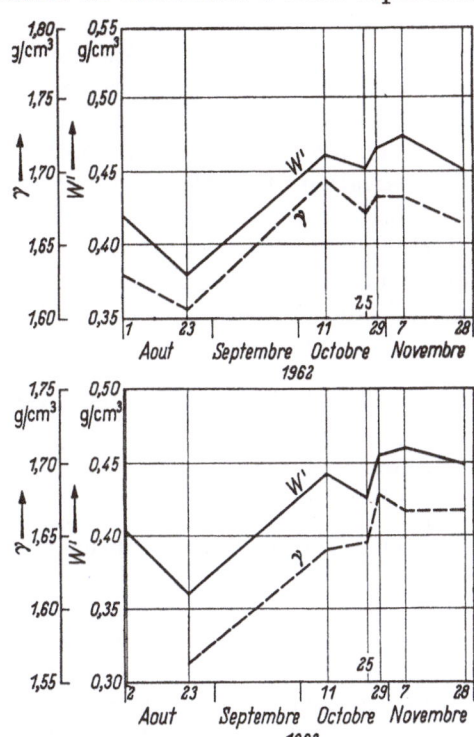

Fig. 7. Variations des valeurs au niveau 75 cm.

Pour ce cas particulier, la mesure est donc influencée en partie par le rayonnement diffusé. Ce résultat était à prévoir, étant donné la longueur de parcours du rayonnement. La précision obtenue avec ce

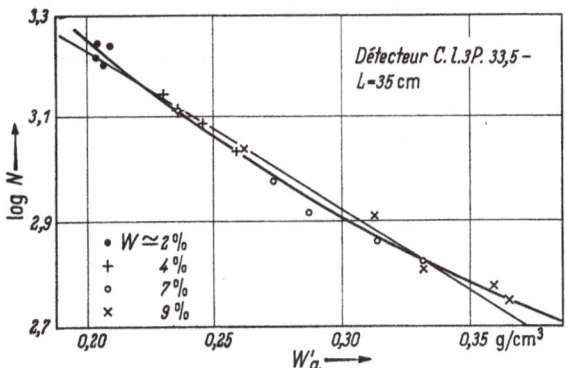

Fig. 8. Courbe d'étalonnage expérimentale.

nouveau principe de mesures est très intéressante. Elle est évidemment moins bonne que celle des mesures radioactives classiques qui permettent des mesures à mieux que 1%, mais il sera certainement possible de l'améliorer. Une étude plus détaillée, tant du point de vue théorique qu'expérimental, paraîtra prochainement.

Cette mesure complète la mesure aux rayons gamma déjà disponible et met ainsi à la disposition du Mécanicien du Sol un nouveau moyen d'investigation utilisable aussi bien en surface qu'à moyenne profondeur. Les appareils basés sur ce principe pourront être utilisés avec des matériaux graveleux, les zones pertubées par l'introduction de l'appareil ayant une influence nettement moindre sur la mesure que dans le cas de la rétrodiffusion.

Bibliographie

[1] MEIGH, A. C., and B. O. SKIPP: Gamma-ray and neutron methods of measuring soil density and moisture. Geotechnique **X**, No. 3, 110—126 (1960).
[2] MARTINELLI, P.: Mesures d'humidité par transmission de neutrons thermiques. 1960, C. E. A., C. E. N. de Saclay. DC/R/AR/60—6/PM/G./M/.
[3] WACK, B.: Mesure de l'humidité des sols par diffusion d'un faisceau de neutrons thermiques. C. R. S. de l'Acad. des Sciences **254**, 1002—1004 (1962).

Discussion

Contribution de R. PELTIER: Dans nos essais effectués au Laboratoire Central des Ponts et Chaussées, notamment dans les essais effectués par M. ROCOPLAN, il semble que les photons de faible énergie donnent des résultats aberrants. Ceci s'expliquerait par le fait que les photons de faible énergie donnent lieu non seulement à l'effet Compton mais aussi à l'essai photoélectrique. D'où la conclusion à laquelle nous sommes arrivés: que pour avoir des essais corrects, il fallait utiliser des filtres laissant passer uniquement les photons de très forte énergie.

Réponse de B. WACK: L'influence de l'énergie du neutron sur le résultat de la mesure de l'humidité est complexe. Elle se traduit principalement sur la valeur de l'équivalent-eau qu'il faut prendre pour la courbe d'étalonnage. Suivant l'énergie des neutrons utilisés l'équivalent-eau peut, d'une part, avoir une valeur plus ou moins grande et, d'autre part, avoir des variations plus ou moins grandes pour différentes compositions chimiques du sol sec. Mais, il n'existe théoriquement aucun domaine d'énergie où l'influence de la composition chimique du sol soit négligeable.

Rapport Général
Relatif à la 4ᵉ Sous-Section

Rapporteur: **P. M. Sirieys**

La quatrième sous-section traite des études sur modèles réduits et des ouvrages «in situ». Elle a fait l'objet d'un volume de communications sensiblement moins important que les trois autres sous-sections.

Le modèle bi-dimensionnel, composé de petits rouleaux métalliques, a toujours la faveur des chercheurs pour résoudre les problèmes de Génie Civil.

Ainsi VERDEYEN a poursuivi, en collaboration avec NUYENS, ses nombreux travaux sur modèle. Ces deux auteurs ont étudié les rideaux ancrés à faible profondeur et proposent une interprétation des expériences basée sur la théorie des caractéristiques. L'analyse de la stabilité de l'ensemble palplanche et ancrage par l'examen de la cinématique du phénomène global les conduit à une méthode de calcul pratique. En Génie Civil ces travaux trouvent une application immédiate dans les problèmes d'ancrage des palplanches ou des ponts suspendus.

D'autre part BIAREZ et un groupe composé de BOUCRAUT, HAERINGER, MARTIN, MONTEL, NEGRE, STUTZ et WIENDIECK ont utilisé ce même modèle dans une double perspective:

Sous un aspect fondamental, le modèle leur permet d'étudier outre les charges limites, la loi d'écoulement plastique. Les auteurs aboutissent à un bon accord avec l'isotropie (coaxialité des deux tenseurs contrainte et vitesse ou déformation) et à une loi de non-variation du volume en régime plastique.

Sous leur aspect pratique, ces études leur permettent de résoudre des problèmes tels que: l'enfoncement des fondations (ils déterminent les composantes normale et tangentielle des efforts à la surface de contact), le soulèvement des fondations (qui s'interprète correctement par la théorie des caractéristiques), la translation et la rotation des écrans en fonction de la profondeur (ils notent l'absence de mise en plasticité parfaite au voisinage de l'écran et estiment nécessaire le recours à une solution Elastique-Plastique ou Rigide-Plastique). Ces recherches s'appliquent aux problèmes de Travaux Publics tels que: Pylones, Murs de soutènement, Fondations excentrées, Murs de quai, Roues et locomotion sur le sable. Ces auteurs concluent, en l'attente de méthode de calcul plus complète, aux besoins d'un calcul semi-empirique.

BRINCH HANSEN aboutit à des conclusions analogues et propose des formules empiriques exprimant les déformations des sables, formules dont a constamment besoin l'Ingénieur de Génie Civil pour la construction d'ouvrages divers.

Le domaine de la reconnaissance des sols par les mesures radio-actives est étudié par WACK qui précise les possibilités d'utilisation des mesures par rétrodiffusion. Dans le compactage d'un matériau graveleux il met en évidence un gradient de densité sèche dans l'épaisseur des couches. Des profils hydriques de divers sites alpins ont pu également être tracés avec une très bonne précision. Il propose enfin un principe nouveau de mesure de l'humidité par transmission de neutrons et pense pouvoir améliorer la précision relative des résultats.

Peltier fait part de recherches effectuées dans le même sens et ajoute que Rocoplan et Legrand ont appliqué ces méthodes d'auscultations aux sols mais également aux bétons.

Des résultats d'essais en vraie grandeur sont présentés par Bjerrum. Les déformations du sol dépendent du cycle de charges qui lui sont appliquées, ainsi sont comparées les déformations obtenues sous charges constantes avec celles qui résultent de grandes variations de charge dans le temps. Dans ce second cas la vitesse de déformation est constante. Ces résultats s'appliquent particulièrement à la réalisation de silos et de réservoirs, en un mot de structures à fluctuation de charges importante. Koizumi et Ito comparent les tassements prévus par le calcul avec ceux, observés depuis huit ans, d'immeubles construits sur une argile volcanique.

Enfin la question des barrages en terre a été minutieusement analysée par Londe qui expose les problèmes de l'Ingénieur-Constructeur. Il conclut, à la suite de nombreuses analyses statistiques, à l'impossibilité d'effectuer des moyennes et à la nécessité de s'intéresser aux détails géologiques mineurs, il souligne l'importance du grand nombre de paramètres qu'il est nécessaire de peser et insiste sur l'utilité d'expériences simples mais nombreuses telles que les limites d'Atterberg. Il a engagé un dialogue entre théoriciens et praticiens en proposant un programme minimum de travail commun comportant en premier lieu une limitation à l'étude des problèmes plans et une analyse de l'influence de l'hétérogénéité (variation des paramètres physico-mécaniques avec l'espace) et du facteur temps. En seconde urgence il suggère l'examen de l'influence de la troisième dimension et surtout la mise au point de méthodes de calcul rapides et peu coûteuses.

En définitive, les problèmes posés par le Génie Civil sont encore résolus par des méthodes semi-empiriques. Les modèles bidimensionnels, qui s'appuient sur la théorie de la Plasticité, et les mesures sur l'influence du cycle de mise en charge, liée à la Viscosité du matériau, apportent des solutions à de nombreux problèmes pratiques. Toutefois le concept du continu homogène et isotrope ne donne pas entière satisfaction aux projeteurs, la voie est ouverte aux études des milieux anisotropes tenant compte des hétérogénéités et discontinuités locales qui jouent en particulier un rôle prépondérant dans le calcul des charges limites.

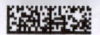